From Cells to Proteins: Imaging Nature across Dimensions

NATO Security through Science Series

A Series presenting the results of scientific meetings supported under the NATO Programme for Security through Science (STS).

The Series is published by IOS Press, Amsterdam, and Springer Science and Business Media, Dordrecht, in conjunction with the NATO Public Diplomacy Division.

Sub-Series

A. Chemistry and Biology (Springer Science and Business Media)
B. Physics and Biophysics (Springer Science and Business Media)
C. Environmental Security (Springer Science and Business Media)
D. Information and Communication Security (IOS Press)
E. Human and Societal Dynamics (IOS Press)

Meetings supported by the NATO STS Programme are in security-related priority areas of Defence Against Terrorism or Countering Other Threats to Security. The types of meeting supported are generally "Advanced Study Institutes" and "Advanced Research Workshops". The NATO STS Series collects together the results of these meetings. The meetings are co-organized by scientists from NATO countries and scientists from NATO's "Partner" or "Mediterranean Dialogue" countries. The observations and recommendations made at the meetings, as well as the contents of the volumes in the Series, reflect those of the participants in the workshop. They should not necessarily be regarded as reflecting NATO views or policy.

Advanced Study Institutes (ASI) are high-level tutorial courses to convey the latest developments in a subject to an advanced-level audience

Advanced Research Workshops (ARW) are expert meetings where an intense but informal exchange of views at the frontiers of a subject aims at identifying directions for future action

Following a transformation of the programme in 2004 the Series has been re-named and re-organised. Recent volumes on topics not related to security, which result from meetings supported under the programme earlier, may be found in the NATO Science Series

www.nato.int/science
www.springeronline.com
www.iospress.nl

Series B: Physics and Biophysics – Vol. 3

From Cells to Proteins: Imaging Nature across Dimensions

edited by

Valtere Evangelista
CNR Institute of Biophysics, Pisa, Italy

Laura Barsanti
CNR Institute of Biophysics, Pisa, Italy

Vincenzo Passarelli
CNR Institute of Biophysics, Pisa, Italy

and

Paolo Gualtieri
CNR Institute of Biophysics, Pisa, Italy

Springer

Published in cooperation with NATO Public Diplomacy Division

Proceedings of the NATO Advanced Study Institute on
From Cells to Proteins: Imaging Nature across Dimensions
Pisa, Italy
12-23 September 2004

A C.I.P. Catalogue record for this book is available from the Library of Congress.

ISBN-10 1-4020-3614-0 (HB)
ISBN-13 978-1-4020-3614-9 (HB)
ISBN-10 1-4020-3616-7 (e-book)
ISBN-13 978-1-4020-3616-3 (e-book)

Published by Springer,
P.O. Box 17, 3300 AA Dordrecht, The Netherlands.

www.springeronline.com

Printed on acid-free paper

Printed in the Netherlands.

In memory of Antonietta Barsanti

CONTENTS

PREFACE

This volume contains the lectures and seminars given at the NATO Advanced Study Institute on "From cell to proteins: Imaging nature across dimensions", held in Pisa, Italy in September, 2004. The Institute was sponsored and mainly funded by the Scientific Affairs Division of NATO. The European Biophysical Societies' Association also supported the School and contributed to its success. It is my pleasant duty to thank these institutions. This ASI offered updated information on how many information we are able to achieve in the exploration of the inner details of biological specimens in their native structure and composition.

How deep we can see inside Nature smallest secrets? Will it be possible someday in a near future to investigate living structures at atomic level? This area of study is very interdisciplinary, since it uses and it is based on the principles and techniques of biology, physics, chemistry, mathematics, and engineering to elucidate the structures of biological macromolecules, of supramolecular structures, organelles, and cells. Scientists from very different backgrounds contributed to the understanding of this fast developing field of research, which has seen considerable progress during the last years. These revelations came not (only) through biochemistry and molecular biology, but through staggering improvements of microscopy.

The widest range of dimensions is covered by the electron microscope (EM): it can image whole cells, their organelles, cytoskeleton and supramolecular assemblies, as well as individual biomolecules, their submolecular structure and, ultimately, individual atoms. Recently, the scanning force microscope (SFM) has opened completely new perspectives for analyzing the surface topography of biological matter in its aqueous environment at a resolution comparable to that achieved by EM. Most exciting, the SFM is the first imaging device allowing direct correlation between structural and functional states of biomolecules at submolecular resolution. The developments of implementation of laser beam and stage scanning systems incorporating confocal optics, the advent of new electro-optical detectors with great sensitivity, linearity, and dynamic range, the possibility of 2D fast image enhancement, reconstruction, analysis and 3D display, and application of luminescence techniques gives the possibility to investigate the chemical and molecular details of life processes, and we can say that this digital approach, comprehending FLIMT and FRET techniques combined with the use of quantum dots, can probe the spatio-temporal organization of cell metabolism. Moreover, through the use of antibody labelling and confocal or multiphoton microscopy joined to algorithms for image restoration, it is now commonplace to distinguish and study structures an order of magnitude below the theoretical limit of resolution of the light microscope (about 250 nm). Also, individual 20 nm particles carrying specific protein probes can be seen, for example tracking across the surface of a living cell.

We can conclude that with the microscope we can explore the microcosm, from tissues to cells and all the way down to molecules and atoms, and that this ASI represented a successful opportunity for carrying on and implementing an interdisciplinary approach to the study of theoretical and practical aspects of this research field.

The first part of the book deals with a general overview of the state of the art of microscopy including theoretical aspects, whereas the second part deals with uses and applications of the different kind of microscopes.

I do hope this book has caught the spirit in which the ASI was conceived.

Paolo Gualtieri

Director of the ASI "From cell to proteins: Imaging nature across dimensions"

OPTICAL MICROSCOPY FOR CELL IMAGING

John M. Girkin
Institute of Photonics, University of Strathclyde, Glasgow, UK

Abstract. Man's curiosity to "see" and hence understand his world has driven the development of a wide array of imaging systems from the telescope to the microscope. As the eye is probably the single most important human detector, optically based systems have played a crucial role in this growth of understanding and the ability to image the very small has lead to a number of significant breakthroughs in knowledge. Allied with the technological developments have come significant advances in the understanding of optical systems in general thus the optical microscope has played a major role in man's comprehension of the Universe. It is the aim of this chapter to enable a user of optical microscopes to select the most suitable system for the specific imaging task from the plethora of instruments and options now available. The basic principles behind optical microscopes used for cellular imaging will be outlined and then the role of each component will be considered in the build up of the final image. With the fundamental concepts established the more advanced imaging techniques will be explained covering confocal and multiphoton microscopy through to sub diffraction far field microscopy.

Keywords: Light Microscopy, Fluorescence Microscopy, Confocal Microscopy, Multiphoton Microscopy

1. INTRODUCTION

Before considering the details of optical microscopes it is worthwhile considering what is needed in a good image from the perspective of cellular biology. Although these considerations will vary considerably from experiment to experiment there are some universal requirements. The overall quality of the image must be sufficient for the user to visualise the desired detail and features and this quality is controlled by three factors. The first, and obvious requirement, is that the microscope must provide a *magnified* image of the object. The magnification causes the image to be spread over a larger area of the detector (frequently the eye) and hence the detail can be seen more clearly. The second factor is the *resolution* of the microscope. This is the ability to separate two closely spaced points and is in the end determined by the quality of the optics and the laws of physics. The third and equally important factor is the *contrast* present between the different

V. Evangelista et al. (eds.), From Cells to Proteins: Imaging Nature across Dimensions, 1–22.

features. In the simplest terms this is the ability for the eye to distinguish between the two points. For example a black dot on a black background can not be seen, but if we place a local white ring or background around the dot then the dot will be rendered visible. In biological samples much of the skill of the experimenter is in determining how to maximise this contrast to enable the features of interest to be brought out above the background, for example through the use of specific labelling and spectral filters to separate the various features and to some extent all developments in microscopes for the last 100 years have been attempts to increase image contrast. All microscopes contain these three essential features and it is the detail in which these are implemented, and the balance between the components, that differentiates the various forms of microscopes.

1.1.1. Light

All of the microscopes considered in this chapter utilise the interaction of light with the sample and thus it is worth considering the properties of light in outline before the details of the optical microscope are discussed. Light forms part of the electromagnetic spectrum and is normally considered to be the portion with a wavelength between 400 and 700nm, defined by the region in which most people's eyes respond. In terms of colour this is from the deep violet to the far red. Most optical microscopes actually have the ability to work beyond this range as ultra-violet light (down to around 350nm) is frequently used for exciting some important flurophores (eg Calcium indicators and DNA labels) and in the near infrared up as far as 1100nm which is now used for optical trapping and for certain specialised imaging modes.

In a vacuum all light, and in fact all electromagnetic radiation, travels at the same velocity defined as being 299,792,458ms^{-1} (for most purposes 300 million ms^{-1}). However, as light enters different materials it is slowed down and the ratio of the speed of light in the material to that in a vacuum is known as the refractive index (n). This slowing depends on the wavelength of the light and as will be shown later can lead to a degradation of images.

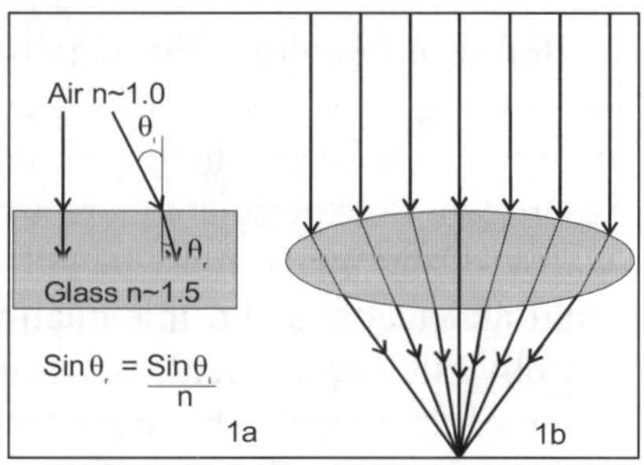

Figure 1. 1a refraction at a plane surface, 1b plane wave refracting at a curved surface (lens)

If a light ray hits the material at 90° then the wave is just slowed down but if it hits at any other angle the light is bent towards the perpendicular (Figure 1a) giving rise to refraction. If the material has a curved surface as a bundle of light rays hit they will be bent at different angles causing the light to come to a focus for a positively curved surface (Figure 1b). By using different shapes of curve and different glasses microscope optics can be designed to compensate for various errors in the optical system.

Two other crucial parameters possessed by light, which are used in microscopy, are interference and polarisation both of which are characteristics of all waves. Waves are said to interfere when the peak of one wave adds with the trough of another wave with the net result of zero (or no light). In the opposite case, when the peaks are matched, then constructive interference takes place giving a bright spot (Figure 2). For such effects the light waves have to be *in phase* meaning they have the same wavelength and that their movements are locked together (they go up and down collectively).

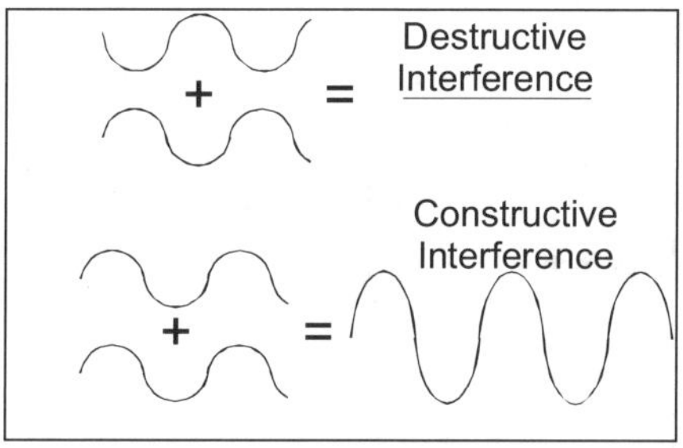

Figure 2. Destructive and constructive interference

Polarisation is the property of the direction of the wave's oscillation. For a wave travelling in the x direction the waves can either oscillate in the y or z directions, or a mixture of the two. Various methods can be employed to distinguish the plane of oscillation but the most common experience is in the use of polarising sun glasses which cut down glare from water by blocking the preferred reflected polarisation from the surface of water.

The remaining feature of light that needs to be considered, as it generally sets the fundamental limit of resolution of microscopes, is *diffraction*. As light passes through a small aperture (one approaching the wavelength of the light) the wave spreads as it passes through the opening. As seen in figure 3 the plane wave front entering the opening is curved after passing through resulting in an intensity pattern with a peak and side lobes. This so called Airy disk limits the resolution of a microscope as will be described later. Further details on any of the topics mentioned here can be found in all standard optics text books several of which are listed [1, 2, 3].

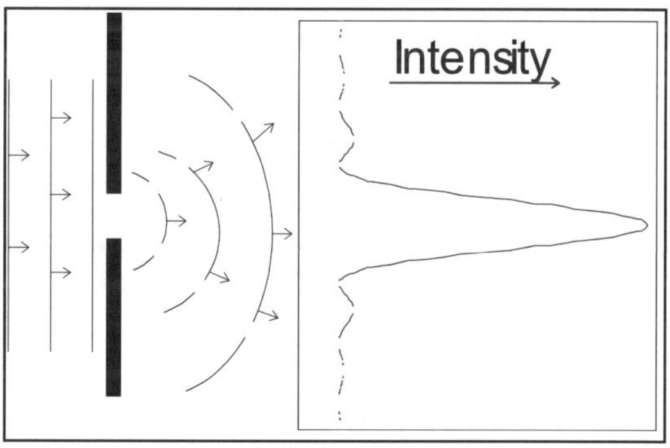

Figure 3. Diffraction through an aperture and the resulting intensity profile

2. THE BASIC MICROSCOPE

As mentioned above the aim of any microscope is to present a magnified image of the sample onto the detector such that fine detail can be resolved. In the basic case a single lens magnifier forms the simplest microscope and in the mid 17th Century a number of fundamental discoveries were made using single lens systems. However, with the limited space available here we will only consider compound microscopes consisting of, in essence, two magnification units. The components of such a system are shown in figure 4 and we will now consider each of these in turn.

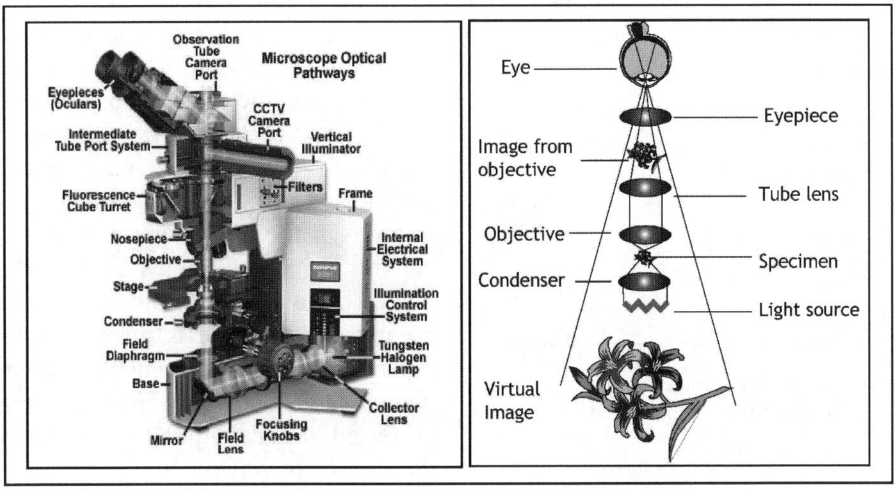

Figure 4. Commercial microscope and the critical components in diagrammatic format

2.1. Objective Lens

Probably the most crucial component in the microscope is the objective lens and financially they can form about 50% of the total cost of an instrument. The aim of the objective lens is to collect as much light from the sample as possible. The fine detail is contained in the peripheral light rays from the sample and therefore in simple terms the wider the collection angle of the lens the higher the resolution of the lens. This collection angle is known as the numerical aperture (NA) of the lens defined as the sine of the half angle. As this collection angle increases then, in general, the working distance of the lens decreases. In normal circumstances the half collection angle can not be greater than 90° in air (due to the refractive index values of air and glass) but by using water rather than air between the first lens element numerical apertures of up to 1.2 are possible and with oil replacing water 1.5NA lenses are available.

The NA of the lens sets the ultimate resolution that is possible with the system. As described above as features approach the wavelength of light diffraction takes place. A definition of the resolution of a system is thus used of the minimum separation between two points at which the two points can each be seen. In effect what takes place is that each point produces an Airy disc and as the points become closer together the finite peaks as the centre of the discs start to overlap until you can no longer distinguish between the two points. The commonly accepted distance at which this happens is given by

$$d = (0.61 \times \lambda) / NA$$

where λ is the wavelength of the light used for the imaging and NA the numerical aperture of the lens. The crucial factor to remember is that the numerical aperture of the lens, and not the magnification, determines the ultimate resolution of a microscope.

However, life is not perfect and no system really achieves the theoretical possibilities. In the case of optics aberrations are induced into the system (Figure 5).

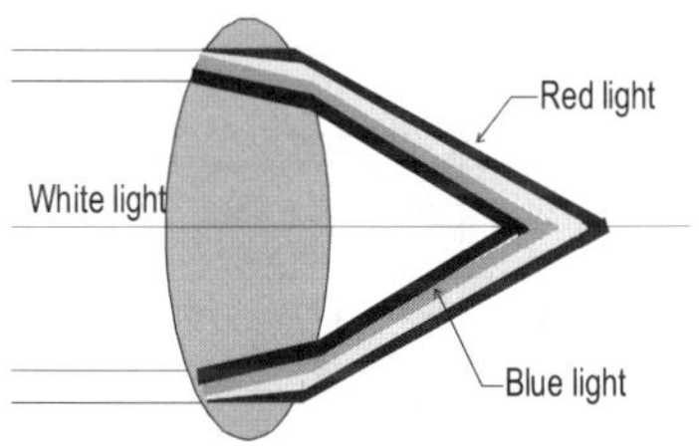

Figure 5 a Chromatic aberration

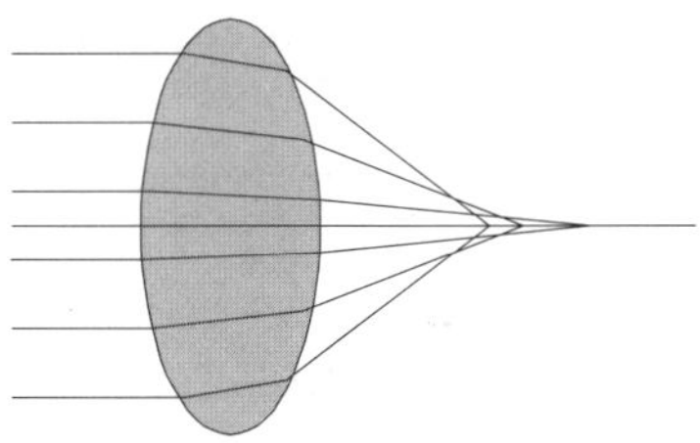

Figure 5b Spherical aberration

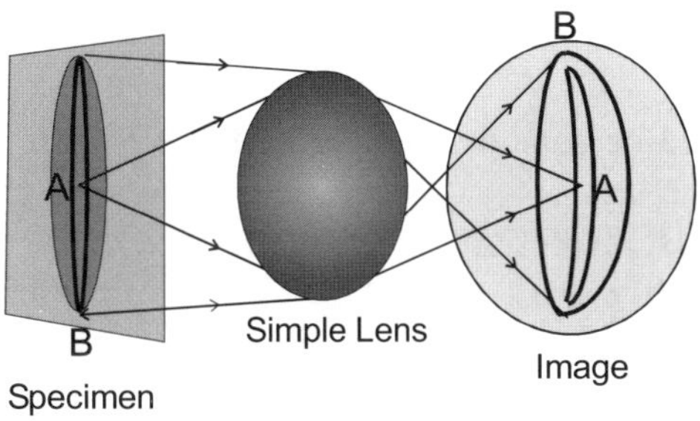

Figure 5c Field Curvature

Figure 5. Diagrammatic representation of the three major sources of aberration in an optical microscope.

a) *Spherical aberration*; In which the light beams entering the lens at the edge of the lens are refracted more than those at the centre leading to an extended focus. For a given optical system this can be largely designed out but any changes in refractive index will cause some spherical aberration and so, if for example, an air objective lens is used for live cell imaging in which the cells are in water one will inherently have spherical aberration degrading the ultimate resolution. However, frequently there is little that can be done as the sample survival must come first.

b) *Chromatic aberration*; The refractive index of materials is dependent on the wavelength of the light and hence a lens will tend to focus the blue rays closer to the lens than the longer wavelength red light. Again this leads to a blurring of the image with colours appearing around subject but by careful lens design this can largely be overcome, but with a financial penalty.

c) *Flatness*; The final aberration considered here is the fact that a flat sample can appear to be curved as one moves away from the centre of the image field. This leads to an overall distortion but this can be corrected should a wide flat field be required.

As indicated all of the aberrations can, in theory be removed through careful lens design and table 1 outlines the way that different lenses are corrected.

Lens Type	**Cost**	**Chromatic Correction**	**Spherical Correction**	**Notes/Use**
Achromat	•	Blue & Red	Green	Green, monochrome, viewing
Fluorite	••	Blue, red, part green	Blue, Green	Colour photography in white light
Apochromat	•••	Deep blue, blue, green, red	Blue, green red	Colour viewing and fluorescence
Plan-fluorite, Plan-apochromat	••••	As fluorite, or apochromat	As fluorite, or apochromat	Flat field, recording (film or digital), scanning

Table 1. Summary of Objective lens types

As a guide the following should be considered in order to help with the correct of the best lens for a specific task.

- Highest NA possible for best resolution
- Correct immersion fluid (water or oil) or air
- Select working distance
- With or without coverslip
- Magnification to obtain as large an image as required on the detector

2.2. Illumination

For transmission microscopy the illumination plays a crucial factor in the ultimate resolution of the system. Again in brief the higher the numerical aperture condensing lens that can be used the better. This illuminates the sample with a wide cone angle of light thus enabling the objective lens to collect peripheral rays adding to the detail of the image. In most practical situations the working distance available for the illumination lens frequently imposes a limitation on lens selection and although it may often be preferred to use a high NA oil lens frequently a longer working distance air lens has to be used.

The alignment and setting of the illumination system is also crucial and this problem was solved by August Köhler (or Koehler) in 1893 when working for the Carl Zeiss company and all microscopes now use this method of illumination (figure 6). The light from the source is collected and imaged through a field diaphragm setting the angle of the illumination cone

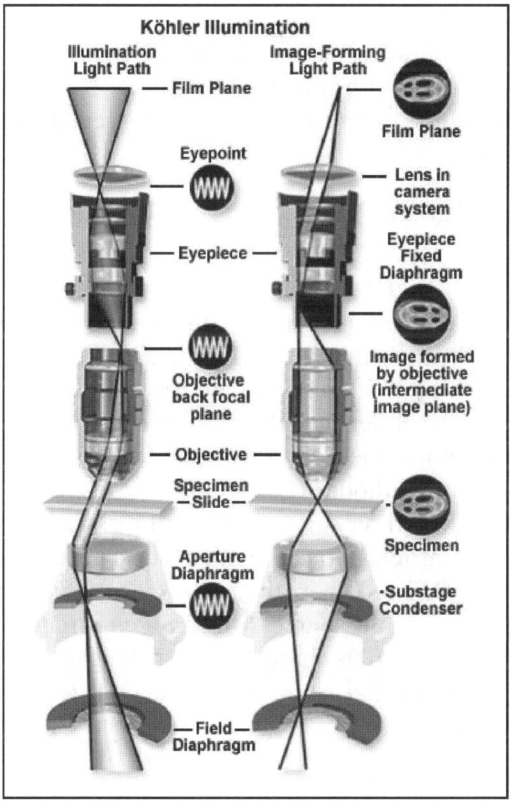

Figure 6. Kohler Illumination (Courtesy Nikon Ltd)

(with the lens NA) and then through a aperture diaphragm setting the size of the field. Through the correct choice of these parameters the sample can be illuminated without excessive light lowering the contrast of the image. The correct setting of the illumination is important and the guide below provides a starting point.

1 Switch on lamp and open field and aperture diaphragms and focus specimen
2 Close field diaphragm to around 90%
3 Raise condenser lens & focus image of field diaphragm onto specimen (polygon)
4 Centre image in field of view then open field diaphragm so it just disappears
5 Remove eyepieces and observe aperture diaphragm with filament or even illumination depending on the source
6 Adjust diaphragm to 2/3 to 3/4 open (best compromise resolution v contrast)
7 Replace eyepiece and observe
8 For a different objective repeat process adjusting both apertures (eg for a x40 close field diaphragm and open aperture diaphragm but this may be a calibration on microscope
9 Adjust intensity with filters
10 **Do not** adjust intensity by adjusting condenser lens or aperture diaphragm.

If this method is adopted then one has an excellent starting point for high quality imaging.

There are no several options for sources for illumination. The most commonly used method is with a conventional tungsten light bulb or a halogen bulb with latter giving a brighter and whiter light but runs at a hotter temperature. With both of these sources the light level on the sample can be controlled with neutral density filters or by adjusting the bulb current which does alter the colour of the light. One recent solution to this has been the use of white light emitting diodes (LEDs) [4]. These are optically very efficient, bright, can be switched on very rapidly and provide a constant white colour over a wide range of intensities For fluorescence imaging in which a narrow spectral band is required then the use of arc lamps (mercury or Xenon) provide the most suitable source with an appropriate filter. With the most recent generations of LEDs these low cost, long lifetime devices may well replace more conventional sources. At the final level of sophistication for most advanced forms of microscopy (confocal, multiphoton) then laser sources are required.

2.3. Detection

The most commonly used detecting system is the observer's eyes. These highly sensitive detectors visualise in colour, have a huge dynamic range and integrated signal processing! Their main limitation being that they can not record a permanent record of the image. The value of the magnification and numerical aperture of the eye pieces in a microscope are again important. If one requires an overall magnification of say 200x this could be achieved either using an objective of x20 or x10 magnification with x10 or x20 eye pieces. Typically for high resolution one would prefer a x20 0.75NA objective with x10 eye pieces rather than a x10 0.5NA objective, x20 eye piece combination. The latter may give a wider field of view but with a lower overall resolution.

In terms of permanent records in a microscope one is generally (in non-scanned systems mentioned below) left with either an electronic camera or more traditional film based system. Film provides the highest possible resolution with a wide dynamic range and a permanent record. However, storage can be an issue, processing is messy and data analysis can be hard. More recent digital images are now challenging film but they still have a lower overall resolution and are generally less sensitive (in relation to signal to noise considerations). They are easier to manipulate and analyse through data storage can quickly become an issue. Digital systems are also capable of very rapid imaging through typically then with a lower dynamic range.

The above should have provided a basis of the understanding of the optics of microscopes and given the reader the tools with which they can make decisions on lens selection and image recording. It is not designed to be a complete guide to microscopy as entire books have been written on what is again a rapidly changing landscape. The equipment suppliers provide excellent supporting literature which, when combined with the understanding provided above, will enable the user to make the correct decision. Some useful guidance can be found on the internet as well as in the references [5, 6, 7].

3. ADVANCED WIDE FIELD MICROSCOPY

As discussed at the start of the chapter the aim of all microscopes is to provide a magnified, high contrast image of the sample. We will now consider some of the advanced techniques used to increase the contrast within a sample.

3.1. Polarisation

Although not widely used for cellular imaging polarisation microscopy does provide a useful way of increasing the contrast in virtually transparent samples. In its simplest form light from the illumination source passes through a polariser which allows through only one polarisation of the light. The light transmitted through the sample is then collected by the objective in the conventional manner but passes through a second polariser before reaching the detector. If the two polarisers are aligned to let through the same polarisation then a conventional image is seen. However, the second polariser, known as the analyser, can be rotated. When this is aligned at 90° to the illumination polariser then no light will reach the detector unless the polarisation of the light is altered as it passes through the sample. For certain samples this can happen as the sub-cellular structure may have a property known as birefringence which rotates the polarisation of the light as it passes through. This is normally caused by the alignment of long chain molecules which lie with their long axis's parallel with each other. Hence by observing the image as the analysing polariser is rotated different parts of the sample can be made visible. The technique in the life sciences has been largely surpassed by Phase Contrast and DIC microscopy

3.2. Phase Contrast Microscopy

Phase contrast microscopy was introduced in 1934 by Frits Zernike with the aim of enhancing the contrast in transparent samples. The main advantage, as with DIC imaging, is that cells can be imaged in their natural state without the inclusion of exogenous labels. A simplified optical diagram is given in figure 7. Light from the illumination source is passed through an annular condenser to produce a ring of illumination. Light passing through the sample is either un-deviated or is diffracted and suffers a phase change due to the gradients of refractive index within the sample (its wave front has been delayed compared to the un-deviated light). In order to see this change Zernike introduced a phase plate which delayed the un-deviated light (and partially blocked it to further enhance the contrast) so that in the image plane the un-deviated light suffered destructive interference, giving no signal, whereas the light delayed by the sample interferes constructively to yield a bright image. Thus a phase, or time delay, in the light has been converted to an intensity variation which optical detectors can visualise. This invention won Zernike a Nobel prize.

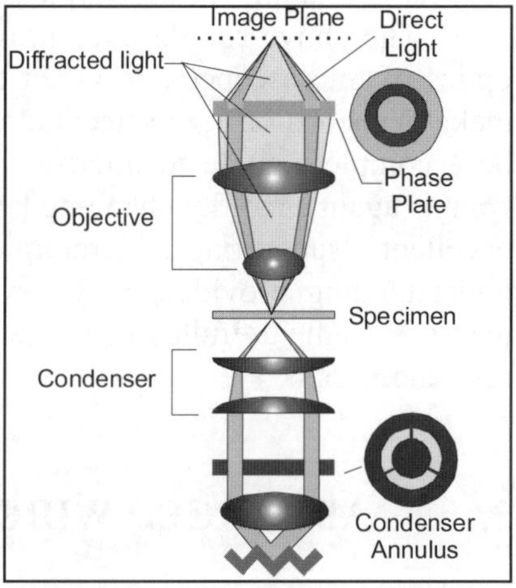

Figure 7. Principles of Phase Contrast Microscopy

Implementing phase contrast on a conventional microscope is straight forward. A pair of condenser and phase rings are purchased for each objective lens and condensing lens and placed into the optical path and once they have been correctly aligned around the optical axis (to obtain a uniformly dark field with no sample present) no further adjustment is required. Thus one has a low cost method of imaging nearly transparent samples. There are limitations as phase images are usually surrounded by halos around the sample boundaries which can obscure the detail. In addition the phase elements do reduce the effective NA of the lenses thus reducing the maximum possible resolution and for obvious reasons it is not suitable for thick samples, however, the method is simple, low cost and should not be overlooked.

3.3. Differential Interference Contrast (DIC) Microscopy

DIC imaging is an advance upon conventional phase contrast techniques but adds a level of sophistication and cost to the microscope. Figure 8 illustrates the simple optical basis of the technique. Light from a conventional source passes through a polarising element before being incident on a Wollaston (or Nomarski) prism. This splits the beam of light into two beams travelling in slightly different directions by using two quartz wedges cemented together. The full operation of Wollaston prisms can be found in any basic optics text book [1, 2, 3]. The two, slightly separated beams now have opposite polarisations hence do not interfere with each other and after the condenser lens pass through the sample to be recombined after the objective lens in a second Wollaston prism and subsequently to the detector.

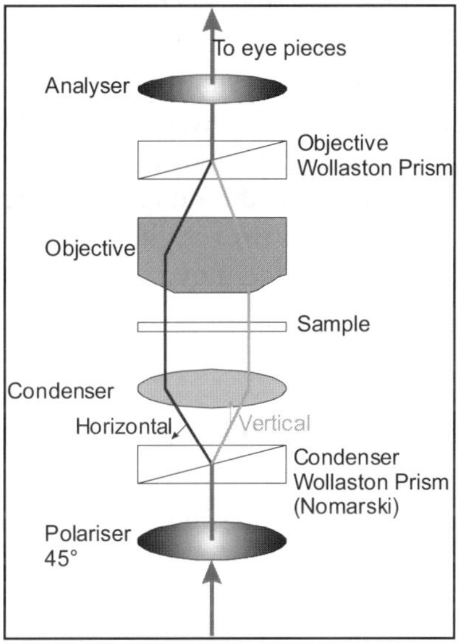

Figure 8. Basic components of DIC Microscopy (beam separation increased to improve clarity)

The two beams traverse the sample where their paths are altered according to the specimen's local thickness and refractive index thus causing slight, relative delays, between the two beams. The polarisation added by the first Wollaston prism is then removed by the second prism so that the beams can now interfere. As the beams pass through the sample so close to each other they are very sensitive to local changes in refractive index and hence the image is a map of the slope of the local path length differences. This design means that on one side of a feature the changes appear to be dark (say going from high to low refractive index), whilst on the other it will be light.

This gives the image a shadow effect providing a feeling of "three dimensionality".

Compared to straight phase imaging DIC has a number of advantages. It is possible to make full use of the NA of the lenses leading to the maximum possible resolution. It is also possible to use DIC on labelled samples as slight absorption of the incoming light will not appear in the final DIC image which can then be superimposed on a fluorescent image. In addition the halos mentioned above are also absent and the method can be used with any achromat or fluorite lens. The disadvantage of the technique is that birefringent sample dishes (ie plastic) can not be used and the pair of Wollaston prisms required for each objective lens are expensive. Additionally the full benefit of DIC imaging can only be obtained when the system has nearly perfect Köhler illumination and all of the optical components are correctly aligned and incorrect adjustment of a single component can significantly degrade the image quality. The resulting image is a result of difference in the optical path, which is determined by both the local refractive index and the sample thickness, and hence some prior knowledge of the sample topology is required for accurate interpretation of the image. However, the technique is very powerful and again should not be forgotten in the move to higher and higher technology solutions to imaging challenges.

3.4. Fluorescence Microscopy

All of the techniques described above can provide detailed images of cells but in order to develop a better understanding of the function of the various components a further, specific, contrast aid is required and this is frequently provided by fluorescence imaging techniques. The fluorescence may come from naturally occurring compounds within the sample or can be provided through the use of specially labelled fluorescent compounds that bind to specific ions, or parts of the cell. In either case the principle of fluorescence microscopy is the same. Light at one wavelength is absorbed by the flurophore and

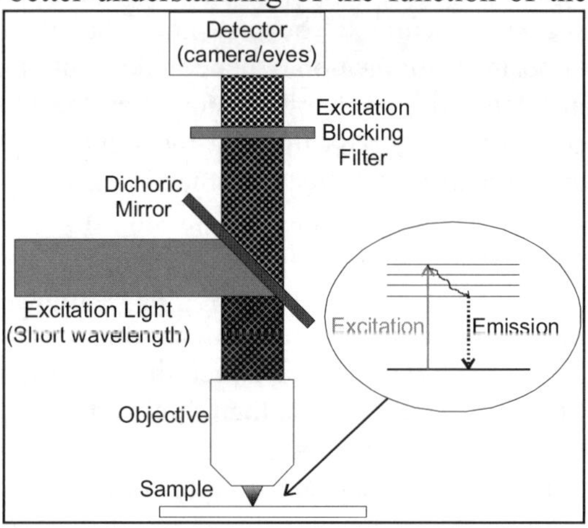

Figure 9. Epi-fluorescence configuration and energy diagram

subsequently re-emitted at a longer wavelength, a short period later, with some of the initial absorbed energy being lost as vibration energy (heat) within the sample. For the rest of this chapter this will form the basis of all of the imaging methods described. But how is this effect used in a practical instrument? The basic principles of all fluorescence microscopy are illustrated in figure 9. Light from a monochromatic source (single wavelength) is directed onto the sample through the illumination lens and the emission detected after a filter, used to block the source light. In most systems the excitation is now through the objective lens (as shown in figure 9) in what is known as the *epi-fluorescence* configuration. The light can be directed through the normal condenser lens, opposite the objective, but this method generally means that without very high quality spectrally separating coatings and filters some of the excitation light leaks through onto the detector and so the nature of the fluorescence emission (light is emitted in all directions) is utilised to improve the rejection of the illumination light. Thus the light from the source is directed into the back of the objective through a dichroic filter. This filter reflects one spectral region (in this case the shorter wavelength) and transmits another (in our case the longer wavelength). The light then passes into the sample, is absorbed and then re-emitted at a longer wavelength and hence passes back through the dichroic element onto the detector. The excitation light is thus blocked from reaching the detector. In certain circumstances a second blocking filter can be employed before the detector to ensure the complete removal of the shorter wavelength light. The detector can be the eye, film, a digital camera or in the systems described below a photomultiplier. In certain circumstances the sample may be labelled with several flurophores that are either excited by one wavelength (but emit at several) or more normally excited at different wavelengths. This means that several different constituent parts of the cell can be imaged in sequence, or through the use of multiple filters and detectors, simultaneously.

In order to achieve this on a conventional microscope a spectrally narrow source, as described earlier is required along with the dichroic beam splitter and filter combination. This is typically provided by the manufacturer in a single cube which can be slide into the optical path of the microscope. Initially the eye provides a very useful detector but again cameras are normally incorporated into the system to record the image. It is worth noting here that the light levels in fluorescence microscopy are significantly lower than in the more conventional methods and hence a higher quality detector is required. In the case of digital cameras this normally means a cooled camera to reduce the background thermal noise. However, standard commercial digital cameras have improved to such an extent in the last five years that even a reasonable priced system, providing you have some control over the shutter and can turn off the flash, can give excellent results.

Fluorescence microscopy thus provides a highly specific method of increasing the image contrast with high resolution. There are, however, some disadvantages. In most cases the sample requires careful preparation with the addition of specific compounds to enable the fluorescence to be generated. This can harm the cell, and without doubt means that one has perturbed its natural state to some extent. Sample preparation is a highly skilled art and selecting the best flurophore and the correct concentration and incubation time only learnt with experience. In many cases the cell membrane has to be altered to allow the flurophore to permeate into the cell. In addition the excitation light slowly degrades the fluorescent compound and the sample becomes dimmer with repeated exposure. This effect is known as photobleaching and although a significant problem in confocal imaging it can also be an issue with wide field fluorescent microscopy. The breakdown products of the flurophore can also be toxic to the cell and many of the interesting labels require excitation in the near ultra-violet which can lead to cell damage directly from the light source. Despite these drawbacks fluorescent microscopy is probably the most powerful and widely used imaging technique in many laboratories.

4. CONFOCAL MICROSCOPY

All of the above techniques provide information on the sample in a two dimensional format. This means that in the case of thicker samples (frequently those of over a cell thick) information from above and below the focal point of the objective lens is superimposed on the image from the plane of interest. In order to minimise this effect the above techniques are generally used on isolated cells or very thin fixed sections. In an ideal world one would like to image the samples with thin sections and build up a three dimensional image of the sample without physically sectioning the tissue. In this section a number of methods are outlined which provide the optical sectioning capability. In a brief chapter only limited details can be given and the reader is referred to several excellent practical books on the subject [8, 9].

The basic principles of confocal microscopy are outlined in figure 10 and apply to both reflection and fluorescence imaging, though nearly all life science confocal imaging is

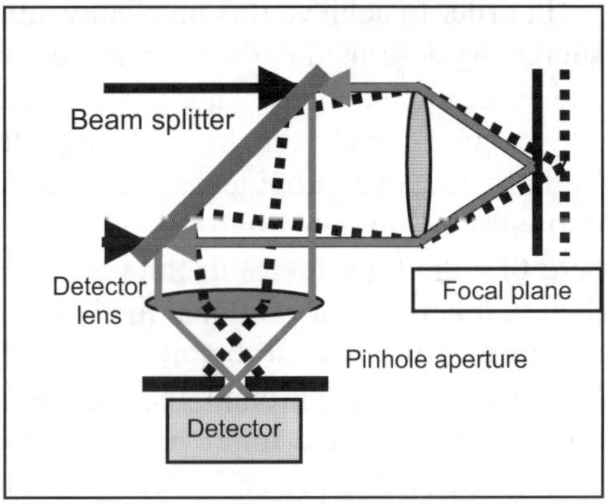

Figure 10. Confocal imaging

undertaken on labelled samples. The light returned from the sample at the focal point of the microscope objective is collimated (in a modern infinity corrected microscope) and then focused down via a second lens such that it passes through a pinhole before reaching the detector. Light from above, or below, the focal point of the objective is not collimated as it leaves the sample meaning that after the second lens most of the light is blocked by the pinhole. One thus has a method of separating the light from different focal planes and one therefore has a method of taking an optical section from the sample and by altering the position of the objective one can obtain the light returned by different levels within the sample. However, as described, this will only return the image from a single column within the sample. In order to obtain a full three dimensional image this point needs to be translated across the sample. One method, as was originally used, would be to translate the sample which although in many respects the simplest option it is not practical for most life science work. Hence an optical scanning method is required.

4.1. Point Scanned Microscopy

The most commonly used method of scanned confocal microscopy is one in which the excitation spot is raster scanned across the sample using a pair of scanning mirrors. Each point in the plane is thus illuminated in sequence. Figure 11 illustrates the basic principles of the technique.

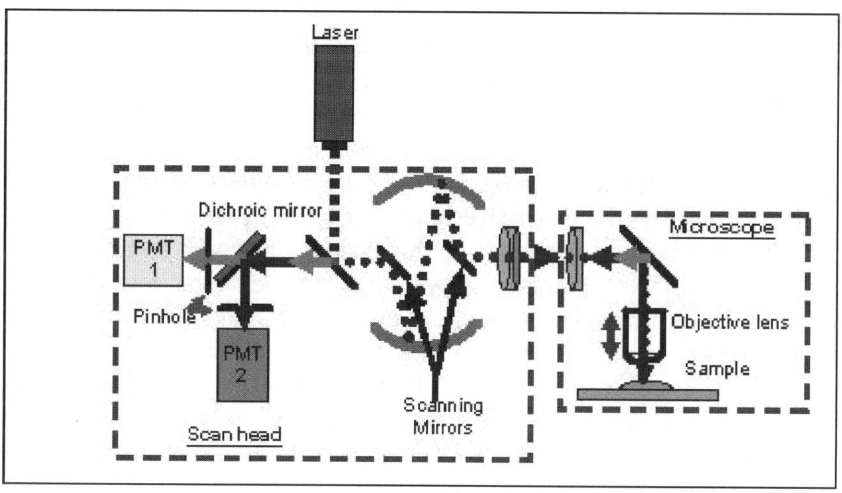

Figure 11. Point scanned confocal scan head using scanning mirrors

Light from the excitation source (almost universally a laser as a high brightness, good quality light beam is required) is directed onto the sample using a dichroic beam splitter (though a partially reflecting optical can be used for reflection imaging). The light then passes onto a pair of scanning galvanometer mirrors before being directed onto the back aperture of the microscope objective. The returned light is then re-imaged by the objective onto the scanning mirrors in which the beam is descanned to pass through the dichroic mirror and the pinhole to reach the detector. Without the descanning on the returned path the beam would only pass through the pinhole in one position of the scan and the descanning thus means that only a single pinhole is required rather than one for each position of the scan. In many systems a multiple series of dichroic mirrors is used with several detectors enabling the fluorescence for a number of flurophores to be collected simultaneously. The returned signal for each point on the scan is stored in a computer and the image thus built up pixel by pixel. After a scan has been obtained for one plane the objective can be moved in or out of the sample to build a full three dimensional image of the specimen. Typically a single image plane can be recorded in around 1 second (for a 512 x 512 pixel image).

The user has control of a number of parameters all of which affect the final imaging quality. As usual the NA of the objective lens plays a crucial role now determining both the lateral (as described above) and axial resolution. In the latter case the resolution is now given approximately by,

Axial resolution $\sim (1.78 \times \lambda)/(NA)^2$

With the use of the confocal pinhole there is an additional improvement in the lateral resolution beyond that of wide field imaging and I would refer the interest reader to reference 8. By adjusting the size of the pinhole the user can reject more of less of the light from outside the focal plane of the objective. Normally the pinhole will be closed down to obtain the best possible resolution (the exact size being set by the optical configuration of instrument and beyond the scope of this chapter) but with feint samples, or ones in which the excitation level has to be kept low to reduce photobleaching, it may be opened. As mentioned above typically a laser is used for the excitation and until a few years ago this was almost universally an air cooled argon ion laser with a lifetime of only a few thousand hours and a source of considerable heat in the laboratory. With the advent of laser diodes, in particular those which operate in the blue compact and reliable sources are now available [10]. Detection of the low light levels emitted is normally via a photomultiplier or avalanche photodiode and the selection of the correct detector is an important decision in the purchase and use of a scanning confocal system.

All of the major microscope manufacturers have confocal imaging systems based upon the principles outlined. They may vary in the way that the scan is implemented, the exact configuration of the microscope but all contain the essential scanning and confocal pinhole. The exact choice of instrument being determined by the exact user requirements such as speed of imaging required, flurophores being used, sample status (live or fixed) and money available! The software to reconstruct the images is also important to the overall performance of a confocal microscope as the presentation, quantification and analysis of three dimensional data sets is really the end result desired of confocal microscopy.

4.2. Multiple Point Scanned Microscopy

A variation on the point scanned system is through the use of a rotating array of small micro-lenses and a matched array of holes (so called Nipkow Disk). As illustrated in figure 12 an expanded laser beam is incident on a rapidly rotating array of micro-lenses. These focus part of the beam down onto a matching array of holes and onto the sample. The returned light then passes back up through the same hole and is reflected back onto a CCD detector (or in fact the eye can be used). By careful arrangement of the lenses in a spiral pattern it is possible to scan the entire sample at high speed. Typically over 7000 lenses and holes will be used rotating at speeds of 1000rpm.

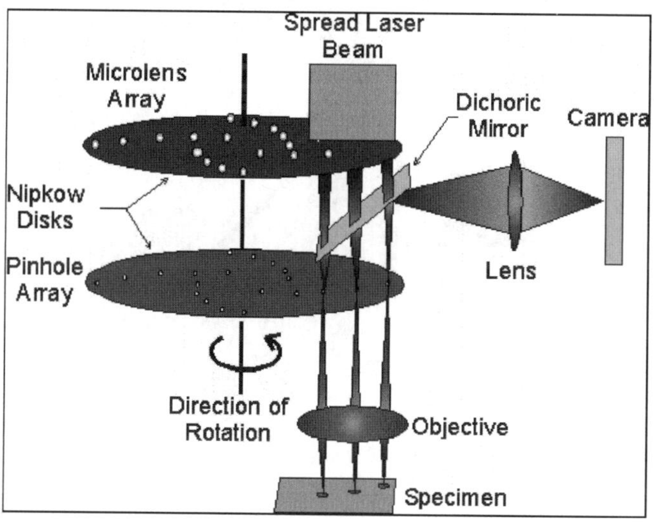

Figure 12. Nipkow Disk Confocal configuration

These systems provide some optical sectioning capability but are not as efficient as rejecting the out of focus light as the more conventional point scanned system [11]. In addition they are wasteful of the laser light as only a small portion of the beam is used at any one time. The main advantage that such a system has is that very rapid imaging is possible with frame rates well in excess of 50fps the limitation mainly being due to the sensitivity of the camera and efficiency of the sample. They are well suited to imaging calcium transients and fluorescence from highly punctuate samples at high speed and are commercially available from several sources.

4.3. Structured Light Microscopy

An alternative approach to obtaining optical sectioning is provided by the method of structured light [12] and a brief description is included here as a commercial "add-on" to a standard microscope is now available. The basic system is shown in figure 13. Light from a non-coherent source, such as a light bulb, is collimated and directed onto a grid with a regular pattern of lines, ideally high contrast sine waves of a single frequency.

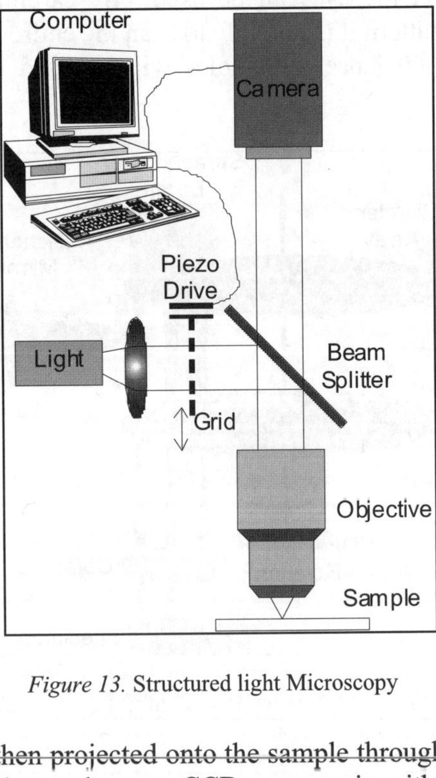

Figure 13. Structured light Microscopy

This grid pattern is then projected onto the sample through the objective and the returned light detected on a CCD camera in either a reflection or fluorescence mode and an image captured. The grid is then translated laterally by a third of its spacing and second image taken before it is advance a further third and the final picture stored.

Each of the images has the grid pattern present and if all the three images are directly added together one just has the conventional wide field image. However, if the images are manipulated in another way,

$$I_{section} = [(I_1 - I_2)^2 + (I_1 - I_3)^2 + (I_2 - I_3)^2]^{1/2}$$

then only the image from the focal plane adds to remove the grid pattern and leave a uniform illumination, whereas the light from outside the focal plane combines to leave a uniformly dark image. This means that after combining the images one only has the image from the focal plane left, and hence an optical section has been obtained.

This method does not provide the high optical sectioning of the pinhole, in practical instruments, but is significantly simpler than any scanned system and hence has a role as a low cost optical sectioning system. One particular role is as an addition to an endoscope where sectioning can therefore be obtained during a minimally invasive surgical procedure.

4.4. Multiphoton Microscopy

Although this subject is covered elsewhere in this book multiphoton microscopy is mentioned here for completeness. In this method light at twice the wavelength normally used to excite a flurophore is directed onto the fluorescently labelled sample. Due to the rules of quantum mechanics there is a probability that two photons well be absorbed simultaneously if sufficient photons are present but the chance is not high. However, at the focal spot of a laser beam a sufficiently high intensity can be produced for enough fluorescence to be generated and subsequently be detected. The effect is not large and generally a laser power of around 1kW is required which would burn most samples. If an ultra-short pulse laser is used in place of a continuous source then the instanteous power may be very high, but the average power low (a few mW) and hence the sample will not be destroyed. The excitation region will be confined to a small volume in space determined by the NA of the objective lens used, but crucially it is confined in three dimensions and thus by collecting the fluorescence generated from this point in space and then scanning the spot one can build up a three dimensional image [13, 9].

In order to achieve such a system a laser producing femtosecond pulses is typically used operating between around 750 and 950nm. This light is directed into a scan head as described above for the point scanned confocal system. The resulting fluorescence, however, does not need to be descanned as it can only have come from one point in the sample and hence can be detected directly behind the objective lens. The long infrared wavelength used for the excitation means that in depth imaging is possible as there will be no photobleaching outside the focal plane and the long wavelength light is scattered less by the sample than the blue or UV light normally used for fluorescence microscopy. There is also a reduced risk of photo-toxic damage to the sample compared to confocal microscopy.

On the downside the systems are very expensive and generally femtosecond lasers are difficult to operate, though the desire of the life scientist for easier to use systems has lead physicists to develop a number of improved methods and techniques for multiphoton instrumentation [14]. Using the technique it is possible to image at depths up to 2mm inside a living sample and it is without doubt the method of choice for in depth, live sample microscopy.

5. EMERGING TECHNIQUES

It is now worth considering two methods of imaging that have emerged in recent years which provide the capability to image beyond the limit set by the diffraction of light and outlined at the start of the chapter.

5.1. Stimulated Emission Depletion Microscopy (STED)

This is a method pioneered by Stefan Hell and lateral resolutions of 28nm have been reported through the use of stimulated emission to deplete the fluorescently excited volume with a second laser pulse [15]. In STED two laser beams are aligned co-linearly through the objective, but with different beam profiles (Figure 14).

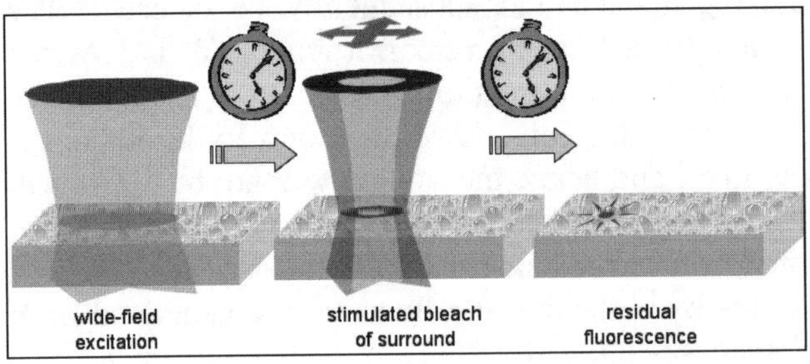

Figure 14. STED microscopy using the dark region at the centre of an annular beam to get sub wavelength resolution

A pulsed excitation beam is used to excite the fluorescence within the sample with pulses up to 100ps. A second pulse between 100fs to 300ps, tuned to the emission wavelength is timed to arrive at the end of the excitation pulse and with a spatial profile that overlaps the excitation pulse, but generally has a dark "hole" overlapping the central point of the original excitation beam. This second beam causes stimulated emission from the sample and hence a gated detector (activated after the end of the second pulse) only receives fluorescent light from the un-depleted, central portion of the initial excitation volume. Increasing the intensity of the stimulating beam effectively reduces the "size" of the central dark regions meaning that the remaining fluorescence comes from a smaller area: which can be sub-wavelength. In the work to date the depletion beam shape has been generated through the use of conventional fixed phase plates which imposes a significant limitation on the imaging options and requires moving the sample under the illumination region. The method has thus not yet been used on a wide range of biological samples, and there are limitations on the flurophores that can be used when considerations of fluorescent lifetimes etc are considered but it an area advances can be expected in the next few years towards practical systems.

5.2. Total Internal Reflection Microscopy

The final method to be explored is known as Total Internal Reflection (TIR) or Evanescent Wave Microscopy. Here high axial resolution (but not lateral) can be obtained through the use of evanescent light waves coupling into the sample. In the simplest configuration light is sent into a glass slide such that it is guided through the slide by total internal reflection and if the slide is viewed normal to the travelling wave nothing is seen. However, if a sample is placed onto the slide a proportion of the light is coupled out from the slide into the sample and can be detected normally to the slide. This evanescent wave only couples 50-100nm at most into the sample and hence one can obtain very high axial resolution. In a more recent variation, through the use of standing waves, lateral resolutions of $1/6^{th}$ the wavelength of light have been achieved [16]. The technique can be used in both reflection and more normally fluorescence. Full details can be found in chapters in reference 9.

6. SUMMARY

This chapter has laid out the main principles behind optical microscopy. The aim was not to present all of the details and complex mathematical formulation present behind the instruments but to provide the life scientist with an understanding of the physical principles so that the microscope can be used to provide high quality images with sufficient unambiguous information to aid research. The practical implementation of standard instruments was then presented with an emphasis on how best each instrument should be used. With increasing complexity the chapter has ended with three dimensional imaging techniques with optical resolutions down to 28nm, approaching that of electron microscopes. Optical microscopes have developed a long way in 400 hundred years and it is to be anticipated that with advances in new technologies new tools will soon be provided to the life scientist for live cellular imaging.

7. REFERENCES

1. Born M. and Wolf E., *Principles of Optics (7th Edition)*. Cambridge University Press, 1999
2. Hecht E., *Optics (4th Edition)*. Adison Welsley, 2001
3. Fowles G., *Introduction to Modern Optics (2nd Edition)*. Dover Publication, 1989
4. Dempster J., Wokosin D. L., McCloskey K. D., Girkin J. M., Gurney A. M. WinFluor - An integrated system for the simultaneous recording of cell fluorescence images and electrophysiological signals on a single computer system. British Journal of Pharmacology 2002; 137: 146P Suppl. S
5. Murphy D. B., *Fundamentals of Light Microscopy and Electronic Imaging*. Wiley-Liss, 2001
6. Inoue K., *Video Microscopy*. Plenum Publishing, 1986
7. http://www.microscopy.fsu.edu provides an excellent interactive tutorial on all aspects of microscopy. 2004
8. J. Pawley, *Handbook of Biological Confocal Microscopy (2nd Edition)*. Kluwer Academic Publishers, 1995
9. Diaspro A ed, *Confocal and Two-Photon Microscopy : Foundations, Applications and Advances*. Wiley-Liss, 2001
10. Girkin J. M., Ferguson A. I., Wokosin D. L., Gurney A. M Confocal Microscopy using an InGaN violet laser diode at 406nm. Optics Express 2000; 7: 336-341
11. Egner A, Andresen V., Hell S. W. Comparison of the axial resolution of practical Nipkow-disk confocal fluorescence microscopy with that of multifocal multiphoton microscopy: theory and experiment. Journal of Microscopy 2002; 206: 24-32
12. Neil M. A. A., Juskaitis R., Wilson T. Method of obtaining optical sectioning by using structured light in a conventional microscope. Optics Letters 1997; 22: 1905-1907
13. Denk W.,. Strickler J. H and W. W. Webb Two Photon laser scanning fluorescence microscopy. Science 1990; 248: 73-76
14. Girkin J. M. Optical Physics Enables Advances in Multiphoton Imaging. Journal of Physics D 2003; 36: R250-258
15. Klar T. A., Engel E., Hell S. W. Breaking Abbe's diffraction resolution limit in fluorescence microscopy with stimulated emission depletion beams of various shapes. Phys Rev E 2001; 64: 066613
16. G. E. Cragg, So P. T. C. Lateral resolution enhancement with standing evanescent waves. Optics Letters 2000; 25: 46-48

INTRODUCTION TO TRANSMISSION AND SCANNING ELECTRON MICROSCOPY

Franco Verni and Simone Gabrielli
Dipartimento di Etologia,Ecologia, Evoluzione, Università di Pisa, via A. Volta 4 56126 Pisa Italy

Abstract: This chapter is a modest attempt to present the basic techniques in both transmission and scanning work. The recommendation outlined in this chapter are likely to be considered standard protocol. As new specific techniques are developed or older once evolved the literature aimed at solving specific problems should be considered.

Keywords: Preparation Techniques; Cpd

1. INTRODUCTION

The goal of the microscopic observations, including electron microscopy (EM), is the examination of morphologically related properties of living or once living specimens. In this chapter will come described the basic treatments of transmission and scanning electron microscopy for biological specimens.

Electron microscopes function exactly as their optical counterparts except that they use a focused beam of electrons instead of light to "image" the specimen and gain information on its structure and composition (Figure 1).
The basic steps involved in all EMs are:
stream of electrons is formed (by the electron source) and accelerated toward the specimens using an electrical potential; this stream is focused using metal aperture and magnetic lenses into a thin monochromatic beam; the beam is focused onto the sample using a magnetic lens.

V. Evangelista et al. (eds.), From Cells to Proteins: Imaging Nature across Dimensions, 23–35.

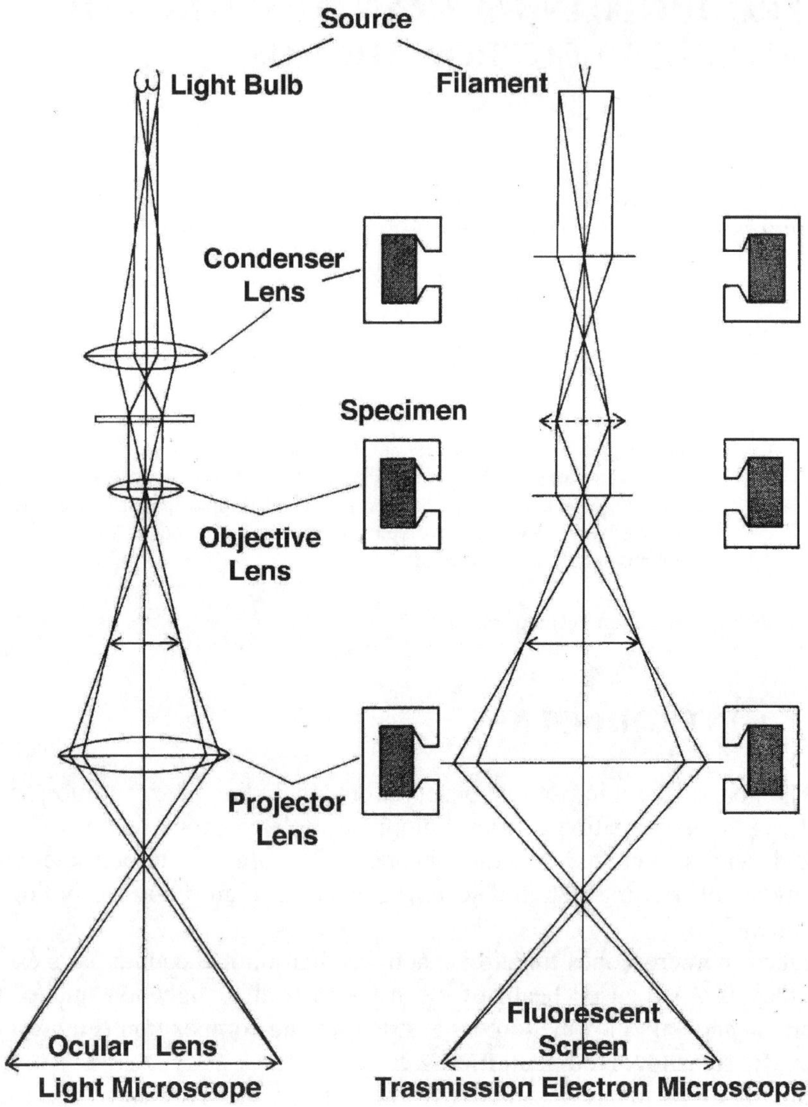

Figure 1. Comparison between lens system of optical and electro microscopy

Today it is possible to use specific approaches to investigate particular detail of biological samples: cytochemistry [1] immunolabel [2-3]; cryosection [4] freeze-fracture [5]. In this chapter are presented the main conventional steps (Figure 2).

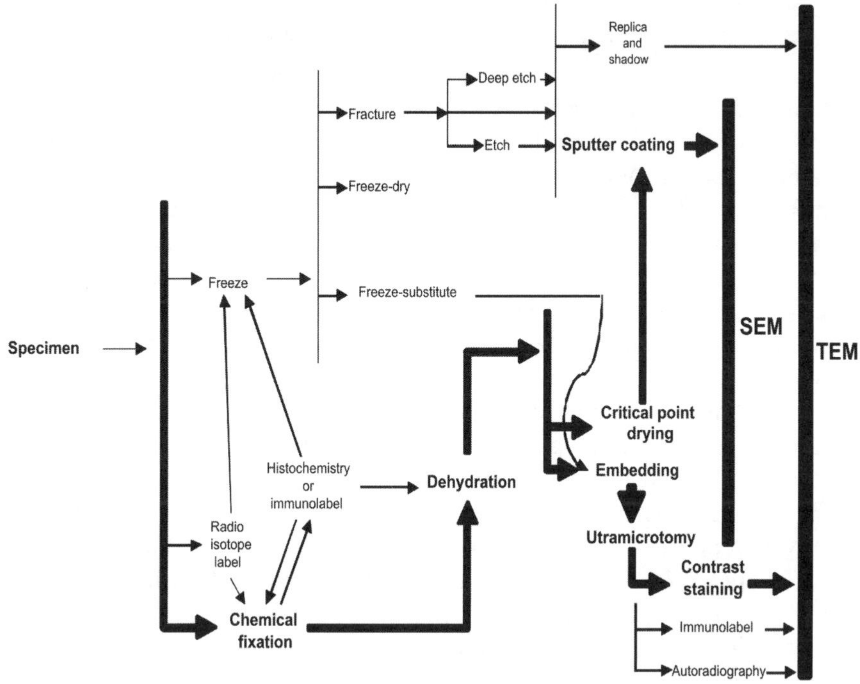

Figure 2. A diagrammatic summary of several approaches to use of electron microscopic techniques (SEM and TEM). The main conventional routine is emphasized. Modified from [6]

2. FIXATION

Observations may include localizations, identifications and analysis of specific chemical entities, but always are related to the specimen's structural components. Since physical principles require specimen observation in conventional electron-beam instruments to a vacuum, is not possible to observe living organism directly by electron microscopy.

In most cases, whichever the microscopical method surveying agrees to apply, is not possible to prescind from the fixation. Infact the fixation is basically the first step of artificial modification of cellular environment and consists in the attempt of instantaneous immobilization of all molecular cell components. Chemical systems (fixatives) are the most common.

Aldehydes, whether glutaraldehyde, formaldehyde, or combinations of the two, are the nearly universal choice as primary fixatives in biological electron microscopy [7] The salient property of glutaraldehyde, introduced as a fixative by Sabatini et al [8], is that it crosslink proteins, including lipo and glycoproteins, into stable gels with very little distortion of their spatial relationships. On the other hand , glutaraldehyde reacts poorly, if at all, with lipids; thus a major of the cellular constituents is not stabilized during initial fixation. Formaldehyde is a useful primary fixative for EM cytochemistry, but not nearly so good as glutaraldehyde for morphology. On the other and it has a more rapid penetration into the sample, in comparison with glutaraldehyde. Karnovsky's fixative [9] is a mixture of glutaraldehyde and formaldehyde. The rationale behind its formulation is that the former component can penetrate rapidly and start the fixation, followed by the slower, more efficient. The purity of glutaraldehyde has been a matter of debate over the years. Products commercially available contain a variety of poorly defined impurities and are complicated by the tendency of glutaraldehyde to polymerize spontaneously. A typical artefact correlated with the use of impure glutaraldehyde, is the explosion of the mitochondrial matrix

The use of the osmium tetroxide for fixation in EM derived directly from its known excellence as fixative for LM. Despite many decades of use as a fixative, its chemical reactions with the tissue components are still poorly characterized. Most importantly, osmium stabilizes lipids by cross linking or otherwise rendering them insoluble in organic solvents. It thus supplies the fixing properties missing in aldehydes. Since glutaraldehyde and osmium, added to issue sequentially have been found to stabilize most cytoplasmic elements, it has not been unreasonable to try applying them simultaneously in the attempt to avoid some of the problems concerning lipid stabilization and slow killing observed with glutaraldehyde fixation. Nevertheless, practically this provides limited benefit for certain purposes and is not to be recommended because the experimenter has not rational way to control the fixation. Infact glutaraldehyde and osmium react chemically with each other, producing at least osmium polymers (Osmium black) and glutaric acid. Good results have been obtained with this method in ciliated protozoa [10]

3. BUFFER VEHICLES

Fixatives almost always are carried into the tissue in a buffer solution. It is not at all clear that buffering *per se* is the most important contribution of the buffer vehicle, since the tissue itself is a buffer. What is clear, however, is that specific ion effects of the buffer vehicle are important to the overall fixation process. Certain buffer solutions of aldehydes and osmium produce fewer artifacts than others, and simple aqueous fixative solutions (i.e., without salts or other additives) produce inferior results.

The role and specific composition of the buffer vehicle are of greater importance during initial fixation with aldehydes than at any other stage in fixation procedure. Buffers most commonly used are phosphate, cacodylate. and organic acid such as PIPES and HEPES [11]. The differences among buffers are largely due to the degree of their interaction with the fixing agent in promoting stabilization, or loss of cell constituents.

Phosphate is the best buffer for a wide variety of specimens. However, it forms insoluble compounds with physiologically important cations such as Ca^{++} and Mg^{++} and with cytochemical reagents such as lead

Buffers for fixatives usually are used at concentration such as 0.05 to 0.2M, the 0.1 is the most common. Fixatives are almost always adjusted according to physiological pHs ranging from 6.8 to 7.4

In the table is represented a baseline fixation protocol

Immerse samples in glutaraldehyde solution for 40 min (extend the fixing time if temperature below 20°C is used) *Glutaraldehyde fixing solution: 3% (purified) in 0.1 M sodium phosphate buffer, pH 7*	Rinse for 5-10 min in 0.1 M sodium phosphate buffer, pH 6.8	Immerse samples in osmium solution for 30 min *Osmium tetroxide fixing solution: 1% OsO4 in 0.1 M sodium phosphate buffer, pH 6.8*	Transfer directly to cold 30% ethanol and proceed through a graded series of cold ethanol

Many substances have been recommended as components of fixative solutions to enhance preservation or contrast, either for general use or for specific applications:

1) Sucrose, or other polymers like dextran, often are added to adjust the fixative solution osmolality.
2) Divalent cations, especially calcium (1.2 mM) may improve the fixation of membranes
3) When used with glutaraldehyde, amine, especially lysine, improve fixation of glycocalyx and cytoplasmic filaments.

4. DEHYDRATION

Dehydration of fixed biological specimens is a critical step in specimen preparation, since, for most epoxy resins, little or no water can be left in the tissue if proper infiltration and polymerization are to take place.

There are a number of dehydrating fluids, acetone and ethyl alcohol being the most common. Propylene oxide, although sometimes considered a dehydrating agent, is more appropriately used as a solvent during infiltration of tissue with epoxy resin. Propylene oxide is not really miscible with water and can only be used in the final stage of dehydration [12]. All solvents commonly used as dehydrating agents cause cell shrinking this is especially obvious in plants or other tissue containing vacuolated cells because vacuoles usually shrink to more than the other cell constituents. To limit the shrinkage of the specimens, dehydration is gradually performed.

A typical dehydration series for animal tissue might be 30 %, 50%, 70%, 90%, 95% and 3x 100% with each step of 10-15 min. The length of each step is dependent on the type and size of sample. The same dehydrating agent doesn't need to be used for the whole sequence; it is possible to start with ethyl alcohol and switch to acetone or propylene oxide at 100% concentration.

5. INFILTRATION AND EMBEDDING

After fixation and dehydration, the structure of the specimens must be rapidly penetrated by an embedding medium appropriate to the investigation, prior to polymerization. This infiltration step involves the progressive replacement of the final dehydrating solvent with an embedding medium which will subsequently be hardened. If the dehydrating solvent is poorly miscible or immiscible with the embedding resin, then a transition medium, miscible with both, must be employed as an intermediate step.

By far the best route to good specimen fixation and infiltration with minimal extraction of cell components is via small specimens less than 1 cubic mm or thin sheet of tissue less than 1 mm in thickness. The rate of infiltration of tissue with resins can be increased markedly by reducing the viscosity of the resin mixture.

For routine EM investigations the embedding medium will be a thermally cured epoxy resin, although a range of embedding media has been developed over the years and is commercially available for specialist preparation techniques such as X ray analysis , cytochemistry and immunocytochemistry (see Rosati Gabrielli in this book)

The more common steps for infiltration and embedding are:

1. Last dehydration step (100%) dry acetone –30 min
2. 50:50 acetone:resin – 45 min
3. 100% resin (in a fresh vial) – overnight
4. 100% resin (at 40°C) – 1h
5. Embedded in gelatin capsules or flat embedding system
6. Polymerized at a temperature appropriate to the resin mixture

Epoxy resins are used for a variety of application, the least economically important is the embedment of tissue for electron microscopy. The first report on practical application for E.M. was by Glauert et al. [13] who used a resin designated as Araldite M® (CIBA corporation). The complete resin formula included an anhydride hardener dodecenyl succinic anhydride (DDSA) and accelerator 2,4,6 –tri(dimethylaminomethyl)phenol (DMP-30)

the latter to bring the reaction to completion in a reasonable period of time. Although the mixture can be stored for up to 10-12 months at -20 °C it is highly recommended that freshly prepared embedding medium always be used. If you choose to store the mixture you should warm it thoroughly prior to adding the accelerator. Many other types of hardners, diluents, catalysts and accelerators are now available. In any event, the introduction of epoxy resins opened a new era for EM. Tissue preservation was excellent, block shrinkage was negligible, sectioning was not too difficult, stability under the beam was good and the contrast of tissue sample was generally acceptable.
Ideal qualities of embedding medium are:

1. Easily available
2. Uniformity from one batch to another
3. Solubility in dehydrating agents
4. Low viscosity as monomer for penetration
5. Uniform polymerization
6. Good sectioning quality that includes homogeneity, hardness, plasticity and elasticity
7. Resistance to heat generated by sectioning
8. Stability in electron beam

6. SECTIONING

The first step of sectioning is the trimming. Trimming is the process of removing excess resin from around the specimen or isolating the area of interest. Inaccurate or improper trimming can introduce sectioning problems. The area for sectioning should be kept as small as possible less than 1 squared millimeter. The upper and the lower edges of the block should be parallel. The sides are trimmed at an angle to produce a block face that has a trapezoidal shape. All embedded tissues are, of course, intended to be sectioned and this often is a major consideration in the choice of an embedding formulation. There are numerous factors that contribute to, or cause artifacts in ultrathin sections including enviromental/physical conditions of the microtome and knife edge, the process itself, and the type and condition of both specimen and resin. Each of these factors can be corrected or, at least, minimized to allow for defect- free ultrathin sections commensurate with the skill and patience of the ultramicrotomist Glass knives for resin sectioning can produce only a finite number of sections. As a rough guide, a ribbon approximately 20 mm long may be cut before the edge causes sectioning defects. These defects often take the form of very small scores which cannot immediately be seen under the stereomicroscope of the ultramicrotome, but only in electron microscope. It is therefore good practice to limit the number of sections cut over each area of knife edge. These problems are partially solved with the diamond knife.

Biological, and to a lesser degree, material specimens must be supported in TEM by some kind of perforated support. The composition of these "grids", as they are universally termed, is nearly as varied as the grid patterns available. Copper is probably still the most popular metal, although nickel is preferred where a less reactive metal is desirable, while for the greatest chemical resistance solid gold or platinum grids are useful

7. STAINING

Although it is quite possible to use just a single stain method, such as lead citrate or uranyl acetate or phosphotungstic acid for routine purposes to obtain a general increase in contrast, most electron microscopists prefer to employ some double staining. Usually ultrathin sections are treated firstly with a solution of uranyl acetate and then with a solution of lead citrate. Double staining is preferred, not only because it produces more contrast, but also because it stains more cell components

En bloc staining with uranyl acetate after osmication and before dehydration can be recommended in cytological studies. The preservation of both membranes and structure containing proteins and nucleoproteins is markely enhanced [14]

8. CRYTICAL POINT DRYING

Fixation and dehydration for the SEM are carried out similarly to those for the TEM. The times in primary fixative, buffers, and alcohol may be extended if the tissue pieces are large. After dehydration in 100% ethanol, the SEM protocol requires drying without introducing surface tension artifacts

The advent of Scanning Electron Microscopy (SEM) in the study of surface morphology in biological applications made it imperative that the surface detail of a specimen was preserved. Specimen air drying can cause severe deformation and collapse of structures; the primary cause of such damage being the effects of surface tension. The specimen is subject to considerable forces, which are present at the phase boundary as the liquid evaporates. The most common specimen medium, the water, has a high surface tension to air. Surface tension could be reduced by substitution with a liquid with a lower surface tension with thereby reduced damage during air-drying. However, the occurrence of what is known as continuity of state suggests a drying technique for which surface tension can be reduced to zero. If the temperature of liquefied gas is increased the meniscus becomes flatter indicating a reduction in surface tension. If surface tension becomes very low liquid surface becomes very unsteady and ultimately disappears.

When this 'critical point' is reached, it is possible to pass from liquid to gas without any abrupt change in state. A specimen that is in the liquid during this process experiences a transition from a ‘wet’ to a 'dry' gas environment without being in contact with a surface, avoiding in this way the damaging effects of surface tension.

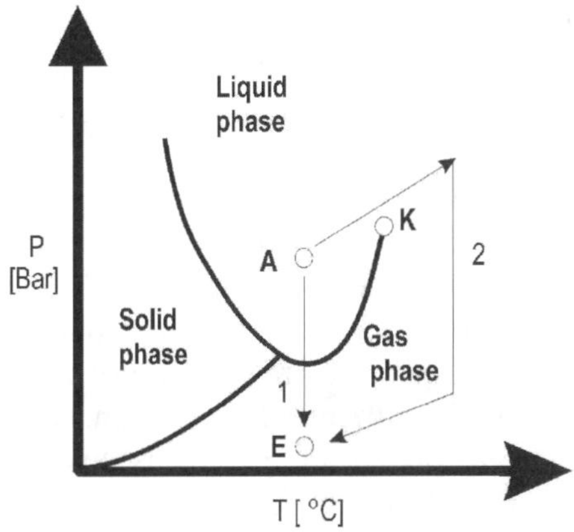

Figure 3. Dehydration of a sample by drying in air (1) or by trough the critical point (2). In both cases the procedures starts from in liquid phase (A). During air drying the liquid where the sample is immersed immediately turns into gas (E). By means of CPD, the step through the critical point (K) allows the liquid to turn into gas without be subjected to surface tension

The critical phenomenon can be used as a drying technique suitable for delicate biological specimens as it achieves a phase change from liquid to dry gas without the effects of surface tension. Carbon dioxide remains the most common medium for the CPD procedure and is termed the 'Transitional Fluid'. However, CO_2 is not miscible with water, therefore water has to be replaced in the specimen with another fluid, which is termed the 'Intermediate Fluid'. If that is able to replace water in the specimen it also serves as the dehydration fluid. Tetramethylsilane (TMS) is an alternative method for preserving excellent surface details [15]

9. MOUNTING

Virtually all specimens need to be affixed or mounted on a stub that can be accommodated in the stage mechanism of the SEM. This is accomplished by glue, silver-or-carbon based paint, double sided adhesive tapes, and other adhesives. Care must be taken in handling delicate biological specimens with forceps or other tools while transferring them to the mounting stub. All order to prevent or minimize charging effects.

10. COATING

The vast majority of dried biological specimens do not possess sufficient electrical and thermal conductivity for the observation with secondary electron imaging (SEI) in SEM. As consequence, various coating techniques have long been used which deposit metal or carbon over the surface of the specimen. Such deposit are normally in the range of 10-25 nm of thickness Figure 4). Most sputter coater use a gold or a gold /palladium (40/60) alloy as target. Both kind give a particularly fine coat when used under the appropriate condition of atmospheric cleanliness, adequate vacuum, and low voltage of the power supply. Metal coating not only aids in thermal and electrical conductivity, but gives somewhat more mechanical stability to many fragile biological specimens. In addition, a stronger secondary electron yield will be obtained from specimens coated with gold or other heavy metal since valence electrons from the metals are weakly bound and therefore easily ejected as secondaries.

While virtually all investigators store specimens for future observation, the best advice that can be given is to observe the material as soon as possible and keep storage time to a minimum.

Figure 4. A sputter coater used to coat non-metallic samples with a thin layer of gold (left) and during the glow discharge (right). Ion bombardment of the cathode will occur, this results in the erosion of the cathode material and is termed plasma sputtering, the subsequent omni-directional deposition of the sputtered atoms forming coatings of the original cathode material.

11. PREPARATION OF FREE CELLS FOR INVESTIGATION OF INTRACELLULAR STRUCTURES

In some cases, it is desirable for SEM investigators to reveal internal structure and organization in the specimens. It is possible observing intracellular structures of free cells by SEM by means of a not stable method. :This method has been attained by using a combination of chitosan embedding and the Osmium- DMSO- Osmium (O-D-O) method [16] Using the O-D-O method, which was devised for the 3-D visualization of intracellular structures, tissue blocks must be cracked with a razor blade and hammer. However, since it is difficult to make a specimen of free cells because they are too small to be cracked, it is better to embed them in some embedding material before cracking. Good results are obtained when chitosan is used as embedding agent. Chitosan, a polysaccharide (Molecular weight about 200,000 two hundred thousand) with a molecular structure similar to cellulose and chitin, can be produced from chitin extracted from crab shells. The embedded specimens are immersed in DMSO and freezed on metallic plate chilled with liquid nitrogen and successively Cracked with blade razor and hammer. In figure 5 are schematized the first phases of this procedure.

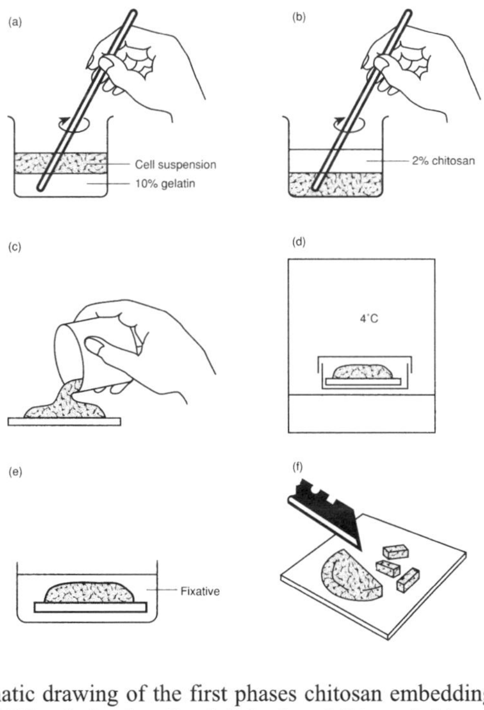

Figure 5. Schematic drawing of the first phases chitosan embedding method. Mixing of cell suspension and 10% gelatine in a Petri dish; the addition of chitosan and mixing; spreading on glass plate; cooling in a refrigerator to harden; fixation; cutting. Modified from [6]

For example with this technique a clear three-dimensional definition of the relationship between the two flagella, the paraxial rod and the photoreceptor in *Euglena gracilis* has been obtained (Figure 6) [17]

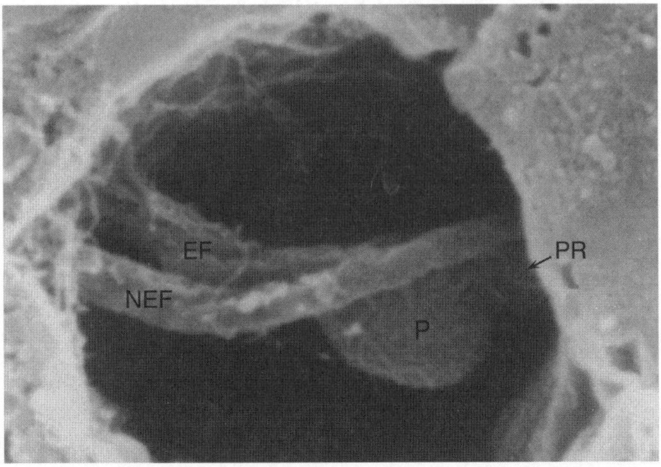

Figure 6. Longitudinal section of reservoir showing the non emergent flagellum (NEF), the emergent flagellum (EF), the flagellar swelling (photoreceptor) (P) and paraxial rod (PR) of *Euglena gracilis* From [17]

12. REFERENCES

1. Knight D.P.,Lewis P:R. "General cytochemical methods" In *Staining methods for sectioned material.* A.M. Glauert ed North Holland Amsterdam 1977
2. Brandtzaeg P. "Tissue preparation methods for immunocytochemistry" In *Techniques in immunocytochemistry vol 1,* G.R. Bullock, P. Petrusz eds Academic Press London 1985
3. Horisberger M. "The gold method as applied to lectin cytochemistry in transmission and scanning electron microscopy In *Techniques in immunocytochemistry vol 3,* G.R. Bullock, P. Petrusz eds Academic Press London 1985
4. Barthel L.K., Raymond P.A. Improved method for obtaining 3 micron cryosections for immunocytochemistry. J Histochem Cytochem 1990; 38:1383-88.
5. Willison J.H.M., Rowe A.J. " Replica, shadowing and freeze-etch techniques" In *Pratical methods in electron microscopy* Glauert A.M. ed Elsevier, Amsterdam 1980
6. Robards A.W., Wilson A.J., *Procedures in electron microscopy* Chicherster, U.K. Wiley & Sons, 1993
7. Lewis P.R, Knight D.P. "Introduction" In *Staining methods for sectioned material.* A.M. Glauert ed North Holland Amsterdam 1977
8. Sabatini D.D., Bensch K.G., Barnett R.J. Cytochemistry and electron microscopy. The preservation of cellular structure and enzymatic activity by aldehyde fixation. J Cell Biol 1963; 17: 19-58
9. Karnovsky M. J. The ultrastructural basis of capillary permeability studied with peroxidase as tracer. J Cell Biol 1967; 35:213-36.
10. Verni F., Rosati G. Peculiar epibionts in *Euplotidium itoi* (Ciliata, Hypotrichida). J Protozool 1990; 37:337-43.

11. Good N.E., Winget G.D., Winter W. Hydr
Biochemistry 1966; 5:467-77
12. Luft J.H. "Embedding media-old and new" In *Advanced techniques in biological electron microscopy,* J.K. Koehler ed Springer-Verlag, Berlin 1973
13. Glauert A.M., Rogers G.E., Glauert R.H. A new embedding medium foe electron microscopy. Nature 1956; 178: 803
14. Knight D.P. "Cytological staining methods in electron microscopy" In *Staining methods for sectioned material.* A.M. Glauert ed North Holland Amsterdam 1977
15. Reville W.J., Cotter M.P. An evaluation of the usefulness of air-drying biological samples from tetramethylsilane in preparation for scanning electron microscopy. J Electron Microsc 1991; 40:198-02.
16. Fukudome H., Tanaka K. A method for observing intracellular structures of free cells by scanning electron microscopy. J Microsc 1986; 141:171-78
17. Rosati G., Verni F.,. Barsanti L., Passarelli V., Gualtieri P. Ultrastructure of apical zone of *Euglena gracilis*: photoreceptors and motor apparatus Electron Microsc Rev 1991; 4:319-42

POTENTIAL AND LIMITATION OF CYTOCHEMISTRY

Giovanna Rosati and Simone Gabrielli
Dip. Etologia, Ecologia, Evoluzione Via A. Volta 4-6 56126 Pisa Italia

Abstract: Selected classic and modern cytochemical techniques are rapidly surveyed. The chemical principles on which they are based are briefly considered, the technical approaches and their potential and limits are analyzed and discussed. Significant examples of application of the different techniques to specific biological problems are also reported.

Keywords: Contrast Methods, Electron Microscopy, Enzyme Cytochemistry, Immunocytochemistry

1. INTRODUCTION

The introduction of the electron microscope as a tool for the biologists brought about a complete reappraisal of the micro-anatomy of biological tissues, organisms and cells. In the early days of its application to biological materials, it was the tool of anatomists and histologists, and many previously unimagined structures in cells were revealed. Successively, specialized techniques to obtain information on the chemical composition of the different cellular structures, the localization and variation of enzymatic activities, the localization of antigens or other specific components within cells, and the presence of single genes were developed. All these techniques, concerned with investigation at the ultrastructural level of chemical composition and activities of tissues and cells, are defined by the term Cytochemistry. A number of these techniques are in use since the beginning of electron microscopy and are now classic whilst others, such as immunocytochemistry, affinity cytochemistry and in situ hybridization, are more recent.

Selected cytochemical techniques are described in the following sections of the chapter and their potential and limits are discussed also with respect to specific examples of application to particular biological problems.

V. Evangelista et al. (eds.), From Cells to Proteins: Imaging Nature across Dimensions, 37–52.

2. SPECIFIC CONTRAST METHODS AND ENZYMATIC DIGESTIONS

2.1. Methods for carbohydrates

The term carbohydrates covers biological material ranging from simple polysaccharides such as stark and glycogen to the complex mucopolysaccharides or the mucoproteins present in many secretory granules. The most used method in light microscopical histochemistry, the PAS method, is thought to depend on the detection of aldehyde groups produced by oxidation with periodic acid. This oxidation step is thought to be moderately specific for sugar residues which contain a pair of free hydroxyl groups on adjacent carbon atoms. Thus, this oxidation step has been proposed also for electron microscopy. Several methods have then been devised for the detection of aldehyde groups at the ultrastructural level, most of them depending upon the reduction of silver ions to metallic silver. The use of osmium tetroxide vapor as developing of the aldehyde oxidation to reveal polysaccharides with periodic acid, introduced by Hanker et al. [1] and then abandoned, has been recently revalued by Guimarâes et al. [2]. With this method, according to the authors, an excellent increment of the general ultrastructural morphology can be obtained.

With these techniques special care must be taken to check the specificity of the staining obtained considering also that they are generally applied on ultrathin sections of fixed material, embedded in resins. For example, to control that the staining does not result from aldehyde or other reducing groups preexisting in the specimen or introduced by fixation, some sections should be treated with aldehyde-detecting reagent without first receiving periodic acid oxidation.

The most known and used of these methods is the "thiosemicarbazide- silver proteinate procedure" [3]. Ultrathin sections, picked up in mylar rings [4], are treated with 1% aqueous periodic acid at room temperature, carefully washed with distilled water, floated on 0.2% thiosemicarbazide (TSC) in 20% acetic acid, washed with decreasing concentration of acetic acid and again with distilled water, and finally treated with the silver proteinate solution in the dark. For the details of the method see [5]). According to Thiéry [3] variations in the length of TSC treatment can be useful in distinguishing simple polysaccharides from mucopolysaccharides or mucoproteins. For the former 1h treatment is sufficient, while for the latter the treatment should be prolonged up to 60-70h. The reaction product with this method is very fine, permitting excellent localization with high resolution.

2.2. Extraction methods

The use of enzyme solutions and other extractive reagents can often improve or confirm the specificity of the selective staining methods like that described above. Before describing the methods it is necessary to emphasize that care must be taken in the interpretation of all experiments with extractive agents. First of all it should be again considered that these methods are applied on specimens fixed and embedded. The length and nature of the fixation procedures, the embedding medium, the thickness of the ultrathin sections can all influence or even prevent the extraction of a particular cellular component. Therefore a negative result should not be considered an absolute answer. Moreover, in specimens preparation, there are uncontrollable factors. For example, the sensitivity of a cellular component to the fixatives does not only depend by its intrinsic nature but also by its accessibility within the cellular context, by its chemical links with other cellular components and by other physic-chemical factors that vary from cell to cell. Variations might even exist in the same cell type according to the functional state at the moment of fixation.

The most used groups of enzymes are carbohydrases and proteases. Glycogen can be removed from sections of material fixed in glutaraldehyde and osmium tetroxide and embedded in epoxy resins by extraction for 1-3 h at 37°C with amylase (5% in phosphate buffer pH7), provided the osmium is first removed from the section with hydrogen peroxide or periodic acid. The effect of glycogen extraction may be verified by the TSC-silver proteinate method. Stark granules in plant need a longer treatment (24h) to be extracted. Other enzymes for sugars, like hyaluronidase, neuraminidase, and chondroitinase have been proposed as valuable cytochemical tools [6-8].

Several enzymes like pepsin, trypsin, chymotripsyn, pronase are available for the extraction of proteins. For each of them a suitable medium at a determined pH is recommended. Standard protocols for their use can be found in the literature. The choice of enzyme must depend much upon the particular biological problem under investigation. Fixation is particularly important here since most fixatives alter the physical, and often the chemical tructure of proteins. Anyway, when osmium tetroxide is used it must be removed from the sections with hydrogen peroxide. Also the nature of the embedding resin has a marked effect on the extraction of different components by proteases. Although hydrophilic resins were originally recommended, epoxy and polyester resins are also suitable provided longer treatment times are used (up to 24 hr). In this case, as control, the incubation of some sections in the medium without the enzyme should be always performed at the same temperature and for the same time of experimental sections. In this way it could be verified whether the obtained extraction is really due to the enzyme or simply depends upon the long permanence of the oxidized section in an aqueous medium (note that the hydrogen peroxide itself can remove some tissue components or render them more soluble).

2.3. Application of combined specific contrast methods for carbohydrates and extraction methods to the study of Polysaccharide reserve in Ciliates (Protozoa)

In the cytoplasm of Ciliates are stored different types of polysaccharides as energy-yielding fuel stores. The most common ones are glycogen and other similar carbohydrates commonly referred to with the term paraglycogen introduced by Bütschly [9]. In some cases, paraglycogen was demonstrated to have an amylopectin like structure [10]. Thus, both polysaccharides would be consisting of alfa-D-glucose residues arranged in chains containing alfa 1-4 linkages with branch points containing 1-6 linkages. The differences would then consist in the relative degree of branching. Each branch ends with a non reducing sugar. When the polysaccharides are used as an energy source, glucose units are removed, by degradative enzymes, one at the time from their non reducing ends. More are the branchings more are the enzymes that can work together, speeding the conversion of the polymer to monosaccharides. This may be indicative of metabolic differences among ciliates according to the type of polysaccharide they form. On the other hand both glycogen and paraglycogen show different features when examined at the electron microscope level. Glycogen is present either in the typical form of rosettes as in mammal liver or as single roughly spherical particles 15-30 nm in diameter like in muscles. Paraglycogen usually appears in the form of granules of variable density and size, ranging from 0.1 to 1.5 μm (figure 1). This variability suggested metabolic differences even in ciliates storing the same polysaccharidic type. So the effects of specific stainings and amylolitic and proteolytic enzymes on the various type of polysaccharides found in ten species of ciliates, different for habitat and feeding habit, were compared [11].

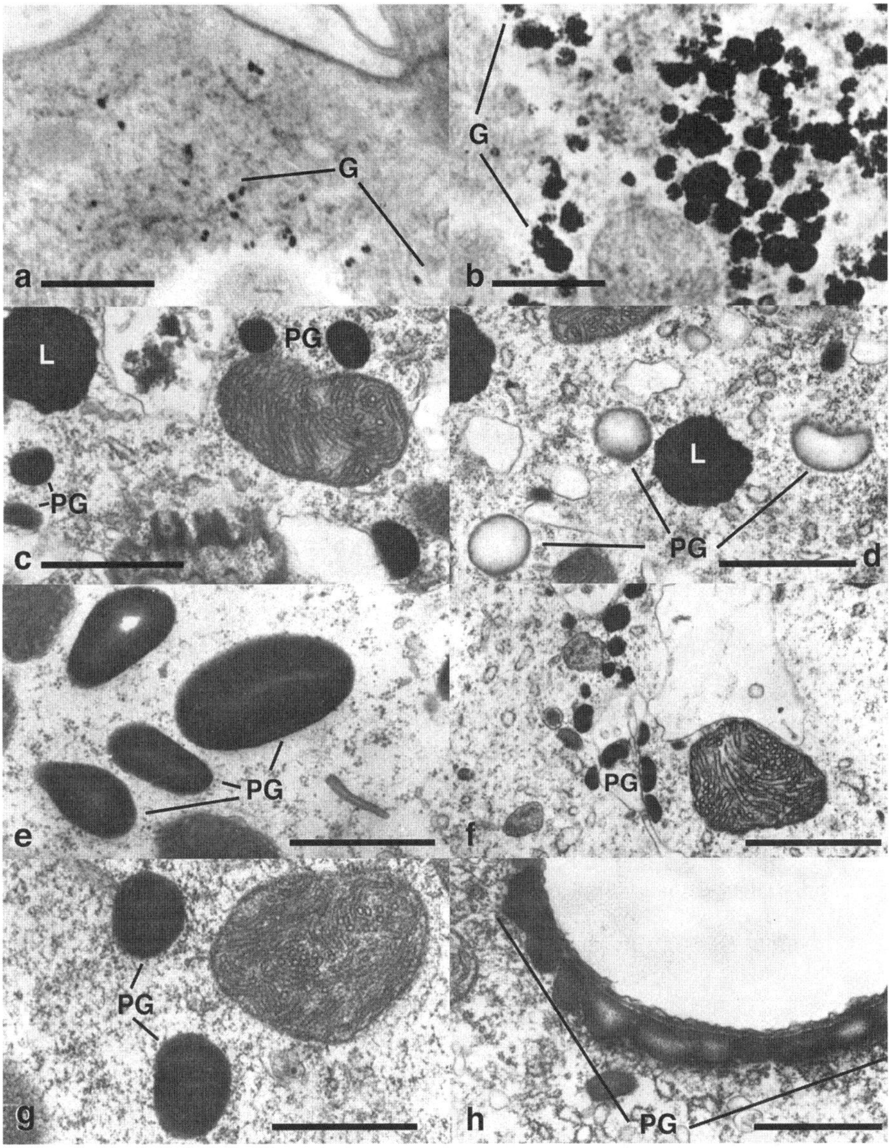

Figure 1. Sections from *Paramecium* (a), *Euplotes* (b), *Oxytricha* (c), *Gastrosyla* (d), *Blepharisma* (e), *Stylonichia* (f), *Dyophrys* sp.(g) and *Dyophrys scutum* (h). Sections a and b are stained with TSC-silver proteinate method. Sections c-h are stained with uranil acetate and lead citrate. The differences in the appearance of polysaccharides are evident. From Verni and Rosati (1980) with permission. Bars represent 1 μm.

Both glycogen and paraglycogen are stained selectively by TSC-silver proteinate method. Enzymatic treatments showed that paraglycogen granules all react in the same way independently of their morphology: they are not affected by oxidation alone, but are extracted by both amylase and pronase. The latter enzyme is thought to act on the proteic substrate to which paraglycogen is linked in the cytoplasm. The polysaccharide, once released, became soluble in the aqueous treatment medium. The results dealing *Gastrostyla* are shown in figure 2.

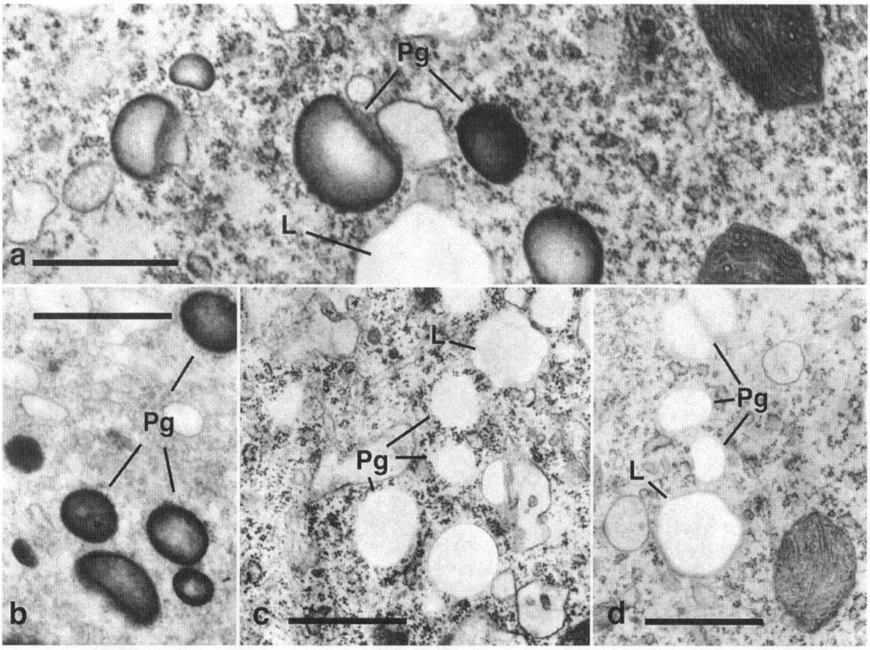

Figure. 2. Micrographs from *Gastrostyla*: a. after oxidation alone paraglycogen granules (Pg) appear not affected while lipid droplets (L) are extracted. b. After oxidation Pg granules are still selectively stained by TSC-silver poroteinate c. Section oxidized and treated with diastase: Pg granules are extracted. d. Extraction of Pg granules following pronase treatment. From Verni and Rosati (1980) with permission. Bars = 1 μm.

It appears reasonable from these results to consider very similar all types of paraglycogen granules in spite of their different appearance. The two types of glycogen, found respectively in *Euplotes* and in *Paramecium*, behaved differently. Glycogen of *Euplotes*, in the form of rosettes, was affected by amylase (although at higher concentration than paraglycogen) and pronase while that of *Paramecium* was not affected by both enzymes even by more drastic treatments (figure 3). It is interesting to mention here that the same differences have been reported between liver and muscle glycogen of mammals [12], whose metabolic differences are well known.

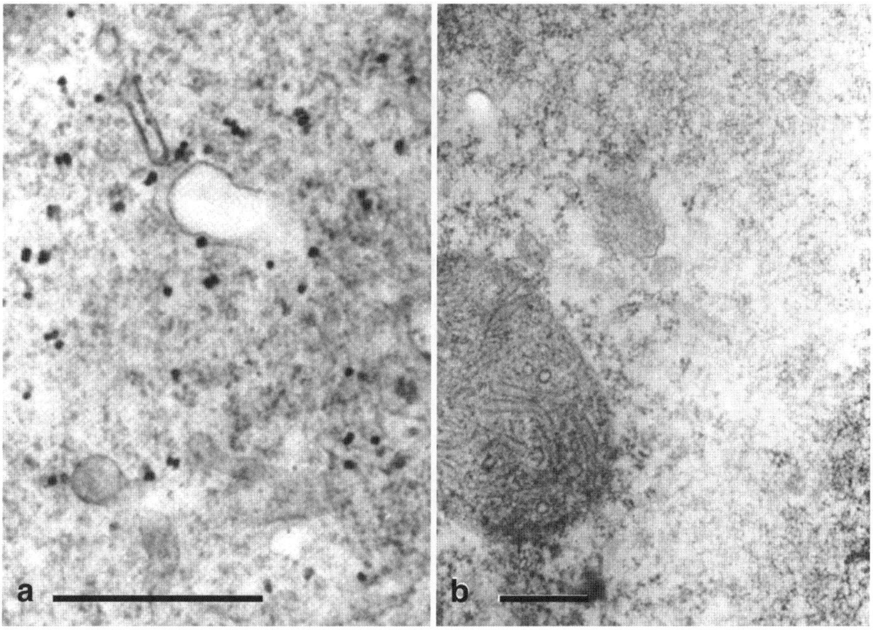

Figure 3. a. Section of Paramecium after treatment with diastase: glycogen granules are still present. B. Section of Euplotes:Action of diastase on glycogen particles of *Paramecium* and *Euplotes*: after treatment with diastase glycogen is completely extracted. From Verni and Rosati (1984) with permission. Bars = 0.5 μm.

3. METAL PRECIPITATION METHODS FOR HYDROLYTIC ENZYMES

Inevitably the enzymes that can be localized cytochemically are a minority. Nonetheless, knowledge of their ultrastructural distribution may be of the greatest value as represents an effective bridge between morphology and biochemistry. Enzyme cytochemistry can also reveal specialized metabolic function of cellular organelles, of different cells within a tissue and provide information about changes in the enzyme activity dependent from different functional states of a cell or tissue or associated with disease conditions. Sensitivity varies from procedure to procedure and also depends upon enzyme concentration and the penetration of reaction components into the cell. Lack of detectability at the EM level should, therefore, be cautiously interpreted.

3.1. Principles of enzyme cytochemistry and technical approaches

All procedures for the demonstration of enzymes at the electron microscopical level center around the incubation stage, in which a precipitate (usually containing a heavy metal) is formed as a result of enzyme activity. During this stage two quite distinct chemical reactions occur in sequence. The first of these reactions is the hydrolysis of the histochemical substrate by the enzyme. The second reaction involves the combination of a hydrolysis product with a capturing agent to form a precipitate. The two reactions are basically independent of each other except that a product of the first becomes a reactant for the second. Hence, modifications introduced to alter one of the reactions may have effect on the other.

The theoretical principles on which the incubation procedures are based is summarized by the following scheme in which the nomenclature used by Pearse [13] is followed:

Substrate (S) ™ Primary Reaction Product (PRP) + Capturing Agent ™ Final Reaction Product (FRP).

In the pre-incubation stages is important the preservation of enzyme activity, possibly without the sacrifice of a good morphological appearance since, otherwise, the products of the enzymatic reactions would be difficult or impossible to localize. This means that the choice of fixatives is extremely important. Oxidizing reagents such as osmium tetroxide cannot be used. Aldehydes like glutaraldehyde, acrolein or paraphormaldehyde are the most recommended. Their concentration, the buffer solutions, the temperature and duration of fixation are all factors affecting the quality of fixation and the maintenance of enzyme activity. Fixation has another important function besides the preservation of ultrastructure: to destroy the selective permeability of the cell membrane. Indeed many components of typical incubation media are only poorly permeable through physiologically intact membranes. Certain additives, of which dimethyl sulphoxide (DMSO) is the best known, are often used to increase membrane permeability.

In the post-incubation stages preservation of the cytochemical precipitate and of its localization is important. In many cases some forms of post-fixation is desirable before dehydration and embedding. The most used is osmium tetroxide that can provide adequate insolubility of the FRP and better overall contrast. On the other hand, osmium may obscure the hystochemical staining. Thus in any new investigation various combinations of fixatives should be tried and, as controls, some samples should be processed by procedures that avoid any step which might introduce extraneous staining. Dehydration and embedding of enzyme histochemical material should present no special difficulties.

For a certain number of enzymes there are protocols for routine methods, in which not only are proposed the substrate and the incubation media but also the most suitable fixation procedures are indicated. For detailed discussion see Lewis [14]

3.2. Application of enzyme cytochemistry to the study of the association between the ciliate *Euplotes magnicirratus* and its bacterial endosymbiont.

A newly discovered, symbiotic relationship between ciliates and bacteria involves the marine species *Euplotes magnicirratus* and an alpha-proteobacterium. The latter, previously identified by molecular analysis as a new species of the *Devosia* genus, is referred to with the provisional name "*Candidatus* Devosia euplotis" [15]. This association is constant in nature (it was found in different ciliate strains collected in different geographical regions), and durable (it is maintained in laboratory conditions for years). A well-defined physiological relationship between the ciliate and its symbiont has been evidenced. Penicillin treated *E. magnicirratus* loose the bacteria. Observed at the interferential contrast microscope, after two days of starvation, they appeared full of the green microalgae they fed on, whilst in the control cells residues of food were no longer visible. The accumulation of the algae in the cytoplasm renders the host swollen, slow in movements, and finally brings it to death [16].

Whether really the loss of symbionts inhibited the digestive process in *E. magnicirratus* was further determined by an *ad hoc* experiment in which 30 treated and 30 control specimens were fed, washed and soon singly transferred on microscope slides. They were then observed every 8h, to check the presence of algae contained in food vacuoles. Twenty-four hours later, the cytoplasm of the majority (81.25%) of the control cells appeared completely transparent and deprived of visible digestive vacuoles. On the contrary, none (0%) of the aposymbiotic *E. magnicirratus* had completed its digestive process, and roundish green bodies, i.e. the food algae, could be still easily recognized inside all the cells. TEM observations performed on aposymbiotic specimens after 10 days of starvation revealed the presence of morphologically intact algae.

The standard protocol to detect acid phosphatase was then applied [14]. *Euplotes magnicirratus* with symbionts and aposymbiotic were fixed in glutaraldehyde at low concentration, washed in buffer, incubated in the following solution: 0.2 M Tris maleate buffer pH 5.2,

0.1M sodium beta-glycerophospate (the substrate), 0.02 M lead citrate (the capturing agent) for 60 min, carefully washed, and postfixed in osmium tetroxide. Controls eliminating the substrate were treated in the same way to exclude an aspecific localization of lead ions. By this method differences were evidenced as concern the activation of acid phosphatase between *E. magnicirratus* with or without bacteria (figure 4). The active enzyme was revealed by a fine lead precipitate on the food remnants contained in digestive vacuoles of the former while no sign of acid phosphatase activation was evidenced in the vacuoles of treated, aposymbiotic specimens containing completely undigested algae. Therefore the ciliate depends on the symbionts for the activation of acid phosphatase, an enzyme essential for food digestion. As a consequence it depends on the bacteria for its survival.

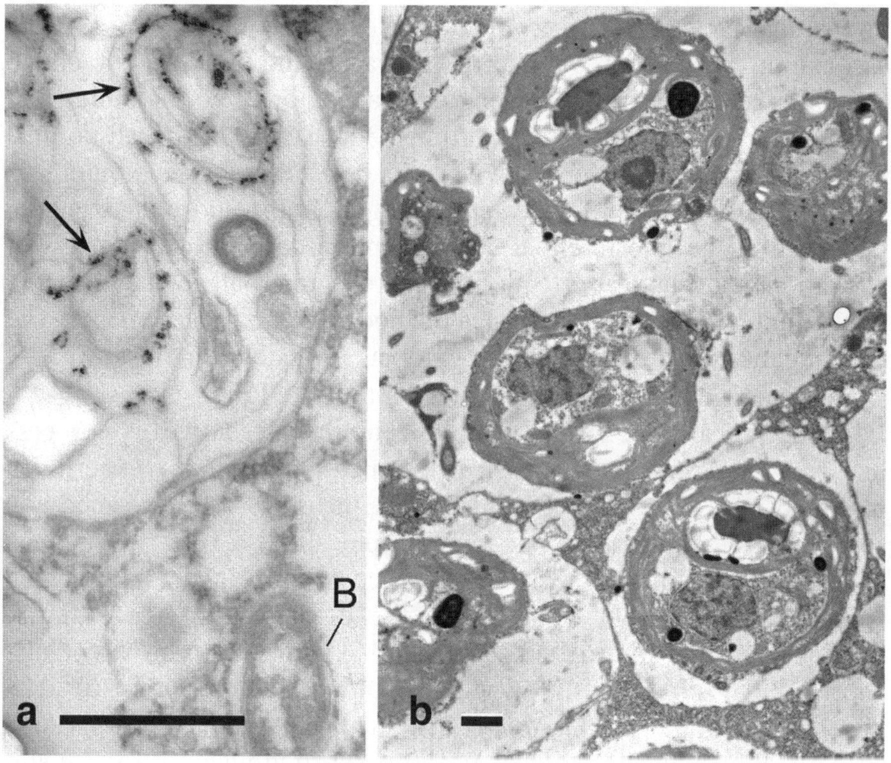

Figure 4. Active acid phosphatase detection. a. In control E. magnicirratus the actve enzyme is evidenced by lead precipitate (arrows) on the remnant of food organisms inside digestive vacuoles. B = bacterium. b. In penicillin treated, aposymbiotic specimens, no active acid phosphatase is detected in vacuoles containing undigested algae. From Vannini et al. 2004b with permission. Bars = 1μm.

3.3. Application of the enzyme cytochemistry in the study of transduction mechanism in Ciliates.

Cyclic AMP is called a second messenger because it transmits and amplifies the cell external chemical signals. It is produced by the enzyme adenylate cyclase (AC) when activated by conformational changes in cell membrane due to external stimuli. To verify the presence and the possible activity of adenylate cyclase-cyclic AMP system in ciliates, cytochemical studies for AC demonstration at the ultrastructural level were carried on throughout the sexual processes (conjugation) in *Euplotes crassus*.[17-18] Adenylate cyclase was localized using 5' Adenylyl imidodiphosphate (AMP-PNP) as substrate and lead as capture agent.[14]. Controls for non specific reactivity were performed by adding to the incubation medium alloxan (an inhibitor of AC) or omitting the substrate.

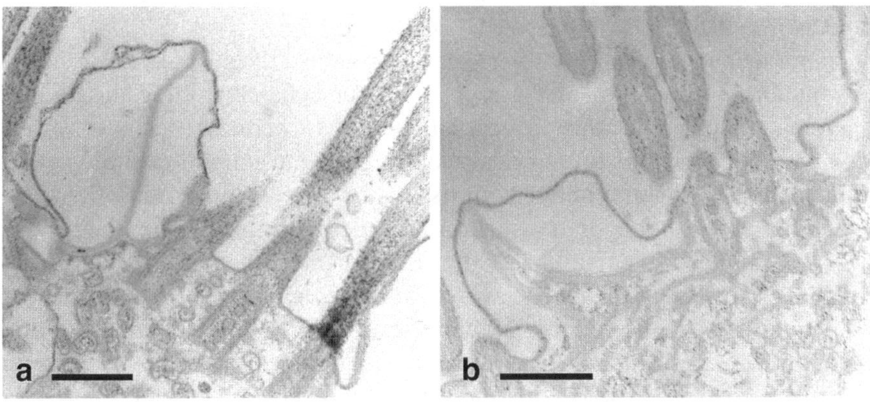

Figure 5. Euplotes crassus preconjugant cells processed for detection of active adenylate cyclase (AC). a. Ten min after mixing sexually competent cells AC is present on the external membrane and inside the ciliary axonemes.b. At the beginning of the visible preconjugant reaction the active AC is almost completely lacking at the ciliary level. From Verni and Rosati (1991) with permission. Bars = 0.5 μm.

The results obtained evidenced a variation in the AC activity during the recognition phases, i.e. when the mixed cells of different mating types, ready for conjugation, exchange specific signals to start sexual processes. Indeed, while active AC was never found in non mixed cells, in about 80% of cells fixed 5-15 min after mixing, an abundant cytochemical reaction indicative of the presence of active AC was observed at the cell membrane level and inside the ciliary structure (figure 5a). On the contrary, when mixed cells were fixed later (at the beginning of the visible preconjugant behaviour) active AC was almost completely lacking at the ciliary level and was weak on the plasma membrane (figure. 5b). These results demonstrate that in *Euplotes* there is a transient activation of AC in the initial part of cAMP system can be considered a possible way for transduction of recognition signals in Ciliates [18].

4. IMMUNOCYTOCHEMISTRY, AND AFFINITY CYTOCHEMISTRY

4.1. Immunocytochemistry

Immunocytochemistry involves the use of antibodies that specifically recognize a cellular antigen and various methods of antibody-labeling to visualize the antigen-antibody complexes. This technique represent a considerable advance in specificity over conventional cytochemical methods which are directed against particular chemical groups shared by classes of molecules.

Immunocytochemical studies can provide the following unique information: 1) Location of the precise cell or cells containing the defined macromolecules. For this aspect immunocytochemical techniques are becoming a major tool for surgical pathologists and immunopathologists in the detection of specific antigens in tissue sections. 2) Detection of changes in cellular distribution of the macromolecules during development. Thus, these techniques have facilitated the assessment of a myriad of complex problems and have decisively assisted in the elucidation of cell lineage of many malignant proliferations. 3) Identification of the subcellular sites of the defined macromolecules. 4) An indication at the ultrastructural level of the particular organelle synthesizing the defined molecule or in which it may be sequestred in the cell.

Part of these information can be obtained by fluorescent-immunocytochemistry in which the antibodies are visualized by fluorochrome labels at the optical level, especially when a confocal microscope is used. However the localization at the EM level, is particularly fine and permits the detection also of molecules present in a low concentration.

4.1.1. Technical approaches

Tissue preparation is probably the major problem to be encountered in immunocytochemistry. Any step of tissue preparation method must necessarily be a compromise between the limitation in obtaining good morphology and the possibility to demonstrate the antigens under study in correct localization and proportions.

Colloidal gold is the most used marker to visualize the antigen-antibodies complexes [19]. In general, to amplify the signal, colloidal gold is not directly coupled to the specific antibody but to an anti-species antibody. Two types of gold labeling are available. 1) Preembedding: the specimens, soon after a mild fixation, are first treated with normal horse serum in PBS (the medium in which the antibody will be diluted) and then incubated in the primary antibody at room temperature in a humidified chamber. Incubation with different concentration of antibody to determine the optimal dilution for immunostaining is recommended. After incubation with the primary antibody, the specimens are washed in PBS, and incubated in the secondary antibody coupled to gold particles. 2) postembedding: the procedure described above is applied on thin sections of specimens, fixed and embedded in resins able to polymerize at low temperature under UV light (the high temperatures necessary for the polymerization of epoxy resins might be deleterious for antigen reactivity).

Besides antibodies, a number of probes displaying highly specific binding properties, can be used in combination with colloidal gold for the detection of ligands bound to target cells. For example lectins are highly specific probes for carbohydrates but this probes are not electrondense. This limitation is circumvented by using lectin-gold complexes. Two different cellular components can be labeled using two distinctive sizes of colloidal gold particles (figure 6)

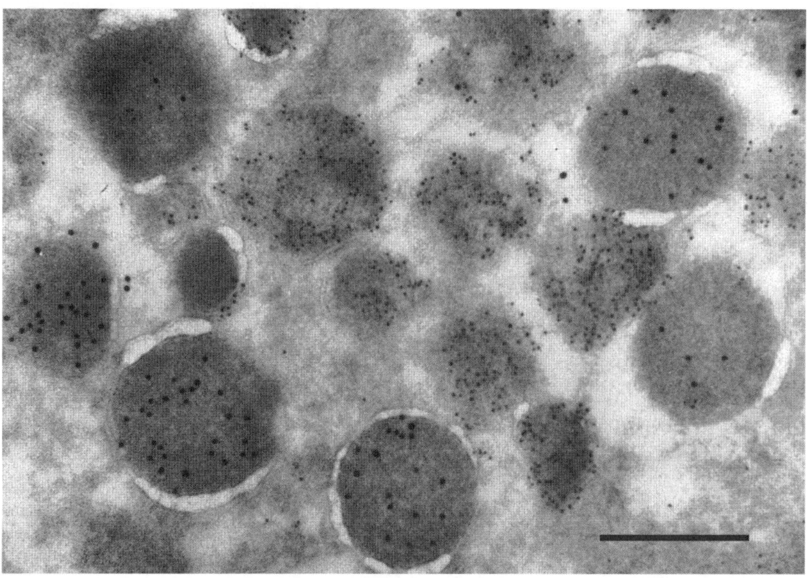

Figure 6. Multiple immunoleballing using different species of primary antibodies on neutrophils.. The binding sites of the rabbit polyclonal antibody were detected with a 5 nm gold probe coated with anti-rabbit IgG, and that of mouse antibody with 15 nm gold probe coated with anti-mouse IgG. From Robard and Wilson 1993 . Bar = 0.5 μm.

4.2. Affinity cytochemistry.

The technique referred to as affinity cytochemistry involves the use of molecules with an high affinity for each other. The most widely used method is the Biotin-Avidin (or streptavidin) System [20]. Avidin is an egg-white derived glycoprotein with an extraordinarily high affinity (affinity constant > 1015 M-1) for biotin. Streptavidin is similar in properties to avidin but has a lower affinity for biotin. Many biotin molecules can be coupled to a protein, enabling the biotinylated protein to bind more than one molecule of avidin. Thus the possibility to visualize proteins present in a very low amount is enhanced. If biotinylation is performed under gentle conditions, the biological activity of the protein can be preserved. By covalently linking avidin with electron microscope markers (for example gold particles) the Biotin-Avidin System, can be utilized to study a wide variety of biological structures and processes. An excellent review on the use of the Avidin-Biotin Complex in bioanalytical applications has been published by Diamandis and Christopoulos [21].

By employing biotinylated nucleic acid probes both DNA and RNA specific regions can be localized by *in situ* hybridization.

4.3. Immunocytochemistry and affinity cytochemistry in studies on human receptors regulation

Actions of adenosine are mediated by four G protein-coupled membrane receptors; A_1, A_{2a}, A_{2b} and A_3. The A_3 adenosine receptor subtype has been implicated in behavioral depression [22] and modulation of ischemic cerebral damage (for review, see [23].The low expression levels of these receptors, together with their coexpression with other receptors subtypes, rendered difficult the study of the mechanisms involved in their regulation in native systems. In a recent study [24] multiple, temporally distinct and sequential processes associated with regulation of the responsiveness of this receptor subtype have been demonstrated by biochemical methods and visualized by means of immunogold electron microscopy. The latter technique was used to directly reveal the localization of A_3 adenosine receptors in human astrocytoma cells in the absence of stimuli and after exposure to an agonist (CI-IBMECA). Thus it has been observed that the rapid impairment of the A_3 receptors caused by a short-term agonist exposure is associated with receptor sequestration into an intracellular endosomal compartment.

By occupying specific surface receptors, adenosine and adenosine analogue modulate neutrophil functions; in particular, functional and biochemical studies have shown that A_1 adenosin receptors modulate chemotaxis in response to chemotactic peptides. For this reason, to visualize these receptors and follow their intracellular trafficking in different situations, could be of interest. The choose of a probe (CHA-biotin XX probe) with good affinity and selectivity for A_1 adenosin receptors allowed Falleni et al. [25] to reveal them, after conjugation with colloidal-gold-streptavidin, as electron-dense gold particles on the human neutrophil surface and inside the cell. The internalization of the ligand-receptor complex was followed in a controlled temperature system. It occurred through a receptor mediated pathway. The kinetics of the intracellular trafficking was fast, taking less than 5 min.

5. CONCLUSIONS

It is evident, from this brief exposition, that cytochemical methods are successfully applicable to a wide variety of biological studies.

For the most common of these techniques there are at present commercial kits and standard protocols available in the literature. Yet, the methods imply different, sequential steps which influence each other. Since for each specimen particular experimental procedures and conditions can be required at each step, in most cases the methods need to be *ad hoc* adjusted. The adjustment often involves preliminary experiments and fruitless attempts. Anyway, positive results reproducible and corroborated by proper controls can be validated as fully significant. On the contrary, a negative result cannot be taken as definitive.

6. REFERENCES

1. Hanker J. S., Seaman A. R., Weiss L. O., Ueno H., Bergman R. A., Seligman A. M. Osmiophilic reagents:new cytochemical principle for light and electron microscopy. Science 1964; 146: 1039-1043.
2. Guimarães E.V., de Carvalho L. Barbosa Santos H.). An alternative technique to reveal polysaccharides in *Toxoplasma gondii* tissue cysts. Mem. Inst Oswaldo Cruz 2003; 98: 915-917.
3. Thiery J. P. Mise en évidence des polysaccharides sur coupe fines en microscopie électronique. J. Microscopie 1967; 6: 987-1018.
4. Marinozzi V. Cytochimie ultrastructurale du nucléole-RNA et protéines intranucleolaires. J Ultrastruct Res 1964;10: 433-456.
5. Knight D.D. "Cytological staining methods in electron microscopy." In *Practical methods in electron microscopy*, Audrey M Glauert, ed. North-Holland publishing Company, 1977.
6. Monneron A., Bernard W. Action de certain enzymes sur des tissus inclus en Epon. J. Microscopie 1966; 5: 697-702.
7. Sorvari T. E., Laurén P.A. The effects of various fixation procedures on the digestability of sialomucins with neuraminidase, Histochem J 1973; 5:405-410.

8. Yamada K., Hoshino M. Digestion with chondroitinases of acid mucosaccharides in rabbit cartilages as revealed by electron microscopy. Histochem J 1973; 5: 195-199.
9. Bütschly O. Protozoa (Infusoria und System der Radiolaria) in Bronns' Klassen und Ordenungen des Thier-Reichs. I.C.F. Winter Leipzig 1889; 3:1098-2035.
10. Grain J. Etude cytologique de quelques ciliés holotriches endocommensaux des ruminants et des équidés. Protistologica 1966; 11:5-52
11. Verni F. Rosati G. Preliminary survey of the morphology and cytochemistry of polysaccharides reserve in Ciliates. Protistologica 1980; 15: 427-434.
12. Rosati G. Enzyme treatments of glycogen particles in rat liver and muscle. J Ultratruc Res 1967; 18:444-455.
13. Pearse A.G.E. *Histochemistry. Theoretical and Applied.* Vol 1. 3rd ed J & A Churchill, 1968.
14. Lewis P.R. " Metal precipitation methods for hydrolitic enzymes." In *Practical methods in electron microscopy*, Audrey M Glauert, ed. North-Holland publishing Company, 1977.
15 Vannini C., Rosati G., Verni F., Petroni G.. Identification of the bacterial endosymbionts of the marine ciliate *Euplotes magnicirratus* (Ciliophora, Hypotrichia) and proposal of 'Candidatus Devosia euplotis' sp. nov. Int. J. Syst. Exp. Microbiol 2004; 54:1165-1170.
16 Vannini C., Schena A., Verni F., Rosati G.. *Euplotes magnicirratus* (Ciliophora, Hypotrichida) depends on its bacterial endosymbiont "Candidatus Devosia euplotis" for food digestion. Aq microb Ecol. 2004; 36:19-28.
17 Verni F., Rosati GAdenylate cyclase and Cyclic AMP in the Ciliate *Euplotes crassus*: involvement in both cell cycle and sexual reproduction. J. Exp. Zool. 1987; 244:289-298.
18 Rosati G., Verni F.. Sexual recognition in Protozoa: chemical signals and transduction mechanisms Zool. Sci 1991; 8:415-429.
19 De Waele M, De Mey J, Moeremans M, De Brabander M, Van Camp B. " Immunogold staining method for detction of cell surface antigens with monoclonal antibodies. " *In Techniques in immunocytochemistry*, Gillian R. Bullock and Peter Petrusz, eds. Academic Press, 1983.
20 Robard A. W. and Wilson A. J. " Immunocytochemical techniques for TEM and SEM" In *Procedures in electron microsc*opy. A. W Robands and A. J. Wilson eds. Wiley, 1993.
21 Guedson J.L., Ternynck T, Avrameas S. The use of avidin-biotin interaction in immunoemzymatic techniques. J Histochem Cytochem 1979; 27: 1131-1139
22 Diamandis E.P., Christopoulos T.K. The biotin-(strept)avidin system: principles and applications in biotechnology. Clin Chem 1991; 37:625-636.
23 Jacobson K. A., Nikodijevic O., Shi D., Gallo-Rodriguez C., Olah M. E.. Stiles G. L., Daly J. W. A Role for central A_3-adenosine receptors. Mediation of behavioral depressant effects. FEBS lett 1993; 336: 57-60.
24 Von Lubitz D. K. Adenosine and cerebral ischemia: therapeutic future or death of a brave concept? Eur J Pharmacol 1999; 371: 85-102.
25 Trincavelli M. L., Tuscano D., Marroni M., Falleni A., Gremigni V., Ceruti S., Abbracchio M. P., Jacobson K.A., Cattabeni F., Martini C. A_3 adenosine receptors in human astrocytoma cells: agonist-mediated desensitation, internalization, and down-regulation. Mol Pharmacol
26 Falleni A., Trincavelli M. L. Macchia M., Salvetti F., Hamdan M., Calvani F., Gremigni V., Lucacchini A., Martini C. A_1 adenosine receptors in human neutrophils: direct binding and electron microscope visualization. J Cell Biochem 1999; 75:235-244

CRYOTECHNIQUES FOR ELECTRON MICROSCOPY: A MINIREVIEW

Pietro Lupetti
Laboratorio di criotecniche per la microscopia elettronica. Dipartimento di Biologia Evolutiva. Università di Siena, Via A. Moro 2, 53100 Siena, Italia

Abstract. Aim of this paper is to present a short overview about the state of the art of the so called cryotechniques for electron microscopy. These protocols have been developed and implemented starting in the 1980's as alternatives, or better to say valid complements, to the standard protocols of chemical fixation routinely used for structural studies of biological samples by electron microscopy. The currently available and more commonly used protocols for rapid freezing such as: plunging, slamming, jet freezing, and high pressure freezing are illustrated and discussed shortly in the first section of this paper.
The second section deals with processing of frozen samples summarizing the standard protocols of cryofracture, it then proceeds with the quick-freeze, deep-etch protocol, and ends mentioning briefly the technique for visualization of macromolecular structures after adsorption on mica, rapid freezing, freeze drying, and rotary shadowing.
This paper ends with a section dedicated to cryo-electron microscopy: the most recent and high resolution protocol for observation of cell organelles and macromolecular assemblies in frozen hydrated conditions.

Keywords: Structural Studies, Electron Tomography, Freeze Fracture

1. INTRODUCTION

Biological samples can be observed by electron microscopy only after treatments designed to preserve and stabilize most of their structural features thus making them compatible with the severe physical conditions imposed during observation: i.e. high vacuum and strong radiation by accelerated electrons.

Since the introduction of first electron microscopes in the 1950's one of the main goals of structural biologists has been represented by the development of protocols designed to improve preservation of samples allowing their observation by the commercially available electron microscopes in a state as close as possible to "suspended life". The identification of chemical fixatives and the subsequent implementations in

V. Evangelista et al. (eds.), From Cells to Proteins: Imaging Nature across Dimensions, 53–70.

the protocols for chemical fixation of living specimens represented a crucial step towards our current knowledge of cell structural biology. A significant number of structural data are still obtained routinely from a wide variety of biological samples after their processing by standard fixation protocols.

Chemical fixation is however characterized by some limitations mainly related to the dynamic nature of living cells: the relatively slow and selective action of water soluble fixatives fails to prevent problems due to water exchanges among cellular compartments and phenomena of membrane fusion and fission may easily occur before fixation is over. In addition to these limitations, the chemical fixation promotes the aggregation and collapsing of some cytoskeletal components and fails to reveal rapid transient mechanisms related to dynamic events such as membrane traffic and cell motility.

Since many years, freezing has been proposed as an alternative to overcome most of the inconvenience related to chemical fixation.
The use of rapid freezing for the immobilization of cell dynamics and the structural preservation of living cells by quick solidification of physiological water has since long been pursued with tenacity by a number of structural biologists. Their efforts allowed the discovery and optimization of protocols for both freezing and observation of biological structures.

2. FREEZING PROTOCOLS

The very first and crucial step along the cryo preparation of biological samples for electron microscopy consists in removing heath from the sample in the fastest possible way. An efficient freezing allows transition of the abundant physiological water contained in all biological samples, from liquid to solid phase, limiting or impeding the formation of ice crystals. These latter if present in the sample as a consequence of slow freezing, are responsible for displacement or even the destruction of structures within the sample thus leading to a consistent loss of ultrastructural information. This concept and its implications on structural preservation of biological samples has been exhaustively studied by Dubochet and Mc Dowall in 1981 [1] and more recently by Lepault et al. [2], Erk et al. [3] and by Ashraf et al. [4]. These studies, together with the pioneer experiences of Dowell and Rinfret [5] on the low temperature forms of ice studied by x-ray crystallography, demonstrated that the size of ice crystals nucleating from liquid water during freezing of samples is inversely proportional to the speed of heath removal. The efficiency of this latter process is in turn influenced by the initial mass of the sample, that in most protocols is therefore contained in the range of microliters; by the surface/volume ratio of it (the wider is the surface in contact with cryogens the better is the freezing rate), and finally by the temperature of liquid cryogens used during freezing (the lowest is the temperature and the boiling point of cryogen, the better are the results).

All freezing protocols available nowadays for EM have been developed under the light of the above considerations and also taking into account the cost/benefit ratio inherent to the selected strategy. Here follows a rapid presentation of the current and more diffuse freezing protocols.

2.1. Immersion or Plunge Freezing

This protocol is the most simple, cheap and therefore more commonly used approach to rapid freezing of biological samples. It consists in the immersion of small tissue blocks or cell pellets in liquid nitrogen, or propane, or freon. When using the latter two cryogens, they are liquefied within metal cylinders that are kept immersed in liquid nitrogen. Samples can be held by tweezers and many adaptations to facilitate sample handling by custom made metal containers have been developed in the labs making extensive use of such protocols. Samples plunging can be done manually or with the help of simple machines that use spring loaded pistons mounted on stable metal stands. Major limitations of this freezing strategy are represented by the tendency of cryogens to boil onto the surface of samples during freezing thus forming a gas layer around them. The gas phase isolates thermally the specimen slowing down the heath removal and thus limiting the ice crystal-free volume to the very surface of sample. The freezing rate follows a steep gradient going deeper into specimens and freezing damage increases rapidly in the core volume making most of the internal volume impossible to preserve structurally. Results are normally improved using faster immersion by mechanical devices and using cryogens with lower boiling points like freon and propane. The use of a cryoprotectant such as glycerol is also quite common and effective in limiting the ice crystal formation during freezing but unfortunately the cryoprotectants themselves may introduce artefacts in the studied structures and can not be use without a preventive chemical stabilization of samples by aldehydic fixation.

2.2. Jet Freezing

This protocol consists in spraying liquid nitrogen, liquid propane, or mixtures of propane-butene onto the specimen. Freezing is performed in commercial machines in which the cryogen pressure and timing of flow can be controlled automatically. Freezing devices are more expensive than those mentioned in section 2.1 and the comparative studies present in the literature [6,7] demonstrate for this method an efficiency similar or slightly better than that of plunge freezing or standard chemical fixation. Also in this case the use of cryoprotection is mandatory since samples must be loaded within gold or copper metal chambers that protect the specimens from being damaged by the strong stream of liquid cryogen but also slow down the freezing rate. Some protocols mention the use of heath conductive solutions (1-hexadecene or Dipalmitoil-lecitine) to improve the freezing efficiency and some custom made machines are proposed and discussed in the literature, allowing a better control on sample humidity [8] or with adaptations to stopped-flow/rapid-freezing with millisecond time resolution to prepare intermediates in biochemical reactions for electron microscopy [9]. Some recent implementations of jet freezing have also been proposed [10] but costs-benefits of this strategy limited significantly its diffusion and only few structural studies mainly on plant tissue or in the field of material sciences have been performed by this freezing protocol.

2.3. Slamming Freezing

Another efficient strategy to remove heath from samples consists in the use of equipments designed to impact small mass samples onto polished metal blocks that have been previously cooled by jet freezing with liquid nitrogen or liquid helium. Specimens are loaded as small hydrated pellets hanging at the tip of spongy holders and slammed by precision pistons, by gravity or gravity and springs, onto a pre cooled metal block hosted in a special chamber in nitrogen or helium atmosphere. Heath transfer is allowed by direct and continuous contact of the specimen surface with that of the metal block for all the duration of freezing. If first impact generates elastic bouncing of sample, freezing can be discontinuous with layers of crystallization alternated to vitrified ones. Most of commercial impact freezers are designed to reduce this inconvenient by built in shock absorbers. A very good efficiency can be obtained using vibration absorbers in series like it happens in the Cryopress. This machine was designed in 1975 and successively implemented by John Heuser and Tom Reese [11,12,13]. In this machine specimen is mounted on a spongy support at the tip of a freezing head composed by a piston whose movement is smoothened by a spring and rubber stops. At impact, the head is captured and secured by a strong magnet onto the furnace hosting the metal mirror that is in turn suspended on springs. All these devices ensure continuous contact and no lateral displacement of specimen with respect to the metal mirror during freezing. In addition, liquid helium at a temperature of 4°K is used for pre cooling the metal block allowing a lower starting temperature for freezing compared to other machines designed to operate at 77°K with liquid nitrogen.

2.4. High Pressure Freezing (HPF)

As briefly discussed above, water increases in volume when passing from the disordered liquid phase to the ordered ice and formation of crystals along this process leads to a dramatic loss of structural information when dealing with biological samples. In alternative to the above methods based on speeding up freezing rate at room pressure, it is possible to freeze the specimen at high pressure to subcool the water before it solidifies thus impairing formation of ice crystals. The principle of HPF is simple [14,15] but putting into practice was more difficult [16]. Almost 30 years after construction of the prototype of a high pressure freezer, appliances of such protocol are still under development [17,18]. Now two factories are producing and marketing similar systems: Baltec and Leica. In these machines a pressure of about 2000 bar is applied in some tens of milliseconds while a double jet of liquid nitrogen is blown on the ~1 mm^3 sample protected by capillary copper tubes, or gold discs, or other custom made metal or cellulose holders. After about one second, the nitrogen flow stops, pressure is released, and the frozen specimen is transferred and stored in liquid nitrogen for further processing. Despite of its consistent costs (both machines are sold at prices around 200 000 USD) the real benefits and advantages of HPF are still matter of discussion. A realistic and impartial position on this matter was expressed by Jaques Dubochet in 1995 [16] and most of his considerations are still actual. This Author claimed that to determine the extent of the improvement that HPF can effect, we need quantitative data and comparisons between specimens in which all the major parameters such as water content and the presence of cryoprotectant are known. It also would be of interest learning more about the depth of water vitrification within the specimen, but obtaining such data is not easy and comparison between the results obtained by normal pressure freezing and HPF is meaningless unless these parameters are defined. The many beautiful micrographs produced on biological specimens after the introduction of HPF may speak for the method but the experience has shown that these products are frequently more indicative of the specific skill of the researcher and the dedication he invests in the work.

3. SAMPLE PROCESSING AFTER FREEZING

3.1. Freeze Fracture

Freeze fracture or cryofracture (the two terms are used as synonymous in the current literature), represent since the 1970's the most widely diffused cryotetchnique for processing of frozen tissue blocks and cell pellets. The protocol consists in fracturing by a cold steel knife frozen samples kept under high vacuum conditions and at temperatures around -140°C. After fracturing, part of the intrinsic water is removed by sublimation in high vacuum, and the resulting surfaces are metal replicated with mixtures of platinum carbon deposited from variable incident angles so that structures in relief are shadowed and cavities receive less metal. At the end of this process shadows accentuate very efficiently local topography of fractured surfaces. After metal shadowing, specimens are thawed at room temperature and pressure, the organic fraction is digested by overnight exposure to strong acidic or bleach solutions and the remaining metal replicas are rinsed in distilled water, recovered on EM grids, and observed by transmission electron microscopy. This treatment is quite selectively devoted to visualization of hydrophobic compartments within cell membrane systems. Fracture planes in fact take the path of least resistance through the frozen block which often carries them through the centre of cellular membranes. As result this technique is particularly suitable for examining the distribution of integral membrane proteins as they span the lipid bilayer. The studies carried out by Daniel Branton and others played a crucial role in the formulation of the fluid mosaic model of cell membranes. The study of structure and function of cell junctions took great advantages by this technical approach in the 1980's and many landmark papers have been published since then, on structure and function of both vertebrate and invertebrate cell junctional complexes. More than 4000 records are selected by Medline when typing "cell junction ultrastructure". See for a review on cell junctions, the paper by Romano Dallai in this volume.

3.2. Quick-Freeze, Deep-Etch QF-DE

Freeze fracture can be made even more informative if the starting material is represented by unfixed tissue frozen in the absence of cryoprotectants by slamming it against a metal block cooled to minus hundreds of degrees, and if including in the protocol a step called "deep-etching". This term refers to vacuum sublimation of ice from a sample freeze-fractured according to the protocol described in section 3.1. The etching is allowed to progress enough to reveal the surface topology at the fractured surface of the sample. This technique, in which greater amounts of water are removed and individual parts of the cell can be seen at high resolution, standing out in deep relief against the background, has been used with great success by its inventor John Heuser, and by the other labs were this technique was disseminated in the 1990's, to capture several different cellular processes that are unusually fleeting, including membrane budding and fusion, synaptic vesicle discharge during neural transmission, movement of cilia and flagella on vertebrate and protozoa cells, and muscle contraction.

In each case with the aim of visualizing the underlying molecular mechanisms occurring.

Here follow some examples of application of QF-DE we have performed in our lab to study the functional morphology of some interesting biological structures.

3.2.1. Pellicular structure in malaric plasmodia

We have studied the pellicular structure of the ookinete stage of *Plasmodium gallinaceum* one of the genuses of protozoa Apicomplexa responsible for malaria [18]. The general organization we have visualized is analogous to that of invasive stages of other members of Apicomplexa. The cytoplasm of this parasitic stage is very reach in organelles (Fig. 1).

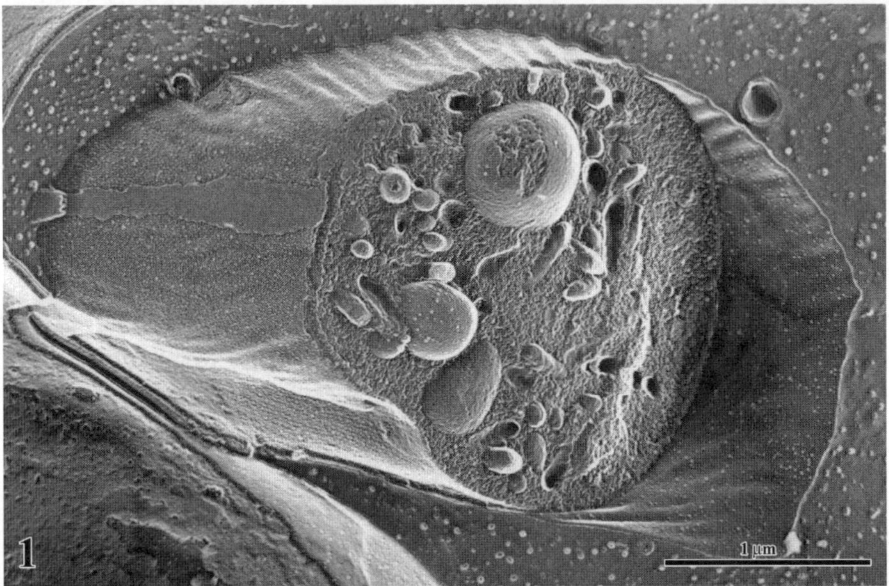

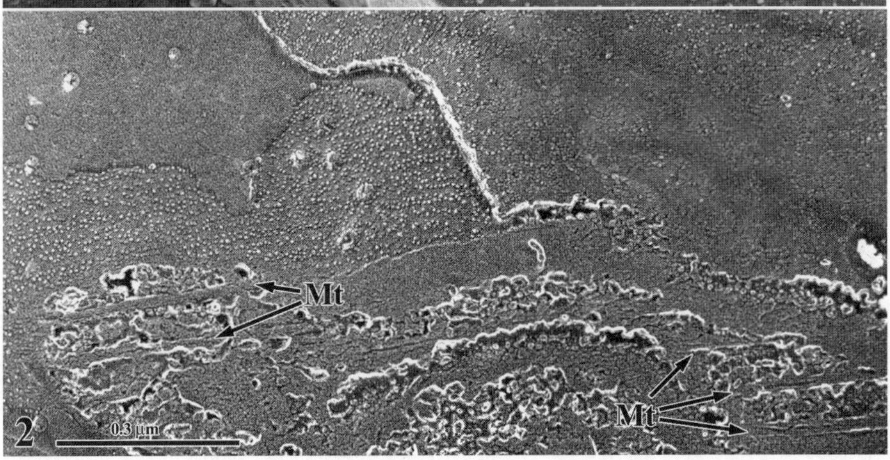

Figure 1. A cryofracture of the mid cell body by quick-freeze, deep-etching (QF-DE) of an ookinete of *Plasmodium gallinaceum* (one of the causative agents of malaria). Etching is deep enough to reveal the presence of numerous cell organelles within the cytoplasm of this stage of the parasite.

Figure 2. QF-DE tangential to the three membrane system characteristic of plasmodial ookinetes. Beneath the innermost membrane of this system are located cytoplamic microtubules (Mt).

The pellicle is composed of three membranes: the plasma membrane, and the two linked intermediate and inner membranes, which in the ookinete form one flattened vacuole that is located beneath the plasma membrane (Fig. 2). Beneath the vacuole is found an array of microtubules that are longitudinally oriented, parallel, and regularly spaced ca. 90 nm apart (Fig. 2). We have also demonstrated by QF-DE the presence of pores crossing the alveolar space. These structures in ookinetes are probably involved in protein trafficking through the alveolar domain and might constitute a novel pathway for the transport of molecules to and from the cortex.

3.2.2. The actin network in insect midgut epithelial cell

We have also studied the midgut epithelial cell from larvae of the beetle *Tenebrio molitor* for the localization and structural organization of actin network [20].

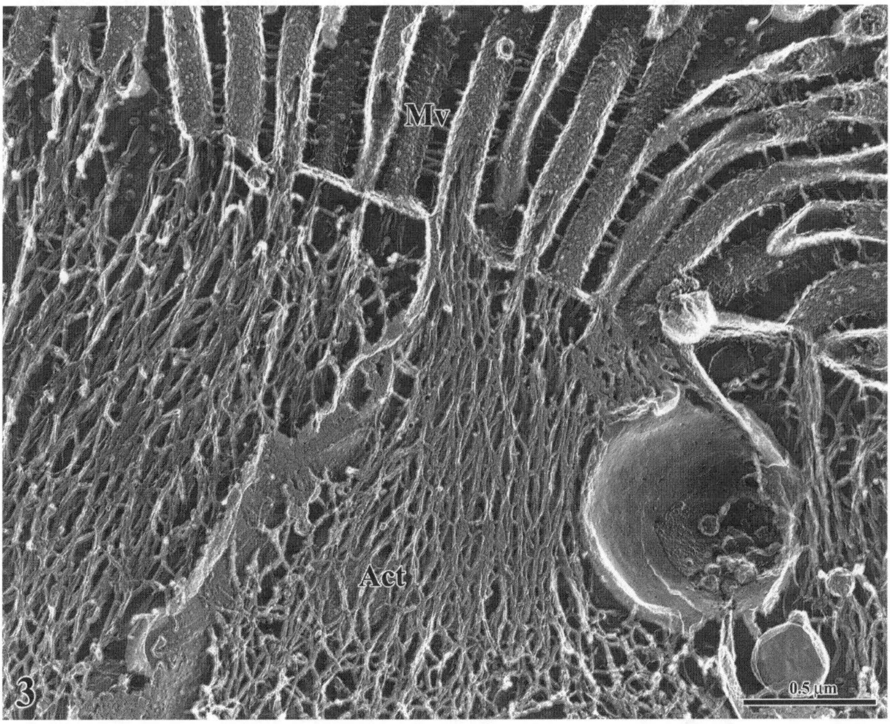

Figure 3. Epithelial midgut cell form larvae of the coleopteran *Tenebrio molitor*. The solubile fraction of cytoplasm has been extracted before processing the sample by QF-DE. A consistent actin network (Act) is visibile emerging from the core of microvilli (Mv) and reaching deeper domains on cytoplasm to contact the lateral plasma membrane.

QF-DE preparations of epithelial cells fixed after dissociation by passing them into a needle show the apical brush border formed by microvilli. When microvilli are fractured longitudinally, a core of filaments can be seen within them emerging towards the basal portion of cytoplam (Fig. 3).

These bundles display the typical longitudinal repeat and thickness of actin filaments; they extend into the cytoplasm for a considerable distance and appear to be inserting into the plasma membranes at the junctional contact (Fig. 4). Hence cytoskeletal elements have an intimate spatial association with the membrane modifications typical of intercellular septate junctions, and may be involved in the positioning of their components and also possibly, of their septal ribbons.

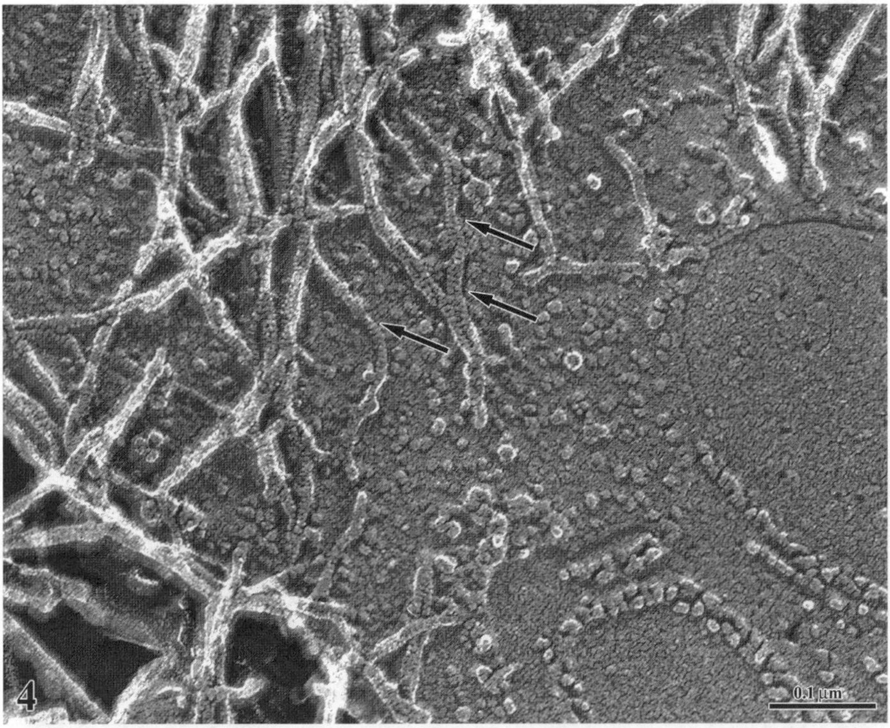

Figure 4. Detail at higher magnification of the sub apical plasma membrane from midgut cell of *Tenebrio molitor*. The preservation of unsoluble cytoplasmic components is so efficient that the typical 7 nm repeat is visible along actin filaments (arrows). The actin network branches and contacts the plasma membrane in correspondence of the intramembrane particles contributing to the formation of intercellular epithelial junctions.

3.2.3. Molecular structure of insect axonemal dyneins

We also have used QF-DE to study the organization of sperm axonemes from the dipteran insect *Asphondylia ruebsameeni* [21,22]. In this species more than 2500 microtubular doublets are disposed in spirals coiled around an axial core of mitochondria; each doublet is provided with outer arms only [23,24]. Such a peculiar axoneme turned out to be a very suitable model for studies on the molecular structure of the dynein arms. Axonemes fractured parallel to their longitudinal axes and close to the cell surface revealed many interesting structural details about this motile system. In such replicas it was possible to resolve tubulin protofilaments along the microtubule doublets, and the outer dynein arms appeared regularly aligned with a longitudinal repeat of 24 nm (Fig. 5).

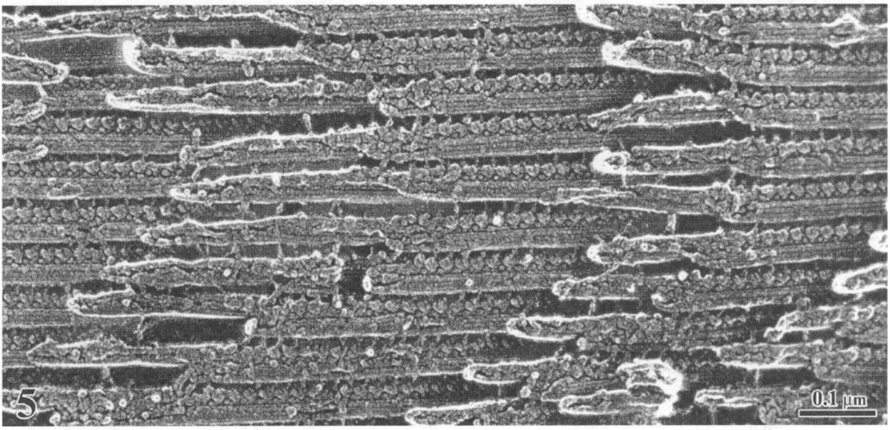

Figure 5. QF-DE of a sperm axoneme from the dipteran *Asphondylia ruebsaameeni.* This tangential fracture to the axoneme shows the parallel rows of axonemal dyneins intercalated among the parallel microtubular doublets.

Dynein arms, at first glance, show a domain organization similar to that previously observed in other species of both invertebrates and vertebrates, consisting of a head region formed by two globular domains in close vicinity of the A subtubule and a thin stalk contacting the B subtubule of the adjacent doublet. However, a more accurate analysis of micrographs revealed the presence of a globular extra-component regularly intercalated between the arms (Fig. 6). We are currently studying by electron tomography the high resolution structure of dynein motor molecules in an insect species closely related to *Asphondylia.*

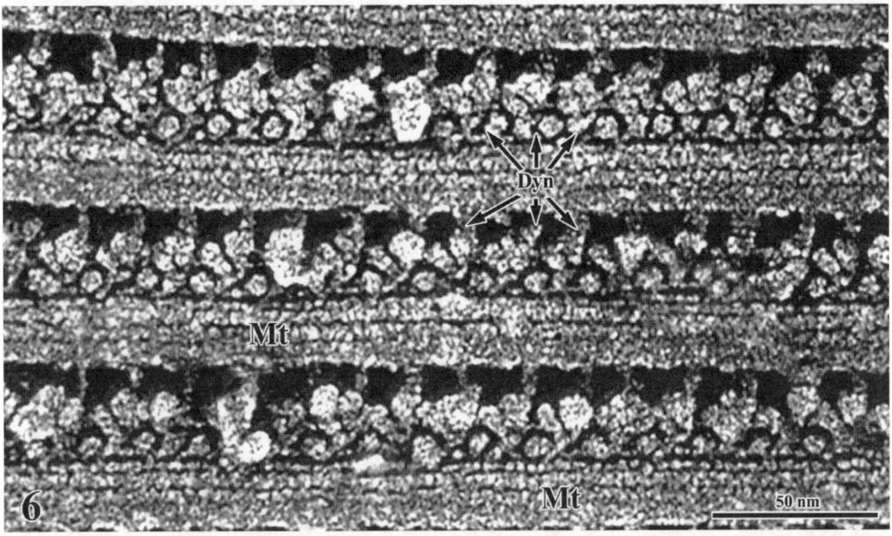

Figure 6. Detail at very high magnification of figure 4. The tubulin protofilaments originate the microtubular wall are visibile as striations on the surface of microtubules (Mt). The molecular subdomains of dyneins (Dyn) are also well visible in this preparation.

3.3. Molecules on Mica

Several techniques have been developed in order to visualize the three-dimensional structure of macromolecules and whole assemblies like viruses and ribosomes. These structures have long been imaged by negative staining and more recently by the frozen thin film technique (see section 3.4 of this paper). However, the contrast characteristic of the images provided by these techniques is so low and there is so much structural overlap that they require analysis by optical diffraction or better an extensive use of computers to model in silico the studied structures. Furthermore negative stains tend to accumulate around intermediate-sized assemblies rendering their finer components invisible. The ideal method of preserving 3D structures would be to maintain a suspension in frozen solvent during electron microscopic viewing [25,26]. Unfortunately, frozen suspensions of molecules are so low in contrast that computer enhancement of the images is mandatory. As a direct consequence the observer has no opportunity to "search" a sample while viewing it in the electron microscope. Such searching is essential when studying molecules that are disoriented relative to each other or when studying fragments of structures that are not completely assembled.

Molecular imaging by freeze-drying of molecules adsorbed to a mica substrate provides images with a great 3D information content, the major advantage of this technique being the improved molecular morphology by avoiding the air drying step involved in other techniques of molecular imaging [27].

The standard preparation consists in the incubation of proteic solutions with mica flakes to allow adsorption of molecules. This step is followed by a gentle centrifugation, the collection of sediment and rinsing in fresh buffer. After a further centrifugation the pelleted mica flakes are rapidly frozen and then treated by QF-DE with a flat angle (10°-11°) of metal rotary deposition.

Conjoined with platinum replication, freeze drying provides topologically realistic images and permits the imaging of all sizes of samples, from individual molecules to complete molecular assemblies, largely because it provides surface views that are free of image superimposition. Furthermore, metal replication provides such high contrast that it permits the visualization of individual metal grains. The size of these grains limits the resolution to around 2-3 nm with platinum, the easiest metal to use, but this is adequate to resolve major surface contours on molecules larger than 50-60 kDa [27].

This protocol has been successfully used to study not only the morphology of molecular complexes but also the mechanisms of macromolecular assembly and disassembly in various processes, including remodelling of cytoskeletons, clathrin-mediated endocytosis, cell-to-cell recognition, and the formation of extracellular matrices.

3.3.1. The molecular structure of the solubile toxin VacA

We present here one example of application of the above protocol for structural studies on macromolecular complexes. We have studied by freeze drying on mica the structure of VacA: the vacuolating cytotoxin from the gram-negative bacterium *Helicobacter pylori* [28,29],. These bacteria are among the causative agents of gastric ulcer and VacA has been extensively studied for its strong vacuolating effects on gastric mucosal cell (one of the first stages of the infection leading successively to gastric ulcer and carcinoma). VacA forms high molecular weight homo-oligomers which contain either six or seven copies of a 95 kDa polipeptide. Electron microscope visualization of VacA has revealed that the oligomers are arranged in flower like structures with 6 or sevenfold radial symmetry, depending on the number of 95 kDa oligomers that they contain (Fig. 7). Images of rotary shadowed oligomers were processed by multivariate statistical analysis to evidence clusters of equivalent and homogeneous oligomers. 3D reconstructions of the replicas so classified were performed by random conical tilt tomography. Computational analyses on tilted series of micrographs allowed to obtain serial sectioning of metal replicated molecules (Fig. 8). This data set was then used to perform the 3D rendering of the oligomerized molecule shown in Fig. 9.

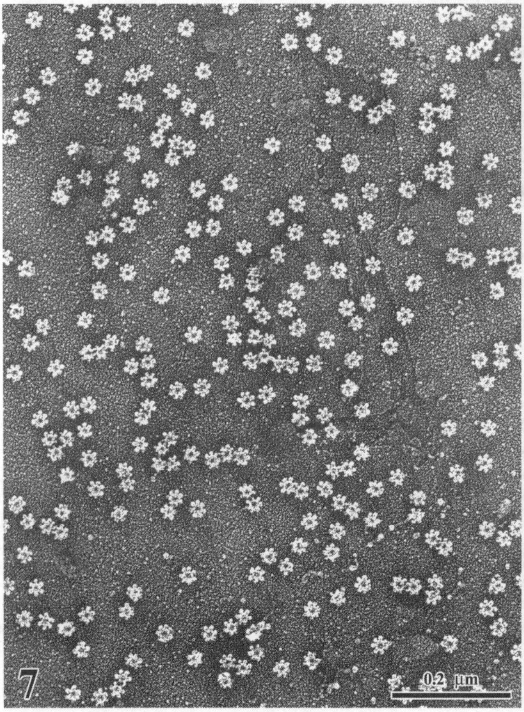

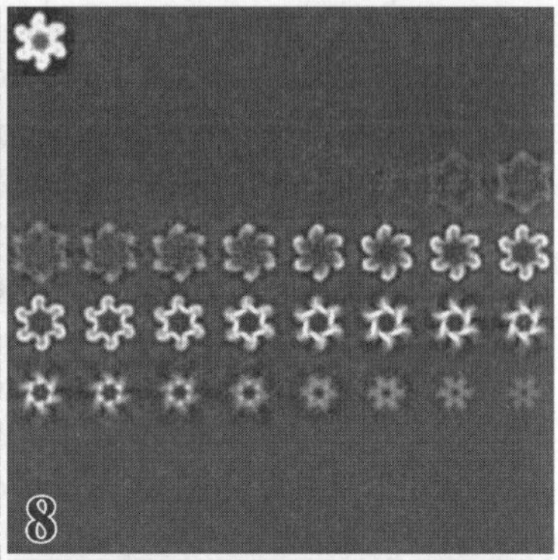

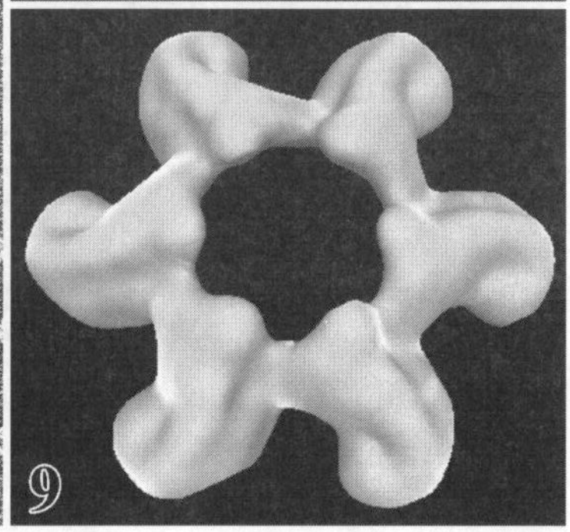

Figure 7. Olygomers of the soluble cytotoxin VacA secreted by the gram-negative bacterium *Helicobacter pylori* (one of the causative agents of gastric ulcer). A suspension of purified molecules has been processed by adsorption on mica, freeze-drying and rotary replication. Many oligomers with the typical hexameric organization are visible.
Figure 8. Tilted series micrographs of field similar to that shown in fig. 7 were used to calculate the tomogram shown in fig. 8. The sequence show the density of metal at different sectioning levels of the molecule. Averaging of numerous oligomers was also performed in order to filtrate the noise introduced with metal sputtering.
Figure 9. 3D model of VacA obtained by computer rendering of the tomogram in fig 8. This toxin molecule has a diameter of about 20 nm and is characterized by the presence of six complexes organized around a central hole. We calculated for this prep a resolution of about 2nm.

The model we obtained demonstrated a cart wheel array of the six–seven peripheral arms around a central ring. This latter displays a central hole with a diameter of about 8 nm . The estimated diameter of the whole molecule is about 20 nm. We further investigated by the same strategy of electron tomography on the structure of subunits composing the whole oligomer [30] and our data together with other evidences from the research group of Dr. J. Telford support the hypothesis that VacA is an AB type toxin and suggest a model in which the smaller of the two subunits is arranged in a uniform ring on the surface of the molecule in such a way as to contribute to the overall stability of the molecule.

3.4. Cryo Electron Microscopy

The potential of this technique and its highly valuable purpose are well summarised by Joakim Frank, one of the fathers of electron tomography. In the summary of a paper he published in 2002 [31] this Authors writes: "...understanding molecular interactions is key to understanding the workings of the cell. Molecules engage in dynamic associations forming complexes that are often too large and too flexible for structural studies by x-ray crystallography. Cryo EM is a new powerful tool that enables us to visualize these complexes despite their unwieldly properties..."

Cryomicroscopy allows in fact to observe by transmission electron microscopy frozen hydrated specimens embedded in vitreous ice by plunge freezing in liquid ethane. The protocol and its first applications in the field of structural biology have been extensively described by the Authors that ideated and successively developed this technique [32,33].

Biological macromolecules, cell organelles or procariotes fractions are suspended in proper buffering solutions and a small droplet of suspension is applied to EM grids pre treated with holey carbon film. The grid is then gently blotted and the remaining thin layer of solution is plunge frozen in liquid ethane. The vitrified ice containing the structures to be studied is then transferred and mounted, under liquid nitrogen, on a sample stage designed to keep the sample frozen during its observation within the transmission electron microscope. This latter is normally equipped with high accelerating voltage to reduce radiation damage to which vitrified samples are particularly sensitive, and a computerized stage able to move on x and y axes as well as tilting the sample so that series of images can be collected at different observation angles from the same area of the grid. Major advantages of this technique are a significantly increased resolution (especially after computer processing of the collected images), and the possibility to observe molecules in their native hydrated stage and in the absence of fixatives, cryoprotectants, and any kind of stain or contrasting agents). There is however a series of limitations and inconveniences that limit the very high potential of this strategy. The electron microscopes designed for this kind of observations are very expensive. Image collection is time consuming and sometimes it is difficult to evaluate the quality of samples before image analyses. Samples are in fact characterized by very low contrast and are quickly damaged by electrons during observation. This makes mandatory the post processing of micrographs by sophisticated protocols of image analysis. This latter approach is becoming more and more powerful but it must be stressed that, with time, it is becoming a science in itself thus frequently making necessary the development of collaborations between structural biologists, and experts of computer science. A very interesting strategy of image analysis that is now being disseminated in many labs is the Electron tomography. This protocol is based on collection of serial projections of the sample from different angles by tilting the specimen within the electron microscope. There are protocols to retro project all the collected images in a 3D space provided that the angles of collection are known. In this way 3D density maps of complex objects can be constructed and these data sets can be visualized by 3D rendering. Despite of this limitations the quality and quantity of data concerning the structure and function of macromolecular complexes as well as virus and bacterial structures has been increasing fast in these last years and most institutions are setting up their own cryo EM facilities. Major results obtained by cryo electron microscopy and electron tomography deal with the structure of viruses [34], bacterial components [35], nuclear pores [36], ribosomes [37,38], membrane channel proteins and many other very important structrures [39]. Further development of this technique are represented by observations of vitrified cell sections and their analysis by electron tomography to visualize the molecular organization of cytoplasm [40] and by the very promising studies on the interaction of molecules and their ligands by fitting atomic structures obtained by x-ray diffraction into the cryo-EM density maps [31]. These last years have been very productive and stimulating for scientists involved in the study of cell and structural biology. The goal of all microscopists of visualizing images of suspended life seems now a closer possibility. However, as Jaques Dubochet, a master in the field of cryo-EM, stated in one of his commentaries "...cryoelectron microscopy helps us to obtain snapshots of "suspended life"...but microscopists must still learn how to get the best out of it. One thing they will not achieve, however, is "suspended life" for themselves. More hard work is needed but potential results will worth the invested efforts and energies.

4. ACKNOWLEDGEMENTS

I wish to thank my mentor Prof. Romano Dallai for introducing me to the challenging and great world of electron microscopy and for his constant tutorage and valuable inputs in our joint researches of functional morphology of invertebrates. My gratitude also goes to Prof. John Heuser for giving me access to advanced cryotechniques for electron microscopy, for his constant help, and for the many stimulating discussions on our data. And last but not least, many thanks to the lab staff: Paolo Salvatici, David Mercati, Fabiola Giusti and Eugenio Paccagnini for their active and highly professional collaboration.

5. REFERENCES

1. Dubochet J., McDowall A.W. Vitrification of pure water for electron microscopy. J. Microsc. 1981; 124, RP3-RP4.
2. Lepault J., Bigot D., Studer D. and Erk I. Freezing of aqueous specimens: an X-ray diffraction study. J. Microsc. 1997; 187: 158-166.
3. Erk I., Nicolas G., Carofe A., Lepault J. Electron microscopy of frozen biological objects: a study using cryosectioning and cryosubstitution. J. Microsc. 1998; 189: 236-248.
4. Ashraf A., Norlen L.P.O., J. Dubochet. Cryo-electron microscopy of vitreous sections of native biological cells and tissues. J. Struct. Biol. 2004; 148: 131-135.
5. Dowell, L. G. and A. P. Rinfret Low-temperature forms of ice as studied by x-ray diffraction. Nature 1960; 188:1144-1148.
6. Swales L.S., Lane N.J. Insect intercellular junctions: rapid freezing by jet propane. J. cell Sci. 1983; 62: 223-236
7. Greene W.B., Walsh L.G. An improved cryo-jet freezing method. J Microsc. 1992; 166: 207-218.
8. Bailey S.M., Chiruvolu S., Longo M.L., Zasadzinski J.A. Design and operation of a simple environmental chamber for rapid freezing fixation. J. Electron Microsc. Tech. 1991; 19: 118-26.
9. Pollard T.D., Maupin P., Sinard J., Huxley H.E. A stopped-flow/rapid-freezing machine with millisecond time resolution to prepare intermediates in biochemical reactions for electron microscopy. J. Electron Microsc. Tech. 1990; 16:160-166
10. Fields SD, Strout GW, and Russell SD. Spray-freezing, freeze substitution (SFFS) of cell suspensions for improved preservation of ultrastructure. Microsc. Res. Tech. 1997; 38:315-28.
11. Heuser, J. E. Quick-freezing evidence in favor of the vesicular hypothesis. Trends in Neurosciences 1978; 1:80-82.
12. Heuser, J. E., T. S. Reese, M. J. Dennis, Y. Jan, L. Jan and L. Evans Synaptic vesicle exocytosis captured by quick freezing and correlated with quantal transmitter release. J Cell Biol 1979; 81: 275-300
13. Heuser J.E. Development of the quick-freeze, deep-etch, rotary-replication technique of sample preparation for 3-D electron microscopy. Prog. Clin. Biol. Res. 1989; 295: 71-83.
14. Müller M., Moor H. Cryofixation of thick specimens by high-pressure freezing. In: The Science of Biological Specimen Preparation, 1984; ed. by Revel, J.P., Barnard T., Haggis G. H.
15. Moor H. Theory and practice of high-pressure freezing. In: Cryo-techniques in Biological Electron Microscopy, 1987; 175-191, ed. by Steinbrecht R. A., Zierold K., Springer Verlag.
16. Dubochet J. High-pressure freezing for cryoelectron microscopy. Trends Cell Biol. 1995; 5: 366-368.
17. Studer D., Graber W., Al-Amoudi A. and Eggli P. A new approach for cryofixation by high-pressure freezing, J. Microsc. 2001; 203: 285-294
18. Walther P. Recent progress in freeze-fracturing of high-pressure frozen samples. J. Microsc. 2003; 212 : 34-43.
19. Raibaud A., Lupetti P., Paul R. E., Mercati D., Brey P. T., Sinden R. E., Heuser J., Dallai R. Cryo-fracture electron microscopy of the ookinete pellicle of *Plasmodium gallinaceum* reveals the existence of novel pores in the alveolar membranes. J. Struct. Biol., 2001; 135: 47-57.
20. Dallai R., Lupetti P., Lane N. The organization of actin in the apical region of insect midgut cells after deep-etching. J. Struct. Biol., 1998; 122: 283 – 292.
21. Lupetti P., Mencarelli C., Rosetto M., Heuser J.,E., Dallai, R. Structural and molecular characterization of dynein in a gall-midge insect having motile sperm with only the outer arm. Cell Motil. Cytoskel. 1998; 39: 303-317.

22. Mencarelli, C, Lupetti P., Rosetto M., Mercati D., Heuser J. E., Dallai R. Molecular structure of dynein and motility of a giant sperm axoneme provided with only the outer dynein arm. Cell Motil. Cytoskel. 2001; 50: 129-146.
23. Dallai R. Lupetti P., Frati F., Afzelius B., Mamaev B.M. Spermatozoa from the supertribes Lasiopteridi and Stomatosematidi (Insecta, Diptera, Cecydomyiidae): Ultrastructure data and phylogeny of the subfamily Cacidomyiinae. Zool. Scripta. 1996; 25: 51-60.
24. Dallai R., Lupetti P., Frati F., Mamaev B.M., Afzelius B. Chracteristics of spermatozoa from five gall-midge species (Diptera, Cecidomyiidae). Acta Zool. (Stockh.) 1997; 78: 33-37.
25. Adrian, M., Dubochet, J., Lepault, J., and McDowall, A.W. Cryoelectron microscopy of viruses. Nature 1984; 308:32-36.
26. Lepault, J., and Leonard, K. Three dimensional structure of unstained, frozen-hydrated extended tails of bacteriophage T4. J. Mol. Biol., 1985; 182: 431-441.
27. Heuser J.E. Protocol for 3-D visualization of molecules on mica via the quick-freeze, deep-etching technique. J. El. Microsc. Tec., 1989; 13: 244-263.
28. Lupetti P., Heuser J., Manetti R., Massari P., Lanzavecchia S., Bellon P., Dallai R., Rappuoli R., and Telford J. Oligomeric and subunit structure of the *Helicobacter pylori* vacuolating cytotoxin. J. Cell Biol., 1996; 133: 801-807.
29. Lanzavecchia S., Lupetti P., Bellon P., Dallai R., Rappuoli R., Telford J. Three dimensional reconstruction of metal replicas of the *H. pylori* vacuolating cytotoxin. J. Struct. Biol. 1998; 121: 9-18.
30. Reyrat J.M., Lanzavecchia S., Lupetti P., de Bernard M., Pagliaccia C., Pelicic V., Charrel M., Ulivieri C., Norais N., Ji X., Cabiaux V., Papini E., Rappuoli R., Telford JL. 3D imaging of the 58 kDa cell binding subunit of the Helicobacter pylori cytotoxin. J. Mol. Biol. 1999; 290: 459-470.
31. Frank J. Single-particle imaging of macromolecules by cryo-electron microscopy. Annu. Rev. Biophys. Biomol. Struct. 2002; 31:303-19
32. Dubochet J., Adrian M., Lepaut J. and McDowall W., Cryo-electron microscopy of vitrified biological specimens. Trends in Biochem. Sci., 1985; 10: 143-146.
33. Cyrklaff M., Adrian M., Dubochet J. Evaporation during preparation of unsupported thin vitrified aqueous layers for cryo-electron microscopy. J Electron Microsc. Tech. 1990; 16: 351-5.
34. Baker, T. S., Olson, N. H., Fuller, S. D. Adding the Third Dimension to Virus Life Cycles: Three-Dimensional Reconstruction of Icosahedral Viruses from Cryo-Electron Micrographs. Microbiology and Mol. Biol. Rev., 1999; 63: 862–922.
35. Berg, H. C. The rotary motor of bacterial flagella. Annu. Rev. Biochem. 2003; 72:19–54.
36. Fahrenkrog B., Aebi U. The vertebrate nuclear pore complex: from structure to function. Results Probl. Cell Differ. 2002; 35: 25-48
37. Frank J., Penczek P., Agrawal R.K., Grassucci R.A., Heagle A.B. Three-dimensional cryo-electron microscopy of ribosomes. Methods Enzymol. 1999; 317:276–91
38. Frank, J, Agrawal, RK. A ratchet-like inter-subunit reorganization of the ribosome during translocation. Nature 2000; 406: 318–22
39. Subramaniam, S., Milne, J.L.S. Annu. Three-dimensional electron microscopy at molecular resolution. Rev. Biophys. Biomol. Struct. 2004; 33:141–55
40. Baumeister, W. Electron tomography: towards visualizing the molecular organization of the cytoplasm. Curr. Opin. Struct. Biol., 2002; 12:679–684.

CELL MEMBRANE SPECIALIZATIONS AS REVEALED BY THE FREEZE-FRACTURE TECHNIQUE

Romano Dallai
Dip. di Biologia Evolutiva, Università di Siena, Via A. Moro 2, 53100 Siena, Italia

Abstract. Two examples of membrane specializations are here reported as revealed by thin sections and freeze-fracture replicas. They are involved in intercellular junctions (tight junctions of urochordates and septate junctions of insects) and ciliary specializations of the apical corona of rotifers. The appearance of these specializations is described and their functional role is discussed.

Key words: Membrane Specializations

1. INTRODUCTION

Cells perform several activities towards the environment or adjacent cells though their limiting plasma membranes. For this purpose, they differentiate structural membrane specializations which perform mechanical and/or functional roles. Numerous examples of membrane specializations have been described in the last few decades and their structure has been well understood after careful studies of freeze-fracture replicas. The method is based upon the principle that by evaporation of platinum and carbon in a vacuum, a high fidelity replica can be made of the surface of frozen hydrated biological material. It was shown [1] that the fracture preferentially follows a path of low resistance through the hydrophobic region of lipid bilayers, thus cleaving membranes in two leaflets and exposing extensive areas of their interior [2]. The two freeze-fractured membranes presented two distinct appearances, the P-face (protoplasmic face) containing numerous globular particles 6-7 nm in diameter, while E-face (external face) was relatively smooth. Freeze-fracture observations provided evidence that protein particles are indeed embedded into the membranes. Moreover, this technique allows to reveal the presence of particular arrays of intramembrane particles (IMPs), 10 nm in diameter, localized at certain regions of the plasma membrane or at the contact between cells which define membrane specializations. These observations provided valid morphological support for

V. Evangelista et al. (eds.), From Cells to Proteins: Imaging Nature across Dimensions, 71–84.

further functional interpretations of the observed specializations. Many of these membrane specializations deal with intercellular junctions and ciliary or flagellar membranes. Here I illustrate two examples of these structures, i.e.: a) the typical intercellular junctions of some tunicates (Urochordata) and insects; b) the ciliary specializations of a rotifer.

2. MATERIALS AND METHODS

2.1. Materials

a) samples of the rotifer *Brachionus plicatilis*; b) fragments of trachea of buffalo; c) midgut of the wingless insect *Orchesella villosa*, of the cockroach *Periplaneta americana* and of the mosquito *Aedes aegypti*; d) salivary glands of the milliped *Pachyjulus* sp.; e) deferent duct of the centiped *Scutigera* sp.; f) branchial basket of the urochordates *Botryllus schlosseri*, *Botrylloides leachi*, *Phallusia ingeria*; g) stomach of the urochordate *Diplosoma listerianum*.

2.2. Methods

Transmission electron microscope (TEM)

The material has been fixed for 3 to 10 h at 4°C in 2.5% glutaraldehyde 0.1 M phosphate buffer (PB) pH 7.2 to which 2% sucrose was added. After rinsing in PB several times the material was post-fixed for 1-2 h in OsO_4 in the same PB. After rinsing, the material was dehydrated with alcohol and embedded in a epoxy resin. After routine staining, ultrathin sections were observed with a Philips CM10 TEM at 80 kV.

For the visualization of the intercellular space, part of the tissues were treated with 1% colloidal (or nitrate) lanthanum added to PB, using the electron-opaque tracer as an extracellular "negative" stain [3].

For freeze-fracture preparations, the material was preliminary fixed for 20 min at 4°C in the fixative as previously described. Samples were then washed and cryoprotected in 10 to 30% glycerol in PB, before freezing by plunging into propane cooled in liquid nitrogen. The material was then mounted in a Balzers BAF 301 freeze-fracturing device and fractured at –100 to –115°C at $2x10^{-6}$ Torr. The preparations were shadowed with platinum-carbon and backed with carbon. The tissue was removed with sodium hypochlorite and the replicas, rinsed in water, were mounted on copper grids for examination with the TEM.

3. RESULTS AND DISCUSSION

3.1. Intercellular junctions

Besides desmosomes, the two main epithelial cell junctions with adhesive functions are *tight junctions*, which apically seal the intercellular cleft and which were thought to be present only in vertebrates, and *septate junctions*, found only in invertebrates. This view, however, is not entirely true: tight junctions are also found in lower chordates and simple types of tight junctions occur also in some arthropods [4-6].

The presence of tight junctions (TJs), or *zonulae occludentes*, in polarized epithelia is commonly demonstrated by ultrastructural observations of thin sections or, better, by freeze-fracture replicas. In sections of tissues prepared for electron microscopy, TJs are morphologically detectable along the cell contact at the points where the intercellular space is obliterated and the two outer leaflets of the adjacent plasma membranes are in close apposition (Figure 1B). This is not always obvious and only the study of replicas made by freeze-fracture allows to identify the presence of TJs. In such preparations, TJs are shown as branching and anastomosing strands or ridges of aligned IMPs on the P-face and complementary grooves on the E-face (Figure 1A,C,D). The interconnected ridges forming a network represent the integral protein *occludin*. The most important role of TJs is to determine an epithelial cell polarity by separating the apical domain from the baso-lateral domain and preventing the free diffusion of lipids and proteins between these two cell compartments through the intercellular space (paracellular pathway barrier). In the specimens incubated with lanthanum, the tracer penetrates the intercellular cleft, but it was halted at the punctate tight-junctional oppositions. The occlusion of the intercellular space is mediated by the interaction between *occludin* and *claudin*, a family of trans-membrane proteins which span the membrane four times [7]. These proteins, together with the single span trans-membrane proteins JAM (junctional adhesion molecules), CRB1 and CAR, interact directly with the cytoplasmic PDZ domain containing scaffolding proteins, which function as adapters to recruit other cytoskeletal and/or signalling molecules [8-9]. As tight junctions participate in the control of ions through the paracellular pathway, they become responsible for the transepithelial resistance.

TJs may be variable in their permeability properties, sometimes being either "tight" or "leaky", according to the number of junctional strands present at the junctional region (Figure 1D). For example, in the renal proximal tubule, TJs may be represented by a single continuous strand; a similar pattern was also described in the branchial basket of ascidians at the level of the intercellular junction between ciliated cells and parietal cells [10] (Figure 1E).

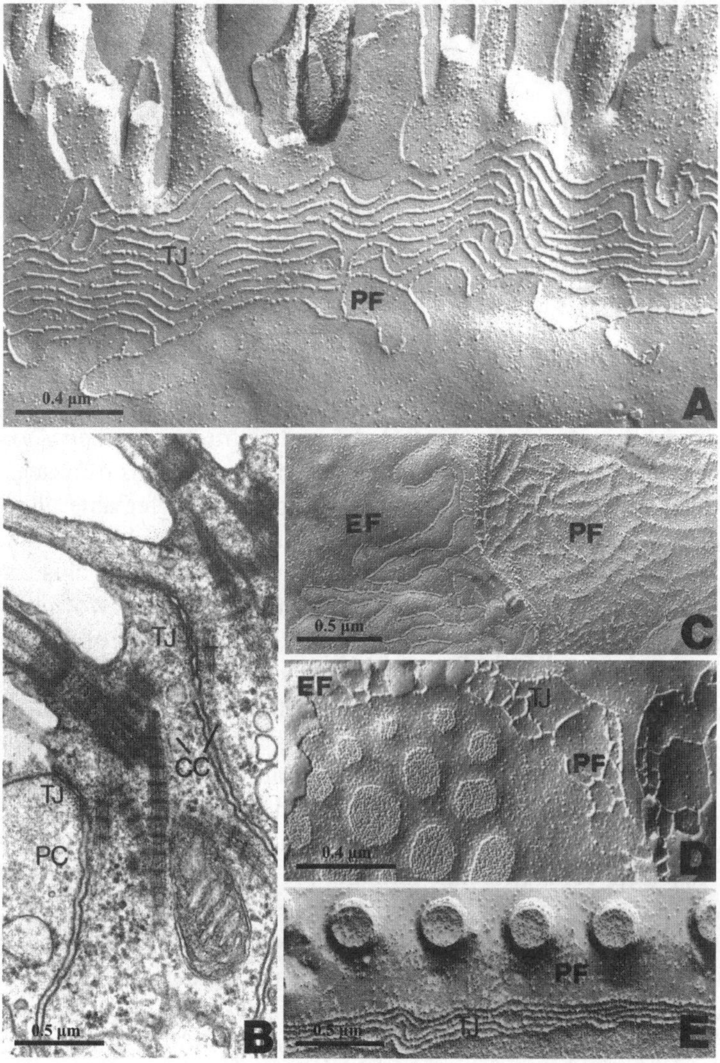

Figure 1. **A)** Freeze-fracture replicas through the buffalo tracheal epithelium. The network of strands associated to the P-face of the membrane (PF) of a typical tight-junction (TJ) is visible at the apical region. **B)** Cross section through the epithelial cells of the branchial basket of *Botryllus leachi*, to show the different extension of the tight junctions (TJ) between ciliated cells and between ciliated (CC) and parietal (PC) cells, respectively. **C)** Freeze-fracture replicas of a tight-junction between epithelial cells of the *Botryllus schlosseri* branchial basket. Note the extensive network of the tight junction strands with IMPs associated with the P-face (PF) and the complementary grooves with the E-face (EF) of the plasma membrane. **D)** Freeze-fracture replicas of the tight-junction between epithelial stomach cells in the urochordate *Diplosoma listerianum*. A slender band of tight junctions (TJ) in a network conformation is visible. **E)** Freeze-fracture replicas of the tight junction between parietal and ciliated cells in the banchial basket of the urochordate *Phallusia ingeria*. The cross-fractured cilia lie to one side of the tight junctional P-face (PF) ridges. Note the few parallel strands of IMPs of the tight junctions (TJ).

Intercellular septate junctions (SJs) are the most common type of cell junctions in invertebrates. They are characterized by a constant intercellular cleft of 15 to 20 nm, crossed by regularly spaced bridges or septa (Figure 2A). The glycoproteic material present in the cleft, however, can obscure the septa. SJs usually form a belt at the apical region of epithelial cells and are often associated with other types of junctions (i.e. *zonulae* and/or *maculae adhaerentes* and gap junctions). SJs from cross sectioned tissues show a rather uniform appearance. Two types of septate junctions are, however, present in tissues of arthropods, the most diverse phylum of the animal kingdom: pleated (PSJs) and smooth septate junction (SSJs). These two types of junctions are clearly distinguishable in tissues after lanthanum tracer treatment. PSJs exhibit distinct undulating septa in oblique sections. The electron-lucent zig-zag strands are visible in a dense background of lanthanum tracer (Figure 2B). On the contrary, SSJs (also known as continuous junctions) are characterized by sinuous electron-transparent strands in tangential sections of lanthanum treated tissue (Figure 2C). In freeze-fracture replicas the PSJs consist of rows of 8-10 nm IMPs which are visible associated with the P-face of the membrane, while complementary E-face exhibits grooves (Figure 2E). The pattern of PSJs can vary according to the tissue and organism studied, but it does not seem to be influenced by the fixation of the material. On the contrary, SSJs of fixed material are characterized by rows of IMPs fused into ridges and associated with the P-face of the membrane, while grooves are evident on the complementary E-face (Figure 2D); in non-fixed material, this pattern of particle distribution is reversed. Little variations on the SSJs are found in different arthropod groups [11-14].

SJs mainly serve as adhesive devices, but they may also represent a partial barrier to paracellular flow, as it was experimentally shown [15].

Cytoskeletal elements are associated with both TJs and SJs and they are localized close to the specialized plasma membrane. These elements are represented by actin filaments, α-actinin, myosin and vinculin [16-20]. Moreover, in the submembrane domain of TJs, several proteins have been described [21].

TJs were initially considered to be exclusive to vertebrates, but from the 1970s they were also found in arthropods and they appear to be the morphological basis to the blood-eye, blood-testis and blood-brain barrier [5]. Moreover, TJs were also described in the lower chordates, such as tunicates [6,10,22].

There is convergence to similar functions between junctions with different structures [6]. In both TJs and SJs, fluorescent dyes introduced into the apical membrane domain and incorporated into the inner membrane leaflet cannot diffuse into the basolateral membrane surface [23-24]. For TJs, this has been explained as an intramembrane "fence", with the TJs strands preventing lateral diffusion of proteins and lipids. On the contrary, the intercellular "gate" of the outer membrane leflet is fused between adjacent cells, thus reducing the cleft and increasing transepithelial resistance [25-26]. The homology found between some proteins detected in the tight junctions of vertebrates and those in the septate junctions of invertebrates raised the interesting suggestion of a common evolutionary origin of these two types of junctions [21].

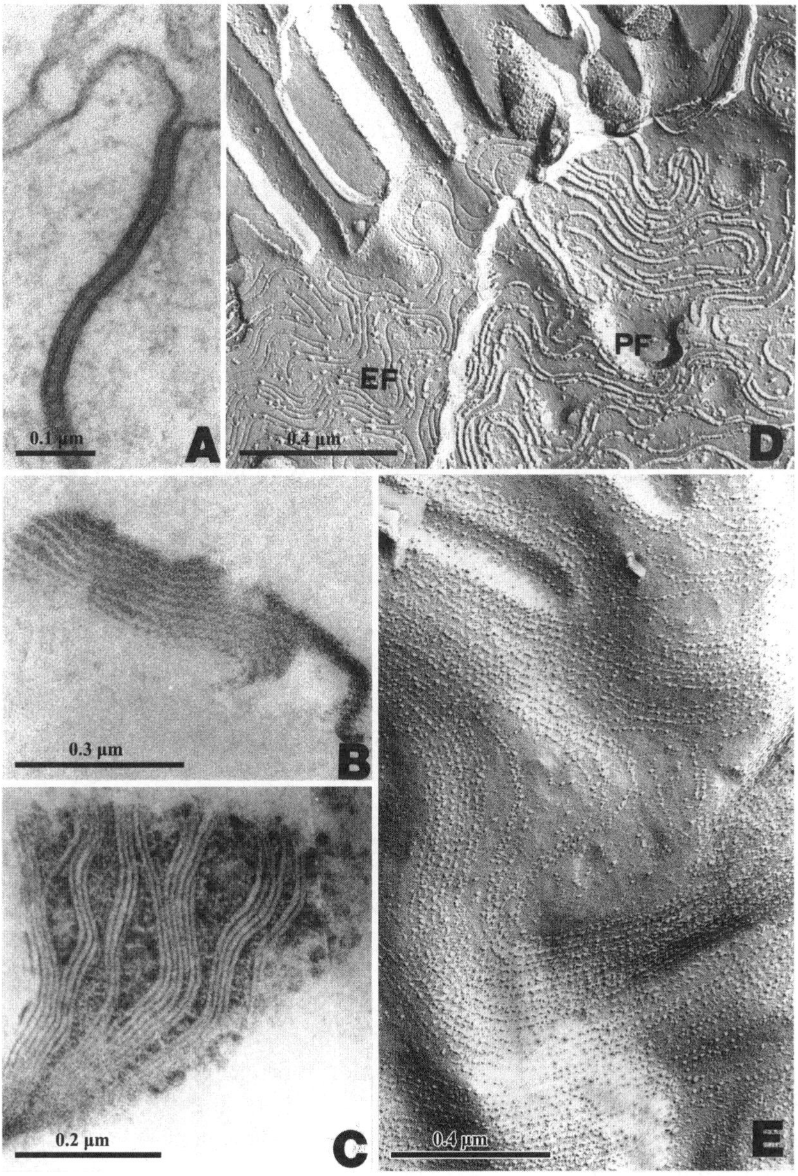

Figure 2. **A)** Cross section through the smooth septate junction between midgut cells in the wingless insect *Orchesella villosa*. Note the parallel array of the opposite plasma membrane and the septa crossing the intercellular space, almost obscured by dense material. **B)** Typical zig-zag translucent strands of the plated septate junction in lanthanum incubated salivary gland epithelium of the milliped *Pachyjulus* sp. **C)** Typical undulating transparent ribbons of the smooth septate junction in lanthanum incubated midgut epithelium of the cockroach *Periplaneta americana.* **D)** Freeze-fracture replicas of the midgut cells in the mosquito *Aedes aegypti* to show the pattern of sinuous ribbons of fused IMPs on the P-face (PF) and complementary grooves on the E-face (EF) of the membrane. **E)** Freeze-fracture replicas of the deferent duct epithelial cells in the centiped *Scutigera* sp. to show the pattern of rows of isolated IMPs.

3.2. Ciliary specializations

Cilia, particularly those propelling mucous or water, show a complex organization due to a series of specializations, which constitute sites for the dynamic anchorage of the axoneme to the ciliary membrane and/or for the transit of ions [27]. Some of these specializations are suggested to be involved in maintaining the integrity of the ciliary structure during ciliary beating [28-30]; others seem to be associated with extra-ciliary structures, as for instance the anchorage of mastigonemes to the membrane in some protozoans ([31] or the bridges connecting adjacent cilia in the ctenophore comb plates [32-33]; others serve for the anchorage of the external network of filamentous material, "the extracellular cuticle" [34], to the ciliary membrane of the apical rotary apparatus (corona) of the rotifer *Brachionus plicatilis* [35]. I will illustrate here the complex organization of the ciliary shaft in this organism.

The basal region of the ciliary shaft of the apical system of *Brachyonus* shows the typical organization of the transition zone described in many Metazoa [36]. A longitudinal section through the ciliated apical cell region shows that a dense foot is visible at the base of the cilium and a striated root extends deeply into the cytoplasm (Figure 3B). The cilium has a basal body and a basal plate at about 0.22 μm from the apical surface of the cell, where the two central microtubules of the axoneme begin. The region comprised between the basal plate and the apical surface of the cell corresponds to the necklace zone [28,37-38] (Figure 3B). For a short region above the plate, the cilium narrows, then it increases its diameter. Beyond this region, the ciliary diameter becomes thinner for about 0.15 μm in length, and then it increases again its diameter. Looking at the whole basal ciliary shaft, three regions, in a linear series, can be observed, with a ciliary diameter of 0.25 μm: the necklace region, an intermediate region, and the region corresponding to the anchorage of the extracellular "cuticle" of fibrous material (Figure 3B).

Serial cross sections through the basal ciliary shaft show the variations in the axoneme organization. In the region corresponding to the ciliary necklace, each doublet is anchored to the ciliary membrane by cup-like connections, previously described as "champagne glasses" [28]. At a distance from this region, where the cilium has a normal diameter, the axonemal doublets are embedded in a dense material which expands in the space between the axoneme and the ciliary membrane. Finally, at the level where the external cuticle anchors to the ciliary membrane, dense large bridges connect the scalloped plasma membrane to the axonemal doublets (Figure 4A, B).

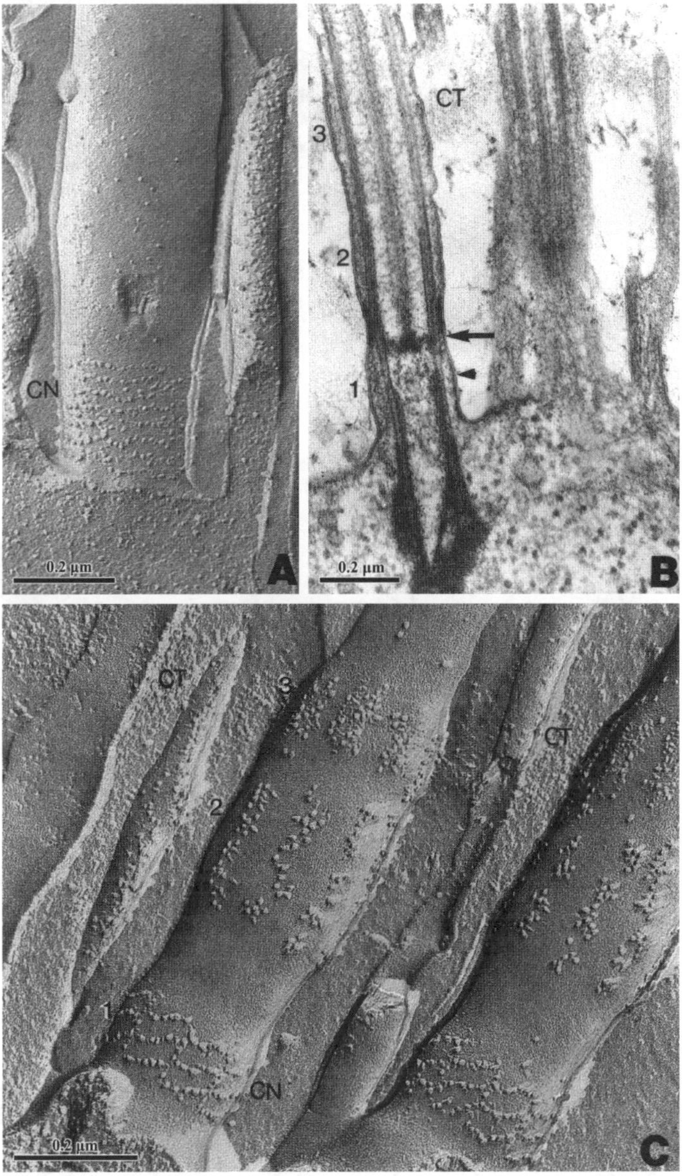

Figure 3. **A)** Freeze-fracture replica of a cilium from the buffalo tracheal ciliary epithelium. Note the several parallel rows of IMPs constituting the ciliary necklace (CN). **B)** Longitudinal section through ciliated cells of the apical corona of the rotifer *B. plicatilis*. Arrow indicates the basal plate. Note the position of the necklace (arrowhead) and the "cuticle" (CT) embedding the cilia. The numbers 1,2,3 mark the position of the three ciliary specializations. **C)** Free-fracture replica of the basal ciliary shaft of the apical corona of *B. plicatilis*. Three dictinct ciliary specializations are recognizable: the basal necklace (CN) (1), the intermediate (2) and the distal (3) aggregates of IMPs. Note the position of the "cuticle" (CT) at the level of the higher (third) specialization.

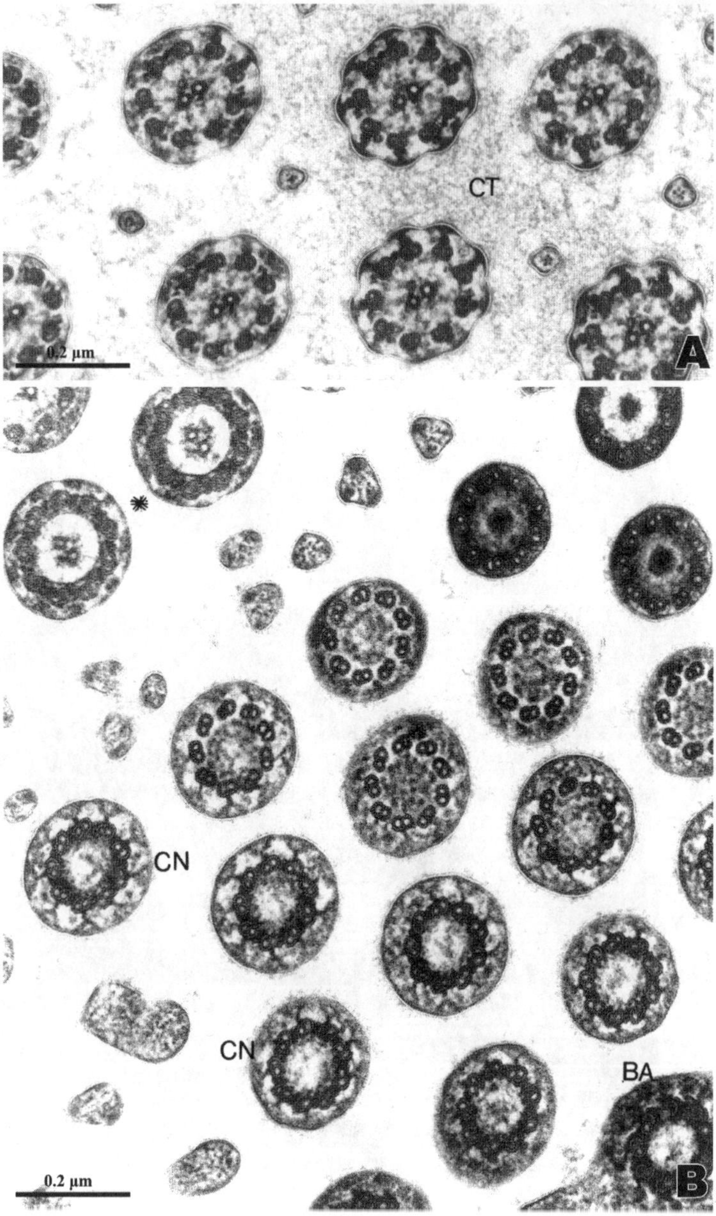

Figure 4. **A)** Cross section through the ciliated cells of the apical corona of the rotifer *Brachyonus plicatilis*. In the bottom right corner, a basal body (BA) is recognizable. The axonemal organization at the level of the ciliary necklace (CN), with the "champagne glasses" structures, is visible. The asterisks (*) mark the intermediate ciliary specialization along the basal ciliary shaft. **B)** Cross section of cilia as in the previous figure A, at a little higher level of the ciliary shaft. The fibrous material intermingled with cilia is the "external cuticle" (CT); the section level corresponds to the position of the higher (third) ciliary specializations.

Freeze-fracture replicas of the basal region of the ciliary shaft clearly show three specialized areas, which can be enlisted as follows: a) starting from the basis, the well known ciliary necklace, consisting of three parallel rows of 10 nm IMPs, disposed in an orderly scalloped fashion. The distance between the rows is 25-30 nm and the period of each row, about 70 nm, corresponds to the spacing of the underlying structures observed in cross section. IMPs are mainly associated with the P-face of the plasma membrane, even though several IMPs also remain on the E-face together with pits; b) a second ciliary membrane specialization above the necklace occurs at the level where the ciliary diameter regains the usual size of 0.25 μm. This specialization consists of longitudinal rows of IMPs, orderly arranged according to the position of the underlying axonemal doublets. IMPs are associated with the P-face of the membrane and complementary pits are on the E-face. Above this second specialization, a further narrowing of the cilium diameter is evident; c) above this region the third distal membrane specialization is visible. This consists of 9 elliptical areas with a few small IMPs and pits, associated with both the P-face and E-face (Figure 3, C) .

The ciliary necklace is a common structure of cilia and usually consists of two or three rows of IMPs, sometimes increasing in the number of rows, as it occurs in the tracheal cilia of some mammals [39] (Figure 3A) or among ciliates [40], although with a greater variability.

The three ciliary membrane specializations described above have a different functional significance. The necklace is clearly the result of the anchorage of the axonemal doublet microtubules to the ciliary membrane by "Y" shaped links. These links take contact with integral membrane proteins that, when fractured, show IMPs exposed on both faces of the membrane leaflets. Thus, the necklace is likely a mechanical device that enables the ciliary axoneme to maintain the proper position within the ciliary shaft. The ciliary specialization placed on the uppermost level along the basal ciliary shaft is the consequence of the anchorage of the fibrous external "cuticle" to the ciliary membrane. Where there is no "cuticle", in fact, as it occurs in the peripheral cilia of the apical corona of *B. plicatilis*, the specialization does not occur. The most interesting specialization is the one placed at the intermediate position along the basal ciliary shaft. It has been shown that this specialization is probably due to the presence of dense material between the axoneme and the ciliary membrane. This specialization does not seem to depend on the presence of external structures and, in fact, it is a constant feature also on cilia of other districts of the apical corona. It could be the structure equivalent to the ciliary plaque of some protists [29,40] retained here as a preferential site for ion passage, possibly Ca^{++} channels. However, these specializations also remind structures described in organisms other than protists, such as ascidians and *Mytilus* [28,41-42], the functional role of which is yet unknown.

In this contest, it seems appropriate to mention that in the space between the peripheral doublets and the surrounding plasma membrane, a process known as *intraflagellar transport* (IFT) has been described [43]. IFT appears to be the mechanism that moves ciliary and flagellar protein precursors to the tip of these structures [44-46]. Particles of variable size move to the tip (anterograde transport) and return from the tip to the base of the flagella or cilia (retrograde transport) [43]. Electron microscopy cross sections indicate that the material interested by IFT consists of particles occurring in groups of variable numbers, called "rafts" [44]; they were formerly observed by Ringo [47] as particles attached by thin connections to the B-subfibers of the outer-doublet microtubules and to the overlying flagellar membrane of *Chlamidomonas* cells [44-45]. The question is whether the raft anchorage to the membrane can be visualized by freeze fracture replicas and, if so, whether some of the specializations described so far in other ciliary systems can be involved in the anchorage of such rafts. This will be matter of future studies.

4. ACKNOWLEDGEMENTS

The paper was supported by a grant from the University of Siena (PAR).

5. REFERENCES

1. Branton D. Fracture face of frozen membranes. Proc Nat Acad Sci 1966; 55: 1048-1056
2. Fawcett D.W. "Junctional specializations." In *The Cell.* Fawcett D.W ed., Philadelphia: W.B. Saunders Comp 1981
3. Revel_J.P., Karnovsky M.J. Hexagonal array of subunits in intercellular junctions of the mouse heart and liver. J. Cell Biol 1975; 33: 3-7
4. Lane N.J., Skaer H.leB. Intercellular junctions in insect tissues. Adv Insect Physiol 1980; 15; 35-213
5. Lane N.J. Tight junctions in arthropod tissues. Int Rev Cytol 1981; 73: 243-318
6. Lane N.J., Dallai R., Martinucci G.B., Burighel P. "Electron microscopic structure and evolution of epithelial junctions." In *Molecular mechanisms of epithelial cell junctions: from the development to disease.* Citi S. ed. R.G. Landes Co., Bionedical Publ. 1994
7. Tsukita S., Furuse M. Claudin-based barrier in simple and stratified cellular sheets. Curr Opin Cell Biol 2002; 14: 531-536
8. Furuse M., Hata M., Furuse K., Yoshida Y., Haratake A., Sugitani Y., Noda T., Kubo A., Tsukita S. Claudin-based tight junctions are crucial for the mammalian epidermal barrier: a lesson from claudin-1-deficiengt mice. J Cell Biol 2002; 156: 1099-1111.
9. Roh M.H., Makarova O., Liu C.-J., Shin K.Y., Lee S., Laurinec S., Goyal M., Wiggins R., Magolis B. The Maguk protein, Pls1, functions as an adapter, linking mammalian homologues of Crumbs and Discs Lost. J Cell Biol 2002; 157: 161-172

10. Martinucci G.B., Dallai R., Burighel P., Lane N.J. Different functions of tight junctions in the ascidian branchial basket. Tissue Cell 1988; 20: 119-132
11. Dallai R., Xué L. Cells junctions in the gut of Protura. Tissue Cell 1992; 24: 51-59
12. Dallai R., Lane N.J., Bigliardi E. Intercellular junctions in myriapods. Tissue Cell 1990; 22: 359-369
13. Lane N.J., Dallai R., Ashhurst D.E. "Structural macromolecules of the cell membranes and the extracellular matrices of the insect midgut." In *The biology of the insect midgut.* Lehane M.J. and Billingsley P.F. eds. Chapman & Hall, London. 1996
14. Noirot C., Noirot-Timothée C. "Cell associations." In *Microscopic anatomy of Invertebrates. Vol. 11A: Insecta.* Harrison F.W. and Locke M. eds. Wiley-Liss Inc. 1998
15. Skaer H. leB., Maddrell S.H.P., Harrison J.B. The permeability properties of septata junctions in Malpighian tubules of *Rhodnius.* J Cell Sci 1987; 88: 251-265
16. Lane N.J., Flores V. Actin filaments are associated with the septate junctions of invertebrates. Tissue Cell, 1988; 20: 211-217
17. Dallai R., Burighel P., Martinucci G.B., Maci R., Camatini M. Actin localization at the tight junction level in invertebrate ciliated epithelia. Tissue Cell 1989; 21: 37-46
18. Dallai R., Lupetti P., Lane N.J. The organization of actin in the apical region of insect midgut cells after deep-etching. J Struct Biol 1998; 122: 283-292
19. Dallai R., Trastullo E., Lupetti P., Mencarelli C. Unusual cytoskeletal association with the intercellular septate junction in the midgut of Collembola. Int J Insect Morphol Embryol 1993; 22: 473-486
20. Colombo A.,Bonfanti P., Camatini M.. Actin, α-actinin, and vinculin are associated with septate junctions in Insecta. Cell Motil Cytoskel 1993; 26: 205-213
21. Knust E., Bossinger O. Composition and formation of intercellular junctions in epithelial cells. Science 2002; 298: 1955-1959
22. Burighel P., Martinucci G.B., Lane N.J., Dallai R. Junctional complexes of the branchia and gut of the tunicate, *Pyrosoma atlanticum* (Pyrosomatida, Thaliacea). Cell Tissue Res 1992; 267: 357-364
23. van Meer G., Simons K. The function of tight junctions in maintaining differences in lipid composition between the apical and basolateral cell surface domains of MDCK cells. EMBO J 1986; 5: 1455-1464
24. Wood R.L. The septate junction limits mobility of lipophilic markers in pasma membranes of *Hydra vulgaris* (*attenuata*). Cell Tissue Res 1990; 259: 61-66
25. van Meer G., Gumbiner B., Simons K. The tight junction does not allow lipid molecules to diffuse from one epithelial cell to the next. Nature (Lond.) 1986; 322: 639-641
26. Mandel L.J., Bacallao R., Zampighi G. Uncoupling of the molecular "fence" and paracellular "gate" functions in epithelial tight junctions. Nature 1993; 361: 552-555
27. Dentler W.L. "Linkages between microtubules and membranes in cilia and flagella." In *Ciliary and flagellar membranes.* Bloodgood R.A. ed. Plenum Press, New York. 1990
28. Gilula N.B., Satir P. The ciliary necklace. A ciliary membrane specialization. J. Cell Biol 1972; 53: 494-509.
29. Plattner H. Ciliary granule plaques: Membrane intrcalated particle aggregates associated with Ca^{++} binding sites in *Paramecium.* J Cell Sci 1975; 18: 257-269
30. Dentler W.L.. Cilia and flagella. Int Rev Cytol (suppl) 1987; 17: 391-456
31. Hill F.G., Outka D.E. The structure and origin of mastigonemes in *Ochromonas minute* and *Monas* sp. J. Protozool 1974; 21: 299-312
32. Afzelius B.A. The fine structure of cilia from ctenophore swimming plates. J Biophys Biochem Cytol 1961; 9: 383-394
33. Tamm S.L., Tamm S. Visualization of changes in ciliary tip configuration caused by sliding displacement of microtubules in macrocilia of the ctenophore Beroe. J Cell Sci 1985; 79: 161-179
34. Clément P., Wurdak E. "Rotifera." In *Microscopic anatomy of invertebrates* Harrison F.W. and Ruppert E.E. eds. Wiley, New York. 1991

35. Dallai R., Lupetti P. Ciliary and microvillar specializations in the corona of *Brachionus plicatilis* (Rotifera, Monogononta)J Submicrosc Cytol Pathol 1994; 26: 497-506
36. Pitelka D.R. "Basal bodies and root structures." In *Cilia and Flagella*. Sleigh M.A. ed. Academic Press, London-New York. 1974
37. Dute R., Kung C. Ultrastructure of the proximal region of somatic cilia in *Paramecium tetraurelia*. J Cell Biol 1978; 78: 451-464
38. Torikata_C. The ciliary necklace. A transmission electron micfroscopic study using tannic acid-containing fixation. J Ultrastruct Mol Struct Res 1988; 101: 210-214
39. Roperto F., Langella M., Oliva G., Restucci B., Varricchio E., Dallai R. Ultrastructural and freeze fracture cilia morphology of trachea epithelium in apparently healthy small ruminants. J Submicrosc Cytol Pathol 1998; 30: 65-69
40. Bardele C.F. Functional and phylogenetic aspects of the ciliary membrane: a comparative freeze-fracture study. Biosystems 1981; 14: 403-421.
41. Dallai R., Burighel P., Martinucci G.B. Ciliary differentiation in the branchial stigmata of the ascidian *Diplosoma listeriam*. J Submicrosc Cytol 1985; 17: 381-390
42. Martinucci G.B., Dallai R., Burighel P. A comparative study of ciliary differentiations in the branchial stigmata of ascidians. Tissue Cell 1987; 19: 251-263
43. Rosenbaum J.L., Cole D.G., Diener D.R. Intraflagellar transport: The eyes have it. J Cell Biol 1999; 144: 385-388
44. Kozminski K.G., Johnson K.A., Forscher P., Rosenbaum J.L. A motility in the eukaryotic flagellum unrelated to flagellar beating. Proc Natl Acad Sci USA 1993; 90: 5519-5523
45. Kozminski K.G., Beech P.L., Rosednbaum J.L. The *Chlamydomonas* kinesin-like protein FLA10 is involved in motility associated with the flagellar membrane. J Cell Biol 1995; 131: 1517-1527
46. Kozminski K.G., Forscher P., Rosenbaum J.L. Three flagellar motilities in *Chlamydomonas* unrelated to flagellar beating. Cell Motil Cytoskel 1998; 39: 347-348
47. Ringo D.L. Flagellar motion and fine structure of the flagellar apparatus in *Chlamydomonas*. J Cell Biol 1967; 33: 543-571

IMAGING, MEASURING AND MANIPULATING BIOLOGICAL MATTER FROM THE MILLIMETER TO NANOMETER SCALE

Daniel Stoffler and Ueli Aebi
M.E. Müller Institute for Structural Biology, Biozentrum, University of Basel, Switzerland

Abstract. Visualizing the structure and dynamics of proteins, molecular assemblies and cellular components is usually key to our understanding of biological function. Here, we discuss the major approaches in imaging, measuring, and manipulating biological matter ranging from the millimeter to the nanometer scale. Relevant biomedical applications at different length scales are chosen to discuss the various aspects of data acquisition with multiple modalities including confocal laser scanning microscopy, electron microscopy and scanning force microscopy.

Key words: Electron Microscopy, Scanning Force Microscopy, Scientific Visualization, Image Processing, Interactive Computer Graphics

1. INTRODUCTION

As the telescope enables us to catch a glimpse of the universe, computerized X-ray tomography and magnetic resonance imaging (MRI) allow us to look inside our body and image its skeleton and organs in three dimensions. Similarly, with the microscope we can explore the microcosm, from cells and tissues all the way down to molecules and atoms. For example, by "cutting" optical sections the confocal laser scanning microscope (CLSM) allows us to look inside living cells and tissues, and to resolve their cytoskeleton and various organelles. Using video-enhanced light microscopy, cell structures can be visualized whose dimensions are one order of magnitude or more below the resolution limit of the light microscope, thus molecular motors can be directly watched at work. The widest range of dimensions is covered by the electron microscope (EM): it can image whole cells, their organelles, cytoskeleton and supramolecular assemblies, as well as individual biomolecules, their submolecular structure and, ultimately, individual atoms. However, to withstand the high vacuum inside the EM, biological specimens have to be dehydrated thus causing them to denature and become subject to preparation artifacts. One way out of

V. Evangelista et al. (eds.), From Cells to Proteins: Imaging Nature across Dimensions, 85–110.

this dilemma has been the introduction of liquid nitrogen- or even helium-cooled cold stages to keep the specimen in a thin film of amorphous ice while inspected or imaged in the EM. Another attempt to keep biomolecules in their native environment has been to embed them in glucose syrup. However, due to the low inherent contrast of biological matter when embedded in ice or glucose, only crystalline specimens (e.g., 2-dimensional (2D) crystalline protein arrays) can be imaged at high resolution by these two methods. Recently, the scanning force microscope (SFM) - a member of the family of scanning probe microscopes - has opened completely new perspectives for analyzing the surface topography of biological matter in its aqueous environment at a resolution comparable to that achieved by EM. Most exciting, the SFM is the first imaging device allowing direct correlation between structural and functional states of biomolecules at submolecular resolution.

2. LOOKING INSIDE CELLS AND TISSUES BY OPTICAL SECTIONING WITH A CONFOCAL LASER SCANNING MICROSCOPE

Confocal laser scanning microscopy (CLSM) was introduced around 1980 by M. Petran and A. Boyde and has quickly found wide applications in the biological sciences [1,2]. The primary value of the CLSM to the biologist

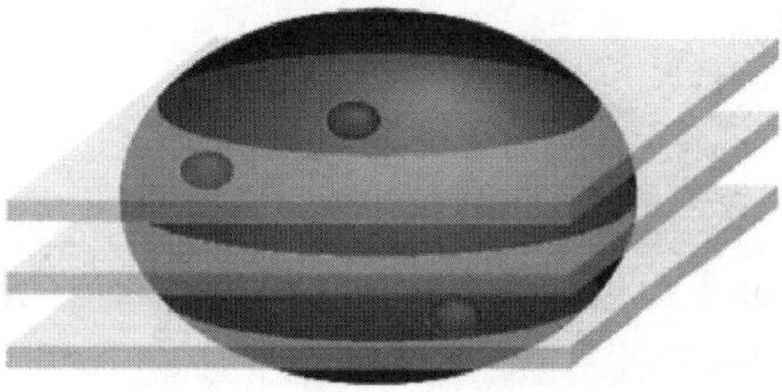

Figure 1. Optical sectioning of a sphere by confocal planes.

is its ability to produce optical sections through a 3-dimensional (3D) specimen - e.g., an entire cell or a piece of tissue - that, to a good approximation, contain information from only one focal plane. Therefore, as illustrated schematically in Fig. 1, by moving the focal plane of the instrument step by step through the depth of the specimen, a series of optical sections can be recorded [3]. This property of the CLSM is fundamental for solving 3D biological problems where information from regions distant from the plane of focus can obscure the image (thick objects). With biological specimens, either the epi-fluorescence or the epi-reflection mode is generally employed. As a valuable by-product, the computer-controlled CLSM produces digital images which are amenable to image analysis and processing, and can also be used to compute surface- or volume-rendered 3D reconstructions of the specimen.

In Fig. 2, the confocal principle is illustrated schematically for the epi-fluorescence imaging mode [3,4]. To image the specimen point by point, a collimated, polarized laser beam is deflected stepwise in the x- and y-direction by a scanning unit (not shown) before it is reflected by a dichroic mirror (beam splitter) so as to pass through the objective lens of the microscope, and focused onto the specimen. The emitted, longer-wavelength fluorescent light collected by the objective lens passes through the dichroic mirror (transparent for the longer wavelength) and is focused into a small pinhole (i.e., the confocal aperture) to eliminate all the out-of-focus light, i.e., all light coming from regions of the specimen above or below the plane of focus. Therefore, the CLSM does not only provide excellent resolution within the plane of section (≥0.25 μm in x- and y-direction), but also yields similarly good resolution between section planes (≥0.3 μm in z-direction).

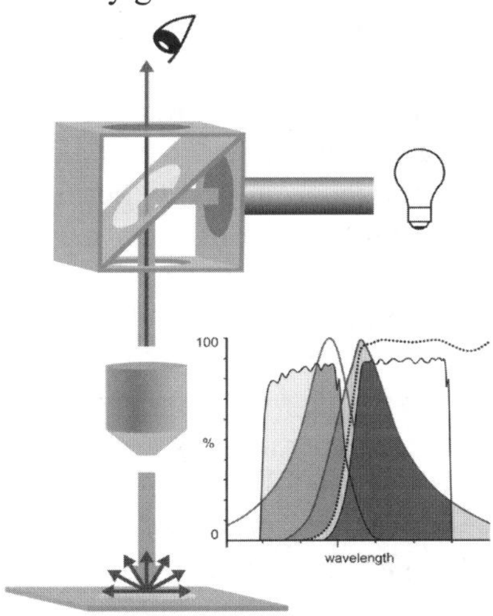

Figure 2. The principle of a confocal laser scanning microscope. The graph depicts excitation and emission spectra of the dye FITC in combination with bandpass filters chosen for sequential visualization of green and red emission. Filters are designed to maximize image contrast and maintain image quality. Image courtesy of Dr. Werner Baschong, M.E.Müller Institute, Biozentrum, University of Basel, Switzerland

The in-focus information of each specimen point is recorded by a light-sensitive detector (i.e., a photo-multiplier) positioned behind the confocal aperture, and the analog output signal is digitized and fed into a computer. At the same time, the analog photo-multiplier signal can be used to generate a TV-like image on a video monitor. The obvious advantage of having a stack of serial optical sections through the specimen pixel by pixel in digital form (Fig. 3, small panels) is that either a composite projection image can be computed (Fig. 3, large panel), or a volume-rendered 3D representation of the specimen can be generated on a computer.

The confocal part of a CLSM consists of an elaborate, highly folded optical bench on which the laser, all the filters, an oscillating-mirror or acousto-optic scanning device, and the detector are mounted. When working in the epi-fluorescence mode, the laser beam is filtered to select the 488 nm, 568 nm or 647 nm wavelength line from an Argon/Krypton laser, and a triple dichroic mirror (to fit all three excitation wavelengths) is used to transmit - rather than reflect - the longer-wavelength fluorescence signal to the detector. For the epi-reflection mode no wavelength filters are needed. Instead, a semi-transparent mirror reflects 50% of the incident laser beam through the objective lens and to reach the specimen, and it transmits 50% of the light reflected by the specimen and collected by the objective lens to the detector. To suppress light reflected by the various optical elements of the microscope, a 1/4-wavelength plate and a polarizing filter are put into the beam path.

Usually, objective lenses with a high numerical aperture (NA) are used to provide good resolution in x-, y- and z-direction (e.g., a 63x/1.4 NA Planapochromat for epi-fluorescence, or a 40x/1.0 NA for epi-reflection). Often, good specimen areas are sparse, so to find them the specimen is screened in the conventional transmission- or epi-fluorescence mode. Once a good region has been located, the CLSM mode is activated, and serial optical sections are recorded at user-selectable depth increments, which can be as small as 0.02 μm. In digital form, each image element yields an 8-bit intensity value in the range 0-255. Typically, the image frame size is 512x512 pixels, and in the best case images are recorded at video rate, i.e., at 25 frames per second in PAL or 30 frames per second in NTSC norm. To improve the S/N ratio of individual image frames, several of them may be recorded in series and averaged. In the case of epi-fluorescence imaging, the total number of scans is generally adjusted to limit specimen bleaching to an acceptable level.

The CLSM allows imaging specimens across a wide range of scales: from bone, cartilage and muscle to the cytoskeleton. In general, thick and opaque specimens that can barely be observed in a conventional light microscope are excellent specimens when it comes to demonstrate the power of a CLSM. For example, 20-25 μm thick sections of bone, cartilage or muscle are ideally suited for 3D imaging in the CLSM. Independent of the thickness and surface quality of such tissue sections, individual confocal planes readily reveal a lateral resolution of 0.3 μm. By recording its auto-fluorescence, even a piece of wood can be optically sectioned to a depth of about 100 μm.

Cultured fibroblasts grown as monolayers can be multiple-labeled with fluorochrome-tagged antibodies against different cytoskeletal components: e.g., actin (labeled with rhodamine = red fluorescence) or tubulin (labeled with fluorescein = green fluorescence). In such cells, actin is primarily in the form of stress fibers (red fluorescence), actin filament bundles adhering to the plasma membrane attaching the cell to the coverslip on which it has been grown. In contrast, the microtubules (green fluorescence) usually radiate outward from the two perinuclear centrioles and reach all parts of the cell. During cell division the microtubules form the spindle apparatus that separates the condensed chromatin (i.e., the chromosomes). The CLSM allows the 3D distribution and relative spatial relationship of these two filament systems to be visualized directly.

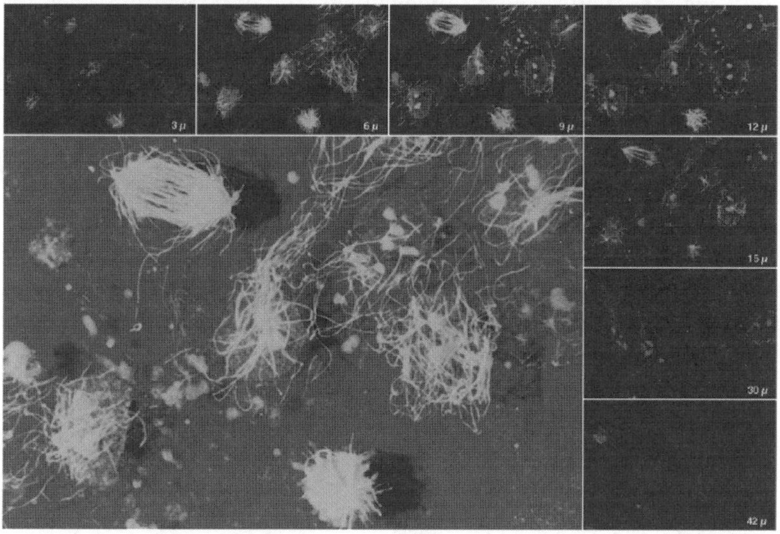

Figure 3. Seven representative optical sections selected from 81 confocal planes (corresponding to a depth of 50 µm) "cut" through a collagen matrix containing growing fibroblasts labeled with fluorescent antibodies to tubulin. Inset, composite shadow-projection image of all 81 confocal sections revealing the spindle apparatus of dividing cells and the regular microtubular network of interphase (i.e., non-dividing) cells.

Exploring the growth and differentiation of cultured cells in 3D collagen matrices - a condition that more closely resembles the natural environment of cells - has been another application of the CLSM (Fig. 3). While difficult to assess by conventional fluorescence microscopy, the spatial relationship of the various cytoskeletal components relative to the different sub-cellular compartments in these matrix-embedded cells is readily determined by CLSM.

3. WATCHING MOLECULAR MOTORS AT WORK BY VIDEO-ENHANCED LIGHT MICROSCOPY

Movement and shape changes of single cells and whole organisms, as well as intracellular transport are fundamental aspects of life. This biological motility results from the specific interaction of motor proteins (e.g., myosins, kinesins or dyneins) with filamentous cytoskeletal substrates: e.g., myosins interact with actin filaments, whereas kinesins and dyneins interact with microtubules. Motor proteins generate force in molecular steps and are fueled by ATP hydrolysis. *In vitro* motility assay systems have been developed that require only a motor protein, its substrate, and ATP to reconstitute motility [5]. Through them, our current view of the biochemistry and biophysics of cyclic chemo-mechanical force generation has been greatly advanced.

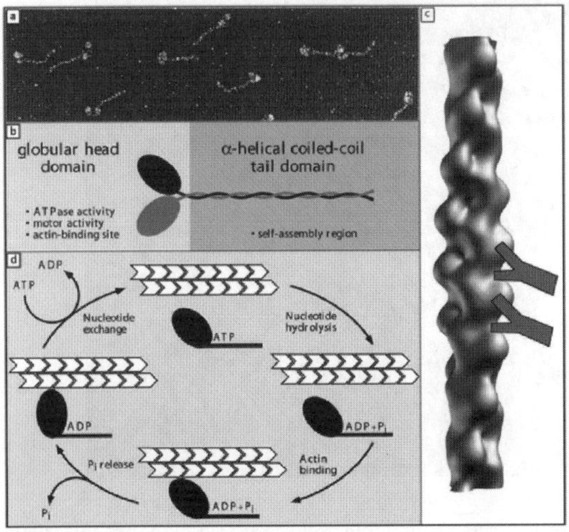

Figure 4. Motor proteins - structure and interaction with their substrate. (a) Electron micrograph of glycerol-sprayed/rotary metal-shadowed muscle myosin molecules which consist of two globular heads and a long, 2-stranded α-helical coiled-coil tail. (b) Schematic representation of the myosin molecule which is composed of two copies each of a heavy chain, and a regulatory and essential light chain [6]. For simplicity, only the two heavy chains are shown here. Myosin polymerization into thick filaments as they are found in the muscle sarcomere proceeds by lateral association of the myosin tails. (c) Interaction of myosin with actin filaments. In this cartoon, two highly schematic myosin heads depicted in light gray illustrate that one head interacts with two adjacent actin subunits [7,8]. (d) Schematic representation of the myosin ATPase/cross-bridge cycle.

Motor proteins transduce the free energy change accompanying the enzymatic hydrolysis of the terminal phosphate of bound ATP into directed movement. For example, myosin is an actin-activated ATPase that rapidly hydrolyzes bound ATP (Fig. 4d: transition from top to right) to yield a ternary complex consisting of myosin and ADP+P_i. This complex binds with high affinity to actin filaments (Fig. 4d: transition from right to bottom). Subsequent release of P_i (Fig. 4d: transition from bottom to left) triggers the "power stroke", i.e., a large conformational change of the myosin head that provides the force for its movement relative to the actin filament. Exchange of bound ADP with ATP (Fig. 4d: transition from left to top) completes the "ATPase/cross-bridge cycle" by lowering the affinity of the myosin head for actin filaments so that they are subsequently released.

The light microscope (LM) allows dynamic biological processes to be imaged in their native (i.e., aqueous) environment with relatively high temporal resolution (i.e., on a sub second time scale). However, the diffraction-limited resolution of the LM is relatively low, i.e., of the order of 250nm. Although the ~9-nm diameter of actin filaments is more than one order of magnitude below this theoretical resolution limit, single fluorescently labeled actin filaments are readily visualized in the LM (Fig. 5).

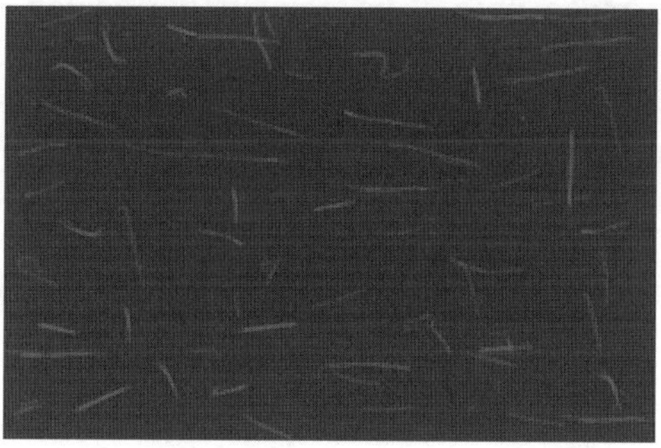

Figure 5. Single fluorescently labeled actin filaments imaged in the LM.

The reason for this apparent discrepancy is that the LM is used to probe whether the area that is imaged contains fluorescent molecules or not, and bright fluorescence can be detected even if the fluorescent object is significantly smaller than the resolution limit. Nevertheless, since the LM cannot overcome the limits set by the laws of diffraction-limited optics, the apparent width of a fluorescent actin filament is defined by and equal to the resolution limit and cannot therefore be interpreted in a straightforward way. However, filament lengths and/or velocities can be determined accurately since these spacings are large compared to the resolution limit.

When working at or beyond the diffraction-limited resolution of the LM, a disadvantage of fluorescence imaging is the relatively low S/N ratio of the images. However, this can be increased significantly by video and computer technology. A computer-aided, video-enhanced LM system such as shown schematically in Fig. 6, serves three different functions: imaging, image processing, and image storage (for a review, see [9]). Images are recorded by a video camera system (i.e., a low-light level CCD or SIT camera for fluorescence LM) and generate signals that can be enhanced in various ways by analog and digital image processing. The camera control unit is usually an "analog enhancer" that allows gray level selection and gain manipulation

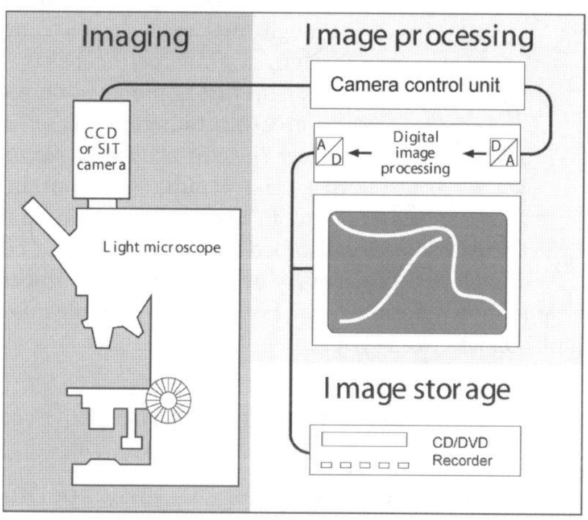

Figure 6. A computer-aided, video-enhanced light microscopy (LM) system performing three different tasks: imaging, image processing, and image storage.

to optimize the image. Digital image processors convert the analog to a digital signal and can store images for (i) background subtraction (e.g., to compensate for uneven illumination), and (ii) to compute moving averages over 2, 4, 8, 16, etc. video frames (e.g., to increase the S/N ratio). To optimize image contrast, most commercially available image processors also allow histogram equalization. Finally, the processed digital images are re-converted into an analog signal that can be displayed on a TV monitor and/or recorded on CD/DVD.

The experimental setup to perform an *in vitro* motility assay is relatively simple. As illustrated in Fig.7 (top), a flow cell (i.e., an open system that allows solution exchange during the experiment) is assembled from a glass slide and a nitrocellulose coated coverslip using silicon grease along the edges of the coverslip as spacer between the two surfaces. A myosin solution is then allowed to flow into the cell. Within seconds, the myosin molecules bind to the coverslip. The nitrocellulose prevents myosin denaturation on the highly charged glass surface. After a washing step to remove excess myosin, free binding sites in the flow cell are blocked to prevent actin filaments from sticking to the glass surface. Fluorescent derivatives of phalloidin (e.g., rhodamin-phalloidin) - a mushroom toxin that tightly binds to actin filaments - are used to fluorescently label actin filaments indirectly. When added to the flow cell, these fluorescent actin filaments bind to the myosin molecules, and the motility assay system is ready for use (see Fig. 5). As depicted schematically in Fig. 7 (bottom), adding "fuel" in the form of ATP produces a sliding movement of the actin filaments powered by the myosin motors attached to the coverslip (see Fig. 4d).

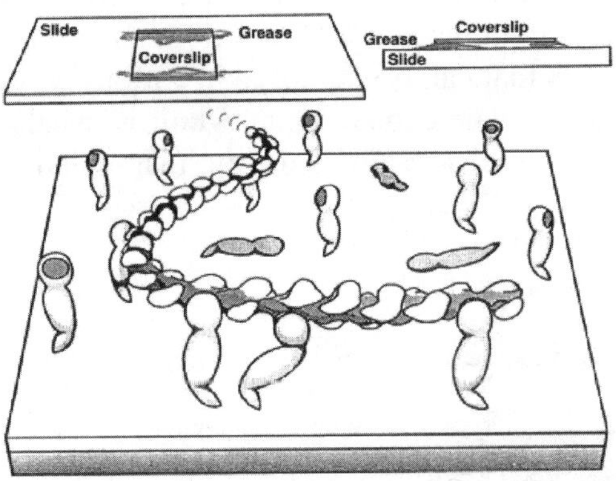

Figure 7. Experimental setup of an *in vitro* motility assay. (Top) Front (left) and side (right) view of a flow cell. Most *in vitro* motility assays are performed using flow cells, i.e., open systems that allow solutions to be exchanged by flowing them through the cell. In its simplest design, a flow cell consists of a coverslip that is mounted on a glass slide using silicon grease as a spacer (i.e., to achieve a separation of ~0.5-1.0 mm). (Bottom) In this cartoon, myosin heads are depicted adsorbed to a nitrocellulose-coated coverslip and are in the process of moving an actin filament. The coverslip is shown upside down, i.e., the myosin molecules and the actin filaments face the slide. Figure adapted from [12].

For a practical illustration, Fig. 8 reveals 3 time frames (i.e., 0, 8 & 16 sec) of an *in vitro* motility assay involving a myosin powered sliding movement of fluorescent actin filaments over a coverslip. In this experiment, individual actin filaments (three of which have been false colored in red, green and blue, respectively) move with an average velocity of about 3 μm/sec relative to the coverslip.

Inspired by successful studies of single ion channels by patch-clamp recordings, a novel *in vitro* assay employing a feedback-enhanced laser trap system has allowed direct measurement of force and displacement that results from the cyclic interaction of a single myosin molecule with a single actin filament [10]. Accordingly, discrete, stepwise movements averaging 11 nm were depicted under conditions of low load, and single force transients averaging 3-4 piconewtons were measured under isometric tension. The magnitudes of the single motor forces and displacements are consistent with predictions of the conventional "swinging-crossbridge" model of muscle contraction [11].

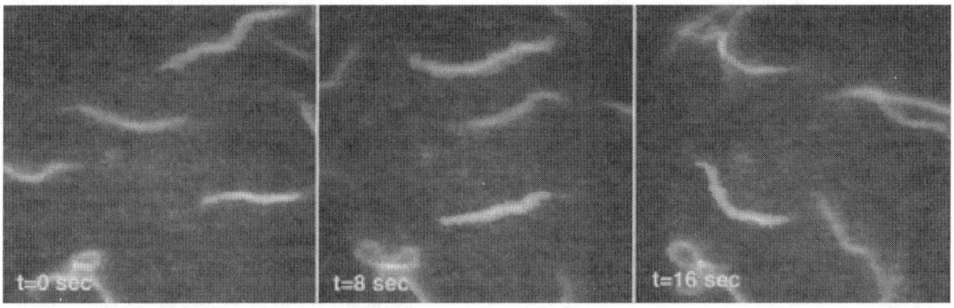

Figure 8. Fluorescently labeled actin filaments. Rhodamine-phalloidin labeled actin filaments were imaged with a Zeiss Axiophot light microscope using a 63x/1.4 NA Planapochromat objective lens. Digital image processing steps included background subtraction, averaging over 4 video frames, and histogram equalization.

4. IMAGING, MEASURING AND MANIPULATING NATIVE BIOMOLECULAR SYSTEMS WITH THE SCANNING FORCE MICROSCOPE

Over the last couple of years the scanning force microscope (SFM) became one of the most powerful tools for determining the surface topography of native biomolecules at subnanometer resolution [13]. Unlike X-ray crystallography, electron microscopy (EM), or light microscopy, which can look “inside” biological matter, the SFM uses no lenses. Instead, SFM works in the same way as our fingers, which touch and probe the environment when we cannot see it. By using a finger to "visualize" an object, our brain is able to deduce its topography while touching it. The radius of the fingertip determines the resolution we can obtain by this method. To achieve atomic scale resolution, a sharp stylus (radius < 10 nm) at the end of a relatively long cantilever is used in the SFM to scan an object line by line thereby contouring its surface while a constant small force is applied to the stylus (Fig. 9). With the SFM the role of the brain is replaced by a computer: the photodiode signal or the z-signal of the position transducer (piezoelectric tube) is used to quantify the surface and the resulting 3D topographic image is then displayed on a computer screen (Fig. 9). This simple technique provides a glimpse into the microscopic world, and it enables us to understand how the "smallest bricks" (i.e., the biomolecules) of biological systems might function. Most excitingly, the SFM allows biomolecules and larger assemblies to be imaged not only under physiological conditions, but also while biological processes are at work [14; reviewed in 15].

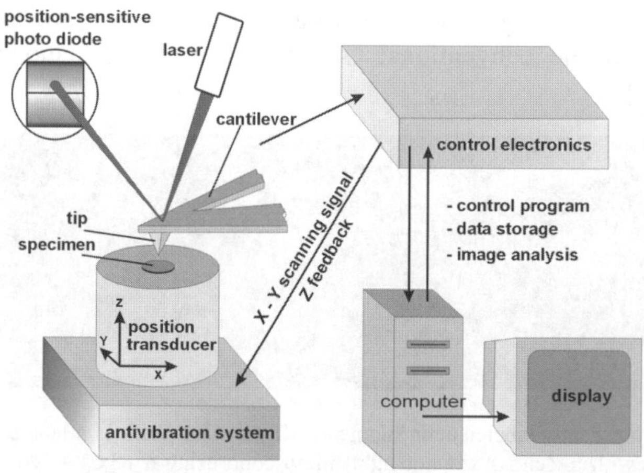

Figure 9. Principle of the scanning force microscope (SFM). A fine tip at the end of a relatively long cantilever is scanned over the surface. At sufficiently small forces the corrugations of the scanning lines represent the surface topography of the sample. The deflection of the cantilever is detected by reflecting a laser beam onto a 2-segment photodiode. The photodiode signal is used to drive a servo system, which controls the movement of the piezo xyz-translator. In this manner the applied force between the stylus and the sample can be kept constant within some tens of a piconewton. The imaging process can be performed in a liquid cell filled with buffer solution. This ensures that the biomolecules remain hydrated. Image courtesy of Dr. Martin Stolz, M.E. Müller Institute, Biozentrum, University of Basel, Switzerland.

Applications with the SFM are manifold. For example, the SFM can provide insight into the binding properties of biological systems. In order to determine the specific interaction between two kinds of molecules (e.g., avidin and biotin) in an SFM, one kind (i.e., avidin) is bound to the tip of a cantilever and the other kind (i.e., biotin) covers the surface of the sample support. The adhesion force upon separation is then a measure of the binding strength. This method allowed the intermolecular forces between individual ligand-receptor pairs [16,17], complementary DNA strands [18], cell adhesion proteoglycans [19], and the specific antigen-antibody interaction [20] to be determined.

As a result of their high S/N ratio unprocessed SFM images provide atomic detail of solids. In contrast, biomolecules such as proteins which are designed to undergo conformational changes and form flexible supramolecular assemblies, are mechanically "soft", i.e., they are best compared with a "sponge". Hence their surface topography cannot be probed at atomic detail by SFM. Nevertheless, state-of-the-art specimen preparation and instrumentation now allow the surface topography of native supramolecular assemblies to be imaged at subnanometer resolution. For example, SFM images of the nuclear pore complex (NPC) -- a large supramolecular assembly -- revealed the distinct surface topographies of the cytoplasmic and nuclear face (Fig. 10a,b; [14]). On the cytoplasmic face the 8-fold rotational symmetry of the NPC could be resolved (Fig. 10a, inset) while on the nuclear face the nuclear baskets were readily visible, as well as "remnants" of the nuclear lamina meshwork in areas devoid of NPCs (Fig. 10b).

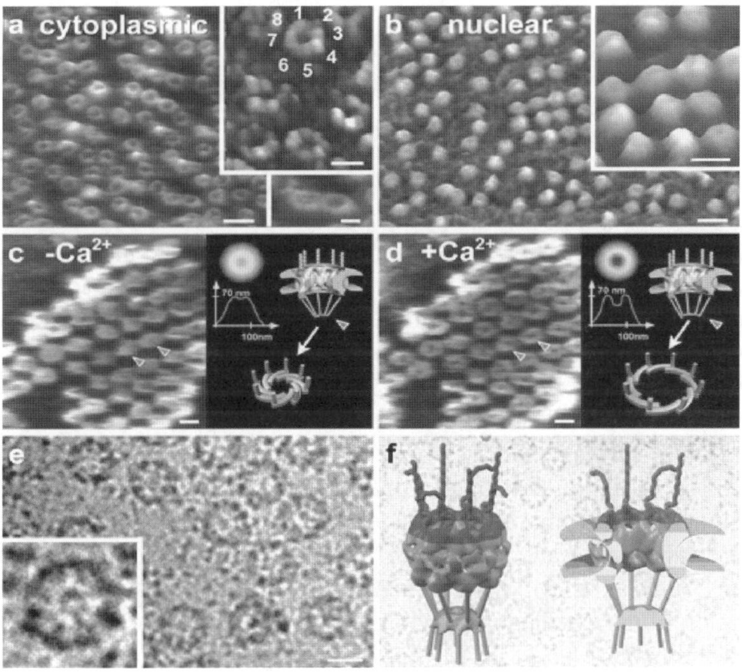

Figure 10. Visualization of native nuclear pore complexes (NPCs) by SFM and cryo-EM. (a, b) Corresponding SFM images of NPCs kept functional in physiological buffer reveal a distinct morphology for the cytoplasmic (a) and the nuclear face (b) of spread *Xenopus* nuclear envelopes (NEs). (c) and (d) reveal reversible calcium-mediated structural changes of the nuclear baskets (i.e. the distal rings) by time-lapse SFM of the same individual NPCs. (e) Appearance of thick-ice embedded chemically unfixed and unstained *Xenopus* nuclear envelope (NE) spread over a carbon support film imaged by cryo-EM. (f) Tomographic 3D reconstruction of the NPC from the type of data shown in (e). All scale bars 100μm.

Furthermore, time-lapse SFM allowed for the first time the direct observation of distinct structural changes at the level of single NPCs under various physiological conditions (Fig. 10c,d; [14]). These experiments revealed the repeated opening and closing of the nuclear basket of individual native NPCs kept 'alive' in physiological buffer at its distal end in response to adding or removing calcium without, however, affecting the overall height and shape of the basket (Fig. 10c,d). These observed 20-30 nm diameter openings may be interpreted in terms of the nuclear basket's distal ring acting as an "iris-like" diaphragm, which is closed in the absence of calcium and opens upon addition of calcium (Fig. 10c,d).

Further biochemical, structural and molecular data suggest that the distal ring might act as a ligand-sensitive docking- or even a gating site for cargo transported in and out of the nucleus [14]. In contrast to the calcium-induced effects observed on the nuclear face of the nuclear envelope (NE), the cytoplasmic face appeared rather unaffected by the same treatment [14]. Further time-lapse SFM experiments [21] provided strong evidence that the elusive "central plug" which can often be observed by EM in the central channel of individual NPCs (Fig. 10e) most probably represents the distal ring of the nuclear basket (Fig. 10e, inset) and/or cargo caught in transit while translocating through the central pore of the NPC. These time-lapse images of functional, native NPCs in their physiological environment by SFM and high-resolution structural data obtained from cryo-EM of NPCs embedded in thick ice (Fig. 10e,f; [21]) provided valuable insights towards a general understanding of the conformational dynamics of macromolecular machines, and a more complete mechanistic understanding of nucleocytoplasmic transport, particularly for the translocation of cargo through the central pore of the NPC.

Indentation type (IT) SFM represents a novel approach for studying the progression of cartilage diseases [22]. More specifically, the IT SFM enables to map, for example, the mechanical properties of healthy and diseased articluar cartilage from the millimeter down to the nanometer scale, thus assessing all important scales of cartilage organization. Pathological situations such as chronic inflammation or acute trauma may lead to the degradation of the cartilage layer coating the articulating bone surfaces in the affected joints. *Ex vivo* IT SFM provides the means to probe representative elasticity measurements of a given piece of cartilage at the nanometer scale – compared to clinical indentation testing devices which measure cartilage elasticity at the millimeter or centimeter scale and therefore cannot resolve its local, hierarchically organized fine structure (see Fig. 11a,b). For example, IT SFM measurements of osteoarthritic cartilage biopsies revealed a significant hardening compared to healthy cartilage (see Fig. 11). This hardening is caused by the loss of glycosaminoglycans (GAGs), which can also be mimicked in a test tube by enzymatic digestion of native articular cartilage with the enzyme Cathepsin D (Fig. 11). In the future, an "arthroscopic SFM" [23] might even be employed for *in situ* diagnosis of osteoarthritic cartilage.

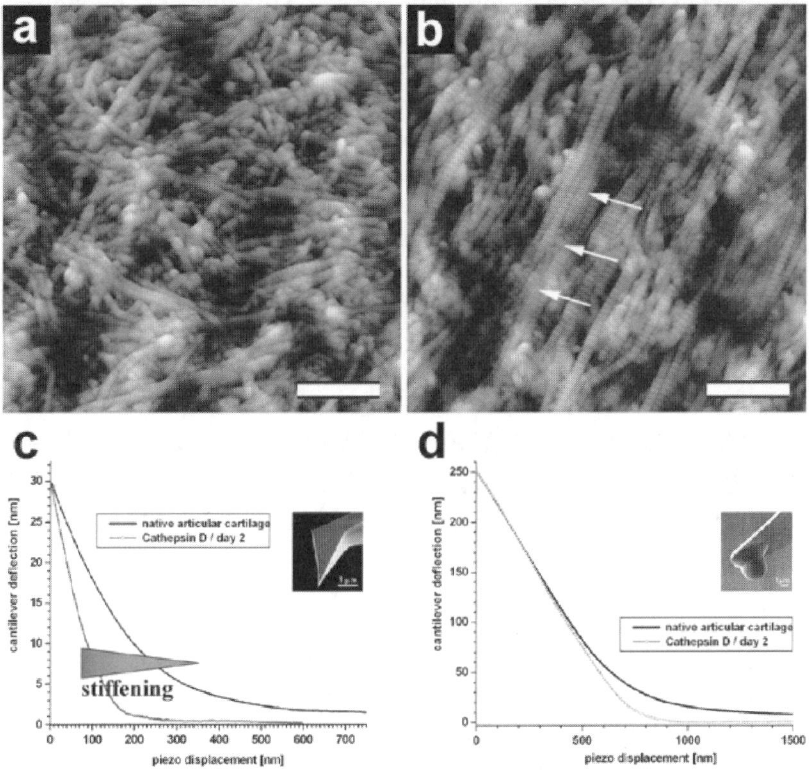

Figure 11. Surface topography of (a) normal articular cartilage and (b) ostheoarthritic articular cartilage, imaged in air, and (c) the corresponding elasticity measurement at the nanometer scale. (a) The 67-nm repeat of individual collagen fibers is clearly resolved by SFM. (b) In contrast to the normal cartilage which exhibits a random orientation of the collagen network, in the diseased cartilage the collagen fibers exhibit a preferred orientation and a more densely packing of the collagen fibers. The orientation of the fibers might follow the directed movement within the joint more easily once the glycosaminoglycans become degraded in the course of disease progression. Elasticity measurements were taken of the same samples by employing two different indenter sizes, i.e. (c) nanometer-sized sharp pyramidal shaped tips (tip-radius = 20 nm), and (d) micrometer-sized spherical tips (tip-radius = 2.5 μm). Each set of the two curves was obtained from normal articular cartilage and from Cathepsin D digested articular cartilage. For digestion the articular cartilage was treated with Cathepsin D at 37°C for two days. Comparison of the two slopes when recorded on the micrometer scale indicates no difference between native and Cathepsin D treated articular cartilage. At the nanometer scale, however, the Cathepsin D treated articular cartilage exhibits a stiffening compared to the healthy articular cartilage. The two icons in (c) and (d) depict at what spatial level the indenters assay the cartilage architecture, i.e. a micrometer scale indenter (d) may simply distinguish between the chondrocytes and the extracellular matrix (ECM), whereas a nanometer scale indenter (c) may reveal the distinct fine-structure of the ECM. Such an indenter may probe, for example, the glycosaminoglycan-filled interstitial space between the collagen fibers that is often altered during disease progression. Scale bars, 1μm (a and b). Image courtesy of Dr. Martin Stolz, M.E. Müller Institute, Biozentrum, University of Basel, Switzerland.

These results emphasize that SFM represents a useful technique in fundamental research as well as for practical applications. Today, SFM images provide information on the surface structure of biomolecular systems, which is complementary to other established techniques such as light and electron microscopy, nuclear magnetic resonance imaging and spectroscopy and X-ray crystallography. The advantage of directly observing biomolecular systems in their native environment opens the exciting possibility to analyze simultaneously their structural and functional properties at the submolecular level. Taken together, SFM provides us with new means to gain structural and functional insights from the millimeter to the nanometer scale, and fosters potential applications in nano-biotechnology, medicine and related fields, which are only limited by our imagination.

5. IMAGING AND MEASURING BIOMOLECULES AND THEIR SUPRAMOLECULAR ASSEMBLIES BY SCANNING TRANSMISSION ELECTRON MICROSCOPY

The scanning transmission electron microscope (STEM) was first introduced in the early seventies. By the end of the decade many of its applications focused on biological research. Although, as may be judged by the wealth of publications (for a review, see [24]), the STEM has become an important tool for the biologist, thirty years later only a few dedicated STEMs are routinely used for biological research: for example, one at the M. E. Müller Institute for Structural Biology, Biozentrum, University of Basel, Switzerland, and one at the Brookhaven National Laboratory, Upton, New York, USA. The instrument's ability to provide both images and quantitative data simultaneously is invaluable for structure determination. Its most successful application is the mass determination and mass mapping of biomolecular structures. Biochemical methods such as gel electrophoresis require dissociation of the complex and only yield the relative amounts of its various structural proteins. Measuring the total mass in STEM allows these data to be expressed in terms of absolute copy numbers. Thus, in many cases the STEM has been instrumental in determining the subunit stoichiometry of oligomeric protein complexes.

While the precision and reproducibility of STEM mass measurements are comparable with those of the analytical ultracentrifuge, the possibility of determining the mass not only of entire supramolecular assemblies but also of their distinct components has opened exciting new avenues which have occasionally been entered [25] but are not yet fully explored. In some cases the two techniques, STEM and analytical ultracentrifugation, complement each other, for example, when the protein content of 2D crystalline protein arrays is required.

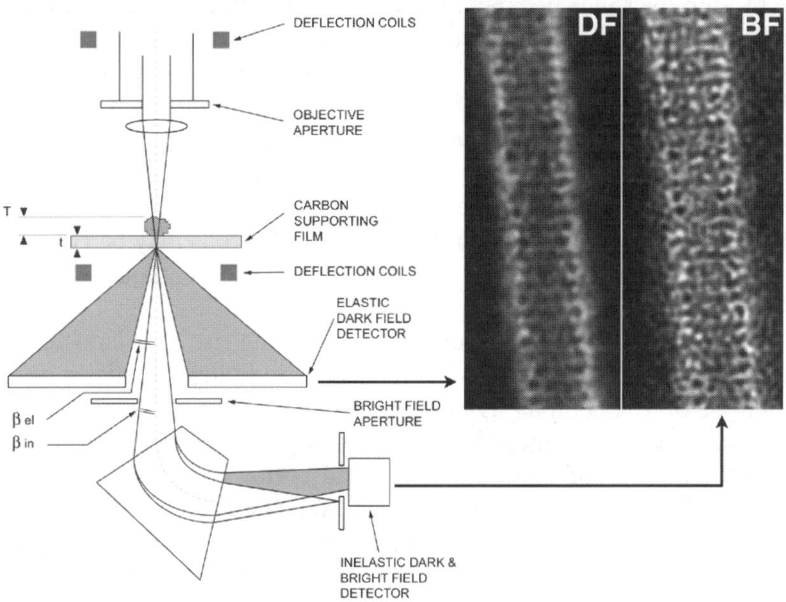

Figure 12. Schematic diagram of a dedicated STEM showing the lens, aperture and deflection systems. Elastically scattered electrons are collected by an annular detector and provide the elastic dark-field signal. A spectrometer deflects those electrons that have lost energy at a larger angle than the unscattered electrons, thus facilitating the acquisition of the inelastic dark-field signal. The coherent bright-field signal arising from unscattered and low-angle elastically scattered electrons is collected through a small aperture placed on the optical axis. The various signals can be collected in parallel and processed on-line as the focused electron probe is scanned over the sample. The simultaneously acquired elastic dark-field (DF) and bright-field (BF) images of a negatively stained bacteriophage T4 giant phage tail are displayed with equal contrast to illustrate the different characteristics of these two imaging modes.

The electron microscope (EM) extracts structural information carried by the scattered electrons. In the fixed-beam transmission EM electrons scattered by the irradiated sample are collected over a narrow solid angle and focused by the objective lens onto the image plane. The elastically scattered electrons, i.e., those that have changed their direction but not lost any energy on interaction with the specimen, interfere with the unscattered electrons to produce a phase contrast image. The inelastically scattered electrons, i.e., those that have changed both direction and energy on interaction with the specimen, generate a diffuse background image that is in some EMs eliminated by an energy filter. As illustrated in Fig. 12, in the STEM the objective lens focuses the electron beam onto an atomic scale sample volume. All scattered electrons can then be collected by a variety of detectors placed behind the specimen and their information exploited to the fullest extent. An image is generated simply by moving the focused beam step by step over the specimen. Hence a STEM image may be considered as a collection of individual scattering experiments. Various types of signals discriminated in scattering angle and/or energy loss yield different structural and chemical information and may be captured simultaneously in different channels. This simultaneous and controlled acquisition of information lends itself to quantitative analyses that are difficult to realize with other instruments. In addition, as there is no limitation of the solid angle and the energy loss interval over which the scattered electrons may be collected, 60-100% of them contribute to the image. This provides a unique opportunity to image beam-sensitive biomolecules at low dose.

The elastically scattered electron signal, giving rise to a dark-field image, is used for mass determination. The mass of individual particles, the mass-per-length of filamentous assemblies, and the mass-per-area of sheet-like structures can all are determined [26]. A large mass range can be covered, e.g., DNA strands of mass 2kDa/nm, and entire viruses of mass 100MDa have both been measured. Although care must be taken to remove all heavy metal salts from the sample, the latter may contain a mixture of biological species since even if not visually distinguishable, these will become distinguishable in terms of their mass as illustrated by the example below (see Fig. 13).

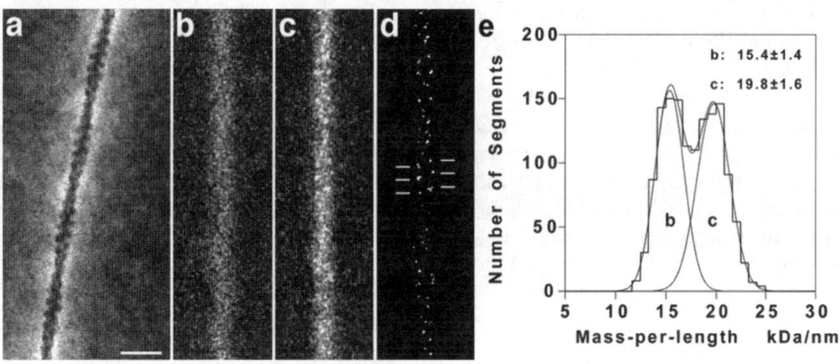

Figure 13. (a) STEM dark-field micrograph of negatively stained, phalloidin stabilized muscle actin filament. (b) As (a) but unstained. (c) Unstained actin filament stabilized with the gold-phalloidin complex. (d) As (c) but contrast adjusted to display the highest intensities only, these correspond to single cold clusters (diameter ~1 nm). The image reveals labeling along the two long-pitch actin strands which are staggered by half a subunit. The spacing indicated between clusters on the same strand corresponds to 5.5 nm which is exactly the axial extent of an actin subunit. Scale bar, 20 nm. (e) Pooled mass data from two experiments: peak b from unstained, phalloidin stabilized actin filaments, peak c from actin filaments when gold complexed phalloidin was employed. Expected experimental value for 1:1 stochiometric bonding, considering the gold contribution at the calibration used and the additional linker protein: 19.2 kDa/nm.

As illustrated by Fig. 13a, significant fine structure is visible on STEM dark-field micrographs of negatively stained muscle actin filaments [27]. Nevertheless, stoichiometric binding of the ~800 Da mushroom toxin phalloidin to these filaments can only be detected in such images after 3-D helical reconstruction [28]. However, the electron scattering power of phalloidin can be significantly raised by complexing the toxin with an 11-gold-atom cluster (gold-phalloidin complex: MW ~7,300 Da), compare Fig. 13b (without gold) with Fig. 13c (with gold) [29]. The high signal-to-noise (S/N) ratio of STEM dark-field images recorded from unstained filaments allowed the gold clusters (diameter ~1 nm) to be directly visualized on the raw data (Fig. 13d). STEM mass measurements revealed the stoichiometry of the phalloidin-actin filament interaction to be unaffected by the presence of the gold label (Fig. 13e). Peak b in the histogram corresponds to phalloidin stabilized filaments as displayed in Fig. 13b, and correspondingly, peak c is from filaments stabilized by the gold-phalloidin complex as shown in Fig. 13c. A significant mass difference is evident arising from the gold cluster and its associated organic linker.

6. ATOMIC SCALE STRUCTURE DETERMINATION OF MEMBRANE PROTEINS BY ELECTRON CRYSTALLOGRAPHY

In spite of major advances in X-ray crystallography, and recent developments in NMR spectroscopy, the structure of most membrane proteins has remained elusive. Of large diversity in function, these proteins share the feature of a hydrophobic belt by which they interact with the lipid bilayer. This attribute explains the slow progress of structural analyses for which membrane proteins need to be solubilized with detergents and purified. Solubilization tends to destabilize the proteins, particularly when short-chained detergents as required for three-dimensional (3D) crystallization are used. Thus, solubilized proteins rarely form crystals suitable for X-ray analysis. A powerful alternative is the reconstitution of 2-dimensional (2D) membrane protein crystals in the presence of lipids. In this approach, the native environment of membrane proteins is restored, as well as their biological activity [30]. Electron crystallography [31] can then be used to determine the protein structure at atomic resolution. In addition, the scanning force microscope (SFM; see above) has demonstrated its potential to directly visualize conformational changes of membrane proteins within such 2D crystals. These developments open the exciting possibility not only to analyze the 3D structure of membrane proteins, but also to assess the relationship between their structure and function directly.

2D crystals assemble from the solubilized protein and mixed detergent-lipid micelles by slow removal of the detergent (Fig. 14a). The protein packing arrangement is controlled by the lipid-to-protein ratio (LPR), by the nature of the detergent and lipid, and by salts, pH, and small hydrophilic solutes. Detergent removal is achieved by dialysis, dilution or adsorption to Biobeads. Solubilization and isolation conditions must be tuned so as to maintain the minimum LPR required for the stability of the particular protein. In addition, the optimum conditions (LPR, salts, pH, temperature, etc.) for the reconstitution of 2D crystals need to be established. Crystallization should be a slow and ordered process where individual units assemble one by one to form a highly ordered array. However, the size of the ternary complexes in a mixture of protein, lipid and detergent exhibits an abrupt change close to the critical micelle concentration (cmc) of the detergent when progressively diluted [32]. The formation of non-crystalline vesicles or proteoliposomes is characterized by a decrease of their size after the phase transition as dilution proceeds. In contrast, the size of the assembled structures remains at the maximum value attained when crystallization occurs. In this case, the assemblies are crystalline just after their spontaneous and very rapid formation (Fig. 14b). Optimization of 2D crystallization is thus intimately related to the abrupt phase transition.

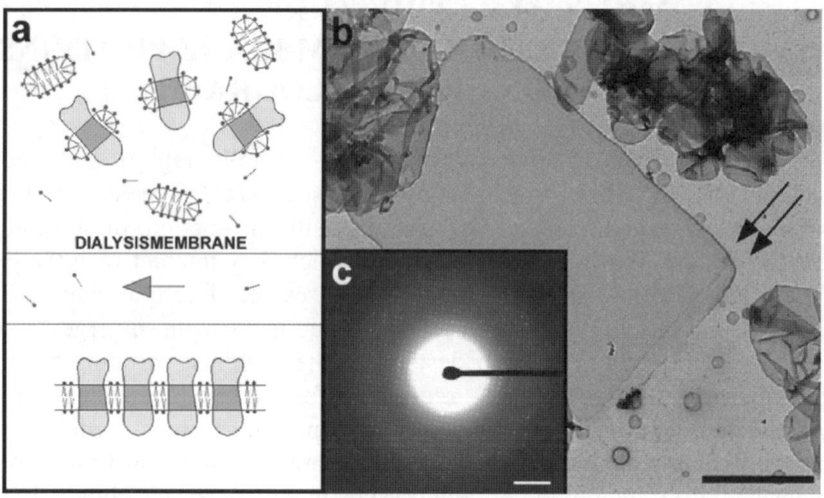

Figure 14. 2D crystallization of aquaporin AQP0 [34]. (a) The hydrophobic belt of the membrane protein (dark shades) which ensures proper integration into the lipid bilayer is shielded by detergent molecules after solubilization. Elimination of detergent molecules by dialysis leads to the exposure of hydrophobic surfaces of both the proteins and the lipids. Hydrophobic forces then drive the assembly of 2D crystals. (b) Reconstitution of lens core AQP0 produces large double-layered two-dimensional crystals. The edges of the two layers can be seen (arrows). (c) Typical electron diffraction pattern of an untilted 2D crystal showing diffraction spots to a resolution of 3 Å. Scale bars correspond to 1 μm in (b) and $(25\ nm)^{-1}$ in (c). Image courtesy of Dr. Thomas Walz, Department of Cell Biology, Harvard Medical School, Boston, MA, USA.

The lens-specific membrane protein Aquaporin-0 (AQP0) depicted in Fig. 14b is a member of the aquaporin superfamily, members of which form pores that are either highly selective for water (aquaporins) or also permeable to other small neutral solutes such as glycerol (aquaglyceroporins). AQP0 water pores are considered essential for the micro-circulation system of the eye lens, proposed to supply deeper-lying fiber cells with nutrients and to clear waste products. AQP0 (and aquaporins in general) must contain a highly specific water channel structure, since the passage of ions and small hydrophilic solutes is blocked, whereas the passage of water is unhindered. Clearly, solving the atomic structure of these proteins is key towards a mechanistic understanding of water transport across membranes.

AQP0 is the only aquaporin known to form membrane junctions *in vivo* [33]. Recently, the structure of this AQP0 membrane junction was solved with electron crystallography [34]. Accordingly, these researchers revealed a junction formed by three localized interactions between AQP0 molecules in adjoining membranes, mainly mediated by proline residues conserved in AQP0s from different species but not present in most other aquaporins (see also Fig. 15). Intriguingly, whereas all previously determined aquaporin structures show the pore in an open conformation, the water pore is closed in AQP0 junctions. The water pathway in AQP0 also harbors an additional pore constriction, not seen in other known aquaporin structures, which may be responsible for pore gating. Based on these findings it was hypothesized that junction formation induces the closed pore conformation in the AQP0 structure. AQP0 and other aquaporins may be in a dynamic equilibrium between an open and a closed water pore conformation, a well-established concept in the ion channel field. Junction formation might then simply stabilize AQP0 in the closed pore conformation, which is otherwise not favored at pH 6 [34]. Its atomic structure (Fig. 15) thus suggests a testable model for the way in which junction formation induces closure of the AQP0 water pore.

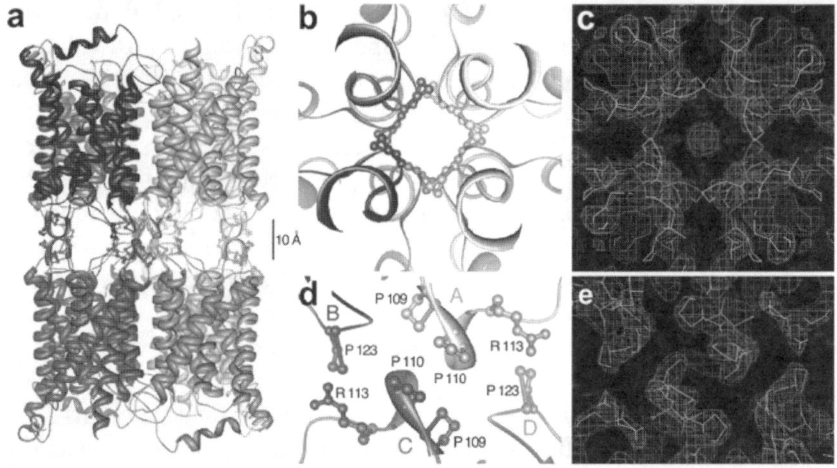

Figure 15. The AQP0-mediated membrane junction determined by electron crystallography [34]. (a) Ribbon representation of the membrane junction with the positions of the two membranes indicated in yellow. (b) The Pro 38 residues contributed by all eight AQP0 subunits in the two interacting tetramers. (c) Corresponding area of the $2F_o – F_c$ map. (d) The C loops connect each AQP0 molecule to two molecules in the opposite membrane. The proline residues (Pro 38) from eight symmetry-related AQP0 molecules in the stacked tetramers come together to form a unique rosette-like structure in the very center of the junction (b, c). The two other interactions involve three AQP0 molecules designated A, B and C. (e) Corresponding area of the $2F_o – F_c$ map. Image courtesy of Dr. Thomas Walz, Department of Cell Biology, Harvard Medical School, Boston, MA, USA.

7. REAL-TIME VISUALIZATION, MANIPULATION AND INTERPRETATION OF BIOLOGICAL OBJECTS BY INTERACTIVE COMPUTER GRAPHICS

Standard molecular modeling and visualization applications include a variety of tasks, ranging from rational design of novel organic compounds to understanding the assembly of biomolecules into distinct supramolecular structures. Inherent to the realization of these aims is the 3D representation of a molecule. At high resolution, as obtained by X-ray crystallography, NMR or computer simulations, this task requires a structural description of the molecule in terms of its atomic coordinates. Atoms are often displayed as solid-rendered spheres, whereas covalent bonds are shown as sticks joining two atoms. Other properties of the molecule, e.g., its hydrophobic or electrostatic potential which governs most of its intra- and inter-molecular interactions, can best be displayed on a solvent-accessible “molecular surface” generated from its atomic coordinates. Provided that structural information in the form of atomic coordinates is available, the concept of the molecular surface has proven to be very useful despite the fact that this representation of the boundary between a molecule and its solvent environment is an abstract formalism which has no simple physico-chemical counterpart.

In contrast, lower-resolution structural information such as obtained by 3D reconstructions from electron micrographs of biomolecules or their supramolecular assemblies cannot be represented by discrete atomic coordinates. These sources of structural information need to be described in the form of volumes or surfaces, and thus are a challenge for real-time visualization on the computer screen. Two different techniques have been developed over the past years for the visualization of volumetric data, extracting different kinds of information from such data sets. The first technique is known as contouring, where all values in the data set that are below a specified threshold are discriminated. This produces a discrete, iso-valued surface called isocontour. The second technique, known as volume rendering, treats the entire data set as a contiguous density cloud that is visualized by modulating the opacity of the object based on the values present in the data set.

The SFM (see above) unveils yet another aspect of biomolecular structure. In contrast to all other experimental methods that explore the 3D architecture of biological matter, the SFM provides a topographical view in form of a surface relief of the specimen.

During the past years it has become clear that structure determination is best achieved by gathering experimental data with different data acquisition methods and under a variety of preparation conditions, because a given method either provides only partial insight or is subject to preparative experimental artifacts due to the size and complexity of the molecular system under investigation. Also, different preparation methods produce structural information at different levels of resolution, and frequently the form of the data obtained is different as well. Such heterogeneous data makes the task of directly comparing, combining and correlating structural information in 3D rather difficult.

Currently, most simulation and analysis tools are being developed for a specific scale, thus hindering the integration of data across scales. An additional challenge for developing new tools is the emerging grid technology. The grid infrastructure enables software applications to integrate instruments and displays, as well as computational and information resources managed by diverse organizations in widespread locations. Such distributed computational resources and data repositories create the need for secure and efficient communication between processes running in distributed hardware environments.

Thus the next generation of scientific analysis and visualization software will have to achieve a high level of flexibility and portability in order to take advantage of the rapidly evolving software, hardware infrastructures, and ever-growing databases. Recent developments in the field of high-level programming languages and increased computational capabilities of typical desktop workstations provide an exciting opportunity for investigating new solutions for these challenges. For example, the Molecular Graphics Laboratory (MGL) at The Scripps Research Institute (TSRI) has spearheaded the development of a novel "component-based" software development strategy [35,36] centered on the Python programming language [37]. The essence of this approach is to use a high-level, general-purpose, interpretive programming language as a computational environment. Rather than writing a standalone program to answer specific scientific questions, the Python interpreter, acting as a framework, is extended with components implementing specific functionality. These extensions to the language can be loaded dynamically at runtime, effectively augmenting the environment with new computational capabilities. Python serves as powerful, yet flexible glue for rapidly assembling components into custom applications. Software components written in pure Python present the additional benefit of being platform independent, i.e. the same source code runs on any computer running a Python interpreter. This approach has to be contrasted with the more traditional "application-centric" approach, where the environment is a program that can be extended by embedding or interfacing new computational methods. In the latter approach, adding a new method usually only augments that particular program.

The benefits of the component-based approach have already been demonstrated by the creation and wide distribution of a number of scientific visualization tools [38-40] and numerous applications [21,41]. For example, the freely available component-based Python Molecular Viewer "PMV" (Fig. 16a; [38,40]) enables the easy integration of computational methods developed by others into PMV while at the same time providing a homogeneous user-interface to operate on these methods.

The component-based visual programming environment "Vision" (Fig. 16b [39,40]) allows scientists to visually and interactively build new "programs" without typing a single line of code. Here, computational methods (represented as "node" icons) can be connected on a canvas with rubber-bands to form computational pipelines which can then be executed like a program. The visual programming paradigm has been successfully employed in various software tools for over two decades, however Vision's implementation provides a number of important innovations including unprecedented levels of flexibility. Vision's unique features and its application-domain independence (i.e., Vision is usable beyond the biomedical realm [42]) enables scientists to efficiently and effectively design and build discovery environments that are tailored for the growing computational needs, and that will integrate into network based computational and collaborative environments [40].

Other scientists developing similar tools have recognized the validity and advantages of a component-based approach and are progressively adopting this software development strategy [42].

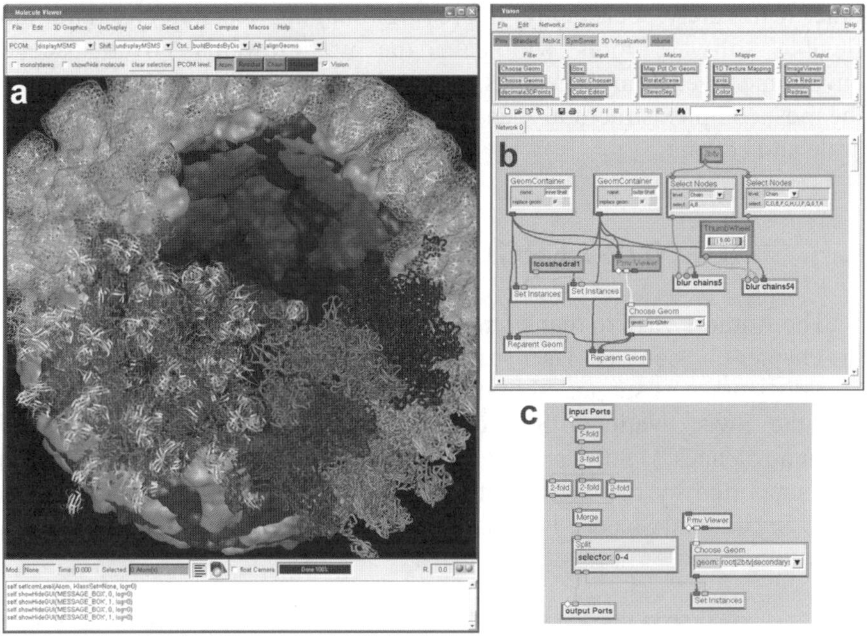

Figure 16. Prototyping advanced visualizations of the bluetongue virus. This viral capsid contains 900 proteins comprising almost 3 million atoms. By exploiting the symmetry information we only need to store molecular data for 15 proteins and less than 50,000 atoms. (a) The Python molecular visualization application "PMV" depicting the bluetongue virus capsid half-shell. Chains A and B forming the inner shell are colored blue and cyan respectively. (b) The visual programming environment "Vision", started from within PMV. The network depicted is used to compute a separate low-resolution surface for each protein, thus providing more flexibility for the graphical representation of the viral capsid in PMV. 5 symmetry matrices (see (c)) corresponding to a pentamer are used to duplicate a ribbon diagram of the secondary structure, another 5 matrices are used to duplicate a Carbon-alpha trace, and matrices corresponding to the upper-half of the capsid are applied to the outer shell container, while 45 matrices (i.e. 60 – 3*5 for the 3 pentamers missing) are applied to the inner shell container. (c) The cyan colored "Icosahedral1" macro node shown in (b) holds a subnetwork depicted here. This network creates 60 nested (4x4) transformation matrices of which 45 are used to build parts of the icosahedral symmetric virus capsid. Image courtesy of Dr. Michel F. Sanner, Molecular Graphics Laboratory, The Scripps Research Institute, La Jolla, CA, USA.

8. CONCLUSIONS

Three-dimensional (3D) structural information is usually key to understanding the function of biomolecules, their molecular interactions and supramolecular assemblies. As discussed in the examples above, a variety of experimental methods have been developed to provide 3D structural insight at various resolutions. Modern graphics computers are powerful tools to visualize these 3D data sets, and to animate their dynamic properties in real-time. Ultimately, constantly improving imaging technologies and more flexible software tools will pave the way towards novel discovery environments enabling scientists to visualize, manipulate and interpret even the most complex structural scenarios – from the nanometer to the millimeter scale and beyond.

9. ACKNOWLEDGEMENTS

The Authors wish to thank Dr. M. Dürrenberger for his helping to prepare the chapter on “confocal laser scanning microscopy”.

This work was supported by the M.E. Müller Foundation of Switzerland and the Swiss National Science Foundation Priority Program NFP 36 (“Nano-sciences”). Software development at The Scripps Research Institute was supported by the Swiss National Science Foundation and grants from the U.S. National Science Foundation and the National Institutes of Health.

10. REFERENCES

1. Pawley, J.B.. *The Handbook of Biological Confocal Microscopy.* New York: Plenum, 1990
2. Boyde A. Bibliography on confocal microscopy and its applications. Scanning 1994; 16: 33-56.
3. Lichtman J.W. Confocal microscopy. Scientific American 1994; 271: 40-45.
4. Wilson, T. and Sheppard, C. *Theory and Practice of Scanning Optical Microscopy.* London: Academic Press, 1984.
5. Kro, S.J., Toyoshima Y.Y., Uyeda T.Q.R. and Spudich J.A. Assays for actin sliding movement over myosin-coated surfaces. Methods in Enzymology 1991; 196: 399-416.
6. Rayment I. and Holden H.M. The 3Dimensional structure of a molecular motor. Trends in Biochemical Sciences 1994; 19: 129-134.
7. Kull F.J., Sablin E.P., Lau R., Fletterich R.J. and Vale R.D. Crystal structure of the kinesin motor domain reveals a structural similarity to myosin. Nature 1996; 380: 550-555.
8. Rayment I., Holden H.M., Whittaker M., Yohn C.B., Lorenz M., Holmes K.C. and Milligan R.A. Structure of the actin-myosin complex and its implications for muscle contraction. Science 1993; 261: 58-65.
9. Inoué, S. *Video Microscopy.* New York: Plenum Press,. 1986.
10. Finer J.T., Simmons R.M. and Spudich J.A. Single myosin molecule mechanics - piconewton forces and nanometer steps. Nature 1994; 368: 113-119.
11. Huxley H.E. A personal view of muscle and motility mechanisms. Annual Review of Physiology 1996; 58: 1-19.
12. Warrick H.M. and Spudich J.A. Myosin structure and function in cell motility. Annual Review of Cell Biology 1987, 3. 379-421.

13. Binning G., Quate C.F., and Gerber C. The Atomic Force Microscope. Phys. Ref. Lett. 1986; 56: 930-933.
14. Stoffler D., Goldie K.N., Feja B., and Aebi U. Calcium-Mediated Structural Changes of Native Nuclear Pore Complexes Monitored by Time-Lapse Atomic Force Microscopy. J. Mol. Biol. 1999; 287:741-752.
15. Stolz M., Stoffler D., Aebi U., and Goldsbury C. Monitoring Biomolecular Interactions by Time-Lapse Atomic Force Microscopy. J. Struct. Biol. 2000; 131:171-180.
16. Moy V.T., Florin E.L. and Gaub H.E. Intermolecular forces and energies between ligands and receptors. Science 1994; 266: 257-259.
17. Florin E.-L., Vincent T.M. and Gaub H.E. Adhesion forces between individual ligand-receptor pairs. Science 1994; 264: 415-417.
18. Lee G.U., Chrisey L.A. and Colton R.J. Direct measurements of the forces between complementary strands of DNA. Science 1994; 266: 771-773.
19. Dammer, U., Popescu, O., Wagner, P., Anselmetti, D., Güntherodt, H.-J. and Misevic, G.N. (1995). Binding strength between cell adhesion proteoglycans measured by atomic force microscopy. *Science* 267, 1173-1175.
20. Dammer U., Hegner M., Anselmetti D., Wagner P., Dreier M., Huber W. and Güntherodt H.-J. Specific antigen/antibody reactions measured by force microscopy. Biophys J 1996; 70: 2437-2441.
21. Stoffler D., Feja B., Fahrenkrog B., Walz J., Typke D., and Aebi U. Cryo electron tomography provides novel insights into nuclear pore architecture: implications for nucleocytoplasmic transport. J. Mol Biol. 2003; 328: 119-130.
22. Stolz M., Raiteri R., Daniels A.U., VanLandingham M.R., Baschong W., and Aebi U. Dynamic elastic modulus of porcine articular cartilage determined at two different levels of tissue organization by indentation-type atomic force microscopy. *Biophysical J.* 2004; 86: 3269-3283.
23. Stolz M., Imer R., Staufer U., and Aebi U. Development of an arthroscopic atomic force microscope. Bioworld (www.bioworld.ch) 2003; 3:2-4.
24. Engel A. and Colliex C. Application of scanning transmission electron microscopy to the study of biological structure. Current Opinion in Biotechnology 1993; 4: 403-411.
25. Reichelt R., Holzenburg A., Buhle E.L., Jarnik M., Engel A. and Aebi U. Correlation between structure and mass distribution of the nuclear pore complex and of distinct pore complex components. J Cell Biol 1990; 110: 883-894.
26. Müller S.A., Goldie K.N., Bürki R., Häring R. and Engel A. Factors influencing the precision of quantitative scanning transmission electron microscopy. Ultramicroscopy 1992; 46: 317-334.
27. Bremer A., Henn C., Goldie K.N., Engel A., Smith P.R. and Aebi U. Towards atomic interpretation of F-actin filament 3-dimensional reconstructions. J Mol Biol 1994; 742: 683-700.
28. Bremer A., Millonig R.C., Sütterlin R., Engel A., Pollard T.D. and Aebi U. The structural basis for the intrinsic disorder of the actin filament: The "lateral slipping" model. J Cell Biol 1991; 115: 689-703.
29. Steinmetz M.O., Stoffler D., Müller S.A., Jahn W., Wolpensinger B., Goldie K.N., Engel A., Faulstich H., and Aebi U. (). Evaluating atomic models of F-actin with an undecagold-tagged phalloidin derivative. J Mol Biol 1998; 276: 1-6.
30. Walz T., Smith B.L., Zeidel M.L., Engel A. and Agre P. Biologically active 2Dimensional crystals of aquaporin CHIP. J Biological Chemistry 1994; 269: 1583-1586.
31. Henderson R., Baldwin J.M., Ceska T.A., Zemlin F., Beckmann E. and Downing K.H. Model for the structure of bacteriorhodopsin based on high-resolution electron cryo-microscopy. J Mol Biol 1990; 213: 899-929.
32. Dolder M., Engel A. and Zulauf M. The micelle to vesicle transition of lipids and detergents in the presence of a membrane protein: towards a rationale for 2D crystallization. *FEBS Letters* 1996; 382: 203-208.

33. Costello M. J., McIntosh T. J. and Robertson J. D. Distribution of gap junctions and square array junctions in the mammalian lens. Invest. Ophthalmol. Vis. Sci. 1989; 30: 975–989.
34. Gonen T., Sliz P. Kistler, J. Cheng, Y. and Walz T. Aquaporin-0 membrane junctions reveal the structure of a closed water pore. Nature 2004; 429: 193-197.
35. Sanner M.F. "Using the Python Programming Language for Bioinformatics." *Encyclopedia of Genomics, Proteomics and Bioinformatics.* Addision-Wesley, John Wiley & sons, Ltd. *(in press)* 2005.
36. Sanner M.F.A component-based software environment for visualizing large macromolecular assemblies. *Structures*, special issue in visualization, simulation and representation of biological complexes. (*in press*) 2005.
37. Lutz M. *Programming Python 2nd Edition.* Sebastapol,CA: O'Reilly & Associates, Inc., ISBN 0596000855. www.python.org 2001
38. Sanner M.F. Python: A Programming Language for Software Integration and Development. J. Mol. Graphics Mod. 1999; 17: 57-61.
39. Sanner M.F., Stoffler D., and Olson A.J. *ViPEr*, a Visual Programming Environment for Python. *In Proceedings of the 10th International Python conference.* 103-115. February 4-7, 2002. ISBN 1-930792-05-0.
40. Stoffler D., Coon S.I, Huey R., Olson A.J., and Sanner M.F. Integrating Biomolecular Analysis and Visual Programming: Flexibility and Interactivity in the Design of Bioinformatics Tools. *Proceedings of the Thirty-Sixth Annual Hawaii International Conference on System Sciences (CD/ROM), January 6-9, 2003, Computer Society Press, 2003.*
41. Stoffler D., Sanner M.F., Morris G.M., Olson A.J., and Goodsell D.S. Evolutionary analysis of HIV-1 protease inhibitors: Methods for design of inhibitors that evade resistance. *Proteins* 2002; 48: 53-74.
42. SciPy Python for Scientific Computing Workshop. CalTech, Pasadena, Sep 2-3 2004. www.scipy.org/wikis/scipy04

INSIDE THE SMALL LENGTH AND ENERGY SCALES OF THE WORLD OF THE INDIVIDUAL BIOLOGICAL MOLECULE

Massimo Sandal, Giampaolo Zuccheri and Bruno Samorì
Department of Biochemistry "G.Moruzzi", University of Bologna, Italy

Abstract. Atomic force microscopy (AFM) has proved to be an essential tool of structural biology, being able not only to image but also to manipulate single biological molecules. These techniques make it possible to investigate the nanometer scale structure of single biological macromolecules and to study how an external force drives single biological molecules towards non-equilibrium conformations, by stretching and breaking bonds and interactions. This chapter focuses on the capabilities of the AFM-based single molecule methodologies to bring us into the nanometer-scale world of the single DNA molecules and into the pico-Newton force-scales of the interactions that sustain the protein folding.

Key words. Atomic Force Microscopy, Dna, Protein Unfolding, Force Spectroscopy, Mechanochemistry

1. INTRODUCTION

After the invention of Binnig, Rohrer and Gerber of the Scanning Tunneling Microscope [1], the emerged probe microscope family has progressively expanded. Within this family of techniques, the Atomic Force Microscopy (AFM) with its capabilities of imaging single molecules has proved to be the most useful to structural biology. In the last decade, also single molecule manipulation techniques based on the same AFM have been developed [2]. These latter techniques make it possible to study how an external force drives single biological molecules towards non-equilibrium conformations, by stretching and breaking bonds and interactions. These single molecule imaging and manipulation methodologies drive structural biology into a new dimension, that of the nanoscale. In fact, structural biology has been so far confined to the Ångström scale having been mostly based on data coming from high-resolution techniques, like X-ray diffraction and NMR. The novel single molecule techniques unveil the strategies and the purposes that nature follows in its "engineering" of complex systems and

V. Evangelista et al. (eds.), From Cells to Proteins: Imaging Nature across Dimensions, 111–137.

of their functions. They are also allowing us to discover how nature switches the self-organization of complex systems among different lengths and energy scales and how these acquire different qualities and functions over those different scales [3].

2. AFM IMAGING OF DNA SUPERSTRUCTURAL PROPERTIES

2.1. The shape of a DNA chain: from the Angstrom to the nanometer scale

The base sequence of a DNA chain encodes its shape and dynamics. DNA is continuously morphing into shapes and structures alternative to the canonical B-form, is coiling in the cell nucleus, is "swilling lazily around in a nourishing molecular soup of transcription factors and other regulatory proteins that are milling around the nucleus."[4]

The shape assumed in space and in time by a particular DNA molecule has been analyzed in terms of the superimposition of the thermal fluctuations upon the intrinsic, lowest energy, structure of a chain with that sequence.[5-7] The average structure of dsDNA depends on the sequence: the stereochemical differences imparted by the different base pairs along the chain give rise to deterministic modulations of the relative orientations of the average planes of the base pairs. These orientations are commonly expressed in terms of the base-step orientational parameters: roll, tilt, twist (see Figure 1).

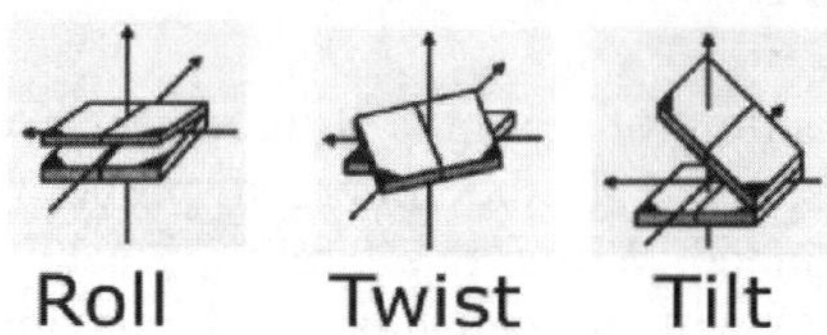

Figure 1. The dinucleotide step orientational parameters: roll, tilt and twist. The composition of the three rotations along the DNA chain gives rise to local and global curvatures.

Considerable effort has gone into defining sets of these parameters corresponding to the lowest energy structures directed by the sequence. Recently, Crothers has critically reviewed the approaches and the sets of values so far reported.[8]

Positive or negative roll or tilt angles give rise to local bending of the double helix axis. These local bends might just lead to a zigzag pattern of the chain axis, which would remain essentially straight, unless they are composed in phase with the repeat of the double-stranded helical winding. In this latter case, they might give rise to extended persistent curvatures that propagate from the Ångström to the nanometer scale.[9] An example of large-scale curvature is the 211 bp segment from the kinetoplast DNA of the Trypanosomatidae Protozoan *Crithidia fasciculata*. This is the most highly curved natural DNA presently known. Its sequence, reported in Figure 2C, is characterized by a periodical recurrence of tracts of 3 to 6 adenines; the centers of most of these tracts are separated by 10 or 11 bp, i.e. the average helical repeat. This distribution of the A-tracts, perfectly phased with the helical winding, makes this short DNA segment to have its lowest conformational energy when wrapped in a circle (see Figure 2A). Experimental evidence of such a large curvature was first provided by Griffith.[10]

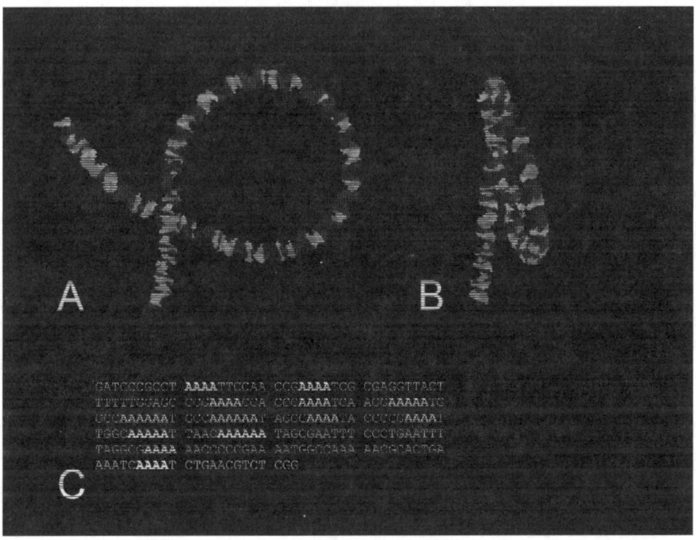

Figure 2. Predicted 3D structure of the curved segment of Crithidia fasciculata (http://archimede.chem.uniroma1.it/webdna.html). A)View along a direction nearly perpendicular to the curvature plane. B)View along a direction almost parlallel to the curvature plane, the segregation of adenines and thymines on the two faces of the curvature plane is clearly visible. C)The base sequence of the molecule shown in A and in B.

Intrinsic curvatures have been monitored and studied by X-ray crystallography on very short double-stranded oligonucleotides, [11] on longer DNA molecules by gel retardation,[12, 13] circularization kinetic,[14-16] electron microscopy (EM),[17, 18] atomic force microscopy (AFM),[19, 20] and have been simulated by molecular dynamics.[21] Commonly, these experiments were carried out with dsDNA constructs with (i) anomalous flexibility sites, like single-stranded stretches,[20] internal loops due to mismatches,[13] a single nick,[22] a double-stranded linker connecting

triple-helix tracts;[23] (ii) segments whose curvature was tailored and controlled by accurate phasing[20] or unphasing,[6, 14] adenine tracts with the helical periodicity. All these approaches to the study of intrinsic curvature are based on the determination of global parameters of the whole chain under investigation like the persistence length, the end-to-end distance, or the cyclization J factor. Information about the sequence encoding and the molecular mechanisms that drive the formation of extended curvatures was inferred from the variations of the values of these global DNA parameters with respect to their values for reference sequences. A combinatorial approach has been proposed for this kind of studies.[14]
The possibility ensured by EM and AFM of seeing individual dsDNA chains deposited on a substrate has made it possible to set up methods to map the curvature along the chain of a natural dsDNA, of any arbitrary sequence.The curvature is computed from the angular deflection at each point along the curvilinear coordinate of the chain [19]. The trajectory of each molecule can be directly gathered from its AFM image, either manually, or through semi-automatic [24] or automatic methods [25 , 26 , 27].

The resolution of both EM and AFM do not allow us to know in an imaged molecule if the sequence goes from one end to the other, or the other way. This information is strictly required if our aim is to map and study how the sequence is directing the molecular shape and drives it towards persistent curvatures. The microscopists have got rid of this uncertainty by labeling one of the two ends of the molecules with a gold bead or a bulky protein [28]. On the other hand, this label drastically affects the entropic response of the chain to thermal perturbation, at least in a proximal section. This is a very bad artifact for an analysis that wants to map superstructures along the chain and to correlate them to the sequence.

Another strategy was envisaged: that of constructing palindromic dimers of the DNA stretch of interest. In a palindromic DNA, the sequence is the same reading from either end. After having chosen the DNA tract whose shape has to be studied, two palindromic dimers of that tract can be prepared with the sequences of these two monomeric units oriented either in a head-to- head (H-H) or in a tail-to-tail (T-T) arrangement. (Figure 3)

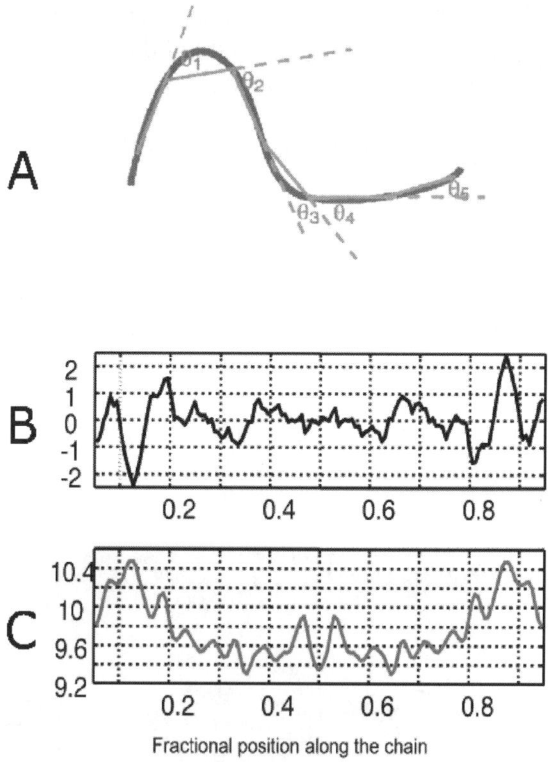

Figure 3. A) Method for quantifying the local DNA curvature from AFM images: the DNA chain axis is digitized in a segment fashion and the local chain direction (and thus the curvature) is determined numerically. B)Plot of the experimental local DNA curvature (evaluated from the AFM images) along the chain of tail-tail palindromic dimers constructed using the EcoRV and the PstI sites of plasmid pBR322 DNA. The average local curvature $\langle C(n,m) \rangle$ in degrees (where n is sequence position, m=2 helical turns in the width of the averaging window) is plotted against the fractional position along the chain, n/N. The experimental plot has been made symmetrical by averaging the two equivalent halves to double the curvature information on the DNA sequence. C)Plot of the experimental local DNA flexibility of the same DNA molecule evaluated as the standard deviation (SD) of the local chain curvature.

After having deposited samples of those dimers on mica their AFM images in air were recorded (Figure 4). Within each sample no two macromolecules have the same shape and conformation, in spite of being totally identical, from any chemical or biochemical point of view. In addition to that, there is no way to discriminate the two dimers just by looking at their images. Contrary to the first impression, the apparently chaotic dynamics that leads to such a variety of shapes is not random. We know that this high variety of molecular shapes comes out from the superimposition of thermal fluctuations upon the lowest energy structure of a molecule with that sequence. We can express this superimposition in terms of curvature by eq.

1, where (C(n)) is the observed averaged curvature at position n on the sequence, ($C_0(n)$) is the intrinsic curvature at the same position n, and <f(n)> the average value of the curvature fluctuation.

$$\langle \mathbf{C}(n) \rangle = \langle \mathbf{C}_0(n) + \mathbf{f}(n) \rangle = \mathbf{C}_0(n) + \langle \mathbf{f}(n) \rangle \tag{1}$$

This latter term can be demonstrated to go to zero if the fluctuations follow the first order elasticity.

On the basis of eq. 1, we thus expect that the intrinsic curvatures and therefore also the lowest potential-energy shape of the molecule with that sequence can be obtained from an equilibrated ensemble of molecular profiles. If more than a thousand profiles are recorded for each of these dimers, the curvature can be computed from the angular chain deflections (4) along all these profiles, averaged over the whole ensemble of profiles, and plotted vs. the position along the chain (fig. 3A).

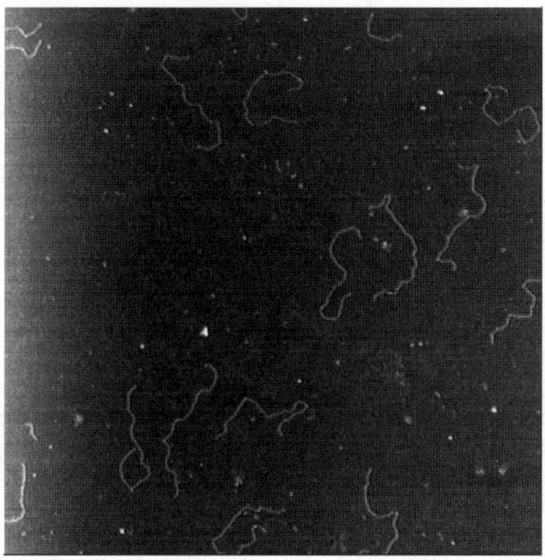

Figure 4. AFM image (2 x 2 μm field) of a spread of DNA palindromic dimers obtained by dimerizing a restriction fragment excised out of plasmid pBR322 DNA. The 1878-base pair long DNA molecules were spread on freshly cleaved muscovite mica from a solution of 4 mM HEPES buffer (pH 7.4), 10 mM NaCl, 2 mM MgCl2 containing approximately 1 nM DNA molecules. Only the molecules completely inside the borders of the image and not presenting any anomalous structure (loops, kinks, bound objects) are used for digitally tracking the helical axis and for the subsequent curvature analysis. Smaller DNA fragments (or residual monomers) are recognized from their measured contour lengths, and subsequently neglected.

From the sequence of the same dimer we can predict a minimum potential energy S-like profile in 2D with two marked curvatures near the end of the molecule (Figure 3B). This theoretically predicted profile came out in the experimental plot: two marked local curvatures with opposite sign (i.e. oppositely oriented) resulted at about 0.1 and 0.9 fractional position along the chain (Figure 3B). In the head-to-head dimer, because of the inversion of the sequence the two curvatures are expected to be near the center of the molecule. Also this prediction came out in the experimental plot (to be published).

In both cases, the profiles of the plots started to reflect the binary symmetry of the sequence only after the ensemble of the molecular profiles contained at least six hundred molecules. The outcome of this binary symmetry is an internal gauge of the statistical significance of the analysis we are carrying out. In fact, when two equal DNA segments are joined to obtain a palindromic molecule the same molecule gains a dyadic axis in the center of inversion of its sequence. Any property of this molecule that is directed by the sequence must reflect this binary symmetry. The statistical significance of the curvature plots is therefore reached only when they exhibit this binary symmetry. So one can implement the set of molecules until this symmetry is obtained. The palindromic symmetry was introduced in order to get rid of the uncertainty about the orientation of the sequence in the AFM images. The palindromic symmetry is at the same time a potential internal gauge of the statistical significance of our analysis.

The intrinsic DNA curvatures can therefore be theoretically predicted and experimentally evaluated either as an average value for a particular DNA chain or as a localized value, mapped along the same chain. Despite these capabilities and many high-resolution NMR and x-ray studies of local bends in oligonucleotides, the description of the origin of large intrinsic curvatures at the atomic level remains disputed. No one doubts that a long-range curvature requires properly distanced A-tracts, and that it depends little on the nature of the sequences that separate the A-tracts,[29] as clearly demonstrated also by the segment from *Crithidia fasciculata* reported in Figure 2C. It is not clear how the A-tracts play a dominant role. The discussion has been mostly focused on the composition of the geometry parameters of the base pairs at the junctions of the A tracts with those of the rest of the chain.[11, 30, 31] It is somewhat surprising that all different theoretical approaches based on nearest-neighbor models,[32-34] despite the different values they utilize for the base step parameters (see also Table 1 and 2 in [8]), reach very similar results in their description of the path of the helical axis in the nanoscale.

We must therefore conclude that AFM imaging and mapping intrinsic curvature in the nanoscale is teaching us as this is a long-range superstructural property that is more determined by the way the double helix selects, composes and phases the local bends over different spatial scales, than by their individual values.

2.2. The flexibility of a DNA chain from the scale length of a dinucleotide step to the micrometer

The sequence is determining not only the lowest energy profile of the molecule but also its local response to the thermal fluctuations. The sequence in this way controls the formation of conformers and superstructures. It should be never forgotten that a conformation, even if poorly populated, can play a crucial biological function. In fact, it can be recognized and selected to switch on processes that the most stable structures might not be able to activate. If our knowledge is limited to the lowest energy profiles, also our possibility to understand functions might be very limited. One of the experimental observables that gives insight on the accessible conformational space of a chain is its local axial flexibility, i.e. the tendency of the long axis of the double helix to deviate both locally and globally from a straight trajectory.

While a significant agreement is found in the literature on the origin and determinants of DNA curvature, the issue of DNA flexibility is still under active debate. The considerations that will follow reflect somehow the point of view of the authors, and are certainly bound to be modified as more results become available.

The axial flexibility of the chains is controlled by the stereochemistry of their sequences, in particular by van der Waals and electrostatic interactions between the adjacent base-pairs.[9, 34-38] The electrostatic interactions are dominated by a large dipole on G-C pairs contrasted with a diffuse distribution of charge on A-T base-pairs.[37, 39, 40] Adjacent A-T base-pairs can thus stack without displacement due to electrostatic forces, whereas the repulsive dipoles in adjacent G-C base pairs lead to a slide displacement with a consequent tendency to a higher positive roll angle.

Another factor that influences the general axial flexibility is the compressibility of both the major and minor grooves and hence the presence of exocyclic groups in the grooves.[41] Recently, De Santis and coworkers showed that the axial flexibility is thermodynamically related to the melting temperature of a DNA tract when a first order elasticity is assumed. This can be easily obtained from the sequence by averaging the formal melting temperature assigned to each dinucleotide step. The results obtained by adopting such a dinucleotide flexibility scale satisfactorily explain the static and dynamic curvature dispersion of DNA images and the sequence-dependent thermodynamic stability of nucleosomes as well.[19, 42-45]

The axial flexibility of a chain in the nanometer scale length can be described in terms of its persistence length (P), a parameter commonly used for defining a polymer bending rigidity. P is defined as the length over which the memory of the polymer axis direction is retained under thermal agitation. A number of techniques including light scattering,[46] rotational diffusion,[47] DNA cyclization kinetics,[48, 49] cryo-electron microscopy[6] as well as conventional electron microscopy[50] and atomic force microscopy[51] have led to estimates for P of around 50 nm for mixed-sequence B-form DNA. The measured persistence length, as determined by most techniques, depends not only on the intrinsic flexibility of the DNA molecule but also on its intrinsic curvature, *i.e.* on the anisotropy of the axial flexibility.[6, 52]

The axial flexibility on the scale length of the dinucleotide steps, or bendability, has been estimated from the range of conformations adopted by the specific base-steps in crystal structures of either DNA oligomers or of DNA-protein crystal complexes.[34, 53-55] The deformations that force the DNA to bend locally (and make it possible to evaluate its bendability) are due to the lattice structure, in the former case. In the latter case, the local bendability of the chain was considered to be tested and evidenced by the tendency of the proteins to bend DNA in the positions where they are bound. On this basis, the pyrimidine-purines steps were found to be more easily deformable than the purine-pyrimidine and purine-purine steps. The bendability of the dinucleotide steps resulted to decrease according to the scale

CG>CA(=TG)>TA>CC(=GG)>TC(=GA)>GC>TT(=AA)>GT(=AC)>CT (=AG)>AT

Assuming that the same spread of the roll and of the other parameters taken into account reflects a Boltzmann distribution in the ensemble, harmonic energy functions were also extracted.[56]

Imaging methodologies that make it possible to visualize the trajectories of the DNA molecules under investigation can therefore map not only the local intrinsic curvatures along the chain (see above) but also the local modulation of flexibility determined by the sequence.

The flexibilities can be evaluated from the dispersion of the curvature values.

$$\left\langle f(n,m)^2 \right\rangle = \left\langle \left| \mathbf{C}(n,m) - \left\langle \mathbf{C}(n,m) \right\rangle \right|^2 \right\rangle = \frac{2RT}{b(n)} m \qquad (2)$$

By evaluating the dispersion of the curvature values, flexibility plots, like that in Figure 3C, were obtained for a population of symmetric molecules obtained by the dimerization of a tract of the DNA plasmid pBR322.[19] In these plots, it was found that the local flexibility is generally higher where the molecule is more curved. This might indicate that the sequence shapes the conformational space of the chain by modulating the intrinsic curvature and the flexibility in the same way. Molecular dynamics simulations (based on all-atom potentials) performed on oligonucleotide duplexes with A-tracts indicate that they are essentially straight and rigid, and that the steps that are more bent and deformed from the canonical B-form are more dynamically deformable.[57]

There are 16 possible dinucleotides, by symmetry only 10 are unique. A good correlation with the flexibility results for the AA(=TT), TA and AT steps. Anti-correlation is obtained instead for the CG, CA(=TG) and also the GC steps. This result is in contrast with the assignments of the bendability of the dinucleotide steps supra reported. In fact, the crystal structures indicated the CG and CA(=TG) steps as the most bendable ones.

Again, like in the case of the curvature, the flexibility assignments in the nanoscale can be hardly reconciled with those made by different techniques on the atomic scale. The following considerations can be made in an attempt to motivate this disagreement.

It can be argued that the dimension and the composition of the basic set of the crystallographic data are not optimal, yet. The dimension is not very large and, as also the authors of these studies point out, the harmonic energy functions evaluated by them can change as new data accumulate.[56] Furthermore, the oligonucleotides analyzed by X-ray diffraction contained tracts systematically GC-rich at the ends, and AT-rich in the central positions.[58] This might have affected the statistical significance of the analysis. Additionally, the spread of the dinucleotide geometric parameters in the crystal structures of the set (about a hundred) of oligonucleotides correlates quite poorly with that of DNA-protein complexes.[8] Nevertheless, both analyses indicate the CG and the CA (=TG) steps as particularly bendable, in marked disagreement with the results obtained in the nanoscale by AFM.[19]

Sensibly, it can be expected that this disagreement might be due to the averaging of the smaller-scale properties along the chain. Okonogi et al. identified the CG step as the most flexible and GC as the most inflexible, while the values of AT and TA were intermediate;[59] the average of the CG and GC flexibility results lower that the average for the combined AT and TA. On this basis, they justify that "sequences containing AT and TA steps are believed to be the most flexible of all dinucleotides." On the other hand, it must be taken into account that the helical structure does not only add and average the different contributions, rather it combines them according to their phase respect to the helical repeat. In this way, a large local flexibility might be even obtained by phasing in an appropriate way a number of steps that might not be the most bendable. Also the DNA curvature is determined more by the way the double helix selects, composes and phases the local bends over different spatial scales, than by their individual values.

One other possible element might be taken into account when the dinucleotide bendabilities obtained for instance from X-ray diffraction analyses (as in references [34, 53]) or from spectroscopic methods (as in reference [59]) are compared with flexibility data obtained by AFM imaging. The molecules imaged by AFM have been transferred from the 3D space of the solution to the quasi-2D space of the substrate surface. This reduction of the degrees of freedom from a 3D to a 2D space could damp out-of-plane motions. Because of this reduction of the dimensionality of the molecules, there is probably a bias towards in-plane flexibility components. This might be another possible source of disagreement between the measurements made over different scale lengths. At the same time, results coming from flattened DNA are particularly valuable from the point of view of structural biology, since they are mimicking the restrictions that are expected to take place also in DNA-protein complexes, where the DNA chain is confined on the protein surface. So this intrinsic limitation of the technique can be useful to shed more light on the topology of these complexes.

Other sources of information about the DNA structural codes in the nanoscale are the structure of the complexes of DNA with architectural proteins, like the histones. These can be analyzed focusing on how conformationally rigid and flexible sequences are combined to drive the positioning of the proteins on the DNA chain that wrap them. The reader is pointed to a very exhaustive review recently written by A. Travers on that topic.[41] His analysis converges with the assignment of AT and TA as pretty flexible steps. In particular, as far as the TA step is concerned, the most recent results by Travers about the positioning of the nucleosomes show that, within their preference to reside in highly curved DNA regions, there is a very strong selection for TA where the minor groove points in (personal communication). The conclusion that TA is likely the most flexible step is supported by the very low stacking energy of this step, and by the data by Zhang & Crothers[14] and by McConnell and Beveridge[57] that point in that direction. This conclusion fits very well with the AFM data.

As far as the AA step is concerned, it seems that for this dinucleotide there is an energetic barrier to overcome for adopting large distortions. On the other hand, it has been noticed that in the very recent structure of the nucleosome core reported by Richmond and Davey,[60] large distortions result at the junctions of this dinucleotide tracts with the flanking sequences. On this basis, as suggested by Travers, it is very likely that the properties of this step depend much on its context (personal communication). The high flexibility assigned to the AA step by AFM might include contributions of the mechanical properties of the flanking positions, due to the nanoscale averaging of the technique.

In conclusion, the capability ensured by AFM imaging of mapping curvatures and flexibility along a DNA chain allows us to increase our understanding of how the DNA double-helical structure acquires different qualities over different spatial scales and how its base-sequence encodes the properties, the superstructures and the recognition mechanisms that DNA exhibits in the nanoscale. Nanoscale structural properties of DNA molecules play important roles in all the functions of nucleic acids, such as replication, transcription, nucleosome formation and, more generally, in all interactions with proteins.

3. MEASURING THE SMALL INTRAMOLECULAR FORCES THAT DRIVE AND SUSTAIN PROTEIN FOLDING

The methodology that allows direct measurement in a single molecule of the forces generated by a molecular interaction, or needed to guide a molecular structure in a new conformation has been termed single-molecule force spectroscopy. The atomic force microscope cantilever is particularly apt to these class of experiments, in being essentially a tiny, mobile dynamometer able to measure forces in the pN scale. The AFM tip is at the same time a probe and a nano-manipulation device: it can be used to pick up single molecules, to stretch them and to follow in real time the subsequent force-induced bond breaking and structural transitions.

Basically, in a force spectroscopy experiment single molecules or couples of interacting molecules are stretched between the AFM tip and a flat surface. The raw data obtained by a force spectroscopy experiment is the "force spectrum": a 2D plot of force *versus* distance. Analysis of the force spectrum gives direct information on the properties of the system under force.

One of the most prominent problems addressed with the help of single-molecule force spectroscopy is the study of the mechanical properties of proteins. Many proteins are exposed to mechanical tension *in vivo*, mainly (but not only) in the extracellular matrix, cell surface and cytoskeleton [61-63] . Protein folding will consequently have evolved to face and take advantage of this. With the help of force spectroscopy, it is possible to simulate the mechanical stresses that molecules can experience *in vivo.* We are beginning to understand how topological and structural features determine proteins mechanical resistance and how these macromolecules can convert a mechanical stress into a biological signal.

Mechanical unfolding of proteins driven by an AFM tip requires a conceptually simple experimental set-up (Figure 5). When pulling the molecule, either the pulling velocity or the applied force can be kept constant. Velocity clamp experiments are simpler, and are the most widely performed at the present state of the art. In this article, we will refer to experiments conducted at constant pulling speed, unless explicitly stated.

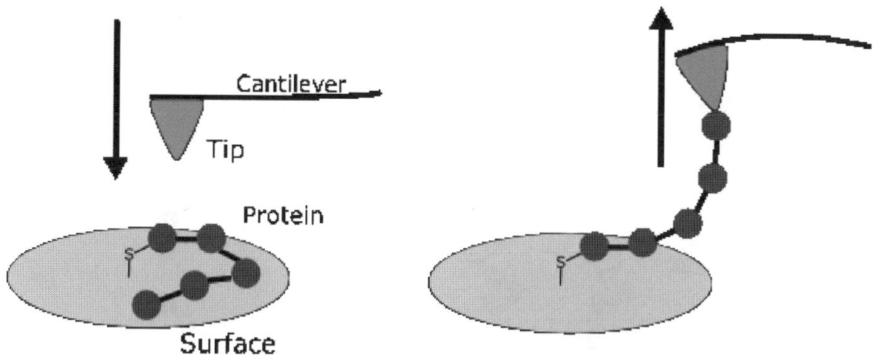

Figure 5. The most common experimental setup for protein force spectroscopy. A sparse layer of proteins must be adsorbed on a flat surface. Attachment of the protein on the surface can be covalent or by simple adsorption. In the image, adding a Cys tail on the N or C-terminal covalently binds the protein on a gold surface via the spontaneous formation of a S-Au bond. The AFM tip is pressed on the surface (left) and then retracted (right), blindly fishing molecules from the surface. A pressure of a few nN by the tip on the protein is usually enough to form a physisorption that can sustain forces up to 1 nN, thus making it possible to stretch the molecule.

Mechanically unfolding proteins composed of tandem modules often results in a remarkable, characteristic saw-tooth pattern in the force-extension graph, with force peaks as high as tens or hundreds of pN. In the apparently simple force-extension graph is hidden complex information about the molecule behavior. To dig out this information, we must before understand what the graph does mean.

3.1. The force spectrum of modular proteins

The saw-tooth pattern is composed of many asymmetrical peaks, as depicted in Figure 6. Every peak of the graph is known to result from the stretch-induced unfolding of a single module (except the last, that usually comes from the detachment of the molecule from the tip or the surface).

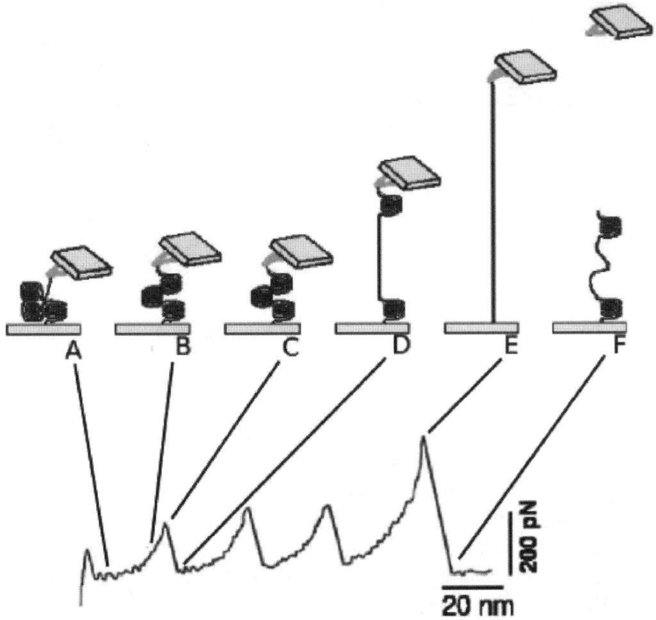

Figure 6. The forced unfolding of a modular protein. The AFM tip picks up the protein (A) and pulls it (B) until the tension breaks critical interactions in a random module (C). The force acting on the cantilever drops (D) following the sudden increase in total contour length, and drawing a single saw tooth on the force-elongation curve. The cycle then repeats until all the modules between tip and surface are unfolded (E) and eventually the protein detachs (F). The distance between a peak and the next approximately equals the length of the unfolded module.

Much of the overall shape of the curve originates from the rising portion of the peaks, and it is mainly the result of the entropic elasticity of the protein chain. Entropic elasticity is an intrinsic property of any polymer, and within the different classes of polymers it depends very loosely from minor details of the molecular structure. When a stretching force is applied as to constrain the polymer, its conformational space is shrinked: at full elongation there would be ideally only one conformation available (a fully straight one). Given that a large majority of conformations are virtually at the same energy, entropy dominates the process, generating an opposing force. It results that small extensions can be reached at a very small force, but approaching the complete elongation the resisting force rises steeply.

The exact profile of the force-elongation curve in this case is often well described by the worm-like chain (WLC) model of elasticity (Figure 7, equation 3) .

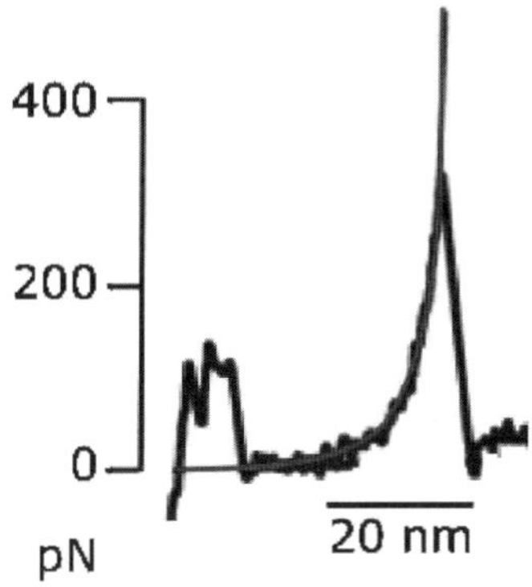

Figure 7. Worm-like chain model fit of the rising section of a titin Ig module unfolding peak.

The model describes the behavior of the polymer as a continuous string of given contour length L. The flexibility of the string is defined as the distance at which bending of the string in a point influences its neighborhood, referred to as the persistence length p. The persistence length and the contour length alone can describe very well the entropic elasticity of a molecule, as long as L >> p [64] :

$$F(x) = \frac{k_b T}{p}\left(\frac{1}{4(1-x/L)^2} - 1/4 + \frac{x}{L}\right) \qquad (3)$$

There are relatively few polymers displaying a perfectly entropic behavior; known examples are polyethylene glycol in hydrophobic solvents [65] and single-strand DNA [66]. More often the elasticity deviates from the WLC profile, implying the existence of an enthalpic component in the response of the molecule to extension. Such deviations can be the result of intramolecular rearrangements and/or of rupture of chemical bonds.

When we begin to stretch a modular protein the intramolecular bonds keep the fold of every module compact, acting as "locks". The contour length measured is that of the folded chain (Figure 6A, 6B). When the action of force breaks critical bonds inside one module, its structure collapses and the force acting on the molecule falls sharply towards zero because the unfolded domain length is suddenly added to the total contour length of the system (Figure 6D). The force value at which the domain unfolds depends both on its structure and topology and on the pulling speed. In the dynamic range of most AFM-based instrumentation (0.1 – 10 μm/s) the most probable unfolding force is in the 100-250 pN range for beta sheet rich structures and almost always lower than 100 pN (even undetectable) for mixed alpha-beta or all alpha structures (As a comparison, a covalent bond can resist forces of about 1000-1500 pN in the same dynamic conditions [67]). The reasons of these differences and the determinants of the mechanical resistance of folds are discussed below.

The modules are unfolded sequentially, so a saw-tooth pattern on the force-extension graph is obtained, where each tooth is due to the unfolding of a single module. Eventually, if pulling continues, the protein detaches from the tip (or the surface), producing a last detachment peak on the graph (Figure 6F).

It was argued that the saw-tooth pattern can be an artifact of desorption of the molecule from the surface. This was ruled out by using artificial multimeric constructs with alternating modules that give peaks of known, different forces (Figure 8). Desorption would follow the linear sequence of the modules, giving alternate peaks of high and low force. Unfolding would instead affect all the weaker modules first, and the strongest last, and this is what happened [68] .

3.2. Tailoring constructs

Why did we talk so far about modular proteins only? Short, globular proteins generally unfold and detach from the tip (or the surface) after short extensions (less than 20-30 nm). It is impractical to study single globular folds in this way, because the unfolding signal is masked by short-range aspecific interactions between the tip and the surface; moreover the bond between the tip and the protein can distort the fold, introducing artifacts and making the pulling direction uncertain. For these reasons almost all protein force spectroscopy experiments are performed either by using multi-modular proteins or more rarely by binding the protein molecule under investigation to the tip or the surface by long polymeric linkers. The former approach has the double advantage of more easily yielding interpretable mechanical signals and to increase data harvesting because multiple unfolding events are obtained per single force curve.

The earliest protein force spectroscopy experiments have been performed with fragments of naturally multimodular proteins, like titin [69] or tenascin [70]. Such experiments gave the first insights about the mechanical behavior of proteins, but they couldn't extend their analysis very far. This is because natural modular proteins are not homopolymers: they are composed of different modules with different mechanical features. Since there is no easy way to assign a specific peak of the force spectrum to a specific module (unless in the sporadic case modules are of clearly different lengths), such experiments give general information on the behavior of the molecule, but can't mine too much in the detail of the mechanical properties of the single folds.

To overcome this problem the first and easiest way is to build a homopolymer of the protein module to study. Since natural modules are thought, by definition, to be practically independent folding units, this should not introduce artifacts and allow to study with precision the mechanical behavior of a given fold. In fact, there are clear hints that spacer sequences and neighbor modules can somehow affect the chemical [71] and mechanical stability of a fold [72]. This problem has only been marginally cited, but it can have serious implication for both unfolding and refolding studies, and it can be of biological importance.

Today, more complex polyproteins are commonly engineered for force spectroscopy experiments. They are often composed of at least two different modules: a number of "marker" modules, of known and simple mechanical behavior (like the I27 modules of titin) that alternate with the "target" module (the one to study) , following a pattern like that depicted on Figure 8. In this way it is possible to investigate the behavior of folds that normally do not belong to multimodular proteins, insulating them between marker modules and preventing potential misfolding issues. The marker modules act as controls and as rulers, allowing for more controlled force and length determination and to know exactly how many target modules have been stretched (this is useful for modules that unfold at low forces, and that therefore can give unclear signals). Notable examples include the study of mechanical unfolding of barnase [73], of PEVK titin segment elasticity [74] and of mechanical anisotropy of protein folds [75 , 76].

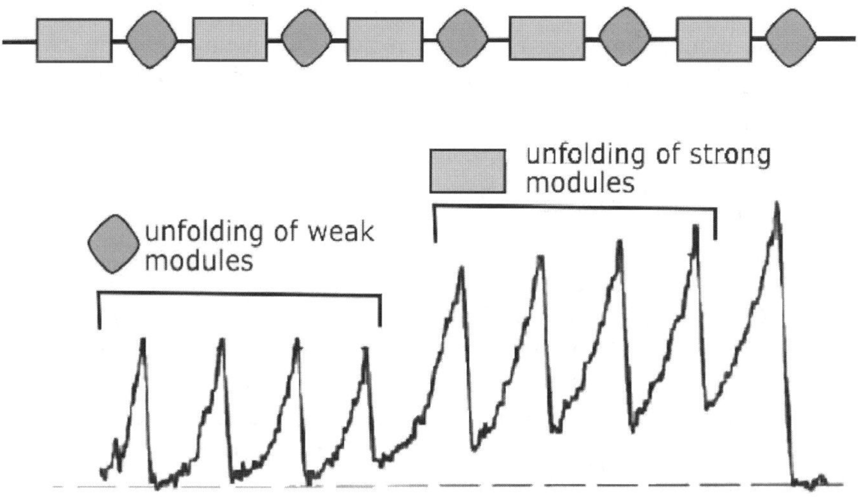

Figure 8. Top, pattern of a generic artificial polyprotein composed of alternating "strong" (rectangles) and "weak" (diamonds) modules. Bottom, force spectroscopy of a construct built with alternating I27 ("weak") and I28 ("strong") modules shows a distinct hierarchy, the weak modules all unfolding before strong ones.

3.3. Mapping the energy landscape of protein unfolding under an external force

In these experiments the rupture of the bonds and of the interactions is normally still thermally driven. The applied force acts somehow just as a catalyst, in promoting the thermal bond disruption by lowering the barrier that confines the bond free energy well. The most probable force of rupture is dependent from the potential energy profile along the force vector direction.

It is therefore clear that there's no absolute value for the rupture force, because it is not only function of temperature but also of the time available to the system to cross the potential barrier. The lower the pulling speed, the more the time is given to the system to thermally cross the barrier. The rupture can therefore take place at a lower force. Every bond -even the strongest covalent one- has a finite lifetime. If we wait for a sufficient amount of time every bond eventually breaks spontaneously. Therefore in the time range of the order of the natural life-time of the bond, the force needed to break the bond becomes zero. On this basis, we can understand how by looking at the relationship between force and pulling speed, we can explore the kinetics of the system far from the equilibrium, and eventually we can also have hints on the energy landscape of the bond under force.

The most generally accepted model [77] predicts simply that force warps the energetic landscape by decreasing it of the mechanical work acting on the system $kv\Delta x$, where k is the system elastic constant, v the pulling speed and Δx the elongation along the force vector. This means that the average force needed to unfold a protein module is linearly dependent on the logarithm of the loading rate kv (Equation 4) :

$$F(kv) = \frac{K_b T}{\Delta x} * \ln\left(\frac{v}{k_0 \Delta x}\right) \tag{4}$$

This relatively simple model follows from the assumption that the barriers keep the same hierarchy under force, but this is not necessarily true. External barriers (at longer Δx) decrease more rapidly than inner barriers. If at zero force the outer barrier is the highest and the inner the lowest, at a given force their hierarchy swaps, and the inner barrier becomes the one kinetically important (sketched in Figure 9). When this happens, the *F/ln(kv)* ratio increases abruptly, reflecting the dependence of kinetics from a new barrier.

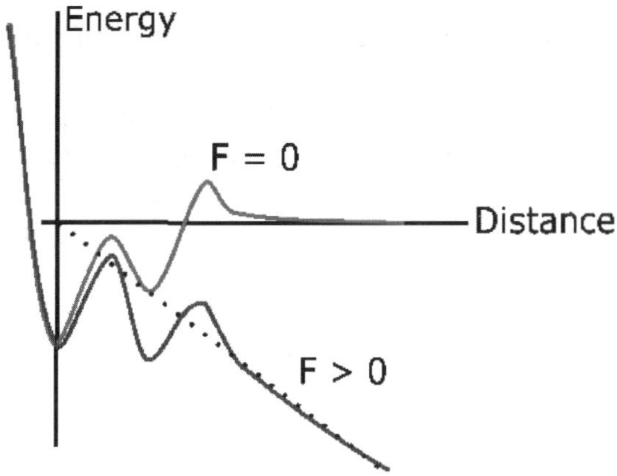

Figure 9. Sketch of how a stretching force modifies the unfolding energy landscape. In presence of an external force, the energy landscape is warped by *kvx* (dotted line), possibly swapping the energetic barrier hierarchy.

From the dependence of the most probable unfolding force from the loading rate it is possible to map the energy landscape of mechanical unfolding. The parameters obtained for each interaction are the bond lifetime at zero force τ_0 and the position X_b of the barrier. The spatial location X_b of the barrier depends only on the slope of the linear relationship, and therefore needs only knowledge of v (if k is assumed to be constant) but τ_0 depends both from the slope and the constant term. The effective loading rate kv must be therefore evaluated precisely where the unfolding takes place. While v can be set at will within the dynamic range accessible to the instrument, k is a global property of the cantilever-molecule system and must be estimated.

Various approaches have been used to evaluate this parameter. Dettmann *et.al.* [78] obtained a value for k measuring the slope of the final part of force peaks. Nevertheless in most cases noise and irregularities of the peak profile often make somewhat arbitrary to define a meaningful "linear slope".

This problem can be circumvented by getting τ_0 and X_b from Monte Carlo simulations of mechanical unfolding. The simulations normally assume a two-state model for the unfolding potential barrier and the WLC elasticity model for protein elongation. The barrier energetic parameters are adjusted to fit the experimental force distribution and dependence of data from pulling speed. With this method the loading rate is implicit in the calculation, being a function of cantilever elastic constant and of molecule persistence length: these parameters must be given to the simulation as explicit data [79]. If there is some knowledge about the interactions involved in mechanical unfolding, more refined models can be used [80].

3.4. What are the determinants of protein mechanics?

The resistance of the modules under force is normally controlled by the rupture of a few critical interactions. After these ruptures the hydrophobic core of the module is exposed to the solvent and its structure collapses rapidly in a cooperative fashion. Chemical or thermal unfolding normally follows different paths in the free energy landscape with respect to the mechanical one. While denaturant concentration or temperature are sensed by the protein as homogeneous and spherically simmetrical fields, an applied force drives the reaction coordinate along the force vector direction. Also from the entropic point of view the chemically / thermally driven and the mechanically driven pathways are markedly different. Thermal or chemical unfolding start from the native folded structure and convert it to increasingly disordered states, until the polypeptide becomes virtually a random coil. Mechanical unfolding instead lead the domain structure collapse abruptly into a transient higher entropy state that still has significant spatial constraints because of its tethering to the tip and the surface. The entropy then decreases again, until eventually the polymeric chain is fully stretched.

Experiments confirm the difference between thermal/chemical and mechanical unfolding pathways. For example barnase (a mainly-beta fold) or calmodulin (an all-alpha fold) are well structured and notoriously thermally stable proteins, but they are easily unfolded by mechanical force, giving low or undetectable unfolding peaks [68, 73]. At the same time the Kringle fold, that lacks significant secondary structure elements and is natively locked by three disulfide bonds, has a significant mechanical resistance even after they have been completely reduced [81], giving force peaks of about 100 pN.

The mechanical unfolding of a protein is also markedly dependent on the force direction in respect to the protein spatial coordinates. This has been demonstrated by pulling E2lip3, an all-beta protein, along two different directions (Figure 9) [75] (An analogous experiment with similar results on ubiquitin unfolding is reported in [76]). The force required for unfolding can vary more than 20-fold when the pulling direction is changed.

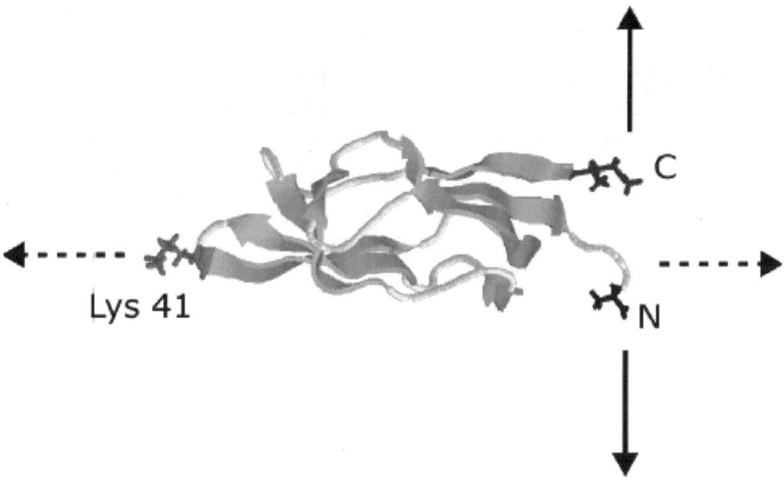

Figure 10. The structure and topology of the E2lip3 protein explain its mechanical anisotropy. When pulling the protein from the N and C termini (solid arrows) the first and the last beta strand are easily peeled apart (no force peaks are observed). When pulling from the N terminus and the Lys41 lipoylation site (dashed arrows) the force vector is almost parallel to the E2lip3 beta strands, and they must shear to break. This latter process requires a much higher average force than the former (force peaks of 150 pN mean height are observed).

This striking result shows very well the main topological determinant that accounts for mechanical resistance of most beta-sheet protein structures. When the E2lip3 protein is pulled between the N and C termini, the force is pulling apart two adjacent beta strands. The bonds between the strands are parallel to the force vector, and they experience the force one by one, therefore being sequentially *unzipped*. When pulling the same protein between the N terminus and a lipoylation site attached to the surface, the force acts on two distant points of the protein, and –most importantly– perpendicularly to the arrays of hydrogen bonds between the beta strands. This means that to the two strands can be detached only through their *shearing*. When the bonds break, they must break all at once. The unfolding through shearing therefore requires a much larger force than unfolding through unzipping.

In principle it is possible to exploit this effect to glean topological –and therefore structural– hints from a beta-sheet protein hard to crystallize or to analyze with conventional means, like membrane proteins or amyloids.

Mechanical properties of alpha helix structures are less well understood. This is because most modular proteins of clear mechanical function share a predominantly beta-sheet structure, receiving therefore much more experimental attention, and that alpha-helix proteins unfold at lower forces and therefore yield harder to record force curves. These facts are related, nature having probably evolved more robust folds for proteins that need to resist mechanical force. Force spectroscopy on spectrin, the only all-alpha protein that has been well investigated, shows a saw-tooth pattern with weak force peaks around 30 pN [82]. The forced extension of the coiled-coil myosin II tail [83] yields an atypical force spectrum with no unfolding peaks but a plateau at 25-30 pN (depending on buffer composition). This indicates an overstretching structural transition. Refolding experiments also indicated that the process can be driven in equilibrium conditions, in striking contrast with the large hysteresis always observed when refolding the Ig-like modules of titin and other multidomain proteins with reentrant topologies (Figure 11).

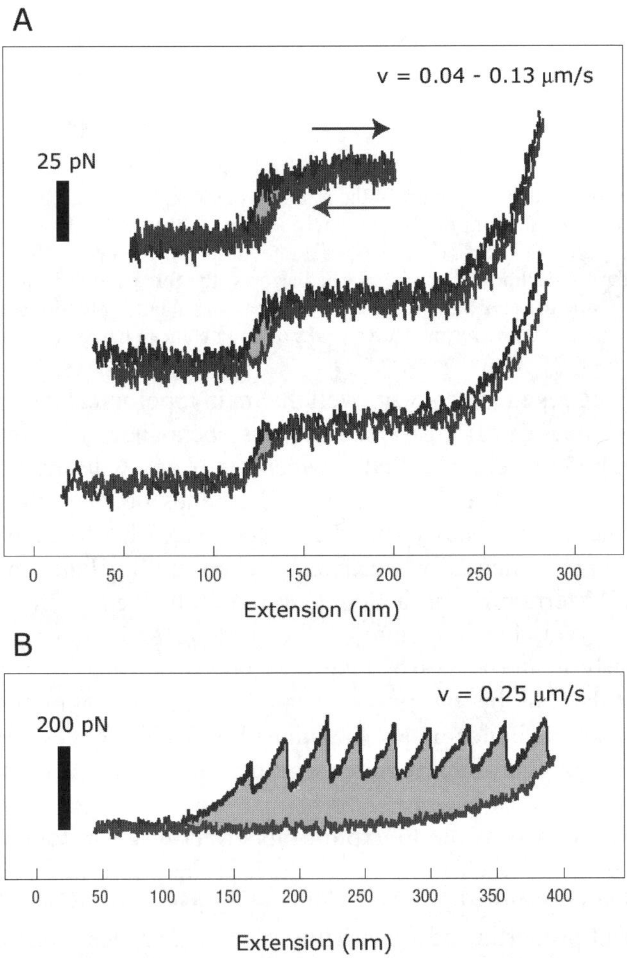

Figure 11. A: series of three pulling and relaxing cycles of a single molecule of rabbit skeletal myosin at extension rates between 0.04 and 0.13μms–1. Myosin behaves as a reversebly elastic structure, and the system works at equilibrium. B: Extension and relaxation of native human cardiac titin with a pulling velocity of 0.38μms–1.The light grey area reflects the energy dissipated during unfolding of titin domains. In this case the system is clearly out of equilibrium. Edited from [83] , with permission.

4. FUTURE PERSPECTIVES: TOWARDS UNIFIED INSTRUMENTS FOR SMALL FORCES, SMALL SCALE AND SHORT TIME STUDIES IN BIOPHYSICS

The realization of instruments that can contemporarily give information on small spatial scale phenomena, involving small forces and short time duration is not too remote. Nowadays, the construction of AFM systems with high performance, fast responding and close-looped piezoelectric positioners allows for the contemporary study of the high resolution topography of a specimen and the accurate measurement of the forces between the probe and the specimen or between the various components of the specimen.

The forthcoming introduction and plausible widespread use of small cantilevers [84] will certainly allow a big step forward in the construction of AFM systems. All this is due to the possibility of small levers of reading the force with a reduced force noise and to respond fast to changes, due to their high resonant frequency [85].

These capabilities will strenghten the developments of the emerging field of single-molecule chemistry. This approach will be increasingly capable of answering questions that traditional ensemble-averaged chemistry cannot even address. AFM studies in the nanoscale are actively yielding fundamental and unique advances on biophysics, biotechnology and nanotechnology.

5. REFERENCES

1. Binnig G., et al. Surface studies by scanning tunneling microscopy. Phys. Rev. Lett. 1982; 49: 57-61.
2. Fritz J., et al. Probing single biomolecules with atomic force microscopy. J Struct Biol 1997; 119(2): 165-71.
3. Samori B. and Zuccheri G. DNA Codes for Nanoscience. Angew Chem Int Ed Engl, 2004.
4. Pearson H. DNA: Beyond the double helix. Nature 2003; 421(6921): 310-2.
5. Schellman J.A. and. Harvey S.C. Static contributions to the persistence length of DNA and dynamic contributions to DNA curvature. Biophysical Chemistry 1995; 55(1-2): 95-114.
6. Bednar J., et al. Determination of DNA persistence length by cryo-electron microscopy. Separation of the static and dynamic contributions to the apparent persistence length of DNA. Journal of Molecular Biology 1995; 254(4): 579-94.
7. Calladine C.R. and Drew H.R. A useful role for "static" models in elucidating the behaviour of DNA in solution. Journal of Molecular Biology 1996; 257(3): 479-85.
8. Crothers D.M., DNA curvature and deformation in protein-DNA complexes: a step in the right direction. Proc. Natl. Acad. Sci. USA 1998; 95(26): 15163-5.

9. El Hassan M.A. and Calladine C.R. Conformational Characteristics of DNA - Empirical Classifications and a Hypothesis for the Conformational Behavior of Dinucleotide Steps. Philosophical transactions-Royal Society of London. Physical sciences and engineering 1997; 355: 43-100.
10. Griffith J., et al. Visualization of the bent helix in kinetoplast DNA by electron microscopy. Cell, 1986; 46(5): 717-24.
11. Dickerson R.E., Goodsell D. and. Kopka M.L MPD and DNA bending in crystals and in solution. J Mol Biol 1996; 256(1): 108-25.
12. Hardwidge P.R., et al. Relating independent measures of DNA curvature: electrophoretic anomaly and cyclization efficiency. J Biomol Struct Dyn 2000; 18(2): 219-30.
13. Kahn J.D., E. Yun and Crothers D.M. Detection of localized DNA flexibility. Nature 1994; 368(6467): 163-6.
14. Zhang Y. and Crothers D.M. High-throughput approach for detection of DNA bending and flexibility based on cyclization. Proc. Natl. Acad. Sci. USA 2003; 100(6): 3161-6.
15. Zhang Y. and Crothers D.M. Statistical mechanics of sequence-dependent circular DNA and its application for DNA cyclization. Biophys. J. 2003; 84(1): 136-53.
16. De Santis P., et al. Sequence dependent circularization of DNAs - a physical model to predict the DNA sequence dependent propensity to circularization and its changes In the presence of protein-induced bending. J. Phys. Chem. 1996; 100(23): 9968-9976.
17. Cognet J.A., et al. Static curvature and flexibility measurements of DNA with microscopy. A simple renormalization method, its assessment by experiment and simulation. J Mol Biol 1999; 285(3): 997-1009.
18. Muzard G., Theveny B. and Revet B. Electron microscopy mapping of pBR322 DNA curvature. Comparison with theoretical models. Embo Journal 1990; 9(4): 1289-98.
19. Zuccheri G., et al. Mapping the intrinsic curvature and flexibility along the DNA chain. Proc. Natl. Acad. Sci. U S A 2001; 98(6): 3074-3079.
20. Rivetti C., Walker C. and Bustamante C. Polymer chain statistics and conformational analysis of DNA molecules with bends or sections of different flexibility. J. Mol. Biol. 1998; 280(1): 41-59.
21. Olson W.K. and Zhurkin V.B. Modeling DNA deformations. Curr. Opin. Struct. Biol. 2000; 10(3): 286-97.
22. Le Cam E., et al. Conformational analysis of a 139 base-pair DNA fragment containing a single-stranded break and its interaction with human poly(ADP-ribose) polymerase. Journal of Molecular Biology1994; 235(3): 1062-71.
23. Akiyama T. and Hogan M.E. Structural analysis of DNA bending induced by tethered triple helix forming oligonucleotides. Biochemistry 1997; 36(8): 2307-15.
24. Rivetti C. and Codeluppi S. Accurate length determination of DNA molecules visualized by atomic force microscopy: evidence for a partial B- to A-form transition on mica. Ultramicroscopy 2001; 87(1-2): 55-66.
25. Sanchez-Sevilla A., et al. Accuracy of AFM measurements of the contour length of DNA fragments adsorbed on mica in air and in aqueous buffer. Ultramicroscopy 2002; 92(3-4): 151-8.
26. Marek J., et al. Interactive measurement and characterization of DNA molecules by analysis of AFM images. Cytometry A 2005.
27. Ficarra E., et al. Automated DNA Fragments Recognition and Sizing through AFM Image Processing. IEEE Transactions on Medical Imaging, in publication.
28. Joanicot M. and Revet B. DNA conformational studies from electron microscopy. I. Excluded volume effect and structure dimensionality. Biopolymers 1987; 26(2): 315-26.
29. Haran T.E., Kahn J.D. and Crothers D.M. Sequence elements responsible for DNA curvature. Journal of Molecular Biology 1994; 244(2): 135-43.
30. Barbic A., Zimmer D.P. and Crothers D.M. Structural origins of adenine-tract bending. Proc Natl Acad Sci U S A 2003; 100(5): 2369-73.
31. Strahs D. and Schlick T. A-Tract bending: insights into experimental structures by computational models. J Mol Biol 2000; 301(3): 643-63.

32. Bolshoy A., et al. Curved DNA without A-A: experimental estimation of all 16 DNA wedge angles. Proc Natl Acad Sci U S A 1991; 88(6): 2312-6.
33. De Santis P., et al. Validity of the nearest-neighbor approximation in the evaluation of the electrophoretic manifestations of DNA curvature. Biochemistry 1990; 29(39): 9269-73.
34. Gorin A.A., Zhurkin V.B. and Olson W.K. B-DNA twisting correlates with base-pair morphology. J. Mol. Biol. 1995; 247(1): 34-48.
35. Calladine C.R., Mechanics of sequence-dependent stacking of bases in B-DNA. J Mol Biol 1982; 161(2): 343-52.
36. Yanagi K., Prive G.G. and Dickerson R.E. Analysis of local helix geometry in three B-DNA decamers and eight dodecamers. J Mol Biol 1991; 217(1): 201-14.
37. Hunter C.A., Sequence-dependent DNA structure. The role of base stacking interactions. J Mol Biol 1993; 230(3): 1025-54.
38. Young M.A., et al. Analysis of local helix bending in crystal structures of DNA oligonucleotides and DNA-protein complexes. Biophys J 1995; 68(6): 2454-68.
39. Hunter C.A., Sequence-dependent DNA structure. Bioessays 1996; 18(2): 157-62.
40. Gallego J., et al. "Molecular electrostatic potentials of DNA base pairs and drug chromophores in relation to DNA conformation and bisintercalation by quinoxaline antibiotics and ditercalinium." in *QSAR and Molecular Modeling; Concepts, Computational Tools and Applications*, Sanz F. Giraldo J. and Manaut F.J.R. Eds. Barcelona, Prous Science Publishers, p. 274-281, 1995.
41. Travers A.A. The Structural Basis of DNA Flexibility. 2004: In press.
42. Boffelli D., et al. A theoretical method to predict DNA permutation gel electrophoresis from the sequence. FEBS Lett. 1992; 300(2): 175-8.
43. De Santis P., et al. Theoretical prediction of the gel electrophoretic retardation changes due to point mutations in a tract of SV40 DNA. Biophys. Chem. 1992; 42(2): 147-52.
44. Anselmi C., et al. Dual role of DNA intrinsic curvature and flexibility in determining nucleosome stability. J. Mol. Biol. 1999; 286(5): 1293-301.
45. Anselmi C., et al. From the sequence to the superstructural properties of DNAs. Biophys. Chem. 2002; 95(1): 23-47.
46. Sobel E.S. and Harpst J.A. Effects of Na+ on the persistence length and excluded volume of T7 bacteriophage DNA. Biopolymers 1991; 31(13): 1559-64.
47. Hagerman P.J. Flexibility of DNA. Annual Review of Biophysics and Biophysical Chemistry 1988; 17: 265-86.
48. Shore D. and Baldwin R.L. Energetics of DNA twisting. I. Relation between twist and cyclization probability. Journal of Molecular Biology 1983; 170(4): 957-81.
49. Taylor W.H. and Hagerman P.J. Application of the method of phage T4 DNA ligase-catalyzed ring-closure to the study of DNA structure. II. NaCl-dependence of DNA flexibility and helical repeat. J Mol Biol 1990; 212(2): 363-76.
50. Frontali C., et al. An absolute method for the determination of the persistence length of native DNA from electron micrographs. Biopolymers 1979; 18(6): 1353-73.
51. Berge T., et al. Structural perturbations in DNA caused by bis-intercalation of ditercalinium visualised by atomic force microscopy. Nucleic Acids Res 2002; 30(13): 2980-6.
52. Scipioni A. et al. Sequence-dependent DNA curvature and flexibility from scanning force microscopy images. Biophys J. 2002; 83(5): 2408-18.
53. Olson W.K. et al. DNA sequence-dependent deformability deduced from protein-DNA crystal complexes. Proc. Natl. Acad. Sci. USA 1998; 95(19): 11163-8.
54. Dickerson R.E. DNA bending: the prevalence of kinkiness and the virtues of normality. Nucleic Acids Res., 1998. 26(8): p. 1906-26.
55. Dickerson R.E. and Chiu T.K. Helix bending as a factor in protein/DNA recognition. Biopolymers 1997; 44(4): 361-403.
56. Matsumoto A. and Olson W.K. Sequence-dependent motions of DNA: a normal mode analysis at the base-pair level. Biophys. J. 2002; 83(1): 22-41.

57. McConnell K.J. and Beveridge D.L. Molecular dynamics simulations of B '-DNA: sequence effects on A-tract-induced bending and flexibility. J. Mol. Biol. 2001; 314(1): 23-40.
58. Olson W.K. NDB Reference Codes of Structures for Naked B-DNA. 1998.
59. Okonogi T.M., et al. Sequence-Dependent Dynamics of Duplex DNA: The Applicability of a Dinucleotide Model. Biophys J. 2002; 83(6): 3446-59.
60. Richmond T.J. and Davey C.A. The structure of DNA in the nucleosome core. Nature 2003; 423(6936): 145-50.
61. Vogel V., et al. Structural insights into the mechanical regulation of molecular recognition sites. Trends Biotechnol. 2001; 19(10): 416-23.
62. Leckband D., Nanomechanics of adhesion proteins. Curr Opin Struct Biol 2004; 14(5): 524-30.
63. Bustamante C., et al. Mechanical processes in biochemistry. Annu Rev Biochem 2004; 73: 705-48.
64. Bustamante C., et al. Entropic elasticity of lambda-phage DNA. Science 1994; 265(5178): 1599-600.
65. Oesterhelt F., Rief M. and Gaub H. Single molecule force spectroscopy by AFM indicates helical structure of poly(ethylene-glycol) in water. New Journal Of Physics 1999; 1(6): 1-11.
66. Smith S.B., Cui Y. and Bustamante C. Overstretching B-DNA: the elastic response of individual double-stranded and single-stranded DNA molecules. Science 1996; 271(5250): 795-9.
67. Grandbois M., et al. How strong is a covalent bond? Science 1999; 283(5408): 1727-30.
68. Carrion-Vazquez M., et al. Mechanical design of proteins studied by single-molecule force spectroscopy and protein engineering. Prog Biophys Mol Biol 2000; 74(1-2): 63-91.
69. Rief M., et al. Reversible unfolding of individual titin immunoglobulin domains by AFM. Science 1997; 276(5315): 1109-12.
70. Oberhauser A.F., et al. The molecular elasticity of the extracellular matrix protein tenascin. Nature 1998; 393(6681): 181-5.
71. Hamill S.J., Meekhof A.E. and Clarke J. The effect of boundary selection on the stability and folding of the third fibronectin type III domain from human tenascin. Biochemistry 1998; 37(22): 8071-9.
72. Li H., et al. Atomic force microscopy reveals the mechanical design of a modular protein. Proc Natl Acad Sci U S A 2000; 97(12): 6527-31.
73. Best R.B., et al. Can non-mechanical proteins withstand force? Stretching barnase by atomic force microscopy and molecular dynamics simulation. Biophys J. 2001; 81(4): 2344-56.
74. Li H., et al. Multiple conformations of PEVK proteins detected by single-molecule techniques. Proc Natl Acad Sci U S A 2001; 98(19): 10682-6.
75. Brockwell D.J., et al. Pulling geometry defines the mechanical resistance of a beta-sheet protein. Nat Struct Biol 2003; 10(9): 731-7.
76. Carrion-Vazquez M., et al. The mechanical stability of ubiquitin is linkage dependent. Nat Struct Biol 2003; 10(9): 738-43.
77. Evans E., Probing the relation between force-lifetime-and chemistry in single molecular bonds. Annu. Rev. Biophys. Biomol. Struct. 2001; 30: 105-28.
78. Dettmann W., et al. Differences in zero-force and force-driven kinetics of ligand dissociation from beta-galactoside-specific proteins (plant and animal lectins, immunoglobulin G) monitored by plasmon resonance and dynamic single molecule force microscopy. Arch Biochem Biophys 2000; 383(2): 157-70.
79. Rief M., Fernandez, J.M. and Gaub H.E. Elastically Coupled Two-Level Systems as a Model for Biopolymer Extensibility. Physical Review Letters 1997; 81(21): 4764-4767.
80. Makarov D.E., Hansma P.K. and Metiu H. Kinetic Monte Carlo simulation of titin unfolding. Journal of Chemical Physics 2002; 114(21): 9663-9673.

81. Bustanji Y. and Samorì B. The mechanical properties of human angiostatin can be modulated by means of its disulfide bonds: A single-molecule force-spectroscopy study. Angewandte Chemie International Edition 2002; 41(9): 1546-1548.
82. Rief M., et al. Single molecule force spectroscopy of spectrin repeats: low unfolding forces in helix bundles. J Mol Biol 1999; 286(2): 553-61.
83. Schwaiger I., et al. The myosin coiled-coil is a truly elastic protein structure. Nat Mater 2002; 1(4): p. 232-5.
84. Viani M.D., et al. Fast imaging and fast force spectroscopy of single biopolymers with a new atomic force microscope design for small cantilevers. Rev Sci Instr 1999; 70(11): 4300-4303.
85. Samori B., Zuccheri G. and Baschieri P. Protein Unfolding and Refolding Under Force: Methodologies for Nanomechanics. Chemphyschem 2005; 6(1): 29-34.

SCANNING PROBE MICROSCOPY ACROSS DIMENSIONS

Ille C. Gebeshuber[1,2], Richard A. P. Smith[2], Hannspeter Winter[2] and Friedrich Aumayr[2]
[1] *Austrian Center of Competence for Tribology AC²T research GmbH, Viktor Kaplan-Straße 2, 2700 Wiener Neustadt, Austria*
[2] *Institut für Allgemeine Physik, Technische Universität Wien, Wiedner Hauptstraße 8-10/E134, 1040 Vienna, Austria*

Abstract. Scanning probe microscopy offers interesting approaches to not only image but also manipulate samples in the micro- and nanoscale regime.
Atomic force microscopy became one of the most versatile microscopy methods in biology, since this type of microscopy works under ambient conditions and in many cases, no extensive sample preparation is necessary.
From imaging single living cells to large protein molecules and even more interestingly protein-protein interactions in real time, to micro- and nanoscale investigations of mechanical parameters such as viscoelasticity, atomic force microscopy has proved a useful technique.
Imaging biomolecules at atomic resolution is a dream which might come true within the next few years. A powerful new technique is magnetic resonance force microscopy, combining three-dimensional magnetic resonance imaging with the excellent force sensitivity of the atomic force microscope. This type of microscopy opens the possibility of performing scanning probe magnetic resonance imaging with a sensitivity more than 10 million times better than the sensitivity of the medical magnetic resonance imaging devices for visualizing of organs in the human body. This improved sensitivity extends magnetic resonance imaging into the nanometer realm.

Keywords: AFM, Diatoms, Frustule

1. SCANNING PROBE MICROSCOPY

In the early 1980ies of the last century, researchers at the IBM Zürich research laboratories invented a new type of microscope: the scanning tunneling microscope (STM, [1]). This instrument, which proved capable of imaging solid surfaces with atomic resolution, revolutionized microscopy and surface analysis in an unprecedented way.

The inventors of the STM, Gerd Binnig and Heinrich Rohrer, were awarded the Nobel Prize in Physics in 1986 (together with Ernst Ruska, who was honored for the design of the first conventional electron microscope).

V. Evangelista et al. (eds.), From Cells to Proteins: Imaging Nature across Dimensions, 139–165

The STM stimulated the development of a whole family of scanning probe methods (SPM) that are all based on instrumental principles very similar to that of the STM:

- atomic force microscopy (AFM)
- scanning near field optical microscope (SNOM)
- magnetic force microscope (MFM)
- magnetic resonance force microscope (MRFM)
- scanning thermal microscope
- scanning potentiometry microscope
- ballistic electron emission microscope (BEEM)
- scanning capacitance microcope
- scanning ion conductance microscope (SICM)

The most popular offspring are the atomic force microscope (AFM) [2], and the scanning near-field optical microscope (SNOM) [3]. For an introduction to AFM for biologists see [4].

A scanning probe microscope basically works like an old fashioned record player, where the up-down movement of the needle sends the recorded impulse through the amplifier and on to the speakers to produce music. For a generalized scheme of an SPM, see Figure 1.

A very sharp "needle", called the probe, scans over the surface of a material. The movement of the needle, due to the interactions between the needle and the atoms in the surface, are detected by a computer and sent to produce an image of the surface on a computer screen. A close examination of any minute changes occurring to the atoms hereby becomes relatively simple.

The mechanical, electrical, magnetic, optical and chemical interactions between the sharp probe and the surface provide a 3D representation of surface parameters at or near the atomic scale. The samples can be in air, vacuum or immersed in some liquid.

The areas and opportunities for SPM based investigation are vast indeed: they cover semiconductor chips, polymers, chemicals, coatings, paper, metals, ceramics, magnetic and recording materials, medicine, biological substances, cells, tissues, films and so on and on through the natural and applied sciences.

STM, AFM and SNOM today represent a set of microscopy techniques that can be applied in many different and highly dedicated modes of operation, such that a variety of physical and chemical properties of a material becomes accessible. This versatility of SPM in general is, apart from the inherent high resolution, its major strength.

Today no scientist can master more than a few of all the microscopy methods based on static electric and magnetic fields, particle beams, electromagnetic radiation, acoustic waves, etc., and presumably would not have access to the instrumentation necessary to apply a number of them. On the other hand the operational principle of SPM is uniform, and often a variety of physical properties can be obtained even by employing only one general-purpose set-up [5].

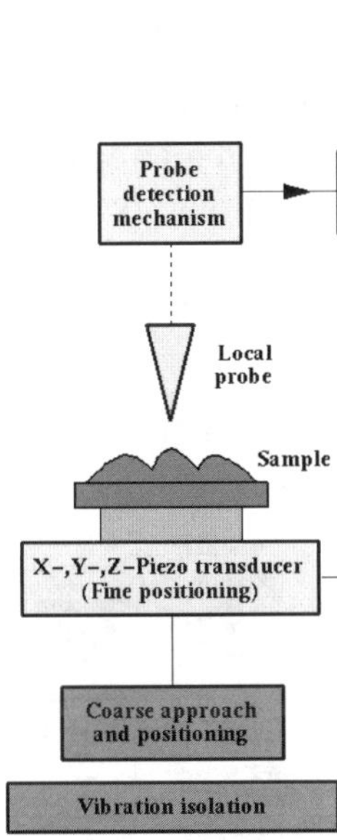

Figure 1. Generalized scheme of a scanning probe microscope (SPM). Image taken from http://spm.phy.bris.ac.uk/techniques/images/SPMschematic.gif.

1.1. Scanning Tunneling Microscopy

The operation of a scanning tunneling microscope is based on the so-called tunneling current, which starts to flow when a sharp tip approaches a conducting surface at a distance of approximately one nanometer. The tip is mounted on a piezoelectric tube, which allows tiny movements by applying a potential difference between sample and tip. Thereby the electronics of the STM system control the tip position in such a way that the tunneling current and, hence, the tip-surface distance is kept constant, while at the same time scanning a small area of the sample surface. This movement is recorded and can be displayed as an image of the surface topography. Under ideal circumstances, the individual atoms of a surface can be resolved and displayed.

It should be noted, however, that STM images not only display the geometrical structure of the surface, but also depend on the electronic density of states of the sample, as well as on special tip-sample interaction mechanisms that are not yet fully understood.

Although the STM itself does not need a vacuum (it works in air as well as under liquids), ultra high vacuum (UHV) is required if one needs to avoid contamination of the samples by the surrounding medium.

A problem in the investigation of metal surfaces is the fact that these surfaces appear very flat to an STM, i.e., the apparent height of individual atoms (corrugation) is 1/100 to 1/10 of an atomic diameter. Therefore, for resolving individual atoms the distance between the tip and the sample must be kept constant within 1/100 of an atomic diameter or less (approx. 0.002 nm). This demands not only very high rigidity of the STM itself but the STM must also be efficiently decoupled from environmental vibrations.

For introductionary texts on STM see [6-8].

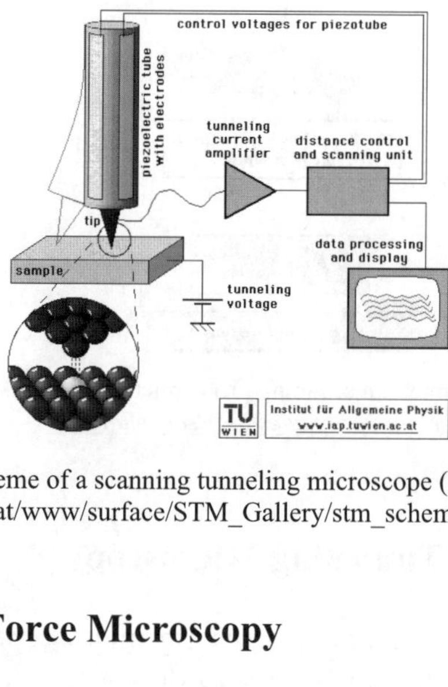

Figure 2. Generalized scheme of a scanning tunneling microscope (STM). Image taken from http://www.iap.tuwien.ac.at/www/surface/STM_Gallery/stm_schematic.html.

1.2. Atomic Force Microscopy

The atomic force microscope measures the interaction force between the tip and the surface. A very sharp tip (cantilever) is scanned across the sample surface using piezoelectric scanners (Fig. 3). The tip may be dragged across the surface or may vibrate as it moves.

The measurement is monitored using the so-called optical beam reflection detection system, in which a laser beam is reflected from the reverse side of the cantilever onto a four quadrant position-sensitive photo sensor. This detector measures the bending of cantilever while the tip is scanning over the sample. The measured cantilever deflections are used to generate a map of the surface topography.

AFM can be used to study insulators and semiconductors as well as electrical conductors, biological samples, magnetic samples, etc.

Three imaging modes, contact, non-contact and intermittent contact (tapping) mode, can be used to produce topographic images of sample surfaces.

In **contact mode**, the tip makes physical contact with the sample and is essentially dragged across the sample surface to make a topographic image. Contact mode imaging can be performed within a liquid environment, which essentially eliminates problems caused by surface moisture such that much lower contact forces can be used.

In **non-contact mode** the cantilever is vibrated near the surface of a sample. The spacing between the tip and the sample is of the order of tens to hundreds of angstroms. Like contact mode, the motion of the scanner is used to generate the topographic image.

Intermittent mode (tapping mode) is like contact mode except for that the vibrating cantilever tip is brought closer to the sample so that it just barely hits (i.e. taps) the sample [9]. This mode is applicable to both imaging in liquids and in air, particularly for soft samples, as the resolution is similar to contact mode while the lateral forces applied to the sample are lower and less damaging.

In **phase imaging** the phase difference between the oscillations of the cantilever driving piezo and the detected oscillations yields information concerning stiffness and viscoelasticiy of the sample.

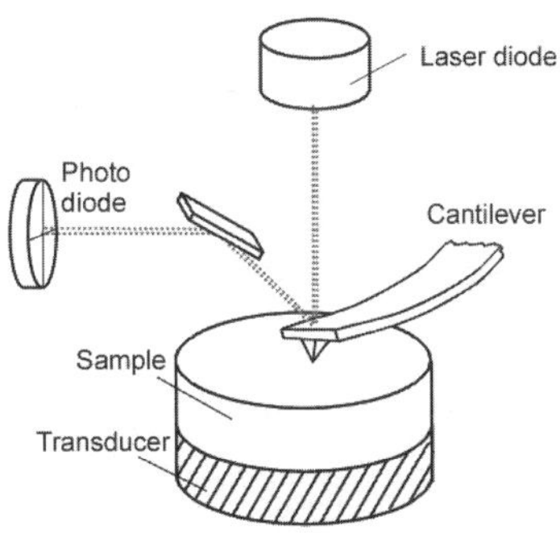

Figure 3. Generalized scheme of an atomic force microscope (AFM). Image by P.K. Hansma (UCSB), taken from http://www.soest.hawaii.edu/~zinin/Zi-AFM.html.

1.3. Scanning Near Field Optical Microscopy

SNOM (Scanning Near Field Optical Microscopy) is a technique that enables work with standard optical tools beyond the diffraction limit which normally restricts the resolution capability of conventional microscopy [3].

It works by exciting the sample with light passing through a sub-micron aperture formed at the end of an optical fibre. Typically, the aperture is a few tens of nanometers in diameter. The fibre is coated with metal to form an aperture and to prevent loss of light, thus ensuring a focused beam from the tip.

SNOM is capable of imaging a variety of fine structures, showing great application potential in life science, material science and semiconductor technology. With spectroscopy, SNOM can realize local spectra, fluorescence sensing and single atom/molecule detection and identification.

1.4. Scanning Ion Conductance Microscopy

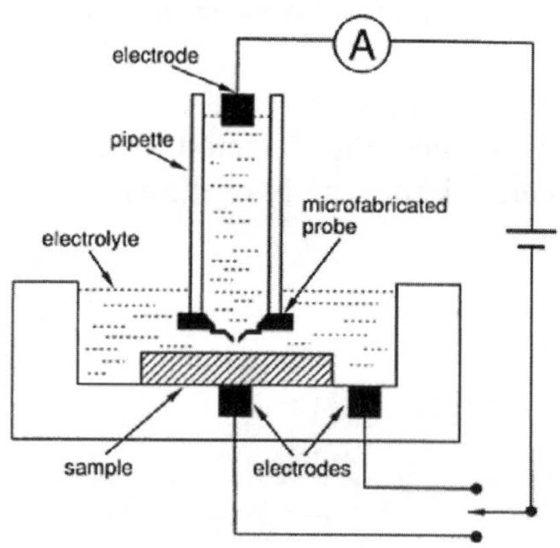

Figure 5. Generalized scheme of a scanning ion conductance microscope (SICM). Reprinted with permission from [10].

The scanning ion-conductance microscope (SICM) can image the topography of nonconducting surfaces that are covered with electrolytes ([11], Figure 5). The probe of the SICM is an electrolyte-filled micropipette. The smaller the opening diameter of the pipette, the better the resolution of the microscope (resolution equals about half the diameter of the pipette).

The SICM measures the ion current between two electrodes, one in the glass micropipette, the other one in the bath solution. The flow of ions through the opening of the pipette is blocked at short distances between the probe and the surface, so limiting the ion conductance.

A feedback mechanism can be used to maintain a given current and in turn determine the distance to the surface. The SICM can also sample and image the local ion currents above the surface.

SICMs are currently not commercially available. We constructed our own SICM [12]. Human red blood cells (sample donated by Stefan Schraml) in Nomarski mode and imaged with our home-built SICM are shown in Figure 6.

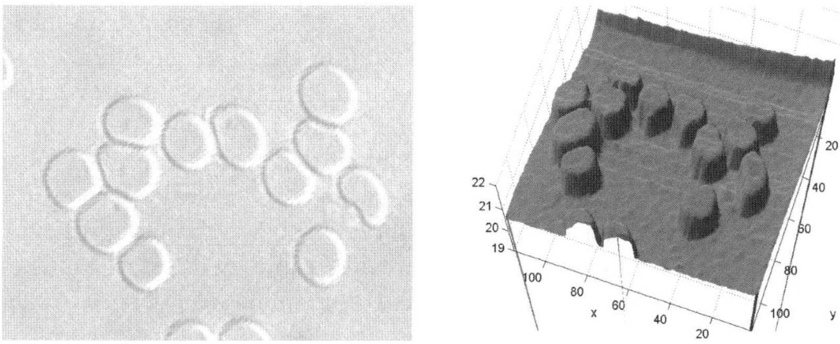

Figure 6. Nomarski light microscopy (left) and Scanning Ion Conductance Microscopy images of human red blood cells in buffer solution on a glass slide [12]. 100μm * 100μm scan size.

2. CELLS

Since many scanning probe microscopes also work in liquids, it is possible to obtain high resolution images of living cells with this type of microscopy.

Below, a couple of interesting examples of SPM investigations on living cells from our own research as well as from other groups are given.

2.1. Diatoms

Diatoms are unicellular organisms with a size of some micrometers. Their surfaces consist of a siliceous skeleton (amorphous glass) enveloped by an organic case.

Individual diatoms range from 2 micrometers up to several millimeters in size, although only few species are larger than 200 micrometers. Diatoms as a group are very diverse, with 12 000 to 60 000 species reported [13]; most of them of exquisite beauty (see Figs. 7 and 8). These naturally nanostructured organisms reproduce via cell division: under ideal conditions, within ten days the offspring of one single cell would number one billion cells (natural "assembly line production" of nanostructures)!

Diatoms live in various wet environments, such as saltwater or freshwater and on wet surfaces. They can either be freely floating (planctonic forms) or be attached to substrates like rocks or plants (benthic forms).

These unicellular organisms are interesting from the point of materials science and biomimetic studies, since they master challenges as diverse as building nanostructured glass shells with high load capacity (a problem interesting for lightweight structures architecture) and engineering strong and robust adhesives that are stable in wet environments (most man-made adhesives fail to bond in wet conditions, owing to chemical modification of the adhesive or its substrate). Furthermore, diatoms excel at preventing dissolution of their silica shells in water owing to a covering layer (up to date technology is currently facing the problem that man made glass fiber reinforced polymers show rapid deterioration when used in water).

Diatoms are small, mostly easy to cultivate, highly reproductive and, since many of them are transparent, they are accessible by different kinds of optical microscopy methods. For an overview of diatoms see [14].

Figure 7. Scanning electron microscope image of a diatom. © W. Oschmann.

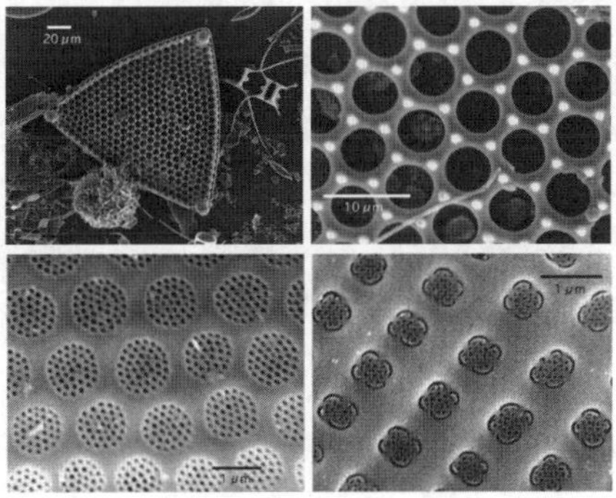

Figure 8. Naturally nanostructured surfaces. The top left image shows an SEM image of a whole cell of *Tricaeratium favus*, the top left image is a zoom into the left image. The bottom images are surfaces of two other species, *Roperia tessellata* and *Achnathes brevipes*. Reprinted with permission from [15].

Currently, human chemical synthesis cannot produce siliceous structures with the hierarchical structural detail of the diatom skeletons nor can ordered siliceous structures be produced synthetically under the benign conditions of diatom biomineralisation [15].

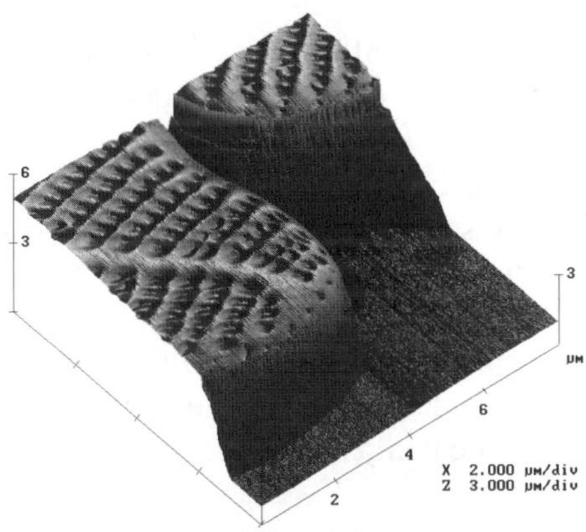

Figure 9. AFM image of parts of two living diatom cells of the species *Navicula seminulum* growing on a glass slide. Note that the flat area does not correspond to the surface of the glass slide, but is determined by the maximum possible extension of the z-piezo of the microscope. Image acquired using AFM contact-mode imaging in water, imaging parameter topography, scan size 8×8 μm^2, scanning frequency 1 Hz. Reprinted with permission from [16].

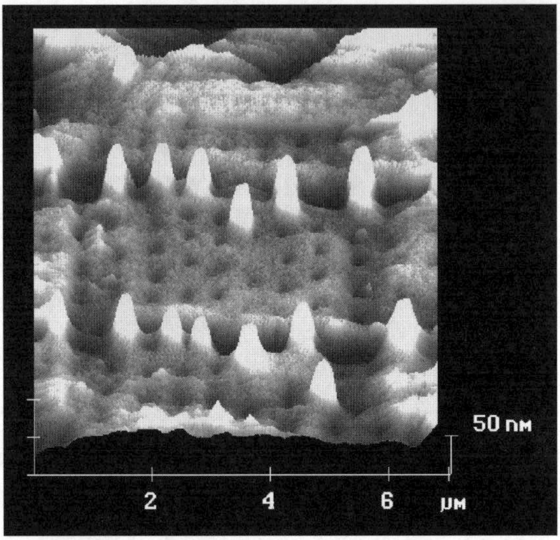

Figure 10. Using the AFM cantilever as a tool. A 5 × 5µm^2 area is repeatedly scanned in AFM contact mode at high feedback force (about 100 nN). After 2 days, a slightly larger area comprising the area where the organic layer had been worn away meanwhile is imaged, revealing a height of the organic layer of about 10 nm. Contact mode, topography, scan size 7 × 7 µm^2, scanning frequency 2 Hz. Reprinted with permission from [16].

Atomic force microscopy offers the opportunity to image these interesting cells alive in ambient conditions, under water, as they continue to grow [16], (see Figure 9).

Since in atomic force microscopy tip is (in contact mode) dragged with ′physical contact′ over the sample, the AFM cantilever can also be used as a tool to mechanically modify the sample. For example, it is possible to determine the thickness of the organic coating enveloping the siliceous skeleton in diatoms by repeatedly scanning (with increased force) a selected area, and imaging a slightly larger area some time later: the altered area is about 10 nm 'lower' than the unaltered surface (see Figure 10).

2.2. Biogenic Adhesives

Most state-of-the-art man-made adhesives fail to bond in wet conditions, owing to chemical modification of the adhesive or its substrate. Engineering strong and robust underwater adhesives that are stable in wet environments is a challenge to current technology. Knowledge of the intrinsic properties of natural adhesives might give valuable information for future engineering approaches.

Three different species of benthic freshwater diatoms (*Eunotia sudetica*, *Sellphora seminulum* and a yet unidentified species) were selected for their strong adhesives by exposure of a greater number of species to highly selective environmental conditions [16].

Eunotia sudetica is very convenient for *in situ* AFM investigation of the diatom adhesive at the cell interface because there is barely any cleft between adjacent cells (Fig. 11a), and valve undulations are less pronounced than in the other species investigated. The diatom adhesive is apparent as small topographic features at the cell interface. The bead-like structures are 10–20 nm high, have lateral dimensions of about 1 μm and are about 1 μm apart. The phase image (Fig. 11b) clearly depicts the altered viscoelastic properties of these structures: the diatom adhesive causes a phase shift of up to 10° compared with the phase difference on the rest of each of the two frustules, where it is within 1° on each, apart from a 2° phase jump between the two adjacent valves. Such a phase jump between cells also appears in the other species where the adhesives are not accessible because of deep clefts between the single organisms.

To access the adhesive *Eunotia sudetica* produces for its attachment to the glass slide on which the cells grow, a *Eunotia sudetica* cell was mechanically removed using a scanning tunneling microscope tip on a micromanipulator stage [15]. Then, the AFM cantilever is brought into contact with the adhesive and lifted upwards again. The force applied after the tip makes contact with the surface can provide a measurement of the stiffness or compliance of the sample while the force required to pull the sample up from the surface is a measure of the adhesion between the tip and the sample. Figure 12 shows such force–distance curves for the diatom surface and the natural adhesive that attaches *Eunotia sudetica* to the glass slide. The diatom surface is less elastic and "sticky" than the adhesive. Several bond-breaking events occur when trying to retract the cantilever from the biogenic adhesive and the pulling force must be applied over nearly one micrometer to completely remove the cantilever. The multiple unbinding events, although not in the monomolecular regime, are similar to the curves found to provide the molecular mechanistic origin of the toughness of natural adhesives, fibres and composites [17].

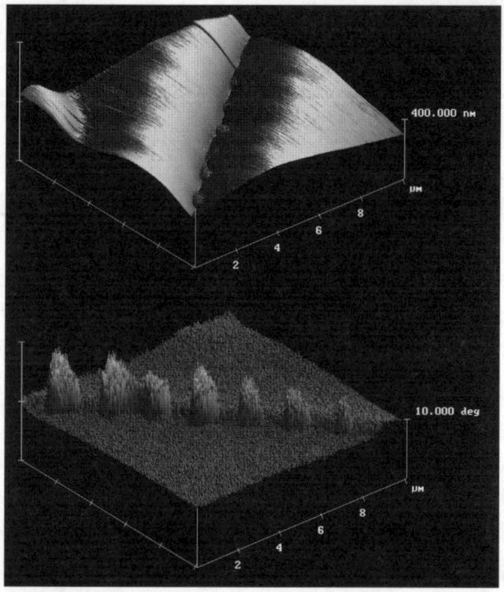

Figure 11: (a) The adhesives in the contact region of two cells of *Eunotia sudetica* are apparent as small topographic features on the slightly undulating cell interface. The corrugation of the bead-like structures is between 10 and 20 nm and their lateral dimension and spacing is about 1 μm. (b) In the phase image these features are far more striking. The diatom adhesive causes a phase lag of about 10° compared with the rest of the frustule surfaces, where on a single frustule it is within 1°. Note the 2° interfrustule phase step, which reveals slightly different viscoelastic properties of the two neighboring valves. Tapping mode, topography and phase, scan size 10 × 10 μm^2, scan rate 5 Hz. Note that for better view (b) is rotated clockwise by 90° as compared with (a). Reprinted with permission from [16].

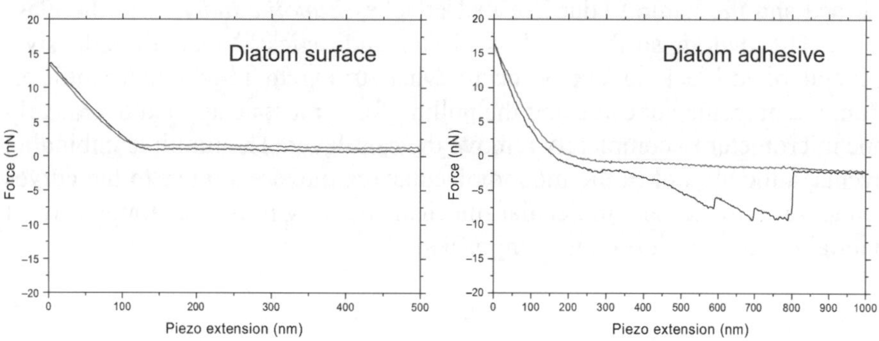

Figure 12: Force–distance curves. Left: No adhesion can be recognized on the diatom surface. Right: Representative data for the diatom adhesive that attaches *Eunotia sudetica* to the substrate. Note that several debonding events occur and that the pulling force must be applied over nearly one micrometer to completely debond the cantilever from the surface. Reprinted with permission from [16].

2.3. Biotribology

Tribology is the study of friction, adhesion, lubrication and wear. Technical micro- and nanosystems often experience failure (e.g. stiction = static friction) due to the lack of reliable concepts, whereas biological systems are reliable also at this scale.

There is an increasing need for a multidisciplinary, system-oriented approach to manufacturing micro- and nanoelectromechanical systems (MEMS/NEMS) that function reliably. Comprehensive understanding of tribology at the micro- and nanoscale can only be achieved through the cross-fertilization of ideas from different disciplines and the systematic flow of information and people among research groups.

Biomicro- and -nanotribology is a new interdisciplinary field of research combining methods and knowledge of physics, chemistry, mechanics and biology. The aim of biotribology is to gather information about friction, adhesion, lubrication and wear of biological systems and to apply this knowledge to innovate technology, with the additional benefit of environmental soundness.

Biological systems with moving parts have optimised their lubrication during evolution. Some diatom species can serve as interesting model organisms for nanotribological investigations:

Bacillaria paxillifer is a motile colonial diatom (Fig. 13). Entire colonies of 5 to 30 cells expand and contract rhythmically and in a coordinated manner. In motion, the cells glide along each other. No wear has ever been detected in this species. Effective biogenic lubricants might have evolved to protect the single cells.

Ellerbeckia arenaria is a colonial diatom which lives in waterfalls. The string-like colonies can be several millimeters long and can be elongated by about one third of their original length. When released, they return to the original position like a spring [18]. This interesting feature could suggest that there are parts in relative motion in this species, coping with friction and wear.

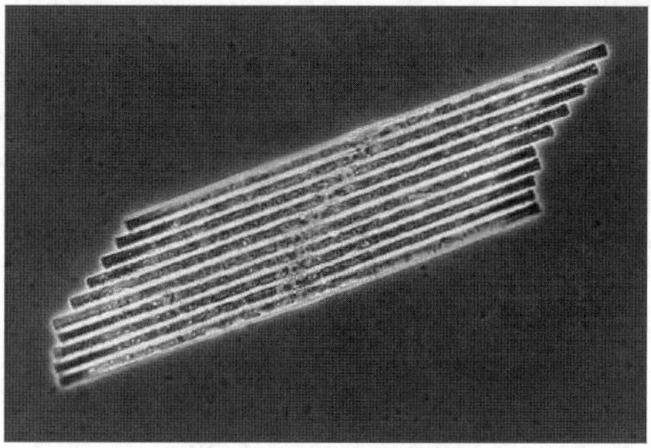

Figure 13: Bacillaria paxillifer colony with 10 cells. In motion, colonies expand and contract rhythmically and in a coordinated manner, with the cells gliding along each other. Biogenic lubricants might have evolved to protect the single cells from wear. © Wim van Egmond, Microscopy-UK (http://www.microscopy-uk.org.uk/).
Eunotia sudetica is a colonial diatom. AFM investigations of living cells reveals bead-like features which might well act as friction reducers, either by means of ball-bearings or a solid lubricant or a strategy which still is completely unknown to engineers.
For an introduction to diatom bionanotribology, see [18 ,19].

3. SINGLE MOLECULES

The AFM has proven to be a useful tool for studying proteins at the single molecule level. In particular, the AFM has been used to measure both structural and mechanical properties of individual proteins in physiologically relevant buffer solutions.

3.1. Chaperonins: Probing Protein-Protein Interactions in Real Time

The *E. coli* chaperonin GroEL and its cochaperonin GroES play important roles in helping proteins to reach their native states. GroEL and GroES work together in a series of adenosine triphosphate mediated steps to capture misfolded proteins, free them from local energy minima and finally provide them with an environment conducive to folding to their native states [20 – 23].
Small cantilever [24] AFMs can be used for making faster and quieter measurements [25]. A prototype small cantilever atomic force microscope [26] was used to observe, in real time, the interactions between individual molecules of the chaperonin protein GroES binding to and then dissociating from individual GroEL proteins, which were immobilized on a mica support [27]

Both GroEL and the GroEL–GroES complex monolayer in buffer solution were repeatedly imaged without the aid of fixing agents (Figs. 15 and 16). The central channel of the protein ring can be resolved on many of the molecules, indicating that GroEL adsorbs to mica in an end-up orientation. The average diameter of the molecules in this image is 14.6 ± 2.2 nm, which agrees well with both X-ray [28] and cryoelectron microscopy [29] data.

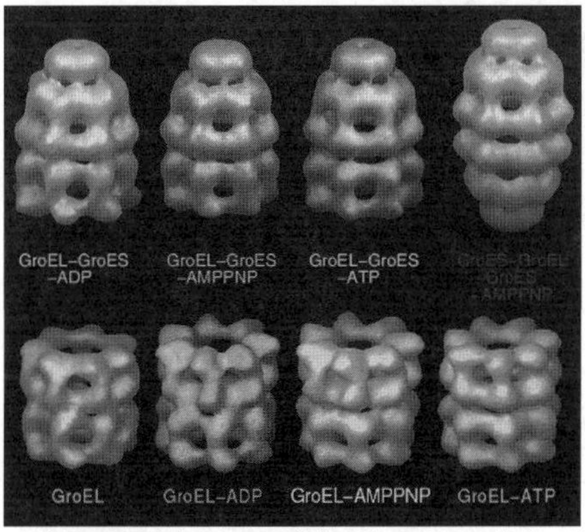

Figure 14. Cryo-electron microscopy images of the GroEL- and the GroEL-GroES-nucleotide structures at 30 Å resolution. Reprinted with permission from [30].

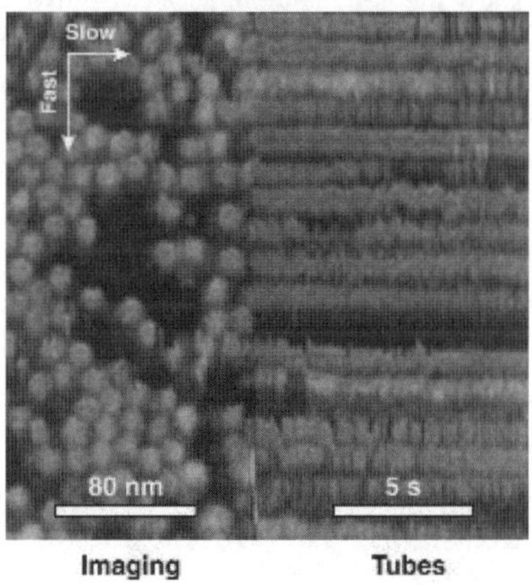

Figure 15. GroEL scanned in two dimensions (left) and in one dimension (right). In this image of GroEL, the slow scan axis was disabled half way through the scan. After the slow scan axis was disabled, the tip repeatedly scanned the same line of proteins, thereby generating protein 'tubes'. Each tube corresponds to an individual protein molecule. Reprinted with permission from [27].

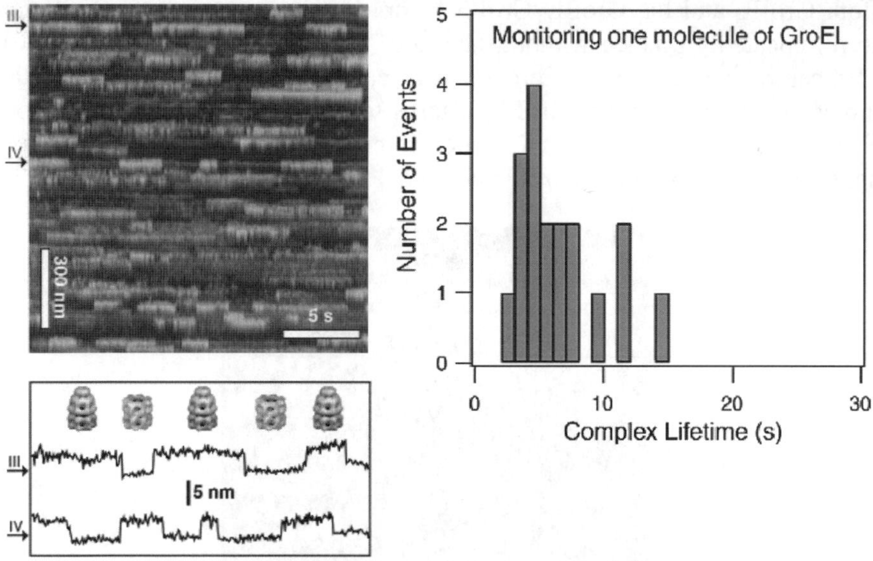

Figure 16. Left: After the addition of GroES and Mg-ATP into the buffer solution, large repeated variations in height along the lengths of many tubes are observed. Left bottom: The cross sections of these tubes show the height stepping between two values that differed by 3.6 ± 1 nm. Cryoelectron microscopy images by [30]. Right: Histogram of measured complex lifetime of an individual GroEL molecule in the presence of Mg-ATP and GroES. Individual GroES molecules attach to and then separate from the same GroEL molecule 18 times during an observation period of ~2 min. Note the absence of events with lifetimes < 2 s. Reprinted with permission from [27].

In order to obtain the temporal resolution required for observing the formation and dissociation of the GroEL–GroES complexes in the presence of Mg-ATP, we scanned the sample in one dimension rather than two. By repeatedly scanning a single line across the sample we were able to monitor the height of a series of protein molecules at rates as high as 20 Hz. This method is illustrated in Figure 16, which shows an image of GroEL molecules in buffer in which the slow scan axis was disabled half way through the scan. From this image it is clear that each 'tube' is associated with a single protein molecule. The height variations along the length of the tubes reflect the time dependent changes in the protein structure.

Upon the addition of both GroES and Mg-ATP, it was observed that many of the tubes displayed repeating well-defined step-like variations in height (Fig. 16, top left). The cross section of the two featured tubes clearly shows the step-like variations in height (Fig. 16, bottom left). The magnitude of these steps is 3.6 ± 1 nm. This is consistent with the height difference measured between structures of GroEL and the GroEL–GroES complex obtained by X-ray crystallography [28] and cryoelectron microscopy [29] studies. We therefore conclude that the observed height variations are caused by a GroES molecule attaching to and then separating from the GroEL molecule associated with the tube.

In one case, we recorded an active molecule for ~2 min and saw the complex form and dissociate 18 times. A histogram of the complex lifetime for this molecule is shown in Figure 16, right. The distribution of complex lifetime has a maximum near 5 s and the average lifetime is ~7 ± 1 s (n = 18).

4. NANOSTRUCTURING ATOMICALLY FLAT SURFACES WITH IONS

Most of the small structures currently used in technology are in the micrometer range. One reason for this is silicon micromachining technology, which works fast and at low cost in this regime. However, needs for increased data-storage density and smaller devices call for nanometer-sized structures.

In the following, a gentle technique to nanostructure surfaces is demonstrated: bombardment with so-called "slow" multicharged ions.

The AFM is the microscope of choice for investigating ion induced nanodefects on flat crystals, because of its great resolution and of the fact that it can image insulating materials.

One important way to produce nanostructures on surfaces involves kinetic sputtering by "fast" ions. However, fast ions unavoidably cause unwanted radiation damage. As opposed to this, potential sputtering (PS), i.e. desorption induced by the potential energy of slow multiply charged ions (MCI), holds great promise for more gentle nanostructuring [31, 32]. It can cause high sputter yields even at such low ion impact energies where kinetic sputtering and defect creation in deeper layers is not possible. While the physical mechanisms of PS have been the subject of extensive investigation [33 – 37], technical applications of slow MCI have so far remained largely unexplored, despite the fact that they provide unique opportunities for etching, ultra-thin film growth and nanostructure fabrication.

4.1. STM and AFM Investigations of HOPG Bombarded with Argon Ions

Highly oriented pyrolytic graphite (HOPG) is a crystalline layered material, that can easily be cleaved with an adhesive tape to obtain atomically flat surfaces (for STM and AFM images in atomic resultion, imaged in ultra high vacuum (UHV), see Fig. 17).

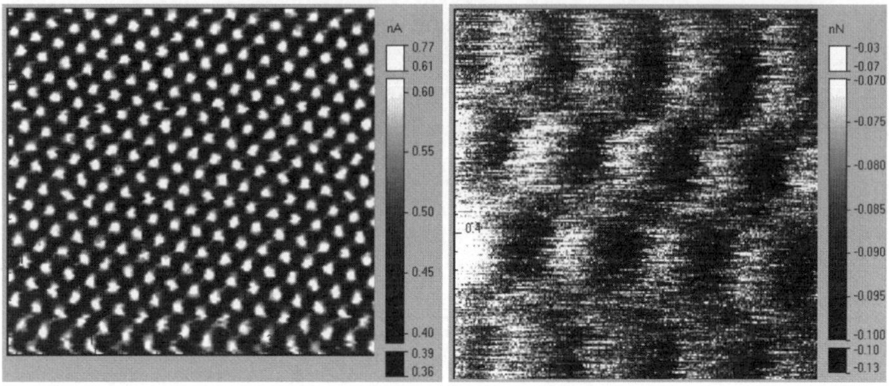

Figure 17. Highly oriented pyrolytic graphite imaged in ultrahigh vacuum with atomic resolution. Left: Scanning tunnelling microscopy image, image size 4*4 nm^2. Right: Atomic force microscopy image, image size 1*1 nm^2.

Figure 18. Highly oriented pyrolytic graphite bombarded with 800 eV Ar^+ ions imaged in ultrahigh vacuum with atomic resolution (right). The fourier transform (left) of the Ar^+ ion-induced defect reveals a $\sqrt{3} \times \sqrt{3}R30°$ surface.

Bombardment with 800 eV Ar^+ ions and subsequent imaging with UHV STM reveals a $\sqrt{3} \times \sqrt{3}R30°$ surface, as is characteristic for interstitial defects or vacancies in HOPG, surrounded by undisturbed surface regions (Figure 18).

Scanning with the AFM at atomic resolution on the irradiated surface did not show any significant topological changes due to ion bombardment. Therefore, it was concluded that the defects observed are mainly due to changes in the electronic density of states of the surface.

For more detailed information on these studies, see [38, 39].

4.2. Ion Bombardement of Atomically Flat Insulator Crystals

We carried out a systematic search for PS with atomically clean monocrystalline insulator targets which have been bombarded by low-energyc ions on the single impact level, searching for resulting surface defects by means of AFM.

The targets remained under UHV conditions during ion irradiation and subsequent AFM inspection.

The insulators studied included LiF(100), SiO_2(0001) α-quartz, muscovite mica and sapphire c-plane Al_2O_3(0001) after their irradiation with low doses of slow singly- and multiply-charged ions.

Ion bombardment was performed at a dose of $5*10^{12}$ ions/cm^2, which is equivalent to five incident ions per 10*10 nm square. Ion bombardment was accompanied by low-energy electron flooding to compensate for surface charging which otherwise would have strongly inhibited the AFM observation.

Resulting nanostructures on the surfaces as investigated by AFM are hillocks a few nanometers high.

Of all crystals investigated, Al_2O_3 showed the most promising results (the size of the hillocks increased with increasing potential energy of the ions with which the samples were bombarded, see Figure 19). For SiO_2 further research might lead to a similar conclusion.

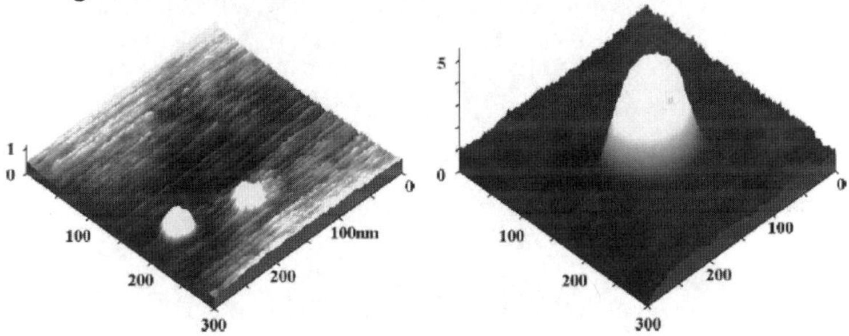

Figure 19. UHV AFM contact mode image of sapphire (Al_2O_3, c-plane 0001) bombarded with 500 eV Ar^+ (left image) and Ar^{7+} (right image) ions. The nanodefects induced by Ar^{7+} ions (same kinetic but higher potential energy than the Ar^+ ions) are considerably higher and wider than those caused by singly charged ions. The defects are real topographic features; all dimensions in nanometers. Reprinted with permission from [39].

Both materials are of great relevance for applications in a rapidly emerging field combining microelectronics and nanotechnology.

For detailed descriptions of these studies, see [39, 40].

4.3. Surface Modifications on Si(111) Produced by the Impact of Slow Multiply Charged Ions

We are currently investigating whether beams of slow MCI can be utilized to produce nanometer-sized surface modifications on silicon substrates. In particular, hydrogen-terminated Si(111) surfaces shall be bombarded in UHV with low fluxes of slow MCI.

Due to MCI - surface interaction the hydrogen atoms should be removed around the MCI impact site and the resultant dangling silicon bonds should than react with oxygen introduced into the UHV chamber so producing ultra-shallow silicon oxide nanodots.

The formation of such nanostructures at ion impact sites is going to be studied with AFM and STM in UHV.

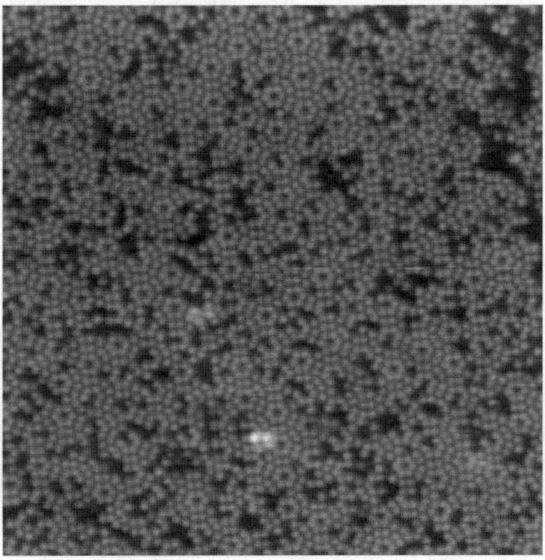

Figure 20. STM image (topography, raw data) of the reconstructed Silicon (111)-(7*7) surface. Scan size 40*40 nm^2, sample bias +1 V (empty-states tunnelling), tunnelling current 1.2 nA, scanning speed 80 nm/s.

Figure 20 shows an UHV STM image of the sample we are going to use for these studies, namely the reconstructed Silicon(111)(7*7) surface. This sample was prepared in a UHV chamber with a pressure of approximately 2.5×10^{-10} mbar. Several steps of heating were used to induce desorption of gases. The first step was radiative heating for 3 hours to remove adsorbed gas from the surrounding components. The subsequent outgassing steps were achieved by passing a current through the sample; the sample details are boron-doped silicon, 0.01-0.02 ohm cm, 13*3*0.5 mm^3 dimensions. The first step was carried out with a current of approximately 1.4 A through the sample; at this point the sample was just below the temperature where glowing becomes visible (approximately 500-600 C). After this step, a series of short heating pulses with increasing current and therefore sample temperature followed, beginning with 5 A, then 6 A and then up to 10 A. These steps were accompanied by pulses of pressure that died away after ~1 minute. The last 10 A step corresponded to a temperature of approximately 1280° C as measured by a pyrometer. After about 10 seconds the current was reduced first to one half and then it was reduced by 0.1 A per 5 seconds. The sample was allowed to cool for about 1 hour before transfer to the AFM/STM unit for studying.

5. SUBATOMIC FEATURES

5.1. Atomic Orbitals in Tungsten

In July 2004, a breakthrough in the resolving power of microscopy was reported by a team of physicists at Augsburg University [41]. An individual tungsten atom was imaged by atomic force microscopy at a resolution of 77 pm. The diameter of a carbon atom is 142 picometers, the diameter of a tungsten atom is 274 picometers. Four distinct peaks that are attributed to highly localized electron clouds can be identified (see Figure 21).

The experiment was performed in UHV at five degrees Kelvin. The microscope was isolated from vibrations by a 30 t foundation and from sound and electromagnetic stray fields by a metal chamber.

Whereas STM only probes the most loosely bound electrons with an energy at the Fermi level, AFM can resolve the charge density variations within a single atom, because the forces between the AFM tip and the sample are of electrostatic nature.

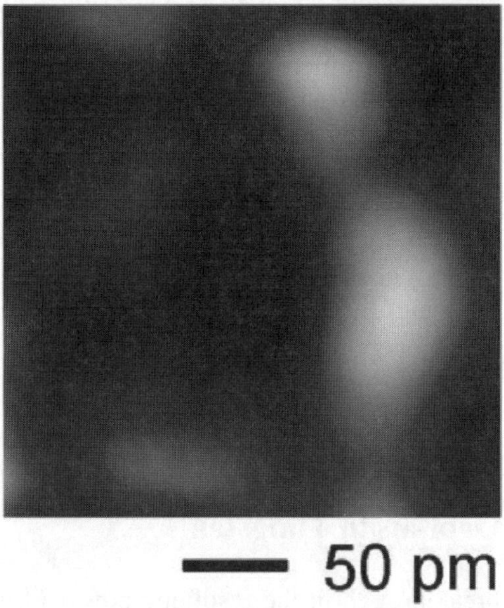

Figure 21. UHV low temperature (5 K) AFM constant-height mode image reveals four-fold symmetry in the amplitudes of the higher harmonics V_{hh} signal (centered close to the maximum of the tunnelling current, data not shown). The symmetry of the V_{hh} signals points to the fourfold bonding symmetry of the W atom viewed parallel to a <001> crystal direction. Reprinted with permission from [41].

The electron structure originates from the quantum-mechanical nature of tungsten bonding. Tungsten develops a body centred cubic crystal structure such that every tungsten atom is surrounded by eight nearest neighbour atoms, causing "arms" of increased charge density which point to the next neighbours. Four of these highly localized electron clouds are visible on surface atoms.

The role of tip and sample was switched in the experiment: the front atom in a sharp tungsten tip was imaged by a light carbon atom of a graphite surface. The reason a light atom was used for probing was to minimize image blurring, since the mapping of one atom with another atom always involves a convolution of the electronic states. The tunnelling current is confined to the top atom because of the sharp increase of tunnelling probability with decreasing distance.

Instead of measuring static deflections or frequency changes, higher harmonics triggered by tip-sample forces (V_{hh}) are analyzed in this technique. The higher harmonics are much more sensitive to short-range interactions than the previously used signals.

5.2. Electron Spins

In 1990 John Sidles set himself the problem of locating the atoms in a single protein molecule. His idea to combine magnetic resonance (MR) with force microscopy in magnetic resonance force microscopy (MRFM) was published as a concept in 1991 [42].

Sidles settled on force microscopy because the performance of induction coils, the detectors in conventional MR, scales unfavourably with size. At least 10^{12} nuclear spins are needed to yield a detectable signal. Shrinking the coil to encompass fewer spins reduces the signal irretrievably below noise. A force microscope, on the other hand, becomes more sensitive the smaller it gets.

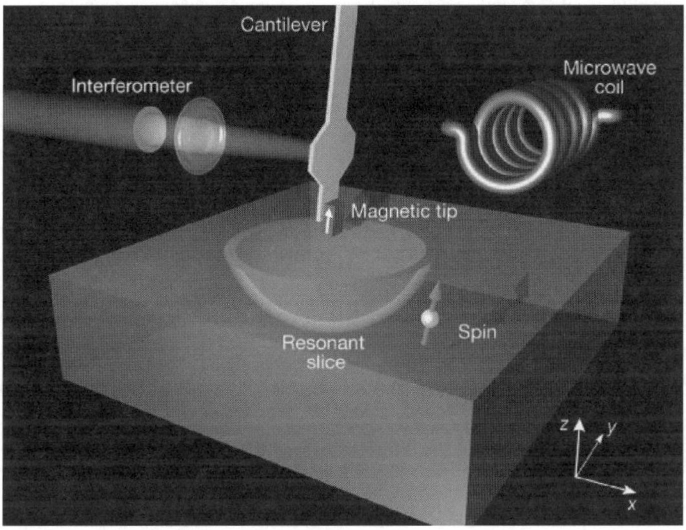

Figure. 22. Configuration of the single-spin MRFM experiment. The magnetic tip at the end of an ultrasensitive silicon cantilever is positioned approximately 125 nm above a polished SiO_2 sample containing a low density of unpaired electron spins. Spins as deep as 100 nm below the sample surface can be probed. Reprinted with permission from [43].

In 2004, Rugar and co-workers improved MRI sensitivity by some 10 million times compared with the medical MRI devices used to visualize organs in the human body. The improved sensitivity extends MRI into the nanometer realm and now allows for single electron spin detection [43].

In their MFRM, a small strong ferromagnetic CoSm dot is mounted on the sharp cantilever, yielding a field gradient of 2 G/nm. Experiments are performed at a temperature of 1.6 K.

The test sample was vitreous silica, which had been bombarded with a 2-Gy dose of Co^{60} γ-rays, yielding 10^{13} to 10^{14} unpaired electron spins (E′ centers) per cubic centimeter.

Unpaired electrons and many atomic nuclei behave like tiny bar magnets. The force exerted by an electron spin on the cantilever is a few attonewtons (10^{-18} N). Such a force is roughly 10 million million times smaller than the force a snowflake exerts resting on your hand. Such a force is even a million times weaker than the forces usually encountered in atomic force microscopy.

Detecting such a small force requires a pliable cantilever. Yet, a pliable cantilever, unlike stiffer AFM cantilevers, can't be positioned close to a surface because the tip would stick to it. Therefore, the cantilever in MRFM is held perpendicular to the sample surface and deflects when a spin lies in front or behind rather than directly below (Fig. 22). A laser interferometer records the cantilever deflections and sophisticated measurement signal acquisition techniques and long measurement times are needed for successful single spin detection.

Several other single-spin detection methods have previously been demonstrated [44 – 50]; see Figure 23 for an STM based method). Yet, MRFM has some attributes that set it apart: The MFRM can sense an electron lying dozens of atomic layers beneath the surface. It is a major milestone toward creating a microscope that can make three-dimensional images of molecules with atomic resolution.

If developed further, the MFRM technique could prove useful for investigating the atomic structure inside materials used in the electronics industry and to image biomolecules – such as proteins - at atomic resolution. However, to reach this goal, nuclear spins have to be detected. Nuclear spins are harder to detect than electron spins, because a proton´s magnetic moment is 658 times weaker than that of an electron.

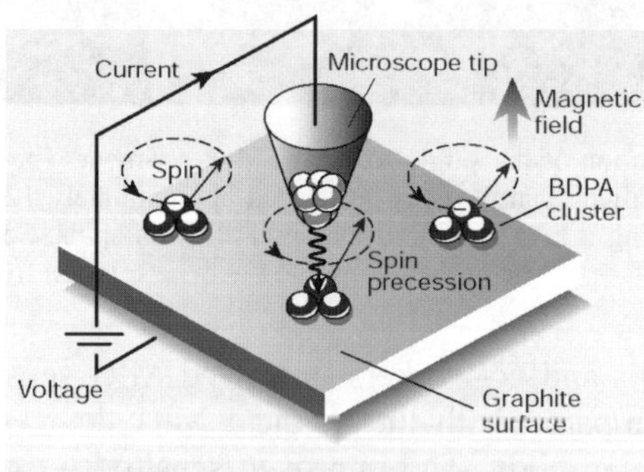

Figure. 23. Spin detection through the union of high-resolution microscopy and resonance techniques. A graphite surface is coated with clusters of organic BDPA molecules. In the applied magnetic field, the electron-spin vectors (red arrows) associated with free radicals in the molecules precess at a certain frequency. When the tip of a scanning tunnelling microscope is brought close to a cluster, the tunnelling current is modulated at the precession frequency; detecting the modulation effectively measures electronic spin in the molecule. Reprinted with permission from [51].

6. CONCLUSIONS AND OUTLOOK

In this review, we presented scanning probe microscopy across dimensions, from large samples like single cells, to single biomolecules and nanometer small ion induced defects on crystal surfaces to subatomic features like orbitals and single spins.

Scanning probe microscopy has even left the surface regime. MFRM promises to be one of the most powerful tools to reach our goal: the 3D-imaging of molecules, at surfaces or in bulk state, at atomic resolution.

7. ACKNOWLEDGEMENTS

Part of this work has been supported by the Austrian FWF (Project No. P-16178) and was carried out at Vienna University of Technology.

Part of this work was funded from the "Austrian Kplus-Program" and has been carried out at the "Austrian Center of Competence for Tribology", AC^2T research GmbH.

8. REFERENCES

1. Binnig G., Rohrer H., Gerber Ch. and Weibel E. Surface studies by scanning tunneling microscopy. Phys. Rev. Lett. 1982; 49: 57-61.
2. Binnig G., Quate C.F., and Gerber Ch. Atomic force microscope, Phys. Rev. Lett. 1986; 56: 930-933.
3. Pohl D.W., Denk W., and Lanz M. Optical stethoscopy: Image recording with resolution $\lambda/20$, Appl. Phys. Lett. 1984; 44(7): 651-653.
4. Morris V.J., Kirby A.R. and Gunning A.P. *Atomic Force Microscopy for Biologists.* World Scientific Publishing Company, 1999
5. Hartmann U. An elementary introduction to atomic force microscopy and related methods, PDF document from the Institute of Experimental Physics, Univ. of Saarbrücken (www.uni-saarland.de/fak7/hartmann/download/afm/afm.pdf)
6. Binnig G. and Rohrer H. The Scanning Tunneling Microscope. Scientific American 1985; 253: 50-56.
7. Quate C.F. Vacuum Tunneling: A New Technique for Microscopy. Physics Today 1986; 39: 26-33.
8. Binnig G. and H. Rohrer In touch with atoms. Rev. Mod. Phys. 1999; 71(2): S324-S330.
9. Hansma P.K., Cleveland J.P., Radmacher M., Walters D.A., Hillner P.E., Bezanilla M., Fritz M., Vie D., Hansma H.G., Prater C.B., Massie J., Fukunaga L., Gurley J. and Elings V. Tapping mode atomic force microscopy in liquids. Appl. Phys. Lett. 1994; 64(13): 1738-1740.
10. Prater C.B., Hansma P.K., Tortonese M. and Quate C.F. Improved scanning ion-conductance microscope using microfabricated probes. Rev. Sci. Instrum. 1991; 62(11): 2635-2638.
11. Hansma P.K., Drake B., Marti O., Gould S.A. and Prater C.B. The scanning ion-conductance microscope. Science 1989; 243(4891): 641-643.
12. Schraml S. Setup and application of a Scanning Ion Conductance Microscope, Diploma thesis, Vienna University of Technology, Austria, 2003

13. Werner D. *The biology of diatoms*. University of California Press, 1977
14. Round F.E., Crawford R.M. and Mann D.G. *Diatoms: Biology and morphology of the genera*. Cambridge University Press., 1990
15. Gebeshuber I.C., Thompson J.B., Del Amo Y., Stachelberger H. and Kindt J.H. In vivo nanoscale atomic force microscopy investigation of diatom adhesion properties. Materials Science and Technology 2002; 18: 763-766.
16. Gebeshuber I.C., Kindt J.H., Thompson J.B., Del Amo Y., Stachelberger H., Brzezinski M., Stucky, G.D. Morse D.E. and Hansma P.K. Atomic force microscopy study of living diatoms in ambient conditions. J. Microsc. 2003; 212: 292-299.
17. Smith B.L., Schäffer T.E., Viani M., Thompson J.B., Frederick N.A., Kindt J., Belcher A., Stucky G.D., Morse D.E. and Hansma P.K. Molecular mechanistic origin of the toughness of natural adhesives, fibres and composites. Nature 1999; 399: 761-763.
18. Gebeshuber I.C., Stachelberger H. and Drack M. Diatom bionanotribology - Biological surfaces in relative motion: their design, friction, adhesion, lubrication and wear. J. Nanosci. Nanotechnol. 2005; 5: 1-9.
19. Gebeshuber I.C., Stachelberger H. and Drack M. "Surfaces in relative motion: bionanotribological investigations." In: *First International Industrial Conference Bionik 2004,* I. Boblan and R. Bannasch, eds. Düsseldorf: VDI Verlag 2004. ISBN 3-18-324915-4, ISSN 0178-9589229-236, 229-236, 2004.
20. Hartl F.U. Molecular chaperones in cellular protein folding. Nature 1996; 381: 571-580.
21. Ranson N.A., White H.E. and Saibil H.R. Chaperonins. Biochem J. 1998; 333(2): 233-242.
22. Sigler P.B., Xu Z., Rye H.S., Burston S.G., Fenton W.A. and Horwich A.L. Structure and function in GroEl-Mediated protein folding. Annu. Rev. Biochem. 1998; 67: 581-608.
23. Fink A.L. Chaperone-Mediated Protein Folding. Physiol. Rev 1999; 79(2): 425 - 449.
24. Walters D.A., Cleveland J.P., Thomson N.H., Hansma P.K., Wendman M.A., Gurley G. and Elings V. Short cantilevers for atomic force microscopy. Rev. Sci. Instrum. 1996; 67(10): 3583-3590.
25. Viani M.B., Schäffer T.E., Chand A., Rief M., Gaub H.E. and Hansma P.K. Small cantilevers for force spectroscopy of single molecules. J. Appl. Phys. 1999; 86(4): 2258–2262.
26. Viani M.B., Schäffer T.E., Paloczi G.T., Pietrasanta L.I., Smith B.L., Thompson J.B., Richter M., Rief M., Gaub H.E., Plaxco K.W., Cleland A.N., Hansma H.G. and Hansma P.K. Fast imaging and fast force spectroscopy of single biopolymers with a new atomic force microscope designed for small cantilevers. Rev. Sci. Instrum. 1999; 70(11): 4300–4303.
27. Viani M.B., Pietrasanta L.I., Thompson J.B., Chand A.,Gebeshuber I.C., Kindt J.H., Richter M., Hansma H.G. and Hansma P.K. Probing protein-protein interactions in real time. Nature Structural Biology 2000; 7: 644-647.
28. Braig K., Otwinowski Z., Hegde R., Boisvert D.C., Joachimiak A., Horwich A.L. and Sigler P.B. The crystal structure of the bacterial chaperonin GroEL at 2.8 A. Nature 1994; 371: 557-559.
29. Xu Z., Horwich A.L. and Sigler P.B. The crystal structure of the asymmetric GroEL-GroES-(ADP)7 chaperonin complex. Nature 1997, 388: 741-750.
30. Roseman A.M., Chen S., White H., Braig K. and Saibil H.R. The chaperonin ATPase cycle: mechanism of allosteric switching and movements of substrate-binding domains in GroEL. Cell 1996; 87: 241–251.
31. Winter HP. and Aumayr F. Slow multicharged ions hitting a solid surface: From hollow atoms to novel applications. Europhys. News 2002; 6: 215-217.
32. Arnau A., Aumayr F., Echenique P.M., Grether M., Heiland W., Limburg J., Morgenstern R., Roncin P., Schippers S., Schuch R., Stolterfoht N., Varga P., Zouros T.J.M. and Winter HP. Interaction of slow multicharged ions with solid surfaces. Surf. Sci. Rep. 1997; 27: 113 – 240.

33. Sporn M., Libiseller G., Neidhart T., Schmid M., Aumayr F., Winter HP., Varga P., Grether M., Niemann D. and Stolterfoht N. Potential sputtering of clean SiO_2 by slow highly charged ions. Phys. Rev. Lett. 1997; 79: 945–948.
34. Aumayr F., Burgdörfer J., Varga P. and Winter HP. Sputtering of insulator surfaces by slow highly charged ions: "coulomb explosion" or "defect-mediated desorption"? Comm. Atom. Molecul. Phys. 1999; 34: 201-219.
35. Schenkel T., Hamza A.V., Barnes A.V. and D.H. Schneider Interaction of slow, very highly charged ions with surfaces. Progr. Surf. Sci. 1999; 61: 23-84.
36. Hayderer G., Cernusca S., Schmid M., Varga P., Winter HP., Aumayr F., D. Niemann, Hoffmann V., Stolterfoht N., Lemell C., Wirtz L. and Burgdörfer J. Kinetically-assisted potential sputtering of insulators by highly charged ions. Phys. Rev. Lett. 2001; 86: 3530-3533.
37. Hayderer G., Schmid M., Varga P., Winter HP., Aumayr F., Wirtz L., Lemell C., Burgdörfer J., Hägg L. and Reinhold C.O. Threshold for potential sputtering of LiF. Phys. Rev. Lett. 1999; 83: 3948-3951.
38. Hayderer G., Cernusca S., Schmid M., Varga P., Winter HP. and Aumayr F. STM studies of HCI-induced surface damage on highly oriented pyrolytic graphite. Physica Scripta 2001; T92: 156-157.
39. Gebeshuber I.C., Cernusca S., Aumayr F. and Winter HP. Nanoscopic surface modification by slow ion bombardment. Int. J. Mass Spectrometry 2003; 229(1-2): 27-34.
40. Gebeshuber I.C., Cernusca S., Aumayr F. and Winter HP. Slow multicharged-ion induced nanodefects on monocrystalline insulator surfaces studied by UHV-AFM. Nucl. Instr. Meth. Phys. Res. B 2003; 205: 751-757.
41. Hembacher S., Giessibl F.J., Mannhart J. Force microscopy with light atom probes. Science 2004; 305: 380-383.
42. Sidles J.A. Noninductive detection of single-proton magnetic resonance. Appl. Phys. Lett. 1991; 58: 2854-2856.
43. Rugar D., Budakian R., Mamin H.J. and Chui B.W. Single spin detection by magnetic resonance force microscopy. Nature 2004; 430: 329-332.
44. Manassen Y., Hamers R.J., Demuth J.E. and Castellano A.J. Direct observation of the precession of individual paramagnetic spins on oxidized silicon surfaces. Phys. Rev. Lett. 1989; 62: 2531–2534.
45. Durkan C. and Welland M.E. Electronic spin detection in molecules using scanning-tunnelingmicroscopy-assisted electron-spin resonance. Appl. Phys. Lett. 2002; 80: 458–460.
46. Wrachtrup J., von Borczyskowski C., Bernard J., Orritt M. and Brown R. Optical-detection of magnetic resonance in a single molecule. Nature 1993; 363: 244–245.
47. Köhler J., Disselhorst J.A.J.M., Donckers M.C.J.M., Groenen E.J.J., Schmidt J. and Moerner W.E. Magnetic resonance of a single molecular spin. Nature 1993; 363: 242–244.
48. Jelezko F., Popa I., Gruber A., Tietz C., Wrachtrup J., Nizovtsev A. and Kilin S. Single spin states in a defect center resolved by optical spectroscopy. Appl. Phys. Lett. 2002; 81: 2160–2162.
49. Elzerman J.M., Hanson R., Willems van Beveren L.H., Witkamp B., Vandersypen L.M.K. and Kouwenhoven L.P. Single-shot read-out of an individual electron spin in a quantum dot. Nature 2004; 430: 431-435.
50. Xiao M., Martin I., Yablonovitch E. and Jiang H.W. Electrical detection of the spin resonance of a single electron in a silicon field-effect transistor. Nature 2004; 430: 435-439.
51. Manoharan H.C. Applied physics: Spin spotting. Nature 2002; 416: 24-25.

IMAGING CELLS USING SOFT X-RAYS

Tom W.Ford
School of Biological Sciences, Royal Holloway, University of London, Egham, Surrey TW20 0EX, UK.

Abstract: A variety of microscopical methods are available to study the structure of cells. However, all of these have certain deficiencies such as low resolution, the inability to image whole cells, the possible introduction of artefacts into the final image due to chemical pre-treatment of the specimen or the restriction of the imaging to specific cell components. X-ray microscopy, especially using soft X-rays, has the potential to avoid most of these deficiencies by providing a method for examining the ultrastructure of intact, hydrated cells at high resolution and without chemical pre-treatment and therefore can provide a complementary method to other established microscopical techniques . Generation of soft X-rays from synchrotron radiation or plasmas produced by the impact of lasers on suitable targets is described, along with imaging systems such as contact imaging and transmission X-ray microscopy. The use of these techniques to study a variety of living cells is discussed along with more specialized techniques such as 3D imaging and elemental analysis.

Keywords: X-Rays, Electron Microscopy, Synchrotron Radiation, Laser Plasmas, Cell Ultrastructure, *Chlamydomonas*

1. THEORETICAL CONSIDERATIONS AND METHODS

1.1. Imaging cells using light microscopy

In order to study cells and their internal organization a degree of magnification is necessary to achieve the required resolution. The human eye has a maximum resolution of around 150μm whilst cells range in size from 1 to 100μm diameter. Although the largest of these cells could in theory be resolved by the human eye alone, in practice it is necessary to use magnifications of several hundred or several thousand times in order to study cells and their internal organization. The development of magnifying lenses e.g. by Leeuwenhoek and the compound microscope e.g. by Hooke in the mid-1600's enabled cells such as sperms and the cellular nature of plant

V. Evangelista et al. (eds.), From Cells to Proteins: Imaging Nature across Dimensions, 167–185.

tissues (cork) to be examined and drawn. Improvements in lens production and in the design of microscopes over the next 300 years, allowed cytologists to view a range of cell types from bacteria to human tissues. With the parallel development of techniques for fixing, sectioning and staining cells, the internal components could be described with increasing clarity. These studies commonly used bright field illumination where the light transmitted by the specimen was viewed by eye or captured on photographic film. Contrast in the image was achieved using a range of coloured dyes, some specific for certain cell components. Later advances in optical design produced microscopes capable of generating phase contrast and differential interference contrast so that cells could be viewed in their natural state and developmental processes such as nuclear division could be filmed in living cells.

Despite all these advances in microscope design and improvements in preparation techniques, the detail visible within cells was limited by the magnification and resolving power of light microscopy. The maximum resolution, defined by the source of the illumination, is approximately equal to half the wavelength of the irradiation and so for light microscopy this is around 250nm.

1.2. Imaging cells using electron microscopy

With the advent of transmission electron microscopy in the 1930's, and the increasing availability of commercial instruments from the 1950's onwards, the resolution limit was improved to around 2-5nm for most biological specimens. However, since electrons are readily absorbed by biological material, cellular detail could only be observed in ultrathin sections of around 50-100nm thickness. In addition, contrast needed to be enhanced by staining with the salts of heavy elements such as uranium and lead. This need for ultrathin sections required the design and development of methods which would ensure that, as far as possible, cells retained the appearance of their natural state and that artefacts were not introduced into the final image. All stages of the preparation process, fixation, dehydration, resin infiltration, sectioning and post-staining have the potential for causing structural distortion. There is well-documented evidence for the loss of cellular material especially during chemical fixation and subsequent dehydration of biological specimens [1, 2, 3, 4]. More recent techniques such as cryo-fixation avoid the need for some of these preparation stages but can cause structural disorganization from freezing damage.

1.3. Imaging cells using X-ray microscopy

At the time that electron microscopy was being developed, there was also interest in using X-rays as a source of illumination since X-rays had a much shorter wavelength (<10nm) than visible light and so should produce images of a much higher resolution. The potential of X-rays for biological imaging had been known since their discovery by Röntgen in 1895 who showed that they could penetrate soft body tissues but were strongly absorbed by dense structures such as bone. This was rapidly exploited in surgery to image bones and to identify foreign objects in biological tissues. However, such systems did not provide a magnified image and optical systems for focusing X-rays were not available. Beams of electrons could be readily focused using electromagnetic lenses and the production of commercial electron microscopes with their very high magnification and resolving power resulted in a rapid expansion of ultrastructural studies using these instruments. X-ray microscopy required two important developments before instruments could be constructed with the potential to image cells with a resolution approaching that of electron microscopy. These were sources of high intensity X-ray irradiation and X-ray optical systems capable of producing a fine focused beam of X-rays.

The main benefit to be gained from using X-ray microscopy, rather than electron microscopy, is that X-rays have a greater ability to penetrate biological specimens and therefore intact cells could be imaged so avoiding the need for chemical fixation and sectioning of the specimen. However, when using so-called soft X-rays the penetration limit is around 10μm so studies are potentially restricted to cells of around this diameter or sections of this thickness. X-ray microscopy also offers the potential for examining cells in a living, hydrated state since the specimen can be enclosed in an environmental chamber for imaging whilst TEM requires the section to be exposed to a high vacuum. The additional benefit of using soft X-rays, especially in the "water window" (2.4-4.4nm), is that a degree of natural contrast is achieved in cells so avoiding the need for staining. Within this water-window, there is high absorption of the X-rays by carbon but very little by oxygen with the absorption coefficient of carbon being around 10x that of oxygen (Figure 1). The degree of transmission of the X-rays through the specimen is therefore dependent on the carbon-density within the specimen. In a living cell, structures with high carbon density, such as membranes, would be expected to absorb more of this radiation than more carbon-dilute areas such as cytoplasm and vacuoles. The image produced is therefore essentially a carbon-density map of the specimen.

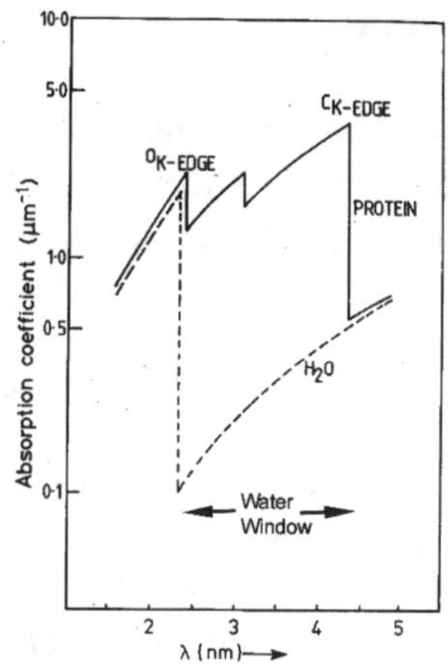

Figure 1. Soft X-ray absorption spectra of water and protein. In the "water window" (2.4-4.4nm) absorption by carbon is almost ten times that by oxygen.

1.4. Sources of soft X-rays

Two principal sources of soft X-rays have been used for microscopy involving biological specimens. These have been the subject of a number of reviews [5, 6, 7, 8, 9,10].

1.4.1. Laser-generated plasmas.

The impact of a laser beam onto a suitable target generates a plasma. The constitution of the radiation within that plasma is dependent on the target material and plasmas typically exist for nanoseconds. The popular target materials for soft X-ray microscopy are gold, molybdenum, tungsten and yttrium which generate plasmas with a high water-window radiation content. Sufficient intensity of soft X-rays must therefore be delivered in this one extremely short pulse and, for this reason, much of the X-ray microscopy using plasma sources has required the use of high-energy laser facilities. However, laboratory (table-top) systems have also been developed [9] which are much more compact and provide easier user access. National high-energy laser facilities are multi-user which can create difficulties for biologists working with labile organisms.

1.4.2. Synchrotron radiation

When electrons are forced into a circular orbit, they emit radiation across the electromagnetic spectrum, including soft X-rays. The electrons are accelerated effectively to the speed of light via a linear accelerator and booster synchrotron before they enter the storage ring where the beam can be maintained for several hours at constant energy. The central cavity of the synchrotron is maintained at ultrahigh vacuum and consists of a series of arcs and straight sections. The arcs contain bending and focusing magnets which maintain the curved trajectory of the electrons and force them into a tight beam of around 50-100μm diameter. As the beam emerges from an arc it emits radiation which travels down a beamline to an experimental station. Along the beamline, a series of optics and monochromators are used to select the required wavelengths and further focus the X-ray beam. Within a soft X-ray beamline, focusing of the beam is achieved using Fresnel zone plates. These consist of a circular grating of alternating soft X-ray-transparent and opaque rings where the spacing diminishes with distance from the centre of the ring. The resolution of the microscope is limited by the width of the smallest zone spacing.

1.5. Soft X-ray imaging systems

Three principal systems have been developed for imaging biological specimens using soft X-rays. Their design is, to some degree, related to the source of the radiation.

1.5.1. Soft X-ray contact microscopy (SXCM)

This was developed principally for use with high-energy, laser-generated plasmas [6] but has also been successfully used with smaller table-top systems. By their nature, these plasmas contain a high intensity of radiation which is mainly soft X-rays providing that a suitable target material has been employed. The radiation is also of extremely short duration, usually only a few nanoseconds. There is therefore a single image capture which freezes any specimen movement but also destroys the specimen. The nanosecond duration of the pulse, and hence image capture, also avoids radiation damage-induced artefacts appearing in the final image. Images were originally captured on photographic film in the first "projection" X-ray microscopes but that has been superseded by photoresists coated with a polymer such as polymethylmethacrylate (PMMA) which forms a higher resolution image around 100x better than photographic film. Exposure of the specimen to the laser-generated soft X-ray-rich plasma takes place inside an evacuated target chamber. To prevent desiccation, the specimen and photoresist are enclosed within an environmental sample holder (Figure 2). The holder is sealed using a silicon nitride window, which is transparent to soft X-rays. Restriction of the radiation, used to form the image, to the water window is the resultant of the plasma composition, silicon nitride window transmission and sensitivity of the PMMA resist.

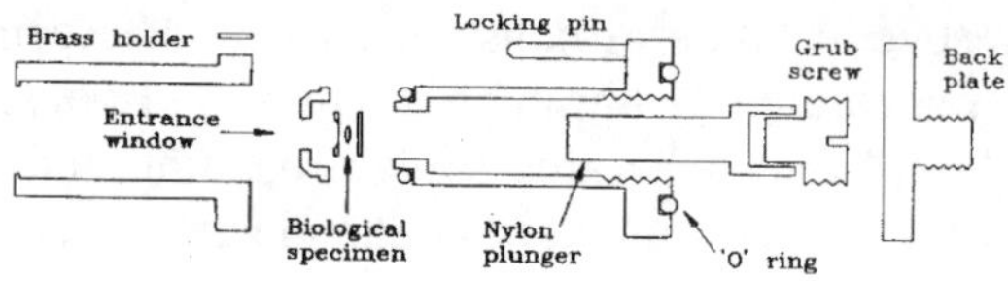

Figure 2. Diagram of an environmental sample holder for imaging living biological specimens using SXCM. The specimen is sandwiched between the silicon nitride window and a photoresist with close contact maintained using a nylon plunger. Once sealed, normal environmental conditions are maintained around the specimen during its period in the evacuated imaging chamber.

The latent image in the photoresist results from the absorbed X-rays generating secondary electrons, which break the local polymer chains such that they are more easily dissolved using organic solvents (Figure 3). The image is revealed following chemical development (methylisobutylketone in isopropyl alcohol) which dissolves away most rapidly areas of the photoresist not covered by the specimen i.e. the higher the soft X-ray fluence received by a particular area of the resist, the more prone that area is to chemical dissolution. Areas covered by increasing density of carbon are progressively more difficult to dissolve.

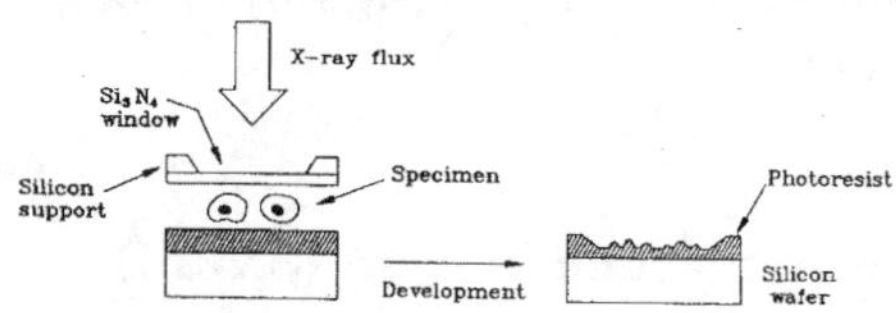

Figure 3. Diagram showing image formation in the photoresist during SXCM and its subsequent development using organic solvents to reveal a carbon map of the specimen.

Examination of the image, which is a 1:1 contact representation of the specimen, can be by incident interference light microscopy though this produces low resolution images and is normally only used to monitor the progress of resist development. Higher resolution final images can be produced by scanning electron microscopy (SEM), though this can cause electron ablation of the photoresist if an area is exposed to the electron beam for too long. The readout method of choice is currently atomic force microscopy (AFM) which is capable of producing the highest resolution readout without damage to the photoresist. In addition, the resist can be subjected to further chemical development following AFM readout which is not possible when the resist is viewed using SEM.

1.5.2. Soft X-ray transmission microscopy (SXTM)

These systems, sometimes called full field microscopy, irradiate the whole of the specimen with monochromatic soft X rays from a synchrotron.

In many ways they are analogous to a typical light microscope or transmission electron microscope (TEM) in their basic construction. The beam of radiation from the synchrotron enters the beamline where it is monochromated to a defined wavelength within the water-window and focused onto the specimen using a condenser zone plate (Figure 4).

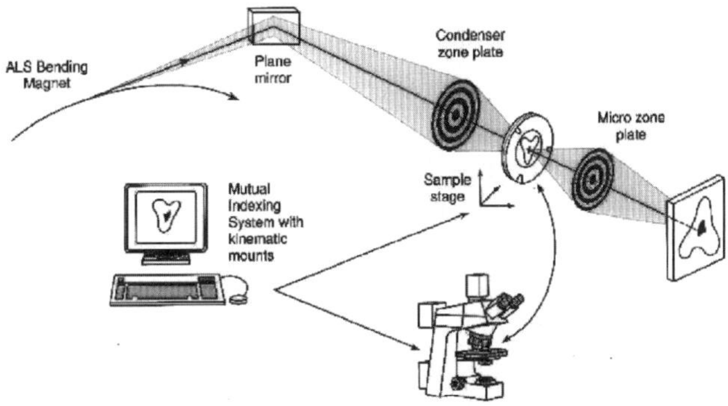

Figure 4. Diagram of the SXTM imaging system on the XM-1 beamline at the Advanced Light Source. Location of suitable specimens using the light microscope, and storage of their co-ordinates, enables accurate location of specimens in the beamline before exposure of the specimen to soft X-rays.

The wavelength for imaging is selected by moving the condenser zone plate relative to the specimen. The specimen, although not in vacuum, is enclosed within an environmental chamber since, in the beamline, it is held within a stream of helium in order to enhance the transmission of soft X-rays onto, and out from, the specimen. The environmental chamber consists of two stainless steel discs each with a central silicon nitride window with the specimen sandwiched between the two windows. The soft X-rays transmitted by the specimen are re-focused using an objective micro zone plate onto the image capture system which is a CCD camera with the image being displayed on a monitor and stored digitally on a computer. The transmission X-ray microscope XM-1 at the Advanced Light Source has a light microscope for selection of fields of view and storage of their co-ordinates. Once the sample holder is placed in the X-ray microscope, it can be driven to these stored co-ordinates [11, 12].

1.5.3. Scanning soft X-ray microscopy

In a scanning soft X-ray microscope, the zone plate focuses the beam to a small spot which is scanned across the specimen in lines with the transmission at each pixel being recorded. In basic construction it is analogous to a scanning electron microscope. Like the SEM, the size of the scanning spot determines the resolution of the microscope. As with the transmission X-ray microscope, the specimen is enclosed in an environmental chamber but in an air gap within the microscope beamline. These instruments are used principally for spectroscopic studies since, compared with SXTM, they have inferior spatial resolution and longer imaging times.

1.6. Specimen damage

As with all types of microscopy, there is always the danger that specimen manipulation and irradiation will introduce artefacts into the image obtained by soft X-ray microscopy. The image obtained will therefore be of little value since it will not be a true representation of the cell in its original state. Problems of chemical damage can be avoided by using living specimens but two major sources of potential damage remain.

1.6.1. Specimen desiccation

Irradiation of a specimen with soft X-rays requires that the specimen is in a vacuum chamber (SXCM) or held in a stream of helium (SXTM). Both of these conditions would result in specimen desiccation especially since the loading, orientation and irradiation of the specimen can take several minutes. In both systems, this is avoided by sealing the sample within a holder, the inside of which is at atmospheric pressure and maintains the normal hydrated environment of the sample.

1.6.2. Radiation damage

X-rays are absorbed by cell components which can result in local heating and chemical disruption. In the long term, chromosomal damage can result in cell and organism death. Over the much shorter time scale of soft X-ray microscopy, damage is largely to cell membranes and it is thought to be mediated via the production of free radicals.

Experiments designed to assess radiation dosages which can cause ultrastructural damage, have been conducted using the unicellular green alga *Chlorella*. Monolayers of cells were exposed to soft X-rays within the water window in a synchrotron beamline before being examined using transmission electron microscopy [13]. Cells left in the environmental holder in the beamline, even for periods of up to one hour, but not exposed to radiation showed no structural damage (Figure 5A). When cells were exposed to radiation of 3.27nm (380eV) it was found that dosages in excess of 10^3Gy caused disruption to chloroplast ultrastructure though the nucleus appeared intact (Figure 5B). Dosages greater than 10^4Gy resulted in complete loss of cellular internal structure.

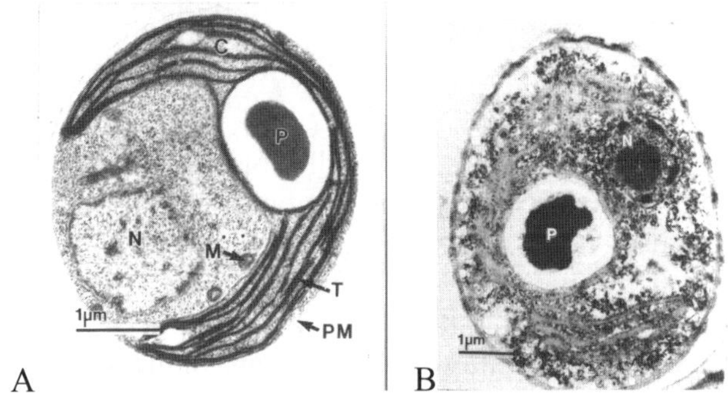

Figure 5. Transmission electron micrographs of Chlorella cells showing the typical appearance of a cell before X-ray irradiation (A) and following exposure to 10^3Gy of 3.27nm radiation (B). C - chloroplast; N - nucleus; M - mitochondrion; P - pyrenoid; PM - plasma membrane; T - thylakoids. Reprinted with permission from Ford et al. (1993).

Dosages typically absorbed by specimens during X-ray microscopy have been estimated to be between 10^4-10^7Gy [14]. This problem of radiation damage has resulted in many of the early studies being carried out using chemically fixed samples. However, this negates the major potential benefit of soft X-ray microscopy which is the ability to using living untreated cells.

Soft X-ray contact microscopy avoids the problem of radiation damage by the rapid exposure and image collection time of only nanoseconds. However, the sample is destroyed in the process so multiple images of the same cell are not possible. Scanning soft X-ray microscopy exposes very small areas of the sample to small doses as the spot is scanned across, so it has been suggested that this would avoid radiation damage. However, during the relatively long time needed to collect an image, any free radicals generated during the first pixel exposure could result in damage to other parts of the cell before they are imaged later in the scan. In transmission X-ray microscopy, radiation damage is potentially a problem due to the soft X-ray fluence needed to expose the whole cell. It can be avoided to some degree by using very short imaging times but this can result in images of poor definition due to the small number of photons collected by the CCD camera. The alternative is to use frozen specimens and this does seem to be a way of avoiding damage but still producing useful, high-resolution images.

2. APPLICATIONS TO THE STUDY OF CELL ULTRASTRUCTURE AND ORGANIZATION

2.1. Introduction

Soft X-ray microscopy offers the cell biologist an alternative method for imaging cell structure which is complementary to well-established techniques such as transmission electron microscopy (TEM). Although the potential spatial resolution of soft X-ray microscopy is not as high as for electron microscopy, soft X-rays have a higher penetration of biological tissues than electrons so allowing specimens of up to 10μm thick to be imaged intact. Even this ability to use relatively thick specimens does have some limitations if cells are to be intact and untreated when examined. Micro-organisms will in general be suitable as will cell cultures obtained from either animal or plant tissues where the specimen can be mounted as a monolayer. Otherwise tissues in either animals or plants will need to be dissected out and sectioned.

Most imaging studies to date have used whole micro-organisms or cultured single cells to compare the images generated by soft X-ray microscopy with those produced by TEM. One potentially important role of soft X-ray microscopy is therefore as a complementary technique to TEM by providing images obtained by an alternative system and one which is using cells in their natural state. Such images can be used to support ultrastructural interpretation of the ultrathin sections used for TEM. This will allow the cell biologist to be more confident in their interpretation of cell ultrastructure seen by TEM since it is unlikely that identical features seen by two quite different techniques are the result of artefacts. However, some unusual features have been seen using soft X-ray microscopy which do differ from images produced using TEM.

2.2. Ultrastructural studies on living cells

2.2.1. *Chlamydomonas*

Chlamydomonas is a motile, unicellular green alga which has been used extensively for at least half a century as a model organism for genetics and research on photosynthesis. It has also been studied at the ultrastructural level using TEM [15]. Cells range in size from around 5-10μm in diameter and are therefore of ideal size for soft X-ray microscopy. Within the cell is a single, large, cup-shaped chloroplast with the nucleus within the cytoplasm of the "cup", whilst in the region of the chloroplast at the posterior of the cell there is a large, dense pyrenoid. The two flagella are inserted at the anterior end of the cell. The plasma membrane of the cell is covered by a glycoprotein layer, rather than a cellulose cell wall as in most plant cells. Their motility is not a problem for SXCM since the nanosecond exposure time "freezes" the image. For SXTM, enclosure of the cells between the silicon nitride windows of the sample holder is usually sufficient to stop movement.

Early studies on live *Chlamydomonas* cells used SXCM due to its extremely short exposure time and its ability to image motile specimens. The relatively long exposure times of scanning and transmission soft X-ray microscopy at that time proved problematical for imaging living cells mainly due to the increased likelihood of radiation damage-induced artefacts. SXCM images of living *Chlamydomonas* cells, as revealed by AFM readout of the developed resist, showed features such as the cell covering, flagella insertion, location of the chloroplast and a number of relatively large, soft X-ray-dense spherical inclusions (Figure 6) [16].

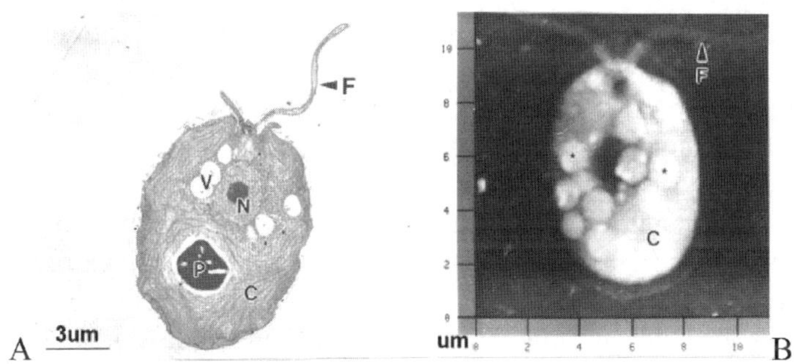

Figure 6. Images of Chlamydomonas cells taken using either TEM (A) or SXCM (B). The numerous soft X-ray-dense spheres visible in (B) (*) have no obvious counterpart in the TEM image except for the highly electron-transparent vacuoles (V). N - nucleus; C - chloroplast; P - pyrenoid; F - flagellum. Reprinted with permission from Ford et al (1995).

These inclusions, or spheres, were unlike any structures seen in conventional electron micrographs and their role is uncertain. It has been suggested that they may be membrane-bound vesicles (which would be supported by later studies using SXTM) possibly involved in the storage of polyphosphates or calcium though these are usually visible in electron micrographs. They are certainly very soft X-ray dense indicating a high concentration of soft X-ray absorbing material. Since it is likely that such material would also result in a high degree of electron scattering, it might be expected that these vesicles would appear dark in electron micrographs. However, vesicles or vacuoles observed in electron micrographs of these cells are often very pale indicating little, if any, electron scattering (Figure 6A). The apparent absence of these spheres in electron micrographs is puzzling since they are membrane-bound and relatively large compared with the overall size of the cell. It is possible that they are destroyed during fixation or the later chemical treatments needed to prepare the cells for TEM. Little else could be identified in SXCM images of *Chlamydomonas* and fine structural detail was probably obscured by the "compaction" of information during the formation of the carbon-density map of the cell.

SXTM is closer in principle to light microscopy and TEM and therefore offered the possibility of images which could be easier to interpret and perhaps show more detail, though never to the resolution of TEM. The main problem with SXTM concerned the possibility of artefacts being introduced into the image by radiation damage during the exposure of the cell to soft X-rays. Since the specimen is not destroyed (at least in theory) when using SXTM (unlike when using SXCM) it should be possible to produce a sequence of images by successive exposures of the same cell. This would allow dynamic processes to be studied in living cells. However, for such studies to be meaningful, the images obtained during the sequence of exposures must not accumulate artefacts resulting from radiation damage. It was therefore necessary to know, not only the maximum exposure of the cell which would produce artefact-free images, but also the number of reliable sequential images which could be produced. Very short exposures are least likely to cause image artefacts though, since the photon count accumulated by the CCD camera will be very low, the image produced will have a poor signal:noise ratio and consequently low resolution and contrast.

Experiments with living *Chlamydomonas* cells have shown that only very short exposures of around 2s or less produced images which were clearly free from radiation-induced artefacts [17]. However, even these short exposures did not permit sequential imaging since damage was clearly visible in the third and fourth images (Figure 7).

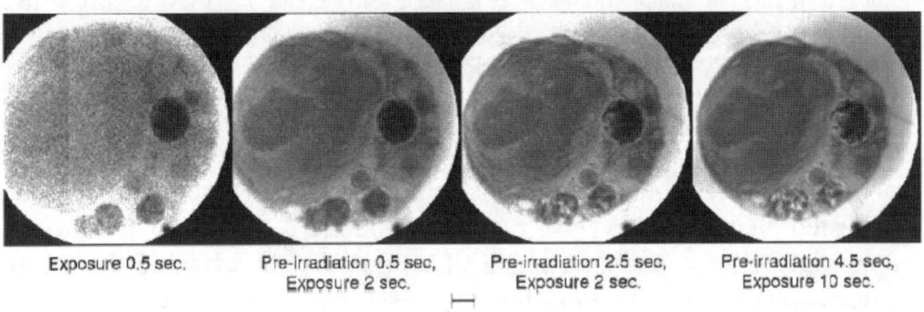

Figure 7. SXTM images of a living Chlamydomonas cell following four sequential exposures. The second image (of longer exposure) shows improved resolution over the first, but the third and fourth images show increasing structural damage, especially to the soft X-ray-dense spheres.

The spheres observed in SXCM images were again clearly visible in SXTM images and proved to be very radiation-sensitive and were therefore good indicators of early structural damage. One possible solution to the radiation damage problem was to use samples which had been subjected to "brief" or "mild" glutaraldehyde fixation. Although such cells are more resistant to radiation damage, the glutaraldehyde treatment itself can cause damage to cell ultrastructure especially the spheres (Figure 8). A more satisfactory alternative is to use frozen samples (see Section 2.3).

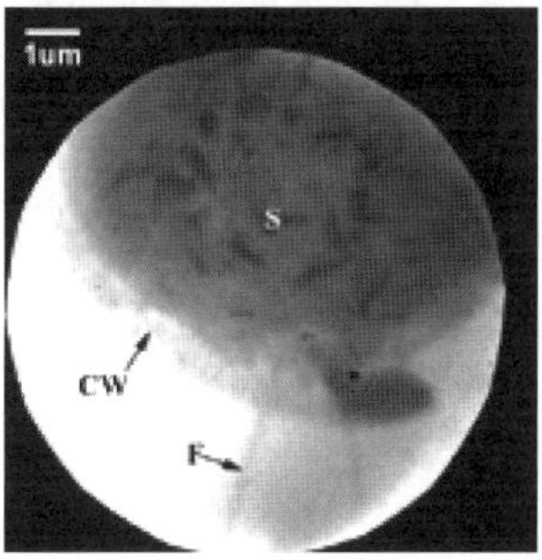

Figure 8. SXTM image of a Chlamydomonas cell following brief (5 minutes) fixation with 1.25% glutaraldehyde. CW - cell wall; F - flagellum; S - spheres

The other benefit of SXTM over SXCM is that image collection and display is immediate allowing rapid evaluation of the image and a higher throughput of exposures. With SXCM, complete development of the image can take tens of minutes even with image evaluation only by light microscopy. On-site access to an AFM can produce a more detailed image evaluation but this adds even more time to the procedure.

2.2.2. Yeast

Yeast is a unicellular fungus consisting of a thick cell wall surrounding a cytoplasm containing nucleus, mitochondria, vacuoles etc. The presence of this very thick wall has caused problems for preparation of cells for TEM and prior removal of the wall is usually required before fixation. Asexual reproduction takes place by budding where, following nuclear division, cell contents bulge through a specific point in the cell wall and, following expansion, the two cells usually split apart.

SXCM studies on living yeast cells have revealed cells with thick walls which have bulges showing incipient budding and cytoplasmic vesicles (Figure 9) [18].

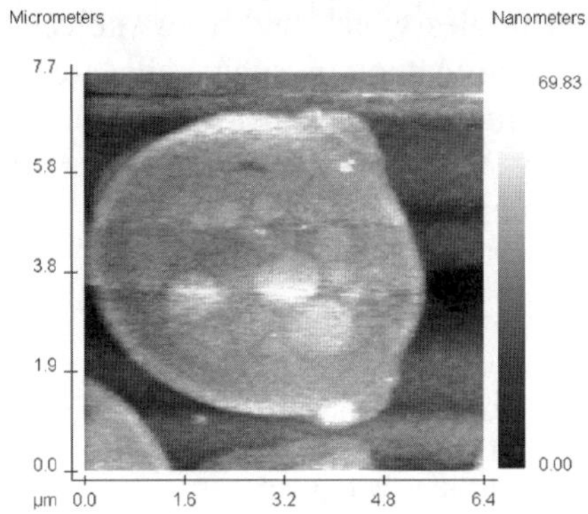

Figure 9. SXCM image of a living yeast cell showing the thick cell wall with two bulges indicating incipient budding. The cytoplasm contains a number of soft X-ray transparent vesicles. Reprinted with permission from Batani et al (2002).

2.2.3. Mammalian sperm

Mammalian sperm consists of a head region containing the nucleus which will eventually fuse with the egg nucleus. The sperm cell swims by means of a long "tail" consisting of a single, membrane-bound flagellum with the power provided by a helical mitochondrion in the "mid-piece" region. Also within this region is a ring of nine outer dense fibres, each around 100nm diameter, which surround the 9+2 arrangement of the microtubules which form the flagellum.

SXCM images of living pig sperm show clearly these three regions (Figure 10A) and, in some images, the outer dense fibres are visible where they join the head (Figure 10B) [18]. SXTM images have shown vesicles in the head and mid-piece region of human sperm along with the helical mitochondrion. These vesicles at the mid-piece region appear spherical and almost transparent to soft X-rays. Changes in the osmotic environment of the sperm causes swelling and contraction in this mid-piece region [19].

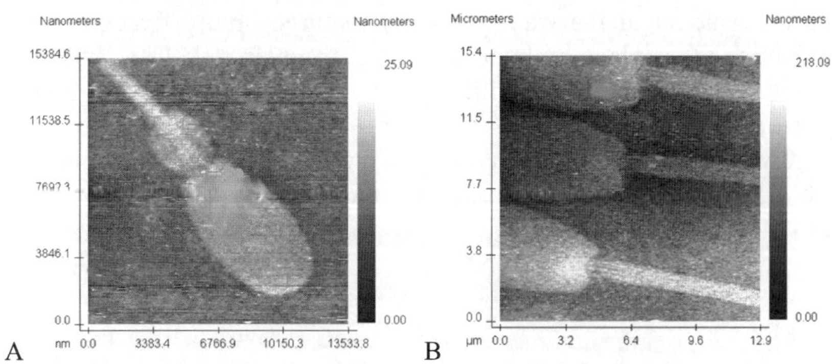

Figure 10. SXCM images of living pig sperm showing the head region, mid-piece region and tail (A). Some of the outer dense fibres surrounding the flagellum microtubules in the mid-piece region are also visible in some of the images (B). Reprinted with permission from Batani et al (2002).

2.3. Imaging frozen specimens

One approach to resolving the problem of radiation damage during soft X-ray imaging has been to freeze the specimen prior to irradiation. It has been claimed that such specimens can withstand cumulative soft X-ray doses of up to 10^9-10^{10}Gy without observable damage to the ultrastructure of the cell [20]. This would allow the use of high fluences of radiation for longer periods so increasing photon collection and improving image quality and resolution. It should also allow multiple images of the same cell, which would be necessary for the construction of 3D images and elemental analysis.

Frozen biological specimens are currently used in electron microscopy and cryopreservation of micro-organisms is routinely used in culture collections. Since water can account for between 40-90% of a biological sample, freezing can cause dehydration, and hence plasmolysis and distortion of cellular contents. It can also result in the formation of ice crystals within the cell, which may disrupt membranous components. Micro-organisms such as unicellular algae tend to have a lower water content than cells of higher plants which are often highly vacuolate. However, the possible hazards associated with freezing still remain. Ultrastructure is preserved best when the water is in a vitreous state, though the formation of this requires rapid freezing at rates up to 10^5C.s^{-1}. Cryo-imaging of *Chlamydomonas* at the ALS on beamline XM-1 used the conventional sample holder where the cells were sandwiched between two silicon nitride windows. The sample was frozen by two jets of helium, cooled by passage through liquid nitrogen, one onto each window. This achieved a freezing rate of 10^3C.s^{-1} and a final sample temperature of -150C, which was maintained during imaging [21]. Cells exposed to soft X-ray radiation doses of up to 2x10^7Gy showed no visible radiation damage These images revealed good internal detail including the insertion of the two flagella, thylakoids within the chloroplast and a number of intact soft X-ray-dense spheres in the cytoplasm (Figure 11).

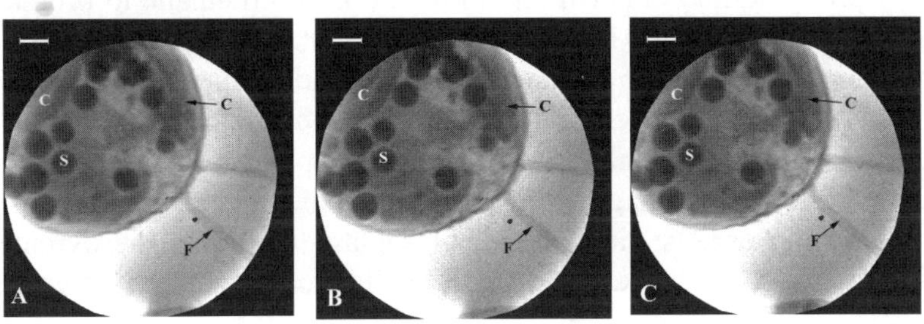

Figure 11. Sequential SXTM images of a frozen Chlamydomonas cell. Accumulated dosage of soft x-rays (A) 4x10^6Gy (B) 6x10^6Gy (C) 8x10^6Gy. C-chloroplast; F-flagellum; S-sphere. Bar = 1.m.

Multiple imaging of the same cell was also possible which enabled montages of larger cells to be produced by moving the cell slightly between exposures. One problem encountered was that ice formation tends to occur on the outside of the silicon nitride windows. To reduce this, a slower freezing rate was used (10-100C.s^{-1}) and the frozen sample was held only at -50C. Spheres in these slow-frozen cells, although still intact after a dosage of 10^7Gy, did have a clear, soft X-ray-transparent halo around them suggesting some shrinkage.

2.4. 3-D imaging

One of the main advantages of using soft X-rays to image cells is that they have a higher specimen penetration than electrons and can be used for whole cells up to ~10μm diameter. This provides the potential for ultrastructural studies on the spatial organization of components within whole cells. Similar studies have been attempted for decades using TEM but this has required the serial sectioning of a single cell followed by image reconstruction. Relatively few studies have been undertaken mainly due to the laborious nature of the technique and difficulties with image reconstruction though reconstruction of serial images of the unicellular green alga *Chlorella* was used to show the presence of a single, convoluted mitochondrion within the cell [22].

3-D or tomographic imaging has been carried out using cells such as *Chlamydomonas* and SXTM using the soft X-ray microscope on the BESSY synchrotron. The sample must be frozen to avoid radiation-induced artefacts since sequential imaging is required and the whole process can take several hours. The cells were enclosed in capillaries of borosilicate glass which has relatively low soft X-ray (2.4nm) absorption [23]. The capillary had an internal diameter of ~8μm, allowing only a single row of cells within the capillary, and was tapered towards one end. Due to capillary action, when the cells were loaded they moved towards the tip. The capillary was cooled rapidly by plunging into either liquid nitrogen or liquid ethane to produce a vitrified state and avoid ice crystal formation. The frozen capillary containing the sample (at -150C) was rotated around an axis perpendicular to the optical axis of the microscope. Images were taken at a number of sequential viewing angles in a 180 rotation. Computer re-construction was used to generate a 3-D image from the individual tilt images. Within this image, the spatial organization of the chloroplast (including the pyrenoid), soft X-ray-dense spheres (vesicles) and flagellar roots could be determined (figure 12).

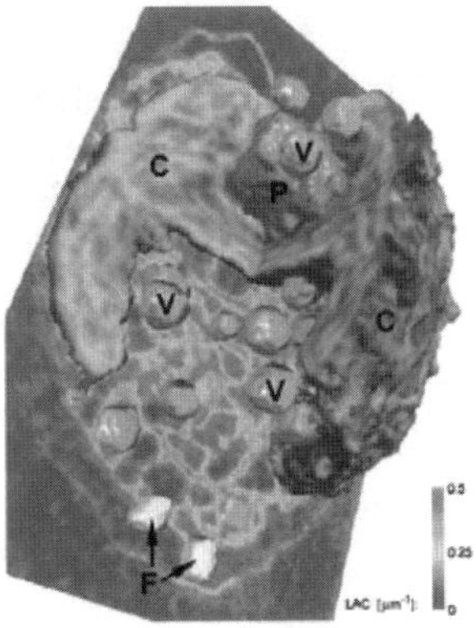

Figure 12. 3-D reconstruction of a single Chlamydomonas cell following imaging of a frozen cell at sequential viewing angles. C-chloroplast; F-flagellar roots; P-pyrenoid; V-vesicles Reprinted with permission from Weiss et al. (2000).

2.5. Elemental analysis

The identification of the location of specific elements at the ultrastructural level is a powerful tool for the cell biologist. X-ray microanalysis is a well-established technique used in both TEM and SEM. The identification of intracellular sites uses TEM and so requires ultrathin sections of the specimen which can make 3-D interpretation difficult. In addition, the possible movement of mobile elements during sample preparation could lead to erroneous conclusions.

SXTM offers the possibility of locating specific elements in whole hydrated cells. Due to their higher resistance to radiation damage, this has involved once again the use of frozen specimens. The cell is imaged either side of the absorption edge of the element being studied and an elemental map can be produced by subtraction of the two images. Cell structures containing large amounts of the element will absorb strongly at its absorption edge and will be highlighted in the resulting image.

Preliminary studies were carried out on the distribution of oxygen in cells of *Chlamydomonas* [21]. The sample was imaged first at 2.4nm and then at 2.3nm (the absorption edge of oxygen). The main structures highlighted in the resulting elemental map were the soft X-ray-dense spheres suggesting that they contained material that was particularly oxygen-rich (Figure 13).

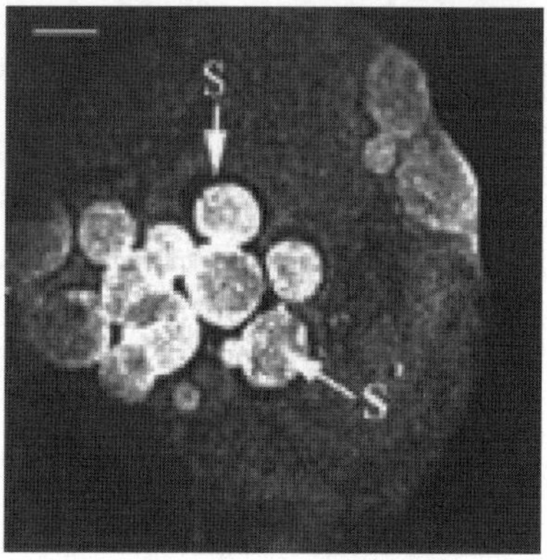

Figure 13. Elemental mapping of oxygen in an SXTM image of Chlamydomonas. Subtraction of the image at 2.4nm from that obtained at 2.3nm highlights areas of high oxygen content which are mainly the spheres (S). Bar = 1.m. Reprinted with permission from Ford et al (2000).

3. CONCLUSIONS

Soft X-ray microscopy provides the cell biologist with a system for studying the ultrastructural organization of cells without the need for sectioning (and its associated chemical pre-treatments). The use of frozen cells is advantageous since it considerably reduces radiation-induced artefacts and also offers the potential for sequential imaging and hence 3-D reconstruction or elemental analysis within intact cells. It is not a technique which will replace conventional electron microscopy, but should be seen rather as a complementary imaging method which can confirm or contradict cellular organization deduced from electron microscopic studies. In addition, it is a technique which requires no specific labelling of cell components and so can be used to study the whole architecture of the interior of the cell rather than a few selected components.

Although small X-ray microscopes are available in several laboratories these are only capable of contact imaging. The main focus of soft X-ray microscopy is currently using transmission mode and such systems are synchrotron-based in large national facilities. The future development of biological soft X-ray microscopy therefore requires continued, and increasing, input from the biological community both nationally and internationally.

4. REFERENCES

1. Coetzee.J. and van der Merwe,C.F. Extraction of substances during glutaraldehyde fixation of plant cells. Journal of Microscopy 1984; 135:147-58.
2. Coetzee.J. and van der Merwe,C.F. Extraction of carbon 14-labelled compounds from plant tissue during processing for electron microscopy. Journal of Electron Microscopy Technique 1989; 11: 155-60.
3. Mersey,B and McCully,M.E. Monitoring the course of fixation of plant cells. Journal of Microscopy 1978; 114: 49-76.

4. Salema,R. and Brandao,I. The use of PIPES buffer in the fixation of plant cells for electron microscopy. Journal of Submicroscopic Cytology 1973; 5: 79-96.
5. Eason,R.W., Cheng,P.C., Feder,R., Michette,A.G., Rosser,R.J., O'Neill,F., Owadano,Y., Rumsby,P.T., Shaw,M.J. and Turcu,I.C.E. Laser X-ray microscopy. Optica Acta 1986; 33: 501-516.
6. Ford,T.W., Stead,A.D. and Cotton,R.A. Soft X-ray contact microscopy of biological materials. Electron Microscopy Reviews 1991; 4: 269-292.
7. Howells,M.R., Kirz,J. and Sayre,D. X-ray microscopes. Scientific American 1991; 264/2: 42-48.
8. Kirz,J. Jacobsen,C. and Howells,M. Soft X-ray microscopes and their biological applications. Quarterly Review of Biophysics 1995; 28: 33-130.
9. Majima,T., Shimizu,H. and Tomie,T. Recent advances in soft X-ray microscopy for living specimens. Bioimages 1999; 7: 59-67.
10. Michette,A.G. X-ray microscopy. Reports on Progress in Physics 1988; 51: 1525-1606
11. Meyer-Ilse, W., Medecki,H., Jochum,L., Anderson,E., Attwood,D., Magowan,C., Balhorn,R. and Moronne,M. New high-resolution zone-plate microscope at Beamline 6.1 of the ALS. Synchrotron Radiation News 1995; 8: 29-33.
12. Meyer-Ilse,W., Denbeaux,G., Johnson,L.E., Bates,W., Lucero,A. and Anderson,E.H. The high resolution X-ray microscope, XM-1. X-Ray microscopy 2000; 129-134 (American Institute of Physics).
13. Ford,T.W., Page,A.M., Foster,G.F. and Stead,A.D. Effects of soft X-ray irradiation on cell ultrastructure. SPIE Proceedings - Soft X-ray Microscopy 1993; 1741: 325-332.
14. Shinohara,K. and Ito,A. Radiation damage in soft X-ray microscopy of live mammalian cells. Journal of Microscopy 1991; 161: 463-472.
15. Harris,E.H. *The Chlamydomonas Sourcebook* San Diego: Academic Press, 1989.
16. Ford,T.W., Cotton,R.A., Page,A.M., Tomie,T., Majima,T. and Stead,A.D. High resolution imaging of the ultrastructure of living algal cells using soft X-ray contact microscopy. SPIE Proceedings - Applications of Laser Plasma Radiation II 1995; 2523: 212-220.
17. Ford,T.W., Page,A.M., Meyer-Ilse,W., Brown,J.T., Heck,J. and Stead,A.D. A comparative study of the ultrastructure of living cells of the green alga *Chlamydomonas* using both soft X-ray contact and direct imaging systems and an evaluation of possible radiation damage. X-ray Microscopy and Spectromicroscopy 1998; II 185-190 (Springer).
18. Batani,D., Botto,C., Moret,M., Milani,M., Lucchini,G., Eidmann,K., Cotelli,F., Lora Lamia Donin,C., Poletti,G., Ford,T. and Stead,A. The use of high energy laser-plasma sources in soft X-ray contact microscopy of living biological samples. European Physics Journal D 2002; 21: 167-179.
19. Abraham-Peskir,J.V. The Aarhus X-ray microscope. Synchrotron Radiation News 2003; 16: 28-31.
20. Schneider,G. and Niemann,B. Cryo X-ray microsocopy experiments with the X-ray microscope at BESSY. X-Ray Microscopy and Spectromicroscopy 1998; I 25-34 (Springer).
21. Ford,T.W., Meyer-Ilse,W. and Stead,A.D. Development and evaluation of cryo-imaging of unicellular algae using soft X-ray transmission microscopy: ultrastructure and elemental analysis. X-Ray Microscopy 2000; 119-122 (American Institute of Physics).
22. Atkinson,A.W. Jr., John,P.C.L. and Gunning,B.E.S. The growth and division of the single mitochondrion and other organelles during the cell cycle of *Chlorella*, studied by quantitative stereology and three dimensional reconstruction. Protoplasma 1974; 81: 77-109.
23. Weiss,D., Schneider,G., Niemann,B., Guttmann,P., Rudolph,D. and Schmahl,G. Computed tomography of cryogenic biological specimens based on X-ray microscopic images. Ultramicroscopy 2000; 84: 185-197.

FROM MICROSCOPY TO NANOSCOPY: HOW TO GET AND READ OPTICAL DATA AT SINGLE MOLECULE LEVEL USING CONFOCAL AND TWO-PHOTON EXCITATION MICROSCOPY.

Alberto Diaspro, Paolo Bianchini, Valentina Caorsi, Davide Mazza, Mattia Pesce, Ilaria Testa, Giuseppe Vicidomini, Giuseppe Chirico, Fabio Cannone, Cesare Usai.
LAMBS-MicroScoBio Research Center - IFOM, Department of Physics, University of Genoa, Via Dodecaneso, 33 16146 Genoa Italy
LAMBS (Laboratory for Advanced Microscopy, Bioimaging and Spectroscopy).
MicroScoBio (Microscopia e Spettroscopia Correlativa per Applicazioni in Biomedicina e Oncologia).
IFOM (FIRC Institute of Molecular Oncology).

Abstract. The application fluorescence to confocal and two-photon excitation (2PE) optical microscopy has led to terrific advances in the study of biological systems from the three-dimensional (3D) micro-spectroscopic level down to single molecule detection (SMD) schemes. Both techniques are particularly relevant for the study of the 3D and dynamic properties of biological molecules within their natural environment, cells or tissues. In particular the advent of 2PE mitigates overall photobleaching and phototoxicity problems, opening new perspectives by providing further attractive advantages. Optical schemes and architectures for confocal and two-photon excitation from microscopic level to SMD will be discussed. Examples of three-dimensional and multiple fluorescence imaging from cells to single fluorescent molecules will be given. Examples in the utilization of confocal and 2PE for specific GFP switching at single molecule level and for monitoring of TPE uncaging will be shown.

Keywords: Confocal microscopy, Two-photon Excitation Microscopy, Single Molecule Detection, Fluorescence.

1. INTRODUCTION

Optical microscopy still occupies a unique niche in studying biological systems due to its ability to allow their examination from macromolecules and organelles to cells and tissues under experimental conditions that are

V. Evangelista et al. (eds.), From Cells to Proteins: Imaging Nature across Dimensions, 187–207.

very close to living or physiological states. This evidence coupled to fluorescence labelling permits the study of the complex and delicate relationships existing between structure and function in biological systems [1-4]. Nowadays, optical microscopy has evolved to a 3D (x–y–z) and 4D (x–y–z–t) image formation and analysis tool, and it is able to offer a mean to probe various processes at submicron and nanometer level [5-7]. The advent of the confocal fluorescence microscope [8-11] has substantially increased our comprehension of the structural and functional organisational motifs of biological systems [12]. Within this evolutive framework, two-photon excitation (2PE) microscopy [13-17] is probably the most relevant advancement in fluorescence optical microscopy since the introduction of confocal imaging in the eighties [18-21]. 2PE microscopy has a three-dimensional intrinsic ability coupled with almost five other interesting capabilities. First, 2PE greatly reduces photo-interactions and allows imaging of living specimens on long time intervals. Second, 2PE produces a unique volume of interaction resulting in a high-sensitivity background free acquisition scheme. Third, 2PE allows a better penetration than 1PE (conventional single photon excitation process) in turbid and thick specimens down to a depth of a few hundreds micrometers. Fourth, due to the broadening of the absorption spectrum of many of the fluorophores 2PE permits simultaneous excitation of different fluorescent molecules. Fifth, 2PE can prime photochemical reactions within a subfemtoliter volume inside solutions, cells and tissues.

Despite of the fact that all far field light microscopes, including conventional, confocal and two-photon microscopes, are limited in the achievable diffraction-limited resolution [22], light microscopy is moving to molecular imaging by means of single molecule detection schemes encouraged by the high signal to noise ratio achievable utilizing confocal and 2PE optical schemes [23-29]. Notwithstanding this, the advances made in the field of biological markers, especially the design of application suited chromophores, the development of the so-called quantum dots [30], visible fluorescent proteins (VFPs) from the green fluorescent protein (GFP) and its natural homologues to specifically engineered variants of these molecules, and the improvements in resolution by means of special optical schemes are moving TPE realizations from microscopy to nanoscopy [6,7].

2. CONFOCAL AND TWO-PHOTON EXCITATION MICROSCOPY

2.1. Three-dimensional fluorescence microscopy

Fluorescence optical microscopy is very popular for imaging in biology since fluorescence is highly specific either as exogenous labelling and endogenous autofluorescence [31]. Fluorescent molecules allow getting both spatial and functional information through specific absorption, emission, lifetime, anisotropy, photodecay, diffusion and other contrast mechanisms. Two-photon excitation of fluorescent molecules is a non-linear process related to the simultaneous absorption of two photons whose total energy equals the energy required for conventional, one-photon, excitation. In any case the energy required to prime fluorescence is the one sufficient to produce a molecular transition to an excited electronic state. Then the excited fluorescent molecules decay to an intermediate state giving off a photon of light having an energy lower than the one needed to prime excitation. This means that the energy E provided by photons should equal the molecule energy gap Conventional techniques for fluorescence excitation use ultraviolet (UV) or visible (VIS) radiation and excitation occurs when the absorbed photons are able to match the energy gap the ground from the excited state. Due to energetic aspects, the fluorescence emission is shifted towards a longer wavelength than the one used for excitation. This shift typically ranges from 50 to 200 nm. For example, a fluorescent molecule that absorbs one photon at 340 nm, in the ultraviolet region, exhibits fluorescence at 420 nm in the blue region. The chance of realizing a three-dimensional (3D) reconstruction of the distribution of fluorescence within a 3D object like a living cell starting from the acquisition of the two-dimensional distribution of specific intensive properties – for example, fluorescence emission - is one of the most powerful properties of the optical microscope. In fact, this allows to achieve complete morphological analyses of living biological specimens through successive volume rendering procedures. The optical microscope is unique in giving the opportunity of optical slicing to get information from different planes of the specimen without being invasive, thus preserving structures and functionality of the different parts of the specimen. In order to get 3D information from the specimen the so called optical sectioning technique is used [32, 33]. Optical sectioning is essentially based on a fine z-stepping either of the objective or of the sample stage, coupled with the usual x-y image capturing. The synchronous x-y-z scanning allows the collection of a set of 2D images, which are somehow affected by signal cross-talk from other planes from the sample, as sketched in figure 1.

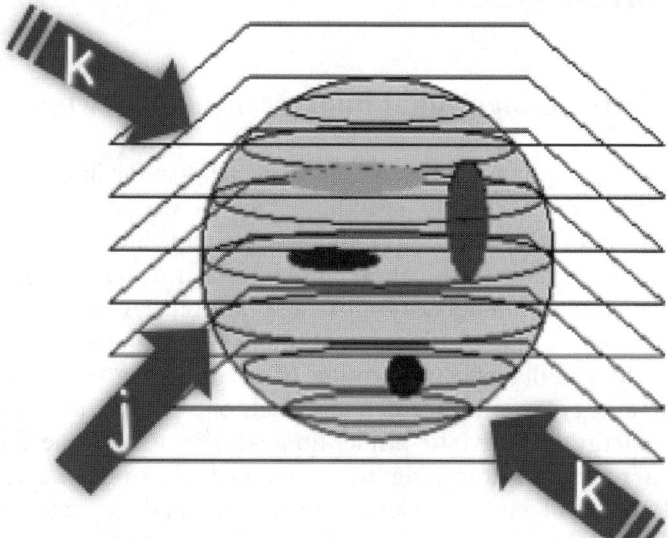

Figure 1. Optical sectioning vignette: current "in-focus" optical plan j can be affected by image contributions comig from the k-adjacent "out-of-focus" planes of a thick real 3D sample.

The observed image O at a generic plane j, coinciding with the geometrical focus position of the lens, is produced by the true fluorescence distribution I at plane j, distorted by the microscope through S, plus contributions from adjacent k planes and noise N:

$$O_j = I_j S_j + \sum_{k \neq j} I_k S_k + N \tag{1}$$

This means that when a set of two dimensional images is acquired at different positions, navigating into the sample by moving the physical optical focus of the lens along the z-axis, in principle one can recover the 3D shape of the object, described by the intensive parameter I, by solving the above written set of equation for very j betwen –k,k. This relationship, usually reported in the Fourier frequency domain [34], links the physical procedure of optical sectioning to some mathematics. In practice, after collecting optical slices one wants to find the best estimate, accordingly to some criterion, of I through the knowledge of the observed images, the distortion or point spread function (PSF) of the image formation system and the additive noise within a restoration scheme classical for space invariant linear systems. So far, this can be computationally done starting from any data set of optical slices. Recently has become available a WWW service, named Power-Up-Your-Microscope that produces the best estimate of I accordingly to the acquired data set of optical slices. Interested readers can find information and check the service trough the webpage http://www.powermicroscope.com, for free [35].

In the next paragraphs it will be evident how part of such image processing can be optically performed using some physical stratagems. The first described here is known as confocal imaging. The second is referred as two-photon or multiphoton imaging.

2.2. Confocal imaging

As reported in a early paper by Minsky [36], an ideal microscope would examine each point of the specimen and measure the amount of light scattered, absorbed or emitted by that point excluding contributions from other part of the sample from the actual or from adjacent planes. This is the situation sketched in figure 1 and simplified in eq.1. Unfortunately, when one tries to obtain images by making many of such measurements at the same time then every focal image point will be clouded by undesired light rays of scattered light deflected from points of the specimen that are not the points we are interested looking at. Most of those "extra" rays would be gone if one could illuminate only one specimen point at a time. There is no way to eliminate every possible such ray, because of multiple scattering, but it is comparatively straightforward to remove all rays not initially wanted at the focal point; just using a sort of second microscope (instead of a condenser lens) to image a pinhole (a small aperture in an opaque screen) aperture on a single point of the specimen. This reduces the amount of light in the specimen by orders of magnitude without reducing the focal brightness at all. Even under this condition, some of the initially focused light will be scattered by out- of-focus specimen points onto other points in the image plane affecting the clarity of the final acquisition, i.e. of the observed image O. But we can reject undesired rays, as well, by placing a second pinhole aperture in the image plane that lies beyond the exit side of the objective lens. We end up with an elegant, symmetrical geometry, shown in figure 2: a pinhole and an objective lens on each side of the specimen.

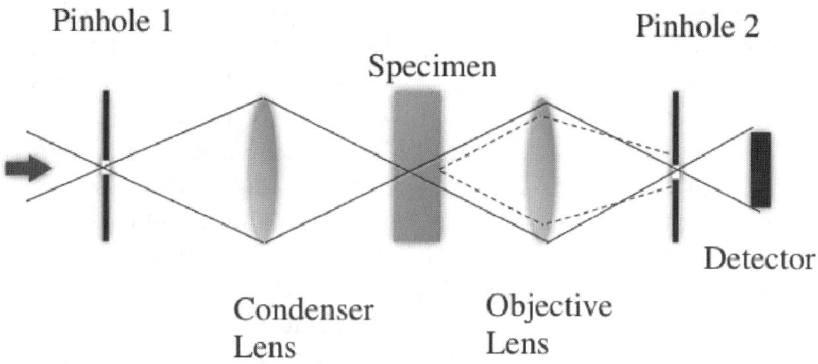

Figure 2. Confocal scheme.

This leads to a situation where one has used two lenses on both excitation and detection sides of the microscope combining two lenses in a unique effect. The final effect is a better resolving power along the three-dimensions as can be demonstrated by imaging two bright spots positioned at a distance close to the optical resolution of the system, figure 3.

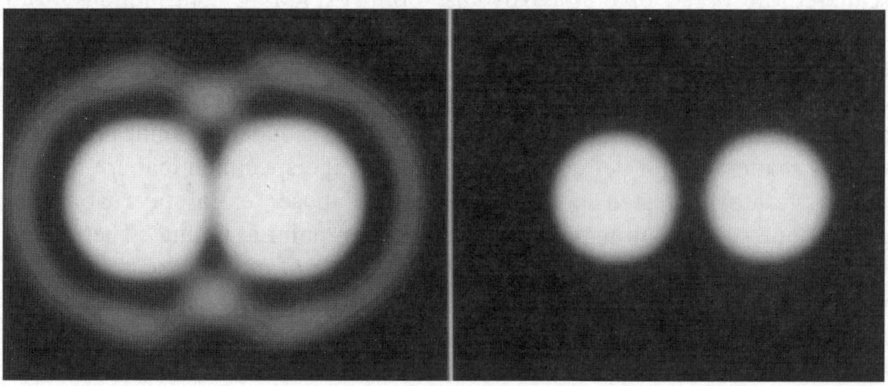

Figure 3. Point-like objects resulting images in the distinct cases of a conventional microscope (left) and of a microscope using the above mentioned stratagem (right).

The price of such a single-point illumination with respect to the classical "one-shot" wide-field illumination-detection case is being able to measure only one point at a time. This is why a confocal microscope must scan the specimen, point by point and that can take a long time because we must add all the time intervals it takes to collect enough light to measure each image point. That amount of time was reduced by using a bright light , i.e. with the advent of laser sources. Unfortunately, the final image cannot be viewed directly by eyes but has to be built point by point on a computer screen. This is a common fate and drawback to all scanning systems, ignoring now fast or real time scanning systems that are out from the purposes of this chapter [37], see figure 4.

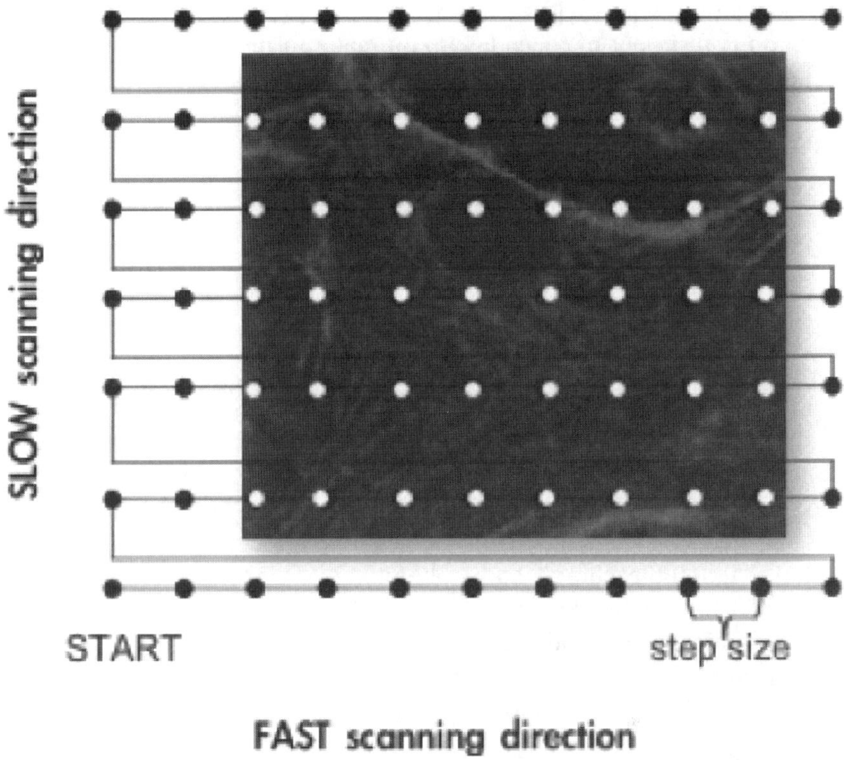

Figure 4. Scanning process.

Moving the stage with the sample or scanning the beam on the sample was the dilemma. When moving the specimen, the lenses of such a system need to be corrected only for the family of rays that intersect the optical axis at a single focal point. Scanning the beam is far more practical and widely used in almost the majority of modern architectures. In any case this stratagem brings fluorescence microscopy into the domain of three-dimensional fluorescence microscopy as well as computational optical sectioning. In such a classical 3D fluorescence optical microscope the fluorescence process is such that the excitation photons are focused into a diffraction-limited spot scanned on the specimen [8]. The three-dimensional ability, i.e. the confocal effect, is obtained by confining both the illuminated focal region and the detected area of the emitted light. So far, the light emitted from the specimen is imaged by the microscope objective lens into the image plane. Here a circular aperture (pinhole) is placed in front of a light detector, as depicted in figure 2. This pinhole is responsible for rejection of the axial out-of-focus light and of the lateral overlapping diffraction patterns. This produces an improvement of spatial resolution of a factor 1.4 along each direction, resulting in a volume selectivity 2.7 times better than in the wide-field case [38]. It is the physical suppression of the contributions from out of focus layers to image formation that produces the so-called optical sectioning effect. Now, the observed image O at a plane j is produced by the true fluorescence distribution I at plane j, distorted by the microscope through S, plus noise N, if considering an ideal situation where contributions from adjacent k planes can be set to zero:

$$O_j = I_j S_j + N \qquad (2)$$

In this case, image restoration is needed only for correcting from PSF distortions – S function - that are less affecting the image formation process than in the conventional case. Unfortunately, a serious drawback occurs. In fact, during the excitation process of the fluorescent molecules the whole thickness of the specimen is harmed by every scan, within a hourglass shaped region [34]. This means that even though out-of-focus fluorescence is not detected, it is generated with the negative effect of potential induction of those photobleaching and phototoxicity phenomena previously mentioned. The situation becomes particularly serious when there is the need for three-dimensional and temporal imaging coupled to the use of fluorochromes that require excitation in the ultraviolet regime. In this context, two-photon excitation of fluorescent molecules provides an immediate practical advantage over confocal microscopy [13]. In fact, reduced overall photobleaching and photodamage is generally acknowledged as one of the major advantages of two-photon excitation in laser scanning microscopy of biological specimens.

2.3. Two-photon excitation imaging

In 2PE of fluorescence, two low energy photons – emitted at a wavelength that is more or less twice with respect to the one needed to prime fluorescence in a conventional way - are involved in the interaction with absorbing fluorescent molecules. The excitation process can take place if these two low energy photons are able to interact simultaneously with the very same fluorophore. Here, the time scale for simultaneity is the time scale of molecular energy fluctuations at photon energy scales, as determined by the Heisenberg uncertainty principle, i.e. 10^{-16}-10^{-15}s [39]. This calculation can be simply done considering a non resonant process and the existence of a virtual state [40]. These two photons do not necessarily have to be identical but their respective wavelengthsthe sum of the energies allows to prime fluorescence emission filling the energetic gap as in a conventional one-photon absorption process. This situation, compared to the conventional one-photon excitation process, is shown in figure 5 using a Perrin-Jablonski-like diagram. It is worth noting that for practical reasons the experimental choice is usually such that such wavelengths are identical. In a 2PE process it is crucial to combine sharp spatial focusing with temporal confinement of the excitation beam. The process can be extended to n-photons requiring higher photon densities temporally and spatially confined. Thus, near infrared (680-1100 nm circa) photons can be used to excite UV and visible electronic transitions producing fluorescence. The typical photon flux densities are of the order of more than 10^{24} photons $cm^{-2}s^{-1}$ that results in light intensities around MW-TW cm^{-2}.

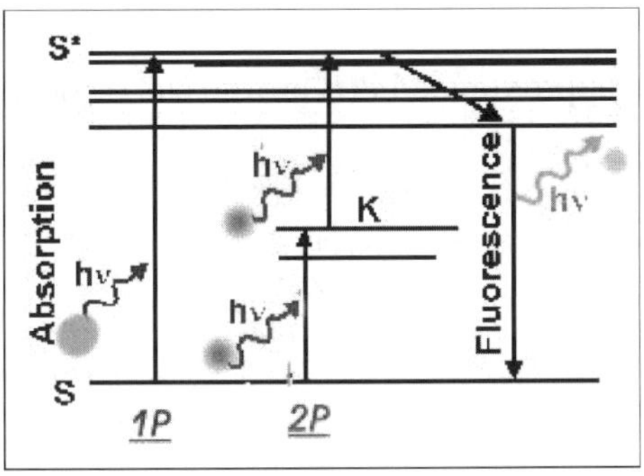

Figure 5. Perrin-Jablonski scheme for conventional (1P) and two-photon excitation (2P) of a fluorescent molecule where K is the virtual state. Operatively this means that excitation of a fluorescent molecule occurring at 360 nm under 1P can be achieved using a radiation at 720 nm under 2P conditions.

A treatment in terms of quantum theory for two-photon transition has been elegantly proposed by [39] using perturbation theory. He clearly described the process by a time-dependent Schroedinger equation, where the Hamiltonian contains electric dipole interactions terms. Using perturbative expansion one finds that the first-order solution is related to one-photon excitation while higher order solutions to n-photon ones [40].

Since the simultaneous absorption of two phonons can be treated as the advent of two independent events each of them related to the emission of fluorescence intensity, 2PE is a process that has a quadratic dependence on the instantaneous intensity of the excitation beam. Introducing the two-photon molecular cross section, as the propensity of the fluorescent molecule to absorb in a 2PE process photons having a certain energy or wavelength, and referring the fluorescence emission as function of the temporal characteristics of the light, I(t), to it, one has:

$$I_f\ (t) \propto \delta_2 * I(t)^2 \propto \delta_2 * P(t)^2 \left(\pi \frac{(NA)^2}{hc\lambda} \right)^2 \tag{3}$$

where $I_f(t)$ is the fluorescence intensity per molecule, $\delta_2(\lambda)$ the two-photon molecular cross section, P(t) is the laser power, and (NA) is the numerical aperture of the focusing objective lens. The last term of eq.(3) simply takes care of the distribution in time and space of the photons by using paraxial approximation in an ideal optical system [41]. Now, for a pulsed laser beam with pulse width, τ_p, repetition rate, f_p, and average power

$$P_{ave} = D * P_{peak}(t) \tag{4}$$

where $D = \tau_p * f_p$, the approximated P(t) profile can be described as

$$\begin{aligned} &P(t) = P_{ave} / D && \text{for } 0 < t < \tau_p \\ &P(t) = 0 && \text{for } \tau_p < t < (1/f_p) \end{aligned} \tag{5}$$

one has:

$$< I_{f,p}(t) > \propto \delta_2 \frac{P_{ave}^2}{\tau_p^2 f_P^2} \left(\pi \frac{(NA)^2}{hc\lambda} \right)^2 \frac{1}{T} \int_0^{\tau_p} dt = \delta_2 \frac{P_{ave}^2}{\tau_p f_P} \left(\pi \frac{(NA)^2}{hc\lambda} \right)^2 \tag{6}$$

One can demonstrate that, in comparison with CW lasers, pulsed lasers operate at the very same excitation efficiency, i.e. fluorescence intensity per molecule, of CW lasers delivering an avergae power higher than the one needed by a pulsed laser of a factor $\frac{1}{\sqrt{\tau \cdot f_P}}$.

The most popular relationship reported below [13] is related to the probability, n_a, that a certain fluorophore simultaneously absorbs two photons during a single pulse, in the paraxial approximation:

$$n_a \propto \frac{\delta_2 \cdot P_{ave}{}^2 \cdot}{\tau_p f_p^2} \left(\frac{NA^2}{2\hbar c\lambda} \right)^2 \tag{7}$$

Introducing 1 GM (Goppert-Mayer) = 10^{-58} [m^4 ·s], for a δ_2 of approximately 10 GM per photon, focusing through an objective of NA >1, an average incident laser power of ≈1-50 mW, operating at a wavelength ranging from 680 to 1100 nm with 80-150 fs pulsewidth and 80-100 MHz repetition rate, would saturate the fluorescence output as for one-photon excitation. This suggests that for optimal fluorescence generation, the desirable repetition time of pulses should be on the order of typical excited-state lifetime, which is a few nanoseconds for commonly used fluorescent molecules. For this reason the typical repetition rate is around 100 MHz. The use of equation (7) indicates how to choose optical and laser parameters that maximize excitation efficiency without saturation. It is also evident that the optical parameter for enhancing the process in the focal plane is the lens numerical aperture, NA, even if the total fluorescence emitted is independent from this parameter as shown by Xu [42]. This value is usually kept around 1.3-1.4. Now, one can estimate n_a for a common fluorescent molecule like fluorescein, that possesses a two-photon cross-section of 38 GM at 780 nm, saturation starts occurring at an average excitation power of 10 mW. The related rate of photon emission per molecule, at a non saturation excitation level, in absence of photobleaching is given by n_a multiplied by the repetition rate of the pulses. This means approximately $5 \cdot 10^7$ photons s^{-1}. It is worth noting that when considering the effective fluorescence emission one should consider a further factor given by the so-called quantum efficiency of the fluorescent molecules. It has been demonstrated that the fluorophore emission spectrum is independent of the excitation mode [42]. Now, even if the quantum-mechanical selection rules for 2PE differ from those for one-photon excitation, several common fluorescent molecules can be used. Unfortunately, the knowledge of one-photon cross section for a specific fluorescent molecule does not allow any quantitative prediction of the 2P trend. The only "rule of thumb" that one could use is related to the possibility of having an effective 2PE cross-section peak at double the wavelength needed for one-photon excitation. However, the cross section parameter is now available for a wide range of dyes including quantum dots [21, 42].

In terms of optical consequences the two-photon effect has the important consequence of limiting the excitation region within a sub-femtoliter volume. This 3-D confinement of the two-photon excitation volume can be understood based on optical diffraction theory [41]. Using excitation light with wavelength λ, the intensity distribution at the focal region of an objective with numerical aperture $NA = \sin(\alpha)$ is described, in the paraxial regime, by [41]:

$$I(u,v) = \left| 2\int_0^1 J_0(v\rho) e^{-\frac{i}{2}u\rho^2} \rho \, d\rho \right|^2 \tag{8}$$

where J_o is the zeroth order Bessel function, ρ is a radial coordinate in the pupil plane, $u = \frac{8\pi \sin^2(\alpha/2) z}{\lambda}$ and $v = \frac{2\pi \sin(\alpha) r}{\lambda}$ are dimensionless axial and radial coordinates, respectively, normalized to the wavelength [8]. Now, the intensity of fluorescence distribution within the focal region has a $I(u,v)$ behaviour for the one-photon case and $I^2(u/2,v/2)$ for 2PE. The arguments of $I^2(u/2,v/2)$ take into proper account the fact that in the latter case one utilizes wavelengths that are approximatively twice the ones used for one photon excitation. As compared with the one-photon case, the 2PE intensity distribution is axially confined [38, 39]. In fact, considering the integral over ν, keeping *u* constant, its behaviour is constant along z for one-photon and has a half-bell shape for 2PE. This behaviour is responsible for the three-dimensional sharp localization of 2PE. Now, the most interesting aspect is that the excitation power falls off as the square of the distance from the lens focal point, within the approximation of a conical illumination geometry. In practice this means that the quadratic relationship between the excitation power and the fluorescence intensity brings about the fact that 2PE falls off as the fourth power of distance from the focal point of the objective. This fact implies that those regions away from the focal volume of the objective lens, directly related to the numerical aperture of the objective itself, therefore do not suffer photobleaching or phototoxicity effects and do not contribute to the signal detected when a 2PE scheme is used. Because they are simply not involved in the excitation process, a confocal-like effect is obtained without the necessity of a confocal pinhole, as sketched in figure 6.

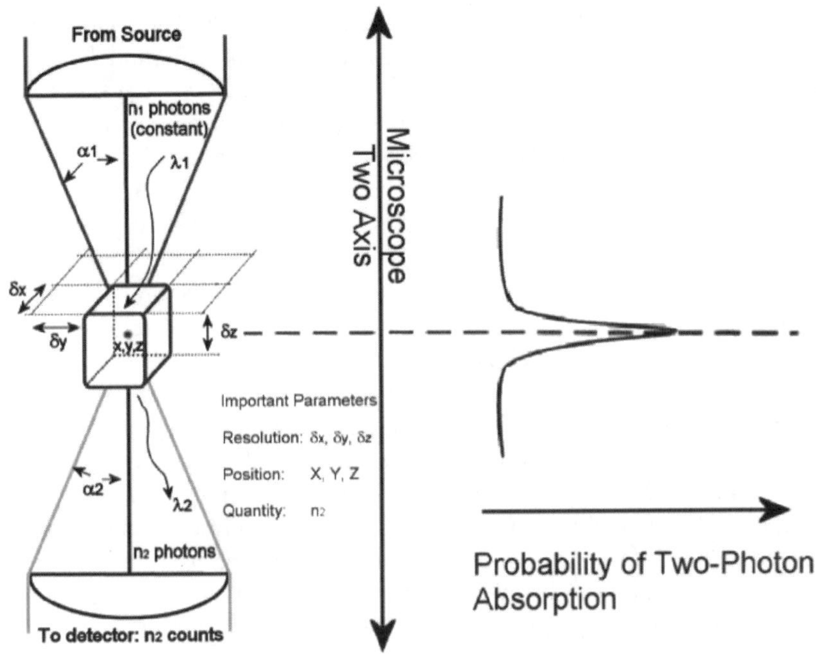

Figure 6. Confinement of excitation volume under 2PE [10].

It clearly appears that also in this case an optical sectioning affect is obtained. In fact, the observed image O at a plane j, produced by the true fluorescence distribution I at plane j, distorted by the microscope through S, plus noise N, holds again the confocal ideal situation where contributions from adjacent k planes can be set to zero as in the confocal situation, see eq. (2). This means that 2PE microscopy is intrinsically three-dimensional. It is worth noting that the optical sectioning effect is obtained in a very different way with respect to the confocal solution. No fluorescence is wasted by removal from the detection pathway. In this case one should be able to collect as much as fluorescence is possible. In fact fluorescence can come only and exclusively from the small focal volume outlined in figure 6 that is of the order of fraction of femtoliter. In 2PE over 80% of the total intensity of fluorescence comes from a 700-1000 nm thick region about the focal point for objectives with numerical apertures in the range from 1.2 to 1.4 [38]. This also implies a reduction in background with consequent increase of the signal to noise ratio that compensates the reduction in spatial resolution due to the wavelength. The utilisation of infrared wavelength instead of UV-visible ones also allows deeper penetration than in conventional case mainly due to the decrease of scattering during the excitation process. This fact implies that deeper targets within a thick sample can be reached. It is of great interest the fact that the scattered emitted light can be completely collected by the photodetectors since it is exclusively originated from the 2PE volume, sketched in figure 6. The main elements needed to realize a 2PE architecture including confocal modality are the following: high peak-power laser delivering moderate average power (fs or ps pulsed at relatively high repetition rate) emitting infrared or near infrared wavelengths (650-1100 nm), CW laser sources for confocal modes, a laser beam scanning system or a confocal laser scanning head, high numerical aperture objectives (>1), a high-throughput microscope pathway, and a high-sensitivity detection system [43, 44]. In typical 2PE or confocal microscopes, images are built by raster scanning the x-y mirrors of a galvanometric driven mechanical scanner [45]. This fact implies that the image formation speed is mainly determined by the mechanical properties of the scanner, i.e. for single line scanning is of the order of ms. While the x-y scanners provide lateral focal-point scanning, axial scanning can be achieved by means of different positioning devices, most popular being: a belt driven system using a DC motor and a single objective piezo nano-positioner. Acquisition and visualisation are generally completely computer controlled by dedicated software. Two popular approaches can be used to perform 2PE microscopy, namely: de-scanned and non de-scanned mode. The former uses the very same optical pathway and mechanism employed in confocal laser scanning microscopy. The latter mainly optimises the optical pathway by minimising the number of optical elements encountered on the way from the sample to detectors, and increases the detector area. Non-descanned mode is not available on confocal "solo" set-ups. Figure 7 illustrates these two approaches also including the conventional confocal scheme with a pinhole along with the de-scanned pathway. 2PE non-descanned mode provides a superior signal-to-noise ratio especially when dealing with strongly scattering samples [46]. In the de-scanned approach pinholes are removed or set to their maximum aperture and the emission signal is captured using excitation scanning device on the back pathway. For this reason it is called de-scanned mode. In the latter, the confocal architecture has to be modified in order to increase the collection efficiency: pinholes are removed and the emitted radiation is collected using dichroic mirrors on the emission path or external detectors without passing through the galvanometric scanning mirrors. The fluorescence emitted is collected by the objective and transferred to the detection system through a dichroic mirror along the emission path. Due to the high excitation intensity, an additional barrier filter is needed to avoid mixing of the excitation and emission light at the detection system that is differently placed depending on the acquisition scheme being used. Photodetectors that can be used include photomultiplier tubes, avalanche photodiodes, and charge coupled device (CCD) cameras.

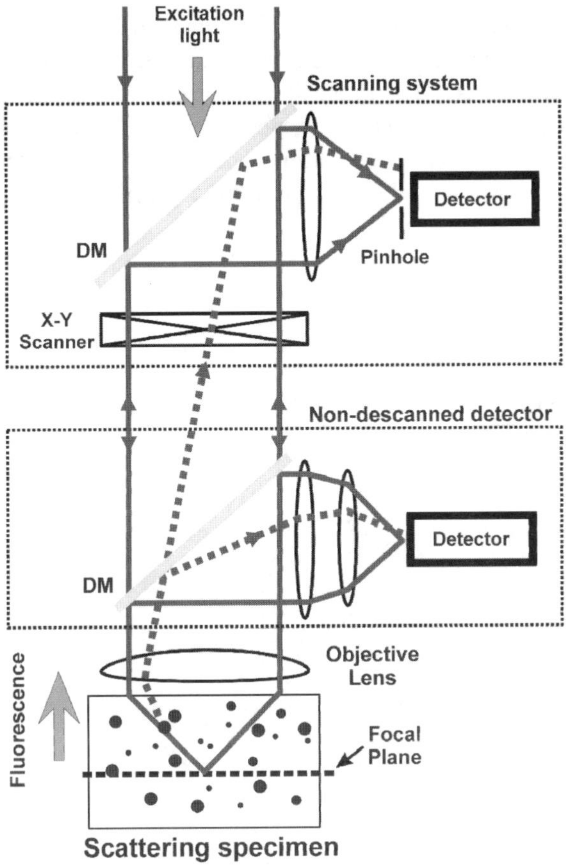

Figure 7. Simplified optical scheme showing de-scanned and no de-scanned modalities (Courtesy of Mark Cannel, adapted from [7]).

3. 2PE IMAGING OF SINGLE MOLECULES

2PE microscopy suffers of the resolution limitation typical in optical microscopy, i.e. 200 nm. As such it does not provide the resolution to image single fluorescent molecules or visble proteins (VP). However, when imaging sparse samples, one gets isolated fluorescent spots that have the size of the point spread function. Then, by means of intensity, photodecay and polarization properties, one can locate single molecules [48]. Here, the focus is on single molecule imaging (far-field) using simple 2PE optical configurations [25, 49, 50]. Following the pioneering work by Sanchez et al. [51] on two-photon imaging of single rhodamine B glass immobilised molecules, spatially resolved applications of ultrasensitive 2PE fluorescence have shown promising results [52]. Two basic issues in single molecule imaging are related to diminish the background signal, either residual scattering or fluorescence, and to discriminate between the signals arising from single molecules and those that correspond to small molecular aggregates. These observations can be performed only by following the time evolution of the fluorescence emission of molecular aggregates, which may be degraded by the prolonged exposure to the exciting radiation as shown in figure 8.

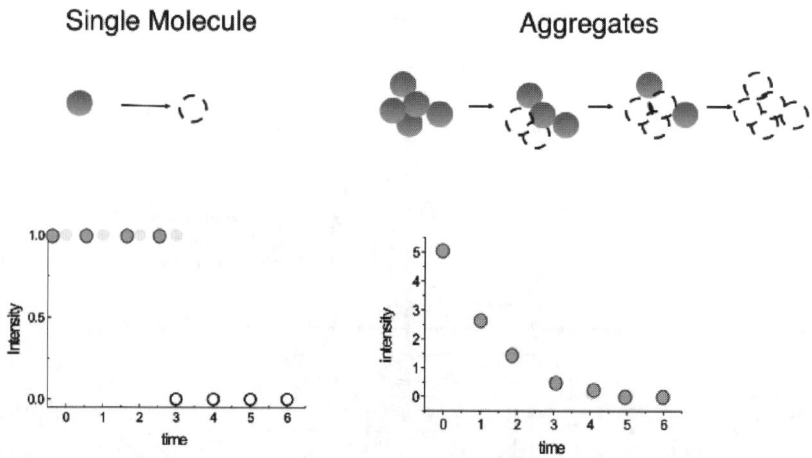

Figure 8. Intensity decay for single molecules and aggregates located on a glass slide within a scanned area of 110 micron. This intensity decay is due to prolonged exposure on a selected area on the sample. This information can be linked with intensity distribution of fluorescence resulting from spot by spot average evaluation, see figure 9.

One can discriminate single molecules from aggregates also considering that the distributions of the pixel content on the images show discrete intensity peaks at specific levels that are found to be multiples of a reference fluorescence level, that corresponds to the dimmest spot revealed in the substrates [48, 52]. This situation is shown in figure 9.

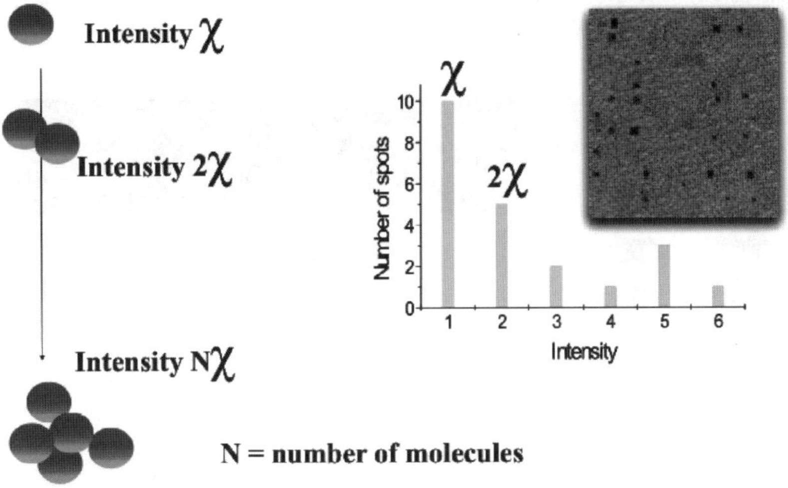

Figure 9. Intensity distribution for single molecules and aggregates located on a glass slide within a scanned area of 110 micron. Quantization of intensity allows to locate single molecules or single fluorescent entities and their higher order aggregates. Photobleaching test can confirm the quantization process. The bar-graph clearly shows the existence of a sinle molecule class of spots. Critical parameters are initial concentration of fluorescent molecules and deposition method [53].

When one is able to localize single molecules from 2PE or confocal images more information can be extracted to study fluorescent molecule properties. For example, a systematic study of single molecule detection by 2PE has been reported on different popular dyes (pyrene, rhodamine 6G, fluoresceine and indo-1) of increasing structural complexity. A second example that we report is the detection of the 2PE fluorescence of a single green fluorescent protein (GFP) mutant, called E2GFP [54] trapped in wet silica gels [55]. The fluorescence kinetics followed for ≅ 110 s showed a bleaching of the E2GFP molecule after ≅ 85 s. After 10 seconds from bleaching, the sample was beam scanned 40 times at 720 nm finding that the E2GFP recovered its own fluorescence, figure 10. This last example points out to interesting applications like 3D optical memories or perennial fluorescent markers for living cells studies.

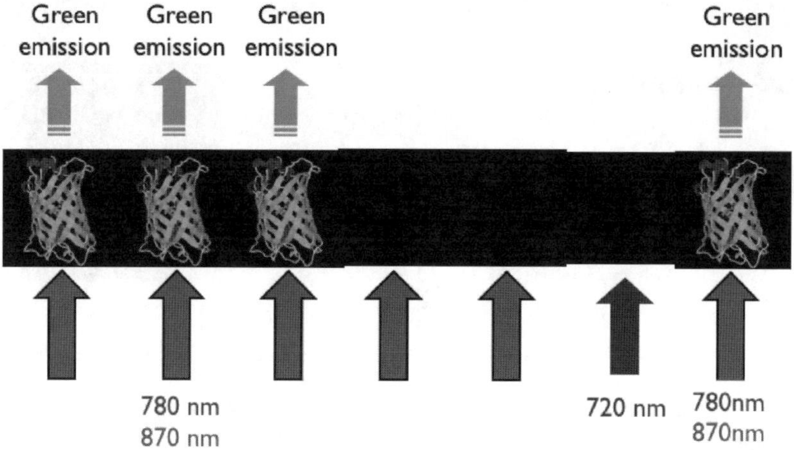

Figure 10. GFP on-off switching under 2PE regime [55].

4. CONCLUSION.

Considering their impact on basic and applied research, the development of confocal and 2PE imaging methods may be viewed as a new deal in optical microscopy towards nanoscopy. In particular, 2PE has generated and still produces a tremendous flourishing of related techniques. New experiments can be designed and performed and comparing one- and two-photon experiments can do a critical reading of the past results. 2PE brings an intrinsic three-dimensional resolution, the absence of background fluorescence, and the charming possibility of exciting UV excitable fluorescent molecules at infrared wavelengths increasing sample penetration. Photobleaching and natural fluorescence are strictly confined in the focal volume of event, and multiple excitations of fluorescent molecules can be more easily and efficiently accomplished than by using conventional excitation. Under a fluorescence acquisition perspective, ballistic and scattered emitted photons can be allowed to contribute to the total acquired signal since they are uniquely produced in a very specific and well-defined 2PE volume. Moreover, a further benefit in establishing a 2PE optical system is the flexibility in choosing the measurement modality favoured by the simplification of the optical design. As a matter of fact, a 2PE microscope offers a number and variety of measurement options without changing any optics or hardware. This means that one can really get multimodal and multifunctional information within the same experiment from the very same sample. Now, one impediment to the understanding and full exploitation of 2PE is given by the difficulty of predicting or measuring two-photon absorption spectra of the fluorescent molecules used. A further extremely practical disadvantage of 2PE fluorescence microscopy resides in the cost of appropriate laser sources. Since there are no physical limitations to the realization of 2PE or multiphoton microscopes, the expectation is that their use will most likely increase as cheaper and more reliable laser sources are developed. As last remark, it has to be noticed that 2PE and confocal microscopy, also as consequence of the high signal to noise ratio achievable, allow the utilization of the optical microscope for studying single molecules without breaking any diffraction barrier. Moreover, recent advances like STED (stimulated emission depletion) and 4PI microscopy, developed within the framework of confocal and multiphoton imaging, brought the resolution limit down to 10-90 nm scale establishing a real bridge between optical microscopy and nanoscopy [6].

5. ACKNOWLEDGEMENTS

This work is dedicated to the memory of Osamu Nakamura that passed away 5:07 a.m. on Sunday, 23 Jan 2005 at Handai hospital. Osamu was an excellent scientist and very bright mind, a very nice person. He brought original and important contributions in microscopy and in particular in 2PE microscopy. His work will be a way for imagining Osamu still with us in our job, at our conferences.

6. REFERENCES

1. Arndt-Jovin D.J., Nicoud R.M., Kaufmann J. and Jovin T.M. Fluorescence digital-imaging microscopy in cell biology. Science 1985; 230: 13330-35.
2. Beltrame F., Bianco B., Castellaro G. and Diaspro A. "Fluorescence, Absorption, Phase-contrast, Holographic and Acoustical Cytometries of Living Cells." In: *Interactions between Electromagnetic Fields and Cells.* Chiabrera A., Nicolini C., Schwan, H.P. Eds. NATO ASI Series, vol.97, New York and London, Plenum Press Publishing, 483-498, 1985.
3. Tsien R.Y. The green fluorescent protein. Ann.Rev.Biochem. 1998; 67: 509-544.
4. Shimomura O. The discovery of aequorin and green fluorescence protein. J.Microscopy 2005; 217:3-15.
5. Amos B. Lessons from the history of light microscopy. Nat Cell Biol. 2000; 2(8): E151-2.
6. Hell SW. Toward fluorescence nanoscopy. Nat Biotechnol. 2003; 21: 1347-55.
7. Bastiaens P.I. and Hell S.W. (eds.) Recent Advances in Light Microscopy. Journal of Structural Biology 2004; **147**: 1-89.
8. Wilson T. and Sheppard C.J.R. *Theory and Practice of Scanning Optical Microscopy.* London: Academic Press, 1984.
9. Wilson, T. *Confocal Microscopy*. London: Academic Press, 1990.
10. Pawley, J. B. (ed.). *Handbook of Biological Confocal Microscopy.* New York: Plenum Press, 1995.
11. Masters, B.R. *Selected Papers on Confocal Microscopy.* SPIE Milestone series, Bellingham, WA: SPIE Press, 1996.

12. Amos W.B. and White J.G. How the Confocal Laser Scanning Microscope entered Biological Research. Biology of the Cell 2003; 95: 335–342.
13. Denk W., Strickler J.H. and Webb W.W. Two-photon laser scanning fluorescence microscopy. Science 1990; 248: 73-76.
14. Nakamura O. Three-dimensional imaging characteristics of laser scan fluorescence microscopy: Two-photon excitation vs. single-photon excitation. Optik 1993; 93: 39-42.
15. Pennisi E. Biochemistry: Photons Add Up to Better Microscopy. Science 1997; 275: 480-481.
16. Diaspro A. Two-photon microscopy. Microsc. Res. Tech. 1999; 47: 163- 212.
17. Diaspro, A. Two-photon excitation microscopy. IEEE Eng. Med. Biol. 1999; 18(5): 16–99.
18. *Masters B.R.* Selected Papers on Multiphoton Excitation Microscopy. *SPIE Milestone Series, Bellingham, WA: SPIE Press, 2002.*
19. Girkin J.M. Optical physics enables advances in multiphoton imaging. Journal of Physics D: Applied Physics 2003; 36: R250-R258.
20. Zipfel W.R., Williams R.M. and Webb W.W. Nonlinear magic: multiphoton microscopy in the biosciences. Nature Biotechnology 2003; **21**: 1369-1377.
21. Diaspro A., Chirico G. and Collini M. Two-photon excitation microscopy. Quarterly Reviews of Biophysics 2005; In press.
22. Abbe E. Beitrage zur Theorie des Mikroskops und dermikroskopischen Wahrnehmung, Arkiv Mikroskopische Anat 1873; 9: 413-468.
23. Mertz J., Xu C.and Webb, W.W. Single molecule detection by two-photon excited fluorescence. Opt. Lett. 1995; 20: 2532-2534.
24. Xie X.S. and Lu H.P. Single molecule enzymology. J.Biol.Chem. 1999; 274: 15967-15970.
25. Sonnleitner M., Schutz G.J. and Schmidt, T. Imaging individual molecules by two-photon excitation. *Chem.Phys.Lett.* 1999; 300: 221-226.
26. Sonnleitner M., Schütz G. Kada G. and Schindler H. Imaging Single Lipid Molecules in Living Cells Using Two Photon Excitation. Single Molecules 2000; 1: 182-183.
27. Diaspro A., Chirico G., Federici F., Cannone F. and Beretta, S. Two-photon microscopy and spectroscopy based on a compact confocal scanning head. J. Biom. Opt. 2001; 6(3): 300-10.
28. Chirico G, Cannone F, Beretta S. and Diaspro A. Single molecule studies by means of the two-photon fluorescence distribution. Microsc Res Techniq. 2001; 55: 359-364.
29. Cannone F, Chirico G. and Diaspro A. Two-photon interactions at single fluorescent molecule level. J. Biomed. Opt. 2003; 8: 391-39.
30. Jaiswal J.K. and Simon S. Potentials and pitfalls of fluorescent quantum dots for biological imaging. Trends in Cell Biology 2004; In press (online August 4th, 2004).
31. Periasamy, A. (ed.) *Methods in Cellular Imaging*, New York: Oxford University Press, 2001.
32. Agard D.A. Optical sectioning microscopy: cellular architecture in three dimensions. Annu.Rev.Biophys. 1984; 13: 191-219.
33. Diaspro A., Sartore M. and Nicolini, C. Three-dimensional representation of biostructures imaged with an optical microscope: I. Digital Optical Sectioning. Image and Vision Computing 1990; 8: 130-141.
34. Bianco B. and Diaspro A. Analysis of the three dimensional cell imaging obtained with optical microscopy techniques based on defocusing. Cell Biophys. 1989; 15: 189-200.
35. Bonetto P., Boccacci P., Scarito M., Davolio M., Epifani M., Vicidomini G., Tacchetti C., Ramoino P., Usai C. and Diaspro A. Three-Dimensional Microscopy Migrates to the Web with "PowerUp Your Microscope". Microsc. Res. Tech. 2004; 64: 196–203.
36. Minsky M. Memoir of Inventing the Confocal Scanning Microscope. Scanning 1988; 10: 128-138
37. Diaspro A. New world Microscopy. IEEE Engineering in Medicine and Biology Magazine 1996; 15 (1) .

38. Jonkman J. and Stelzer E. "Resolution and contrast in confocal and two-photon microscopy." In *Confocal and Two-photon Microscopy: Foundations, Applications and Advances.* A. Diaspro ed. New York: Wiley-Liss Inc., 2001.
39. Nakamura O. Fundamentals of two-photon microscopy. Microsc. Res. Tech. 1999; 47: 165-171.
40. Esposito A., Federici F., Usai C., Cannone F., Chirico G., Collini M. and Diaspro A. Notes on theory and experimental conditions behind two-photon excitation microscopy. Microsc. Res. Tech. 2004; 63: 12-17.
41. Born, M. and Wolf, E. *Principles of Optics*, 6th ed., Oxford: Pergamon Press, 1993.
42. Xu C. "Cross-sections of Fluorescence Molecules in Multiphoton Microscopy." In *Confocal and Two-photon Microscopy: Foundations, Applications and Advances,* Diaspro A. ed. New York: Wiley-Liss, Inc. 2001.
43. Diaspro A. (ed.) *Confocal and two-photon microscopy : foundations, applications, and advances.* New York: Wiley-Liss, Inc. 2001.
44. Girkin J. and Wokosin D. "Practical Multiphoton Microscopy." In *Confocal and Two-photon Microscopy: Foundations, Applications and Advances,* Diaspro A. ed. New York: Wiley-Liss, Inc. 2001.
45. Webb R.H. Confocal Optical Microscopy. Rep. Progr. Phys. 1996; 59: 427-471.
46. Centonze V.E. and White, J.G. Multiphoton excitation provides optical sections from deeper within scattering specimens than confocal imaging. Biophys.J. 1998; 75: 2015-2024.
47. Soeller C. and Cannel M.B. Two-photon microscopy: imaging in scattering samples and three-dimensionally resolved flashed photolysis. Microsc. Res. Tech. 1999; 47: 182-195.
48. Cannone F, Chirico G and Diaspro A. Two-photon interactions at single fluorescent molecule level. J. Biomed. Opt. 2003; **8**: 391-395.
49. Mertz J., Xu C. and Webb, W.W. Single molecule detection by two-photon excited fluorescence. Opt. Lett. 1995; **20:** 2532-2534.
50. Diaspro A , Chirico G, Federici F, et al. Two-photon microscopy and spectroscopy based on a compact confocal scanning head J. Biomed Opt. 2001; 6 (3): 300-310.
51. Sanchez E. J., L. Novotny G. R. Holtom and X. S. Xie. (). Room temperature fluorescence imaging and spectroscopy of single molecules by two-photon excitation. *J. Phys. Chem. A.* 1997; 101: 7019 –7023.
52. Chirico G., Cannone F. and Diaspro A. Single molecule photodynamics by means of one- and two-photon approach. J. Phys D Appl. Phys. 2003; 36(14): 1682-1688.
53. Chirico G, Olivini F, Beretta S and Diaspro A. Identification of single molecules by analysis of the intensity distribution. Biophys. J. 2001; 80(1): 655.
54. Cinelli R. A. G., Pellegrini V., Ferrari A., Faraci P., Nifosi R., Tyagi M., Giacca M. and Beltram, F. Green fluorescent proteins as optically controllable elements in bioelectronics. Appl. Phys. Lett. 2001; 79(20): 3353-3355.
55. Chirico G., Cannone F., Diaspro A., Bologna S., Pellegrini V., Nifos R. and Beltram F. Multiphoton switching dynamics of single green fluorescent proteins. Physical Review E 2004; 70: 030901.

FLUORESCENCE RESONANCE ENERGY TRANSFER (FRET) AND FLUORESCENCE LIFETIME IMAGING MICROSCOPY (FLIM)

Thomas M. Jovin[1], Diane S. Lidke[1], and Elizabeth A. Jares-Erijman[2]
[1] *Department of Molecular Biology, Max Planck Institute for Biophysical Chemistry, D-37077 Göttingen, Germany*
[2] *Departamento de Química Orgánica, Facultad de Ciencias Exactas y Naturales, Universidad de Buenos Aires, C1428EHA Buenos Aires, Argentina*

Abstract. Current approaches and developments in the field of microscopy based on determinations of Fluorescence (Förster) Resonance Energy Transfer (FRET) and Fluorescence Lifetime (FLIM) are discussed.

Keywords: Donor quenching, Acceptor Sensitization, Emission Anisotropy, Quantum Dots

1. INTRODUCTION

Fluorescence Resonance Energy Transfer (FRET), or more exactly Förster (the name of Theodor Förster, the discoverer of the phenomenon) Resonance Energy Transfer is applied extensively in all fields of research science and technology, generally - but not exclusively - as a "nanoruler" with a dynamic range of ~0-10 nm.

"What, where, when" ⇒ spatio-temporal relationships are important
Thus, the need to image living cells with specific chemical and physical probes and adequate spatial and temporal resolution

molecular biology/genetics

cellular biology ⇔ *cellular biophysics*

What are the causes?
What are the effects?
What is (un)important?
What is (non)essential?

Figure 1. Performing "biochemistry" on and in the cell.

V. Evangelista et al. (eds.), From Cells to Proteins: Imaging Nature across Dimensions, 209–216.

Biological cells function in a coordinated manner because the underlying molecular interactions are orchestrated with an extensive degree of spatio-temporal regulation (Figure 1). Despite the extensive and highly complex networking involved, the fundamental physicochemical basis corresponds to a limited set of reactions: binding, conformational transition, covalent modification, and transport. FRET can provide information about all of these molecular processes with high specificity and sensitivity via appropriate probes expressed by or introduced from the external medium into the cell, tissue, or organism (an incomplete list is given in Table 1).

Table 1. Strategies for specific protein fluorescence labeling *in vivo*.

Visible Fluorescent Proteins (VFPs) fusions; include:
"cameleon" analogs for sensing ions, pH, covalent modification,…
photoactivatable and photo-interconvertible VFPs
bimolecular complementation (half VFP molecules)
FlAsH and ReAsH tags (biarsenical dyes for tetracysteine motifs)
dye-specific single chain antibodies (as fusion proteins)
oligohistidine tags
nano-tags (streptavidin-binding peptide fusion tags)
aptamer peptides with bound or binding fluorophore
snap-tags (DNA alkyltransferase fusion protein)
N-terminal cysteine (targets for thioester permeable dyes)
Quantum Dot ligands (factors, peptides, modifiers)[a]
VFPs: intrinsic probes; others: extrinsic
Modifiers of cellular expression: RNAi, inteins, transposon shuffling

[a] see chapter by D. J. Arndt-Jovin, T. M. Jovin, and Diane S. Lidke.

2. FRET

Excellent introductions to the spectroscopic principles forming the basis of this chapter are provided by three monographs [1-3] and a very comprehensive (> 1400 pages) Website devoted to "Multi-Probe Microscopy" [4]. We have proposed elsewhere [5] a somewhat unconventional representation of a *fluorophore* as a "photophysical enzyme", i.e. a catalyst that converts a photon (absorbed) of a given energy to another photon (emitted) of lower energy, and repeats this process until it is degraded photochemically ("photobleached"). FRET as a photophysical phenomenon is generally regarded as *competitive* with other forms of radiative and non-radiative decay from the first excited singlet state of a fluorophore (Figure 2A). That is, it provides an additional pathway for deactivation, without affecting the inherent probabilities, i.e. the rate constants, for other preexisting processes. Thus, introduction of FRET into a system leads to a reduction in "fluorescence lifetime" (a misnomer for what is more accurately denoted as "fluorescence-detected excited state lifetime") because the decay to the ground state is altogether more efficient. FRET provides the motivation for a large fraction of FLIM (Fluorescence Lifetime Imaging Microscopy) experiments. The lowered lifetime implies a reduced quantum yield (fractional decay in the form of fluorescence) and thus FRET is often perceived as quenching, i.e. a reduction in the steady-state rate of emission. That is, the corresponding saturation curve (fluorescence vs. irradiance), is shifted to the "right" (Figure 2B). Somewhat paradoxically the maximal rate at saturation levels of irradiance is unchanged, and is simply given by the radiative rate constant k_f (which may not be as constant as is generally assumed [6]).

FRET is applied extensively both in solution and in imaging systems. We have reviewed the methodology applicable to implementations of FRET in optical microscopy [5, 7, 8] with extensive references to the literature (that are not reproduced here). A classification scheme for FRET imaging methods was proposed and is represented in simplified form in Table 2.

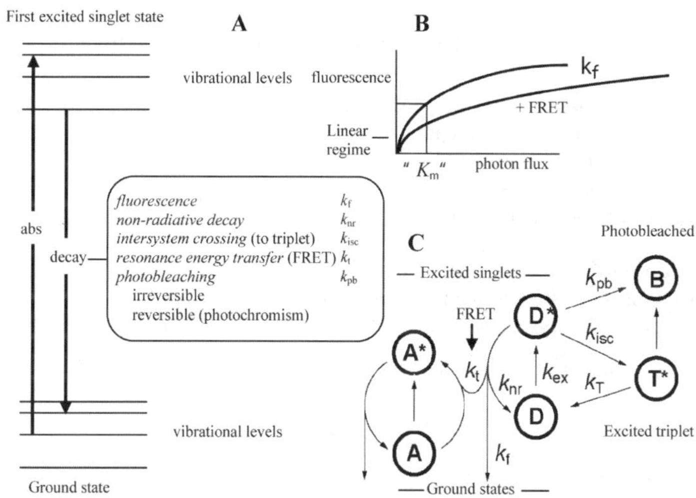

Figure 2. Photophysical states and interconversions associated with FRET. (**A**) Transitions between the ground state and first excited singlet (S_1*) state of a fluorophore. The fluorescence lifetime is given by the reciprocal of the sum of all first-order rate constants leading to deactivation of S_1*. Resonance energy transfer between a donor (D) and acceptor (A), a process which proceeds with an efficiency E dependent on the 6th power of the D-A separation and the degree of overlap between the donor emission and acceptor absorption; it also requires a non-orthogonal orientation of the donor emission transition moment and the acceptor absorption transition moment. FRET leads to a reduction in the donor fluorescence lifetime τ and thus (**B**) to a displacement of the saturation curve to higher photon fluxes (irradiances, intensities). The "K_m", the photon flux required for attaining 1/2 the maximum fluorescence emission rate, k_f, is given by $(\sigma\tau)^{-1}$, where σ is the donor absorption cross-section. The excited singlet states can also undergo photobleaching (shown in **C** for the donor) either directly or via the generally long-lived excited triplet state (T*), with which it is in equilibrium (for more explanation and formulas, see [5]).

Most of these methods derive from very simple two (ground, excited singlet) or three (ground, excited singlet, excited triplet) state depictions of the photophysical scheme for light absorption and emission (Figure 2C). Many of the FRET techniques ([5]; Table 2), some of which have yet to be implemented, exploit the inherent and FRET-sensitive non-linearity of the fluorophore saturation curve.

FRET can be utilized in both a qualitative or quantitative manner. That is, certain reporter systems by design provide a distinctive FRET signal in the determinations of fractional association or other conversion reactions. Thus, much of the recent FRET literature deals with alternative methods for deriving E from the spectral data and a number of publications appeared in 2004 devoted to this topic (Table 2). These and other developments closely reflect the ongoing improvements in the available scanning and wide-field instruments, particularly in the optical (light sources and detectors, objectives, filters) and computational aspects. The utility of any particular approach to FRET (and FLIM) will depend on the technical resources available and the inherent nature of the system under study.

Table 2. Measurement of FRET in imaging systems (adapted from ([5])

	Refs[a]
Donor quenching and acceptor sensitization	
combined donor (D) and acceptor (A) emission signals	6, 9-15
fluorescence-detected excited lifetime(s) (FLIM)	6, 7, 9, 10, 15-26
donor depletion kinetics	27
acceptor depletion (adFRET)	28, 29
Emission anisotropy	
steady-state anisotropy (donor or acceptor)	6, 7, 30-33
homotransfer, energy migration FRET (emFRET, P-FRET)	6, 7, 30, 31, 33, 34

[a] Publications that have appeared since those cited in [5].

3. FLIM

The many virtues of fluorescence lifetime determinations in general cell biological studies and for FRET-based probes in particular have been extensively documented, one is referred to a recent review on the subject [6]. The techniques employed in the imaging context are quite varied (Table 3) and can be classified according to whether they are based on point (one-dimensional) detectors or two-dimensional detectors such as CCD cameras. A second level of classification refers to the use of time-domain and frequency-domain methodology. Often the source of great controversy [6], this distinction is being increasingly blurred by the convergence of the technologies in terms of speed and resolution.

Table 3. Fluorescence Lifetime Imaging Microscopy (FLIM)

"Point" detection ± scanning
time-correlated ± space-correlated single photon (TCSP)
gated boxcar integration or windows (≥ 2)
multiphoton, pump-probe, double-pulse (depletion, upconversion)
white-noise excitation & power spectrum analysis [35]
"Area" detection
gated intensifier + CCD camera
frequency-domain: phase-modulation (source & camera)
combination with optical sectioning, spectral & *r* encoding
indirect techniques: photobleaching kinetics, *r*, …

[a] Publications that have appeared since those cited in [5]: [6-12, 15-26, 36-38]

4. ANISOTROPY

Fluorescence anisotropy, *r*, a measure of emission polarization, is a sensitive reporter of molecular rotational diffusion and rearrangement and of FRET. We have discussed elsewhere the formalism and application of dynamic and static fluorescence anisotropy determinations in imaging systems ([5, 7, 8, 34, 38, 39]; see other references in Table 2). These include confocal laser scanning as well as wide-field microscopes, in the latter case adapted for anisotropy-based dynamic frequency domain fluorescence lifetime imaging microscopy (FLIM), a method we denote as rFLIM [38]. The manifestations of FRET reflected in changes in steady-state or time-dependent *r* are very useful in studies of the association, and proximity of cellular proteins *in vivo.* Both heterotransfer FRET (involving a distinct D and A pair) or homotransfer FRET (in our nomenclature [38], energy migration or emFRET) are feasible (Table 2). EmFRET is particularly useful in studies of activation and processing of molecules engaging in self-association, such as the transmembrane receptor tyrosine kinases involved in signal transduction and expressed as fusions with VFPs [7, 8].

5. ACKNOWLEDGEMENTS

We thank many colleagues for valuable suggestions and discussions. The Max Planck Society and the EU 5th Framework Program (Grant No. QLG2-CT-2001-02278), through which D.S.L. was awarded a fellowship, are acknowledged for financial support. T.M.J. and E.A.J.-E. were the recipients of a joint grant (No. I/77 897) from the Volkswagen Foundation. E.A.J.-E. is also indebted to the Agencia Nacional de Promoción de la Ciencia y Tecnología (ANPCyT), Fundación Antorchas, Consejo Nacional de

Investigaciones Científicas y Técnicas (CONICET), Secretaría de Ciencia, Tecnología e Innovación Productiva (SECyT), German-Argentine DLR-BMBF-SECyT, and the Universidad de Buenos Aires (UBA) for financial support.

6. REFERENCES

1. Lakowicz J.R. *Principles of Fluorescence Spectroscopy*. Kluwer Academic/Plenum, New York, 1999,.
2. Periasamy A. and Day R.N., ed., Molecular *Protein Imaging in Cells and Tissues: FRET Microscopy*. Oxford University Press, in press.
3. Valeur B., *Molecular fluorescence: principles and applications*. Wiley-VCH, Weinheim, 2002.
4. McNamara G., Multi-Probe Microscopy, http://home.earthlink.net/~mpmicro/, 2005.
5. Jares-Erijman E.A. and Jovin T.M. FRET imaging. Nat. Biotechnol. 2003; 21: 1387-1395.
6. Suhling K., French P.M. and Phillips D. Time-resolved fluorescence microscopy. Photochem. Photobiol. Sci. 2005; 4: 13-22.
7. Jovin T.M., Lidke D.S. and Post J.N. Dynamic and static anisotropy in biological microscopy (rFLIM and emFRET). SPIE Proc. 2004; 5323: 1-12.
8. Lidke D.S., Nagy P., Barisas B.G., Heintzmann R., Post J.N., Lidke K.A., Clayton A.H.A., Arndt-Jovin D.J. and Jovin T.M. Imaging molecular interactions in cells by dynamic and static fluorescence anisotropy (rFLIM and emFRET). Biochem. Soc. Trans. 2003; 31: 1020-1027.
9. Zal T. and Gascoigne N.R. Using live FRET imaging to reveal early protein-protein interactions during T cell activation. Curr. Opin. Immunol. 2004; 16: 418-427.
10. Wahl M., Koberling F., Patting M., Rahn H. and Erdmann R. Time-resolved confocal fluorescence imaging and spectrocopy system with single molecule sensitivity and sub-micrometer resolution. Curr. Pharm. Biotechnol. 2004;5: 299-308.
11. Schneckenburger H., Wagner M., Kretzschmar M., Strauss W.S. and Sailer R. Laser-assisted fluorescence microscopy for measuring cell membrane dynamics. Photochem. Photobiol. Sci. 2004; 3: 817-822.
12. Neher R.A. and Neher E. Applying spectral fingerprinting to the analysis of FRET images. Microsc. Res. Tech. 2004; 64: 185-195.
13. Gu Y., Di W.L., Kelsell D.P. and Zicha D. Quantitative fluorescence resonance energy transfer (FRET) measurement with acceptor photobleaching and spectral unmixing. J. Microsc. 2004; 215: 162-173.
14. Ecker R.C., de Martin R., Steiner G.E. and Schmid J.A. Application of spectral imaging microscopy in cytomics and fluorescence resonance energy transfer (FRET) analysis. Cytometry A 2004; 59: 172-181.
15. Dumas D., Gaborit N., Grossin L., Riquelme B., Gigant-Huselstein C., De Isla N., Gillet P., Netter P. and Stoltz J.F. Spectral and lifetime fluorescence imaging microscopies: new modalities of multiphoton microscopy applied to tissue or cell engineering. Biorheology 2004; 41: 459-467.
16. Trugnan G., Fontanges P., Delautier D. and Ait-Slimane T. FRAP, FLIP, FRET, BRET, FLIM, PRIM...new techniques for a colourful life. Med. Sci. 2004; 20: 1027-1034.
17. Peter M. and Ameer-Beg S.M. Imaging molecular interactions by multiphoton FLIM. Biol. Cell. 2004; 96: 231-236.
18. Peter M., Ameer-Beg S.M., Hughes M.K., Keppler M.D., Prag S., Marsh M., Vojnovic B. and Ng T. Multiphoton-FLIM quantification of the EGFP- mRFP1 FRET pair for localization of membrane receptor-kinase interactions. Biophys. J., in press.

19. Becker W., Bergmann A., Hink M.A., Konig K., Benndorf K. and Biskup C. Fluorescence lifetime imaging by time-correlated single-photon counting. Microsc. Res. Tech. 2004; 63: 58-66.
20. Clayton A.H., Hanley Q.S. and Verveer P.J. Graphical representation and multicomponent analysis of single-frequency fluorescence lifetime imaging microscopy data. J. Microsc. 2004; 213: 1-5.
21. Van Munster E.B. and Gadella T.W. Jr. phiFLIM: a new method to avoid aliasing in frequency-domain fluorescence lifetime imaging microscopy. J. Microsc. 2004; 213: 29-38.
22. Ulrich V., Fischer P., Riemann I. and Konigt K. Compact multiphoton/single photon laser scanning microscope for spectral imaging and fluorescence lifetime imaging. Scanning 2004; 26: 217-225.
23. Requejo-Isidro J., McGinty J., Munro I., Elson D.S., Galletly N.P., Lever M.J., Neil M.A., Stamp G.W., French P.M., Kellett P.A., Hares J.D. and Dymoke-Bradshaw A.K. High-speed wide-field time-gated endoscopic fluorescence-lifetime imaging. Opt. Lett. 2004; 29: 2249-2251.
24. Niggli E. and Egger M. Applications of multi-photon microscopy in cell physiology. Front. Biosci. 2004; 9: 1598-1610.
25. Elson D., Requejo-Isidro J., Munro I., Reavell F., Siegel J., Suhling K., Tadrous P., Benninger R., Lanigan P., McGinty J., Talbot C., Treanor B., Webb S., Sandison A., Wallace A., Davis D., Lever J., Neil M., Phillips D., Stamp G. and French P. Time-domain fluorescence lifetime imaging applied to biological tissue. Photochem. Photobiol. Sci. 2004; 3: 795-801.
26. Rolisnki O.J., Mathivanan C., Mcnaught G. and Birch D.J.S. Structural information on nanomolecular systems revealed by FRET. Biosensors & Bioelectronics 2004; 20: 424-430.
27. Clayton A.H., Klonis N., Cody S.H. and Nice E.C. Dual-channel photobleaching FRET microscopy for improved resolution of protein association states in living cells. Eur. Biophys. J. Published on line: 30 June 2004.
28. Jares-Erijman E.A., Giordano L., Spagnuolo C., Kawior J., Vermeij R.J. and Jovin T.M. Photochromic Fluorescence Resonance Energy Transfer (pcFRET): formalism, implementation, and perspectives. SPIE Proc. 2004; 5323: 13-26.
29. Jares-Erijman E.A., Giordano L., Spagnuolo C., Lidke K.A. and Jovin T.M. Imaging quantum dots switched on and off by photochromic Fluorescence Resonance Energy Transfer (pcFRET). Molec. Crystals & Liq. Crystals, in press.
30. Squire A., Verveer P.J., Rocks O., and Bastiaens P.I. Red-edge anisotropy microscopy enables dynamic imaging of homo-FRET between green fluorescent proteins in cells. J. Struct. Biol. 2004; 147: 62-69.
31. Rizzo M.A. and Piston D.W. A high contrast method for imaging FRET between fluorescent proteins. Biophys. J. , in press.
32. Mattheyses A.L., Hoppe A.D. and Axelrod D. Polarized fluorescence resonance energy transfer microscopy. Biophys. J. 2004; 87: 2787-2797.
33. Rocheleau J.V., Edidin M. and Piston D.W. Intrasequence GFP in Class I MHC molecules, a rigid probe for fluorescence anisotropy measurements of the membrane environment. Biophys. J. 2003; 84: 4078-4086.
34. Jares-Erijman E.A., Spagnuolo C., Giordano L., Etchehon M., Kawior J., Mañalich-Arana M., Bossi M., Lidke D.S., Post J.N., Vermeij R.J., Heintzmann R., Lidke K.A., Arndt-Jovin D.J. and Jovin T.M. 'Novel (bio)chemical and (photo)physical probes for imaging live cells."In *Supramolecular structure and function Vol. 8*, G. Pifat-Mrzljak, ed., pp.99-118, Amsterdam: Kluwer,. 2004,
35. Schenkeveld V.M.E. and Young I.T. Simulations of measurements of fluorescence lifetimes using noise-modulated light. J. Fluoresc. 1997; 7: 55-58.
36. Neher R. and Neher E. Optimizing imaging parameters for the separation of multiple labels in a fluorescence image. J. Microsc. 2004; 213: 46-62.

37. Krishnan R.V., Masuda A., Centonze V.E., and Herman B. Quantitative imaging of protein-protein interactions by multiphoton fluorescence lifetime imaging microscopy using a streak camera. J. Biomed. Opt. 2003; 8: 362-367.
38. Clayton A.H.A., Hanley Q.S., Arndt-Jovin D.J., Subramaniam V. and Jovin T.M. Dynamic fluorescence anisotropy imaging microscopy in the frequency domain (rFLIM). Biophys. J. 2002; 83: 1631-1649.
39. Subramaniam V., Hanley Q.S., Clayton A.H.A. and Jovin, T.M. Photophysics of green and red fluorescent proteins: some implications for quantitative microscopy. Methods Enzymol. 2003; 360: 178-201.

QUANTUM DOTS, A NEW TOOL FOR REAL-TIME *IN VIVO* IMAGING

Donna J. Arndt-Jovin, Thomas M. Jovin, and Diane S. Lidke
Department of Molecular Biology, Max Planck Institute for Biophysical Chemistry, D-37077 Göttingen, Germany

Abstract. Semiconductor nanocrystals, or quantum dots (QDs), are exciting new fluorescent probes useful in imaging at the single molecule to the whole animal level. Some recent applications of QDs are reviewed here and the reader is directed to the original literature for further details.

Keywords: Single Molecule, Semiconductor Nanocrystal, Fluorescent Probes

1. PROPERTIES OF QUANTUM DOTS

Structure, localization and dynamics can be visualized in the live animal, in the single cell, or even at the single molecule level, thereby yielding information about physiological state and activity. Fluorescence imaging microscopy complements the wide variety of biochemical, genetic and physiological techniques for studying cellular function. However, organic and biomolecular fluorophores have limitations in their applications. Such fluorophores generally exhibit only moderate Stokes shifts between their excitation and emission spectra, have relatively broad emission spectra (making it difficult to discriminate multiple colors), and undergo photobleaching when monitored over extended periods of time. Promising alternatives to conventional fluorophores are quantum dots (QDs). The properties and advantages of these new fluorophores have been reviewed in a number of articles in this past year [1-3]. Some of their unique properties are listed in Table 1.

V. Evangelista et al. (eds.), From Cells to Proteins: Imaging Nature across Dimensions, 217–225.

Table 1. Properties of Quantum Dots.

- High photostability
- High quantum yield, Q
- Narrow emission band (size dependent), FWHM ~20 nm
- Size tuneable wavelength
- Broad excitation spectrum
- Act as FRET donors
- Long fluorescence lifetime (FLIM)
- Exhibit blinking
- Can be conjugated to ligands and biomolecules
- Non-toxic
- Electron dense, can be visualized in TEM, EFTEM

The core of the QD consists of a semiconductor nanocrystal, typically CdSe, surrounded by a passivation shell of ZnS (Figure 1). An electron-hole pair is generated by the absorption of a photon. Recombination of the electron-hole pair (in ~ 10-20 ns) leads to the emission of a less energetic photon whose wavelength is dependent on the size of the core (smaller → lower wavelength). The size of the quantum dots can be adjusted via the synthesis conditions [4, 5].

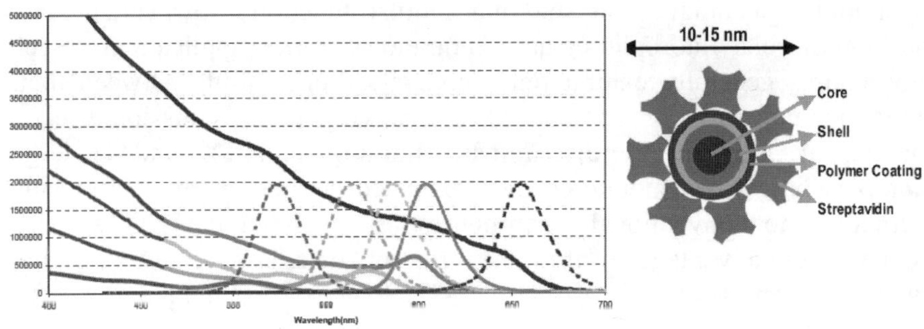

Figure 1. Excitation and emission spectra and composition of quantum dots. Adapted from information from Quantum Dot Corporation. Left: typical spectra of monodisperse Quantum Dots showing continuum excitation and discrete emission spectra. Right: Quantum dots consist of a CdSe nanocrystal surrounded by a ZnS passivation shell. The surface is coated with a polymer that protects the QD from water and allows for chemical coupling to biomolecules. Depicted here is a streptavidin coated QD.

QDs provide many advantages over conventional labels. The excitation spectra of QDs are a continuum, rising into the UV, while the emission spectra are narrow and slightly red-shifted to the bandgap absorption. Thus, multiple QDs with different emission bands can be excited with a single excitation wavelength [6,7]. The large extinction coefficient and the relatively high quantum yield of QDs, and their extraordinary photostability, permit the use of a low sample irradiance and prolonged imaging with a detection sensitivity extending down to the single QD level.

Despite the many advantages of QDs as fluorescent probes for microscopy they have only recently found applications in biology, as a consequence of improvements in the shell and surface coating, that provide isolation of the core from water, increasing solubility in biological buffers, and reactive chemical groups for protein conjugation. QDs are commercially available (see www.qdots.com and www.evidenttech.com) with a variety of conjugated or reactive surfaces - e.g. amino, carboxyl, streptavidin, protein A, biotin, and IgGs (reviewed in [1, 3, 6, 8, 9]). The recent introduction of near infrared (IR) and far-red ternary cadmium selenium telluride quantum dots extends deep tissue, *in vivo,* imaging applications [10-12].

2. Live Cell Imaging

The dynamics and localization of biomolecules in living cells must be qualitatively and quantitatively determined in order to understand cellular function. New insights in such studies have come from the application of QDs. For example, QDs have been employed for specific multiplex protein target labeling [13], multi-color tracking over long-term imaging regimes [14], and monitoring signal transduction [15].

Wu et al. [13] labeled multiple compartments and specific antigens such as the membrane protein erbB2 (Her2), microtubules, actin and nuclear antigen with QDs in both live and fixed cells, and in tissue sections. Both Wu et al. [13] and Jaiswal et al. [14] demonstrated the dramatic photostability of the QDs compared to organic dyes.

In our laboratory, Lidke et al. [15] used a combination of QD-labeled ligands and visible fluorescent protein (VFP)-fusion proteins to study the ligand-induced activation as well as homo- and heterodimerization of members of the erbB family. The QDs were shown to act as bona fide receptor ligands when coupled to a biotin derivative of the naturally occurring epidermal growth factor (EGF) (utilizing the Streptavidin coated QDs). The QD-ligand complex (EGF-QD) was physiologically active, i.e. capable of inducing RTK activation and internalization of erbB1 (epidermal growth factor receptor). Due to the photostability of the QDs, their signal could be monitored continuously for long periods (> 60 minutes). Activated receptors could be tracked from the cell surface through endocytosis, on to vesicular trafficking and fusion in living cells. As little as 50 pM ligand-QDs were easily detectable as single and small clusters of QD-bound receptor, and a new retrograde mechanism for the transport of the receptors on cellular filopodia was discovered.

By studying the behavior of transgenic cells lines expressing VFP-erbB proteins in the presence of QD-EGF we were able to determine that erbB2, but not erbB3, can influence the endocytosis of activated erbB1 as seen in Figure 2. Cell membrane fluorescence and internalized fluorescence were separated by image processing segmentation of data acquired during physiological activation of cells expressing erbB receptors. The bar graph shows values for the internalized signals at 30 minutes after activation by QD-EGF for lines expressing different fluorescent erbB family members. Upon EGF-QD activation of cells expressing erbB1 and erbB2 both homodimers and heterodimers form and are internalized whereas in cells expressing erbB1 and erbB3 heterodimerization does not occur and only erbB1 is internalized. The different behavior of the erbB1 and erbB3 receptors is clearly seen in the cell images shown in Figure 2 as well as in the colocalization histograms of the VFP and the QD signals (see Lidke et al. for details [15]).

Many of the problems in cellular physiology require information about the exact localization of a macromolecule in time, its local environment and the partners with which it may interact. Förster Resonance Energy Transfer (FRET) is a means to assess the direct interaction of two or more molecular species and is covered in a separate chapter in this book (Jovin). The unique spectral properties of QDs with their broad excitation spectra make them ideal donors for fluorescence energy transfer experiments as has been demonstrated *in vitro* [16-18], and can easily be extended to *in vivo* studies.

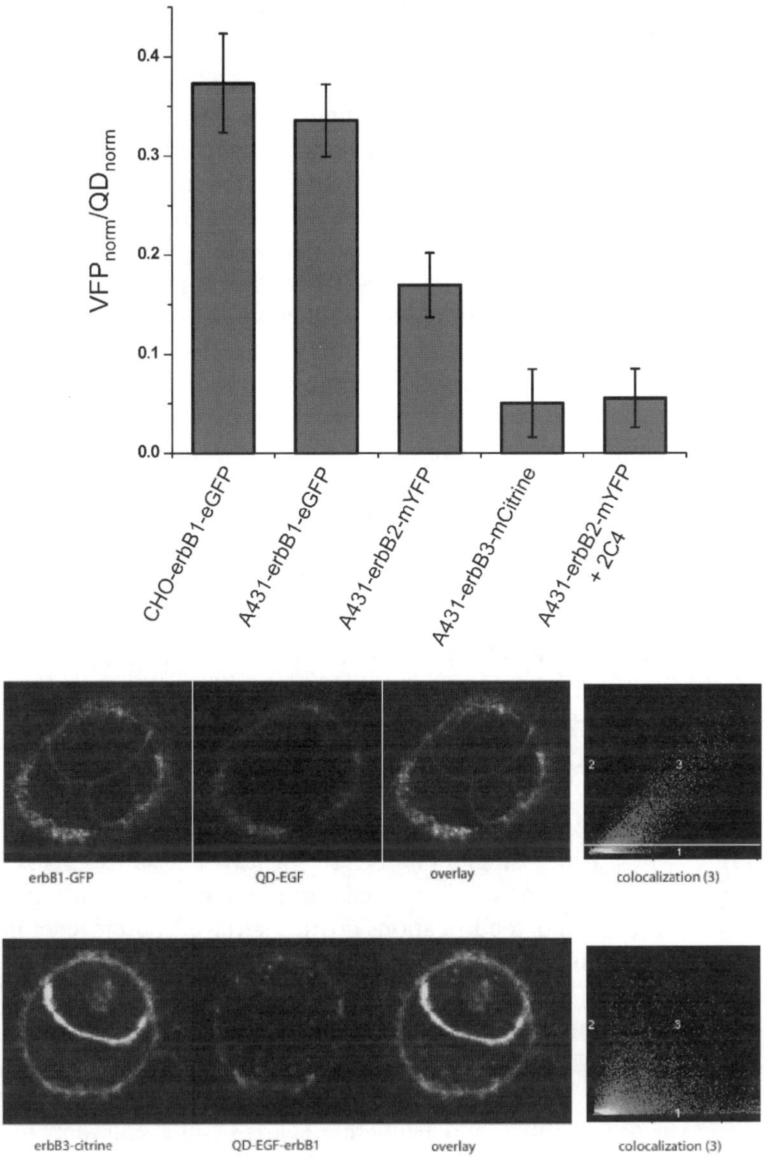

Figure 2. Quantitative analysis of EGF-QD- induced homo- and heterodimerization and internalization. Left: Comparison of the amount of erbB-VFP internalization (VFP_{norm}/QD_{norm}) determined by quantitative image processing. Below: Upper panel, erbB1-GFP binds QD-EGF, homodimerizes and is endocytosed resulting in a high colocalization of both signals in all pixels. Lower panel, erbB3-citrine does not heterodimerize with erB1 so that the citrine and QD-EGF signals are not colocalized after EGF stimulation and endocytosis of the erbB1-QD-EGF (figure taken from [15]).

3. SENSITIVITY AT THE SINGLE MOLECULE LEVEL

Tracking single molecules on the cell membrane has generally involved the use of particles (nanogold of ~ 40 nm or latex beads of ~500 nm) that may interfere with protein dynamics, or small fluorescent labels that suffer from photobleaching. Very small beads have led to underestimation of diffusion constants due to low signal/noise ratios that precluded very fast data acquisition [19]. QDs are smaller (10-20 nm) than traditional beads and, as stated above, are more photostable than conventional dyes. Dahan et al. [20] exploited these qualities for tracking the glycine receptor at the single molecule level in living cells using QDs conjugated to antibodies. We have recently visualized and tracked individual membrane tyrosine kinase erbB1 receptors determining their diffusion constants and velocities of transport upon activated.

Most QDs emitting in the visible wavelength range have a CdSe core that is electron dense and thus amenable to detection by transmission electron microscopy, TEM, including energy filtering techniques [21]. The comparison of light and TEM images provides the means for integrating dynamic measurements and high-resolution localization.

4. WHOLE ANIMAL IMAGING

The photophysical properties of QDs make them excellent probes for *in vivo* imaging when using single photon IR or two-photon sources that penetrate deep into tissues or whole animals. In particular, IR emitting QDs [8] are now available and modifications to the surface coating chemistry have increased the circulating half-lives [10] as well as provided a means for organ targeting [22].

Larson et al [23] found that multi-photon excitation of QDs produced an order of magnitude greater signal than the best two-photon absorbing fluorophores. They used QDs to dynamically image blood flow in capillaries through the skin of live mice and from line scans of QD emission in the capillaries. Both the rate of blood flow as well as the ripples caused by the mouse's heartbeat could be detected.

Deep tissue imaging is optimized in the near IR (NIR) by reducing tissue-induced light scattering. Only 400 pmol NIR QDs and low irradiance were used by Kim et al to successfully track QDs through sentinel lymph nodes in live mice and pigs, providing an improved sentinel lymph node mapping, a prerequisite for cancer surgery resection [12].

Gao et al. [9] linked an amphiphilic polymer to tumor-targeting ligands and coupled these to QDs, thereby effectively targeting human prostate cancer cells grown in nude mice. The QD probes accumulated at tumors both by their enhanced permeability and retention at tumor sites and by antibody binding to cancer-specific cell surface biomarkers. The ability to 'fine-tune' QD targets based on surface coating chemistry [24] will increase their use in both diagnostics and, eventually, therapeutics.

5. STABILITY AND TOXICITY

We and others have found that tissue culture cells loaded with QDs survive for weeks without diminished growth or division and that the QDs are visible over this entire period. In the live animal studies cited here, some of the mice were observed to maintain the QDs over months without obvious deleterious effects [12]. These observations indicate that QDs are safe for long-term imaging in live cells or animals. However, the available data on stability and toxicity is still anecdotal and more studies are required before QDs can be considered safe for therapeutic uses. More information related to biosafety issues is also required.

6. MULTIPLEXING SIGNALS USING QDS

By mixing QDs of various colors in different ratios and capturing them in microspheres, an almost infinite number of specific probes can be constructed and targeted to cells, genes or tissues, thereby establishing a "barcode" identity [25]. Mattheakis et al. coded living cells using peptide delivery of the QDs of various color mixtures. This method has the potential for > 100 different codes [7]. The individual cell types within a mixture of the cells can be recognized by fluorescence imaging or by flow cytometry, opening the possibility of tracking tumor cells in animals or for doing multiplexed physiological assays.

7. SUMMARY

From single molecule studies to tracking in whole animals, QDs are revolutionizing fluorescence imaging. The rapid advances in surface chemistry and reduction in overall size of the particles along with improvements in conjugation strategies continue to broaden their applicability and provide the biologist with a new toolbox of selective, highly fluorescent bioprobes for the future.

8. ACKNOWLEDGEMENTS

We thank many colleagues for valuable suggestions and discussions. The Max Planck Society and the EU 5th Framework Program (Grant No. QLG2-CT-2001-02278), through which DSL was awarded a fellowship, are acknowledged for financial support.

9. REFERENCES

1. Jaiswal J.K., and Simon S.M. Potentials and pitfalls of fluorescent quantum dots for biological imaging. Trends Cell Biol 2004; 14: 497-504.
2. Lidke D., and Arndt-Jovin D. Imaging takes a quantum leap. Physiology 2004; 19: 322-325.
3. Smith A., and Nie S. Chemical analysis and cellular imaging with quantum dots. Analyst. 2004; 129: 672-677.
4. Dabbousi B.O., Rodriguezviejo J., Mikulec F.V., Heine J.R., Mattoussi H., Ober R., Jensen K.F., and Bawendi M.G. (Cdse)Zns core-shell quantum dots - synthesis and characterization of a size series of highly luminescent nanocrystallites. J Phys Chem B 1997; 101: 9463-9475.
5. Alivisatos A.P. Perspectives on the physical chemistry of semiconductor nanocrystals. J Phys Chem 1996; 100: 13226-13239.
6. Bruchez M., Moronne M., Gin P., Weiss S., and Alivisatos A.P. Semiconductor nanocrystals as fluorescent biological labels. Science 1998; 281: 2013.
7. Mattheakis L.C., Dias J.M., Choi Y.-J., Gong, J. Bruchez M.P., Liu J., and Wang E. Optical coding of mammalian cells using semiconductor quantum dots. Anal Biochem 2004; 327: 200-208.
8. Bailey R., Strausburg J., and Nie S. A new class of far-red and near-infrared biological labels based on alloyed semiconductor quantum dots. J Nanosci Nanotechnol 2004; 4: 569-574.
9. Gao X., Y C., Levenson R., Chung L., and Nie S., In vivo cancer targeting and imaging with semiconductor quantum dots. Nat Biotechnol. 2004; 22: 969-976.
10. Ballou B., Lagerholm B.C., Ernst L.A., Bruchez M.P., and Waggoner A.S. Noninvasive imaging of quantum dots in mice. Bioconjug Chem 2004; 15: 79-86.
11. Dubertret B., Skourides P., Norris D.J., Noireaux V., Brivanlou A.H., and Libchaber A. In vivo imaging of quantum dots encapsulated in phospholipid micelles. Science 2002; 298: 1759-1762.
12. Kim S., Lim Y.T., Soltesz E.G., De Grand A.M., Lee J., Nakayama A., Parker J.A., Mihaljevic T., Laurence R.G., Dor D.M., Cohn L.H., Bawendi M.G., and Frangioni J.V. Near-infrared fluorescent type II quantum dots for sentinel lymph node mapping. Nat Biotechnol 2004; 22: 93-97.
13. Wu X., Liu H., Liu J., Haley K.N., Treadway J.A., Larson J.P., Ge N., Peale F., and Bruchez M.P. Immunofluorescent labeling of cancer marker Her2 and other cellular targets with semiconductor quantum dots. Nat Biotechnol 2003; 21: 41-46.
14. Jaiswal J.K., Mattoussi H., Mauro J.M., and Simon S.M. Long-term multiple color imaging of live cells using quantum dot bioconjugates. Nat Biotechnol 2003; 21: 47-51.
15. Lidke D.S., Nagy P., Heintzmann R., Arndt-Jovin D.J., Post J.N., Grecco H.E., Jares-Erijman E.A., and Jovin T.M. Quantum dot ligands provide new insights into erbB/HER receptor-mediated signal transduction. Nat Biotechnol 2004; 22: 198-203.
16. Clapp A., Medintz I., Mauro J., Fisher B., Bawendi M., and H M. Fluorescence resonance energy transfer between quantum dot donors and dye-labeled protein acceptors. J Am Chem Soc 2004; 126: 301-310.

17. Medintz I., Clapp A., Mattoussi H., Goldman E., Fisher B., and Mauro J. Self-assembled nanoscale biosensors based on quantum dot FRET donors. Nat Mater 2003; 2: 630-638.
18. Grecco H., Lidke K., Heintzmann R., Lidke D., Spagnuolo C., Martinez O., Jares-Erijman E., and Jovin T. Ensemble and single particle photophysical properties (two-photon excitation, anisotropy, FRET, lifetime, spectral conversion) of commercial quantum dots in solution and in live cells. Microsc. Res. Technique 2004; 65: 169-179.
19. Murase K., Fujiwara T., Umemura Y., Suzuki K., Iino R., Yamashita H., Saito M., Murakoshi H., Ritchie K., and Kusumi A. Ultrafine membrane compartments for molecular diffusion as revealed by single molecule techniques. Biophys. J. 2004; 86: 4075-4093.
20. Dahan M., Levi S., Luccardini C., Rostaing P., Riveau B., and Triller A. Diffusion dynamics of glycine receptors revealed by single-quantum dot tracking. Science 2003; 302: 442-445.
21. Nisman R., Dellaire G., Ren Y., Li R., and Bazett-Jones D.P. Application of quantum dots as probes for correlative fluorescence, conventional, and energy-filtered transmission electron microscopy. J Histochem Cytochem 2004; 52: 13-18.
22. Akerman M., Chan W., Laakkonen P., Bhatia S., and Ruoslahti E. Nanocrystal targeting in vivo. Proc Natl Acad Sci U S A 2002; 99: 12617-12621.
23. Larson D.R., Zipfel W.R., Williams R.M., Clark S.W., Bruchez M.P., Wise F.W., and Webb W.W. Water-soluble quantum dots for multiphoton fluorescence imaging in vivo. Science 2003; 300: 1434-1436.
24. Ruoslahti E. Vascular zip codes in angiogenesis and metastasis. Biochem Soc Trans 2004; 32: 397-402.
25. Gao X., and Nie S. Quantum dot-encoded mesoporous beads with high brightness and uniformity: rapid readout using flow cytometry. Anal Chem 2004; 76: 2406-2410.

IMAGE ACQUISITION AND ITS AUTOMATION IN FLUORESCENCE MICROSCOPY

Michal Kozubek
Laboratory of Optical Microscopy, Faculty of Informatics, Masaryk University, Botanická 68a, CZ-60200, Brno, Czech Republic

Abstract: Fluorescence microscopy is presently the main tool for the investigation of structure and function of cells and their components. Unfortunately, this type of microscopy suffers from low light levels and consequent high noise levels. This poses big challenge for the image acquisition process. The selection of proper hardware and software parts of the whole set-up (microscope, light detector, driving computer) as well as optimization of the imaging conditions are highly non-trivial tasks that require deep understanding of the laws of physics and mutual dependences of a number of phenomena. This chapter brings insight into these laws and dependencies and provides the reader with a number of hints for the selection of appropriate system components as well as for the optimization of their performance. Special attention is paid to automation in fluorescence microscopy that is absolutely indispensable if multi-dimensional data are to be acquired.

Keywords: Image Acquisition, Fluorescence Microscopy, Automated Microscopy, Image Cytometry, Fish

1. INTRODUCTION

At first glance, image acquisition is a simple task of transforming the “real world” (real biological objects) into the “digital world” (image data stored in computer memory) using a certain optical set-up coupled to a suitable digital camera. As a matter of fact, things are not that easy. Image acquisition is a very responsible task because it prepares the data for the subsequent image analysis step and the quality of the acquired image data is crucial for the successful analysis of its content. Therefore, attention should be paid to choosing both the proper hardware and the software for image acquisition. This is especially true in fluorescence microscopy where the light level is low and, consequently, noise must be dealt with. On the other hand, fluorescence microscopy is well suitable for automation.

Let us have a look at the specifics of fluorescence microscopy. In this field we have to deal with a number of limitations caused by the laws of physics.

V. Evangelista et al. (eds.), From Cells to Proteins: Imaging Nature across Dimensions, 227–270.

These limitations are often in contradiction with the demands of users-biologists.

The most important limitations are the following:

1) Low light level (limited power of light source, photobleaching of fluorochromes, light waste in confocal systems)
2) High noise level (imperfect biological staining, unwanted auto-fluorescence, dust in optical path, electronics noise, photon shot noise due to the low light level)
3) Low optical resolution (hundreds of nanometers)
4) Low image detector dimensionality (0D, 1D or 2D)
5) Mechanical imprecision caused by moving parts of the microscope

Typical demands of the users are as follows:

1) High quality of the images (high signal-to-noise ratio, which means a lot of photons and negligible noise level)
2) High optical resolution (resolving small targets inside biological objects with small mutual distances)
3) High image source dimensionality (up to 5D – x, y, z, λ, t)
4) Precise alignment of optical set-up and precise steps while scanning a particular dimension

Consequence: Selection of suitable hardware and software for image acquisition is a non-trivial and very responsible task.

How can automation help to solve the above-mentioned problems? Why shall we think of the automation in fluorescence microscopy? The reasons may be:

1) Motorized microscope components are more precise and exhibit better reproducibility of the results
2) At least partial automation is required in multi-dimensional image acquisition (at least autonomous capturing of one field of view)
3) Fully automated image acquisition allows obtaining statistically significant data sets, which is especially required in rare event detection or precise topology studies

Consequence: Automation is very desirable in fluorescence microscopy. Fortunately, fluorescence microscopy is very suitable for automation due to independently stained channels (each fluorochrome specifically stains one target type); each acquired channel corresponds to exactly one fluorochrome.

In this chapter, it will be shown how to choose proper components for image acquisition and its automation in fluorescence microscopy:

- Microscopy hardware including automation components (motorized stages and objective movers – stepper motors versus piezo-electrical solution, filter wheels versus filter cubes, acousto-optical elements, confocal units – spinning Nipkow disks versus classical CLSM systems, etc.)

- Light detection hardware (photomultipliers versus cameras, their properties and camera selection guide)
- Computer hardware (suitable for driving microscopes, additional motorized components and light detectors)
- Image acquisition algorithms (strategies and hints for multi-dimensional image acquisition)
- On-line image analysis methods (quick image quality assessment including quick object detection)
- Software for automated image acquisition (commercial versus own)

It will also be shown how to optimize imaging conditions in fluorescence microscopy, more specifically how to properly:

- Prepare the specimen (choosing the proper coverglass thickness, refractive index of oil, refractive index of embedding medium, choosing the proper combination of fluorochromes based on the available light source, optical filters and fluorochrome excitation/emission spectra)
- Suppress noise and artifacts (dark charge noise, readout noise, photon shot noise, monochromatic and especially chromatic aberrations)
- Set acquisition parameters (sampling frequency, bit depth, light intensity, resolution, gain, readout speed, etc.)

Finally, FISH imaging (and its automation) as one of the possible applications in fluorescence microscopy will be discussed.

2. HOW TO CHOOSE PROPER COMPONENTS

2.1. Microscopy hardware

In fluorescence microscopy we can acquire images in two basic modes: wide-field and confocal. Confocal mode yields better optical resolution (depends on the strength of the confocal effect) whereas wide-field mode offers faster image acquisition (short exposure times thanks to the unreduced light throughput). Some instruments allow switching between wide-field and confocal modes or even combining images of the same object acquired in different modes [1]

. The degree of confocal effect (so-called optical sectioning strength) of state-of-the-art commercial microscopes is compared in Table 1. White light source means that mercury or xenon lamps are used as the light source. White light sources are cheap and provide freedom in choosing fluorochromes. On the other hand, the intensity of laser light sources is typically higher (if the power of the laser is sufficiently high). Camera light detectors are better than photomultiplier tubes (PMTs) but cannot be used with single point scanning systems. Direct viewing means that the image can be observed in the ocular. More details about individual microscopy techniques can be found in [2, 3].

Table 1. Comparison of different commercially available microscope types that can be used in fluorescence microscopy. The following microscopy techniques are compared: conventional wide-field (WF), confocal based on structured light illumination using a grid (Grid), confocal based on a Nipkow disk (Nipkow), confocal based on Nipkow disk with microlenses (Microlens Nipkow), confocal laser scanning (CLSM), and confocal laser scanning using two-photon excitation (TPE). Reprinted from (Kozubek et al., 2004).

	WF	Grid	Nipkow	Microlens Nipkow	CLSM	TPE
Producer and model	Many	Zeiss Apotome	Atto CARV	Yokogawa CSU-10	Many	Many
Price	Very low	Low	Low	Moderate	High	Very high
Degree of confocality	None	Medium (fixed)	Medium (fixed)	Medium (fixed)	Low-high (variable)	High (fixed)
Light source	White	White	White	Laser	Laser	Laser (pulsed)
Light detector	Camera	Camera	Camera	Camera	PMT	PMT
Direct viewing	Yes	No	Yes	Yes	No	No
Photo-bleaching	Medium	Medium	Low	Low	High	Low
Photo-toxicity	Medium	Medium	Low	Low	High	Low
Excitation wavelengths	UV-IR	UV-IR	UV-IR	A few	A few	A few

2.1.1. Automation in fluorescence microscopy

There are two stages of automation in microscopy. The first stage is to motorize the required parts of the microscope and make them remotely controlled with a special keypad. The second stage is to replace the manual keypad with a computer interface so that computer can drive the motorized parts of the microscope. Already the first stage is useful because motorized parts of the microscope are more precise. The second stage is useful in the case of automated or semi-automated acquisition of the microscopic images. In this case, special software is required.

Most producers of the optical microscopes offer partially or fully motorized versions of their instruments. Besides, there are companies on the market specialized in manufacturing accessories for optical microscopes. However, although motorized optical microscopes are commercially available,

Table 2. Automation of microscope components. The table summarizes the available hardware possibilities and the corresponding response times. Updated version of the table published in (Kozubek et al., 2001c).

Microscope component	Usage	Response time [$]
Motorized objective revolvers	Slow but precise exchange of objectives	≈1-2 s
Motorized filter-cube revolvers	Slow but precise exchange of filter-cubes	≈1-2 s
Motorized filter wheels	Fast exchange of excitation/emission filters	≈100 ms
Motorized focus drives [#]	Stepper motor (50-100 nm step) control Or piezo (10 nm step) control	≈200 ms / step ≈1-10 ms / step
Motorized stages (x-y or x-y-z)	Stepper motor (50-100 nm step) control	≈500 ms / 100 μm
Electronic shutters	Slow or fast shutter possibility	≈1-100 ms
Lamp intensity controllers	Voltage control via a D/A converter	≈1-10 ms
Acousto-optical elements	Quick wavelength selection	≈25 μs

[#] Although stepper motors are more common, piezo-electrical control is a better choice because of the better resolution, precision and response time.
[$] The times include so-called settling time (time to settle down at the new position); lateral and axial stage movements are not linear and practical values are shown.

they are not widely spread because of the unavailability of suitable software systems for automated image acquisition and analysis from commercial vendors. Hence, the development of own software is often the only possibility.

Table 2 summarizes the accessories for microscope automation available on the market. *Microscope stages* are usually motorized using stepper motors (one stepper motor for each direction). There are X-Y motorized stages (2 stepper motors) and X-Y-Z motorized stages (3 stepper motors) available. Important features of the motorized stages are travel range (maximum travel dimensions), resolution (the shortest step) and repeatability (how precisely one can re-allocate a specific position after moving away from it). For *focus control*, either a stepper motor can be used (e.g. the X-Y-Z motorized stage can be purchased) or a piezo-electric nano-positioning device can be bought (best of all a closed-loop one). Piezo-electric devices usually move objectives up and down instead of the stage. They are quicker and much more precise than stepper motors but have got only a limited travel range (typically 100-

300 μm). *Filters* can be automated in two ways: either two motorized filter wheels are purchased (one for excitation and one for emission) or the whole filter cubes are moved using a motorized revolver (simultaneous exchange of dichroic mirror, excitation and emission filters). In the first case, a dual- or triple-band-pass polychroic mirror is needed (because one mirror must serve for several different dyes). This may slightly reduce image quality but, on the other hand, filter wheels are much quicker than filter cube revolvers. Also *objectives* can be changed using a motorized revolver. If a computer-controlled *shutter* has to be placed into the optical path, either an integrated filter wheel with shutter can be purchased or even a separate electronic shutter. The control of *light intensity* is usually provided by the microscope manufacturer (either the laser power is controlled or the lamp voltage in the case of halogen lamps). The intensity can also be controlled using a filter wheel carrying different neutral density filters instead of filters selecting a certain wavelength range.

The majority of the commercially available microscopes can be automated or is even sold with automated parts. If a specific part of the microscope is not automated, it can usually be replaced by an automated one supplied either by the microscope manufacturer or by a third-party company specialized in automated microscope accessories, such as:

- Ludl (`www.ludl.com`),
- Märzhäuser (`www.marzhauser.com`),
- Prior (`www.prior.com`),
- Sutter (`www.sutter.com`),
- Physik Instrumente (`www.physikinstrumente.com`).

2.2. Light detection hardware

In order to analyze images produced by optical microscopes using computers, the optical image information must be converted into a digital representation and stored into computer memory. In principle this means that photons must be converted into electrons and then the electrons must be counted. In earlier times these two tasks were performed using two independent devices: the conversion of photons into electrons was performed using analog detectors and the conversion of the analog signal into the digital representation was performed using so-called frame grabbers. Frame grabbers were in principle A/D (analogue-to-digital) converters produced as special ISA or PCI cards inserted into computer. Nowadays, mostly digital detectors are used that directly generate the digital signal (the A/D converter is integrated into the detector).

There exist 0D, 1D and 2D light detectors. 3D detectors are unfortunately not available. The description of the light detection hardware will be concentrated on the 2D light detectors. The 0D detectors (photomultipliers) and 1D detectors (line scan cameras) will be mentioned only shortly. Because the vast majority of the used 2D detectors are CCD or CMOS arrays, other types of 2D detectors will not be described. As we are talking about fast imaging devices for the purposes of fluorescence microscopy, „photography and scanner“ method of getting image into computer memory will also be omitted.

2.2.1. 0D detectors: Photomultipliers

Photoelectric cell or *phototube* is an electron tube in which electric current is started by photoelectric emission. Light falling on photosensitive material frees electrons, which flow as a current through the circuit. The photoelectric cell is used in burglar alarms, traffic light controls and door openers. In a modification called the *multiplier phototube*, or the *photomultiplier (PMT)*, metal plates amplify the photoelectric emission, making the tube sensitive to low levels of light. The analog signal (electron current) is finally transformed into the digital signal (usually using an 8-bit A/D converter) for further computer processing. Photomultipliers, however, detect only 0D intensity information.

In optics, photomultipliers are frequently used to detect light originating from a laser light source. By scanning the laser beam column after column and row after row, 2D information is often recorded (e.g. confocal scanning light microscopy). Besides scanning microscopy, photomultipliers are used also in flow cytometers where they record 1D information (the only dimension here is the direction of flowing stream) corresponding to a profile of a cell or a chromosome as it passes the detector (see Section 4.2.1).

The main drawback of the PMTs is low quantum efficiency (typically 20-30%), i.e. only a small number of photons are really detected (the majority of them are lost). Also the response of PMT is usually not linear (in contrary to cameras), which means that k-fold increase in intensity of incoming signal does not correspond to k-fold increase of PMT output. Non-linear output severely complicates quantitative measurements.

2.2.2. 1D detectors: Line scan cameras

Line scan cameras are 1D arrays of light sensitive pixels (usually 512, 1024 or 2048). The pixel size is comparable to the 2D arrays (7 - 25 μm) and mostly also the same technology (CCD) is used. However, the full well capacity (i.e. the number of electrons which can be accumulated) of each pixel can be much higher (up to 100 times). Thus, these devices can accumulate much more signal.

Usually, line scan cameras are used together with line (slit) illumination for the acquisition of 2D data by scanning in the perpendicular direction. This is slower compared to the 2D CCD arrays but at each time only the appropriate object points are illuminated and there is no light from the other points contributing to the image. This is a similar effect as with the confocal point scanning microscopy where only one point at one time is illuminated and acquired. With line scan cameras and slit illumination, the confocal effect is not as high as with point detectors and point illumination, but is still visible and can be sufficient for some purposes. Besides, line scan cameras can be advantageous in some other specific applications such as screening of objects (scene) that continuously move in one direction (e.g. goods moving on belt conveyer).

Line scan cameras are usually not used in microscopy.

2.2.3. 2D detectors: Cameras

Nearly all available cameras nowadays are based on the CCD (charge-coupled device) technology. Recently also CMOS (complementary metal oxide semiconductor) cameras have appeared on the market, however their properties are similar to CCD cameras, therefore a common description for CCD and CMOS will follow. A variety of different camera models are available, from low-cost video-rate cameras up to high quality cooled slow-scan cameras capable of integration of the falling light for several seconds, minutes or even hours. In fluorescence microscopy, the integration feature is very important because of the low light intensity produced by fluorochromes.

Camera consists of an array (two-dimensional matrix) of picture elements (pixels) where each pixel is realized as an *electron well,* which accumulates charge (photoelectrons) just like pools accumulate water during rain. The total accumulated charge is then read out of the chip and digitized by means of an A/D converter for computer processing. Thus, the acquisition consists always of two parts: the exposure (integration of photoelectrons) and the *readout*. The readout procedure depends on the type of the chip. Two basic types are distinguished: frame-transfer and interline-transfer chip.

In *frame-transfer camera*, the whole chip area is light sensitive. After exposure, camera shutter is closed and the chip is read out row by row to a serial register (1D array of an appropriate width) and pixel by pixel out from the serial register to the A/D converter. After readout, the shutter is opened and the camera is ready for the next exposure. The shutter is necessary because of the smearing effect that would occur if the chip was illuminated during the shifts of rows. The newer models of frame-transfer cameras have got two identical arrays: a light sensitive one for the exposure and a light insensitive one for the storage of a freshly acquired image. After the image is exposed, the shutter is closed and the image is transferred quickly using row shifts from the exposure array to the storage array. Then, the shutter is opened and next image can be integrated while the slow readout electronics transfers the content of the storage array into the computer memory. Thus, the storage array enables a sort of parallelism but does not eliminate the problem of shuttering the camera.

On the contrary, *interline-transfer cameras* do not require shutter. Each light sensitive column of pixels is associated a neighboring light insensitive column of pixels into which the exposed picture is transferred. Thus, there are also two arrays of pixels (one light sensitive and one light insensitive) similar to two-array frame-transfer cameras but they are merged together so that the corresponding columns touch each other. This enables to transfer the whole image in one step (instead of a series of shifts) between the arrays and eliminates the problem with shuttering (smearing effect cannot occur). The insensitive array is read out in exactly the same way as in the frame-transfer case. The drawback of this approach is 2 times smaller electron well capacity and 2 times worse light sensitivity (because of the insensitive columns of the chip). The second problem can be, however, eliminated using recently introduced lens-on-chip technology which enables putting a microlens above each pixel which gathers the light into the sensitive half of the pixel.

Light efficiency can further be improved using an *intensified CCD (ICCD)* camera where an intensifier is attached to a normal CCD chip. The intensifier consists of a photocathode, a microchannel plate (MCP) electron multiplier and a phosphor screen. Electrons are ejected from the photocathode in response to incident photons and accelerate toward the MCP input. The number of electrons is increased by the gain of the MCP. The resulting electron image is incident on the phosphor screen that emits many photons for each incident electron. For each photoelectron many photons are thus available to the CCD. The drawback of this technology is a so-called photon shot noise produced by an undesired boost of electrons, which decreases the spatial resolution. ICCDs are, therefore, used mostly only for the observation of fast events when integration is not possible.

If color information is to be acquired, *color cameras* are available: either one-chip or three-chip. One-chip color camera has got masked pixels by red, green and blue filters. The different filters form a certain pattern that is repeated many times over the entire chip. This reduces, of course, the resolution and light sensitivity. A better choice is a three-chip camera where three identical chips are used for the red, green and blue colors. The light is divided using a prism and high-precision optics. Three-chip cameras are, however, quite expensive. In fluorescence microscopy, the monochrome cameras and sequential acquisition of different fluorochrome images are preferred because of the better characteristics of the monochrome cameras.

Cameras can also be divided into analog and digital. *Analog cameras* produce an analog signal (such as TV signal) which must be digitized using a special computer board with A/D converter called *frame-grabber*. The latest trend is to put the digitization electronics right into the camera head so that it produces digital output directly. This type of camera is called *digital camera* Digital cameras produce images of better quality because digital signal can be transported easier and exactly the same information is sent by camera and received by computer.

Choosing the appropriate camera type is a difficult task and a lot of parameters must be considered. A detailed description would take a lot of space, so only a summarizing table (see Table 3) and a couple of general advises (next section) are presented. More detailed description of different cameras and their features can be found, for example, at the address: `www.roperscientific.com` (Library/Encyclopedia).

Table 3. Basic features of cameras.

Camera feature	Units	Description
Resolution	pixels x pixels	Horizontal times vertical number of pixels
Pixel size	μm x μm	Horizontal times vertical size of one pixel
Grade		Quality of the chip (depends on the maximal number of defective pixels)
Full well capacity	electrons	Maximal number of electrons in the well
Read-out noise	electrons/pixel	Preamplifier noise root-mean square (RMS)
Dark current	electrons/pixel/s	Noise produced by thermally generated charge (can be reduced by cooling)
Dynamic range	bits	Depth of digitization (bits in A/D converter)
Readout rate	pixels/s	Number of pixels read out per second
Binning	{pixels x pixels}	Possibility of grouping pixels together
Cooling	no/air/water/LN	Medium used for cooling the chip
Spectral response: curve of quantum efficiency	QE versus nm	Plot of quantum efficiency against wavelength (sensitivity of the chip to different colors)

2.2.4. Choice of an appropriate detector

While PMTs cannot be chosen (they are supplied by the manufacturer of e.g. CLSM system), the CCD (or CMOS) cameras are usually purchased separately from the microscope and the user has to choose from a large number of models produced by a variety of manufacturers. Here are some hints how to select a suitable camera for low-light fluorescence imaging that is the subject of this chapter.

- It is best to choose a *monochrome (black-and-white) camera* because of the sensitivity (fluorescence is often faint, so high sensitivity is advantageous). Sequential scanning of individual fluorescent dyes using a monochrome camera is slower than parallel acquisition of all dyes using a color camera but the quality of the result is much better.
- It is very important to look at the *spectral response* of the chip, i.e. the plot of *quantum efficiency (QE)* against wavelength. This plot tells how many photons are really converted into electrons in average for a given wavelength. The best cameras reach up to 60-70% (back-illuminated cameras even 90%). This is a measure of the *sensitivity* of the chip. One should pay attention to those wavelengths that are emitted by the fluorochromes used or planned for the particular application.
- Further, the camera should be capable of integration of the image for up to several tens of seconds in order to be able to increase the exposure time if necessary. Cameras with the integration possibility are called *slow-scan cameras*. They have slower readout time than normal cameras (e.g. 3 frames per second instead of 30 frames per second) but they are optimized for low-light imaging and yield much better results.
- It is also advantageous if the camera is cooled. For the exposure times of up to several seconds, *air cooling by Peltier element* is sufficient. Water cooling or even liquid nitrogen cooling is not necessary.
- If high-resolution studies are performed (such as precise gene position determination), then the camera should have sufficiently *small pixel size* (in order to increase the sampling frequency). The pixel size varies approximately from 6 to 24 μm. The best cameras have got the pixel size of 6-7 μm.
- The *dynamic range* of the camera should be 12 bit. Lower values (8 bit or 10 bit) are usually insufficient for fluorescence imaging. Higher values do not improve the image quality any more because of low light levels (for low light levels the photon shot noise does not allow for the accurate determination of fluorescence level).
- The *camera resolution* (dimensions of the chip in pixels) is application dependent. Usually, 1300x1000 pixels are adequate. For some applications, larger chips have to be purchased (e.g. to capture a large part of a tissue with high resolution).
- If precise fluorescence intensity measurements have to be performed, it is also advantageous to buy a chip of a *better grade*. Grade says how many defective pixels the chip contains. The lower the number of defective pixels, the more expensive the chip is. Chips with no defects (usually marked as grade 0) are the most expensive. A good compromise is to buy chips with no cluster defects (usually marked as

grade 1), which means that the chip contains defective pixels but they do not occur in clusters.

- Last but not least, the *programmability* of the camera is important if own software is to be written. Especially useful is the *binning* feature that allows to group pixels into superpixels. This is extremely important for software focusing (see Section 2.4.2).

2.2.5. Camera categories

State-of-the-art cameras can be divided into three main categories:

1) Digital cameras for taking *still pictures* (photos)
   - 200-2000 USD, 1-6 Mpixels, up to 1 GB memory on card
   - Slow readout from chip to card, low noise level, high quality
   - Possibility of taking short video sequences (for example 15 Hz, 320x200 pixels, 30s)
   - USB interface for data transfer (from card to computer)

2) Digital cameras for taking *video sequences*
   - 500-2000 USD, 0.5-2 Mpixels, 60/80 min tapes, up to 1 GB memory on card
   - Fast readout from CCD chip to miniDV tape (25/30 Hz, 720x576 pixels)
   - Enable taking low-resolution pictures to card (up to 2 Mpixels)
   - FireWire (IEEE 1394) interface for data transfer (from tape to computer)
   - USB interface for data transfer (from card to computer)

3) Digital cameras for *scientific purposes*
   - 5000-50000 USD, 1-6 Mpixels, no memory cards or tapes
   - Slow readout (1-20 Mpixels/sec), very low noise, very high quality
   - Enable taking pictures or video sequences at low rate (1-10 Hz)
   - Possibilities of cooling, binning, readout speed selection
   - Fully computer controlled using supplied libraries
   - FireWire (IEEE 1394) or own PCI interface for data transfer to computer

Note that the first two categories tend to merge as the quality of still pictures in the second category improves. At the beginning, the second category was sold with 0.5 Mpixels, then with 0.8 Mpixels, in 2003 with 1 Mpixels and in 2004 with 2 Mpixels. As soon as it reaches 3-4 Mpixels, the pictures will be comparable to the first category. Hence, one will have to carry just one instrument capable of acquiring both nice still images to a memory card and digital video sequences to a tape.

2.3. Computer hardware

Let us attend to the question how to choose proper computer hardware. There exist several computer types as well as several operating systems (see Table 4). Each of the three platforms provides a whole range of various models ranging from very cheap ones for text processing up to the top ones for graphics and scientific computation. It is impossible to decide which of the three is generally the best. Every user has own preferences that are influenced mostly by his colleagues and computer types used in his or her institution. For hard scientific computing, UNIX workstations are generally thought to be the best because they have the advantage of very open architecture with a high data throughput: data transfers are fast and more processors can easily be added in future.

Unfortunately, both computer type and operating system are mostly determined by microscope and/or camera producer, especially by the drivers supplied with the instrument. Most microscope and camera manufacturers provide drivers only for PCs, which makes it very complicated or impossible to drive this hardware from Macs or UNIX workstations. Typically a PC with PCI, USB, FireWire, serial or parallel interface and Microsoft Windows operating system are required.

Table 4. Types of computers and corresponding operating systems

Computer type	Manufacturer	Processors used	Operating system
Personal Computer (PC)	Too many	Intel, AMD, Cyrix, etc.	MS-Windows, Linux, etc.
Apple Macintosh (Mac)	Apple Computer	Motorola	Mac-OS
UNIX workstation	Sun, SGI, HP, Compaq, etc.	RISC processors	UNIX (clones)

Other factors that should be considered when buying a computer for image acquisition and/or image processing are as follows:

- Sufficient main (RAM) as well as external (HDD) memory capacity are crucial; also memory speed is an important factor
- Two-processor computer can increase the reliability of the system, especially in Microsoft Windows operating system
- Automatic back-up of the data (e.g. to digital tapes) in regular intervals (such as each second night) is a good idea

- UPS (uninterruptible power supply) for computer as well as microscope is also a good idea

2.4. Image acquisition methods

The image information to be recorded can be 0D (one light intensity value), 1D (illuminated light track), 2D (illuminated area) or 3D (illuminated volume). Some authors speak of 4D or even 5D imaging denoting light wavelength and time as the other dimensions. As discussed earlier, light detectors can be 0D (photomultipliers), 1D (arrays of light sensitive pixels) or 2D (different kinds of cameras).

Let us suppose that we want to record n-D light information using m-D light detector, where m≤n. If m=n, one record is sufficient. If m=n-1, a series of records must be acquired. If m=n-2, scanning in two dimensions is necessary. Thus, for example, 2D information is usually recorded using a 2D light detector, but can be recorded using a 0D light detector combined with scanning (e.g., in point scanning microscopy). 3D information in optical microscopy can be recorded using axial scanning with a 2D detector (e.g., in Nipkow disk microscopy) or both lateral and axial scanning using a 0D detector (e.g., in confocal laser scanning microscopy). A summary of the possible types of multi-dimensional image acquisition can be found in Table 5.

Table 5. Multi-dimensional image acquisition. The table summarizes the possibilities of recording different types of image data. For each type, available detectors and corresponding number of dimensions to be scanned are presented. Reprinted from (Kozubek et al., 2001c).

Information type	Variables	Image type	Detector/ Scanning	Typical application
0D		single value	0D:0D	single measurement
1D	x	1D grey-scale	0D:1D 1D:0D	flow cytometry material inspection
2D	x, y	2D grey-scale	0D:2D 1D:1D 2D:0D	scanning microscopy not used all cameras
3D	x, y, z	3D grey-scale	0D:3D 1D:2D 2D:1D	scanning microscopy not used wide-field microscopy
4D	x, y, z, λ	3D color	0D:4D 1D:3D 2D:2D	scanning microscopy not used wide-field microscopy
4D	x, y, z, t	time series of 3D grey-scale	0D:4D 1D:3D 2D:2D	scanning microscopy not used wide-field microscopy
5D	x, y, z, λ, t	time series of 3D color	0D:5D 1D:4D 2D:3D	scanning microscopy not used wide-field microscopy

Thus, the first decision in image acquisition is which imaging device (light detector) to choose, whether 0D, 1D or 2D. Because 1D devices are not commonly used in microscopy, the decision is reduced to the choice between PMT-based acquisition and camera-based acquisition. There is no general consensus which choice is better. Both systems have got their advantages and disadvantages. The advantage of using a camera detector is quicker and simpler acquisition, linear response, precise spacing of the pixels and higher sensitivity. The disadvantage is fixed sampling frequency (given by the pixel size of the chip). In commercial systems, the manufacturer does the choice: usually in laser-based systems PMTs and in non-laser-based systems cameras are used. However, there are exceptions to this rule: for example, microlens Nipkow disk systems use laser light source and camera detector.

The second decision in image acquisition is how to handle the scanning in individual dimensions. As stated above, if the dimensionality of the detector is less than the dimensionality of the image to be acquired, a method of scanning in the other dimensions must be chosen. The following sections will describe how to handle the different types of scanning.

2.4.1. Scanning in X and Y dimensions (lateral scanning)

There exist 2 types of lateral scanning: *primary* (within one field of view in single point scanning systems) and *secondary* (used for slide scan across many fields of view).

Primary lateral scanning is used in PMT-based systems (such as CLSM). This type of lateral scanning is required to form a 2D image of the observed object. This scanning is always automated by the manufacturers of the point scanning microscopes at hardware level. Usually, two mirrors are used: one tilts the light beam in x-direction and the other one tilts the beam in y-direction. The tilting of the two mirrors is synchronized so that the beam travels in a meander way (TV-like way) across the image.

The reason for *secondary lateral scanning* is that each optical setup (PMT or camera based) has got a limited field of view, i.e. only a fraction of the objects to be acquired can be recorded during one image capture. For example, a 2D camera is often used for the acquisition of 2D information. But the camera sees only a fraction of the specimen and scanning is necessary in order to capture the whole specimen. The final image consists then of many fields of view acquired sequentially. The scanning here does not increase the number of dimensions of the acquired image (if compared with axial scanning or the lateral scanning described in the previous paragraph) but virtually enhances the field of view. In optical microscopy, the lateral scanning cycle is usually implemented by moving the stage with the specimen according to a certain pattern. The most used pattern is a *meander pattern* (i.e. the same pattern as in the previous paragraph), which means that the stage moves horizontally forward from one end to the other, then backward along the neighboring line, then again forward one line further and so forth. In this way, the whole slide can be scanned (e.g. overnight). It is a good idea to provide a small *overlap* of the neighboring fields of view. The overlap should be roughly equal to the dimensions of the observed objects (cells). This enables to analyze also objects which are incomplete (and therefore excluded from the analysis) at the given stage position. On the other hand, the overlap should not exceed the smallest object dimension, otherwise one object can be found in two neighboring images (and counted twice into the resulting statistical summary of the whole slide).

2.4.2. Scanning in Z dimension (axial scanning)

Axial scanning is used in both PMT- and camera-based systems for three purposes:

- *3D imaging:* to obtain 3D image of the observed object
- *2D-Max imaging:* to extend the depth of focus in 2D imaging
- *Software focusing:* to evaluate the sharpness of the image at different z-positions

The first type of z-scanning is required *to record 3D images*. The majority of microscope suppliers usually automate axial scanning so that the whole 3D image can be recorded by pushing one button. There are two parameters that the user must supply: axial step and number of slices. The *axial step* determines the voxel size in z-direction. Its value usually corresponds to the size of the smallest object that we want to record (so that no object is missed). For high-resolution imaging, the axial step should be roughly equal to the depth of focus (i.e. typically about 0.3μm for high NA objectives). The *number of slices* determines the z-dimension (number of voxels in z-direction) of the acquired 3D image. It is usually set so that the axial scanning range (the number of slices multiplied by the axial step) covers the axial dimensions of the largest objects on the slide. Both the axial step and the number of slices are usually constant for the whole slide.

2D-Max imaging is usually not supplied in commercial systems but it is not difficult to implement using the following algorithm: 1) Collect a stack (z series) of xy-images similarly to 3D imaging but with an intermediate resolution (z-step can be slightly larger); 2) Compute a new artificial xy-image as pixel-wise maximum over the acquired z-stack of xy-images (the value at each pixel of the new image is equal to the maximal value at the very same lateral position among all xy-images in the z-stack); 3) Throw away the z-stack and save only the maximum image. Note that sometimes also sum is used instead of maximum but maximum usually produces better results.

Software focusing is indispensable prerequisite of each automated system because the depth of focus in fluorescence microscopy is usually small (typically about 0.3μm for high NA objectives). Therefore, it is reasonable to focus the specimen before acquisition and to refocus the image each time the stage moves (e.g. in an automated lateral scanning cycle) because even the best microscope slides are not ideally flat. Automated focusing is possible both in direct (non-fluorescence) light microscopy (where even hardware autofocusing is available) and fluorescence microscopy using appropriate software. The focusing algorithms take several image samples at different axial stage positions and compute a so-called *focus function* for each image. The focus function has its absolute maximum in the focal position and decreases in both directions. The stage is finally moved to the axial position corresponding to this maximum. Several functions can be chosen for the focusing and their comparison can be found in [4, 5, 6].

The time for focusing one field of view can be reduced to 2-3 seconds using state-of-the-art cameras with binning feature [7]. If focusing is time-consuming, *interpolation or extrapolation techniques* can be used in the scanning cycle so that focusing is performed only on each third field of view or even more rarely. Both techniques assume that the objects (mostly cells) lie within one plane that is tilted relative to the optical axis of the microscope. In the interpolation case, several points (usually corner points plus some inner points) are focused and the algorithm interpolates among them. In the extrapolation case, the algorithm computes the correct axial position according to several previous neighboring measurements. In both cases, proper focusing is applied from time to time (usually at each third image) in order to correct the prediction obtained by interpolation or extrapolation.

2.4.3. Scanning in wavelength dimension

In fluorescence imaging, the wavelength dimension is always present and several different dyes (stains) have to be acquired and combined into one image. Either all dyes are acquired simultaneously (using a color camera or several PMTs) or individual dyes are acquired sequentially through the appropriate highly specific filters (using a monochrome camera). The second approach is slower but gives better results because of the high specificity of the filters. It requires sequential scanning in wavelength dimension and combining (merging) of the individual monochrome images in computer memory into one pseudo-colored image for visualization purposes. This can easily be implemented provided a 2D acquisition procedure is available.

If more than three different fluorochromes (more than three channels) are imaged, then their excitation/emission spectra overlap and cross-talk between channels occurs. In this case *proper spectral imaging* is desirable to get rid of the cross-talk numerically (by spectral unmixing of mixed channels) and recording of full spectrum is required. In such case a *prism* or *grating* is used to disperse the light into individual wavelengths, some microscopes include one of these two dispersion elements:

2000: Leica TCS SP2 (prism)
2001: Zeiss LSM 510 META (grating)
2004: Olympus FV 1000 (grating)

The λ-step (step in wavelength dimension) depends on how many channels one wants to distinguish (using linear unmixing). In Leica and Olympus one can choose the λ-step, in Zeiss the λ-step is set to 10.7 nm and the number of samples is 32.

2.4.4. Scanning in time dimension (time-lapse imaging)

The time step in time-lapse imaging can vary from fractions of a second to minutes or even hours. The choice of the time step depends on the speed of observed objects. If quick events are monitored, then fast image capture and fast readout are crucial. For this purpose intensified CCDs in camera-based systems are often used. Alternatively, one can employ the binning feature of a common scientific grade camera [3]. If slow events are observed and the time step is large (e.g. minutes), then monitoring multiple stage positions can be performed (only few systems offer this possibility but it is not difficult to implement).

Sometimes the problem in time-lapse imaging is to keep the living objects within the field of view (or volume of view if z-scanning is employed). In this case, on-line tracking algorithms are required that keep the object of interest centered within the field (or volume) of view.

2.4.5. Scanning order

There are two main possibilities how to choose the scanning order:

1) primary X, primary Y, λ, Z, secondary X, secondary Y, T
2) primary X, primary Y, Z, λ, secondary X, secondary Y, T

Usually not all dimensions in the list are scanned (only some of them but in the above-mentioned order). Most frequently used is the first possibility. The alternative scanning order (second line) is better if the following criteria are satisfied:

- The time to change filters is large as compared to the time for axial movement (time to make one z-step)
- The reproducibility of axial positions is good
- The number of samples in λ-dimension is small (e.g. 2-4)
- The total acquisition time for one pseudo-color 3D image is short as compared to instabilities of optical set-up (e.g. thermal drifts)

2.5. On-line image analysis methods

On-line processing of the acquired image data is used for a quick inspection of its content, especially for the purposes of deciding whether to store the data to the hard disk or throw it away (or store only its part). On-line image analysis uses quick methods to obtain the results in real-time (seconds or fractions of a second). The computed results are used for the decisions what to do next, i.e. results of the on-line computations influence the software behavior.

On-line processing can be divided into:

- Filtering (image enhancement) methods (used for the purposes of visualization on computer monitor)
- Image segmentation methods (used to find objects in image data)
- Object measurements (size, intensity, etc.)
- Object classification (e.g. positive or negative)
- Quick statistical evaluation (histograms, scattergrams, etc.)

On-line image segmentation is the most important but the most difficult task to do in short time. It can help in the following respects:

1) To assess image quality if followed by object measurements (whether the image contains any objects of expected size, shape, etc.; if not, then the image data is not worth further attention or saving to disk)
2) To decide whether to acquire the given stage position or not (first only one 2D image at coarse resolution is acquired, e.g. during focusing, then it is analyzed and it is decided whether to perform proper multi-dimensional acquisition)
3) To compress the image data and save space on disk (only objects with certain properties are found and saved to disk, alternatively the uninteresting image parts are set to black, which significantly helps to improve the compression ratio

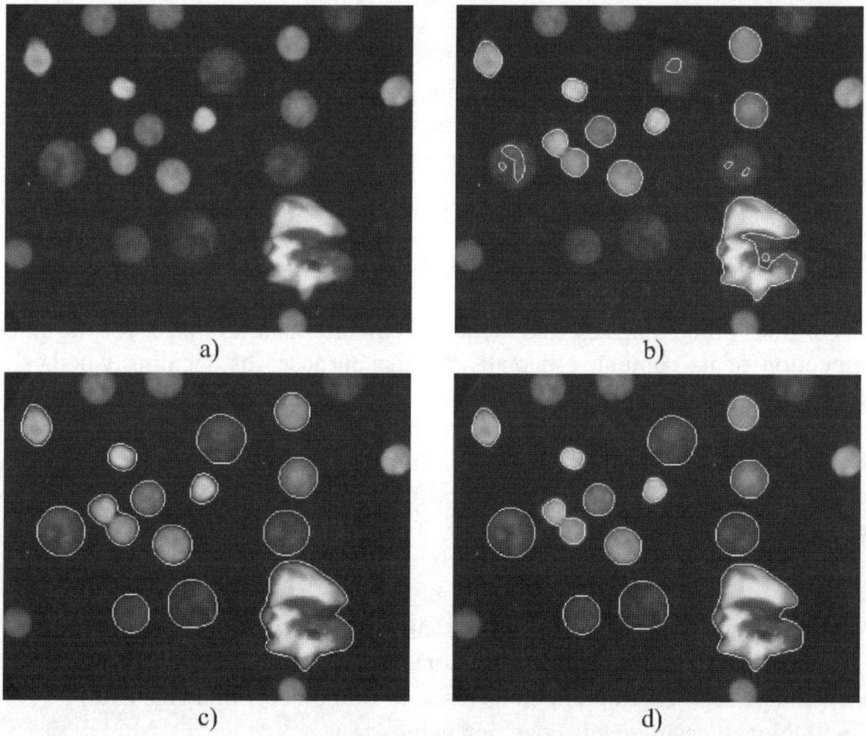

Figure 1. An example of on-line image segmentation using global and local intensity thresholding of cell nuclei stained with DAPI. Cells are at different stages of the cell cycle and, therefore, have different DNA contents and consequently different fluorescence intensities. a) Original image. b) Boundaries obtained by global thresholding with high threshold appropriate for bright cells. c) Boundaries obtained by global thresholding with low threshold appropriate for faint cells. d) Boundaries obtained by local thresholding with individual threshold for each cell. Reprinted from (Kozubek et al., 2001c).

Typically thresholding or local (adaptive) thresholding is used for on-line segmentation [8, 9]. It is quick and relies on the global or local intensity histograms, respectively. The difference between global and local thresholding is illustrated in Figure 1.

2.6. Software

All major microscope manufacturers (Zeiss, Leica, Olympus, Nikon) have their own software developers. Typically they provide, however, only basic microscope automation for the automatic acquisition of one field of view using their microscope. This basic software does not contain algorithms for software focusing, automatic stage scanning, on-line image analysis, etc.

Fortunately, there exist companies that attend to the development of software for microscope automation. Typically these companies provide complete solutions including motorized microscope, light source, light detector, computer and software. The number of supported applications is, however, limited (usually only 2D images can be acquired, not 3D). The most famous companies active in the field of microscopy automation are the following:

- Applied Imaging, UK (`www.aii.co.uk`)
- Imstar, France (`www.imstar.fr`)
- Loats Associates, USA (`www.loats.com`)
- MetaSystems, Germany (`www.metasystems.de`)

The advantage of buying *commercial software* is that it is ready and usually supplied as part of the whole system, thus the supplier should guarantee the proper performance. Disadvantages are the high price and limited extension possibilities (new algorithms can not be easily added).

Another possibility is to develop *own software*. In this case one gains the freedom of gradual extensions and modifications (quick response to the demands of the users) but the development of the first version takes quite a long time (months or even years). Moreover, communication with the hardware suppliers is required in order to obtain necessary drivers and manuals for programming individual hardware components. Unfortunately, not all hardware suppliers are willing to provide assistance.

3. OPTIMIZATION OF IMAGING CONDITIONS

The performance of each optical system depends on many factors such as imaging mode, numerical aperture of the objective, type of objective correction, refractive index of the embedding medium (refractive index mismatching), coverglass thickness, depth of the observed objects below the coverglass, emission wavelength, detector pinhole size in confocal microscopy, ratio between emission and excitation wavelengths in fluorescence microscopy [10, 11, 12]. These factors influence the basic features of the optical system such as point spread function (PSF), optical resolution, optical sectioning (confocality), light throughput, degree of monochromatic and chromatic aberrations, etc.

In fluorescence microscopy, all the above-mentioned features are of great importance. The most important feature of an optical system is the *optical resolution* (further referred to as resolution). It is mostly defined as the shortest distance between two theoretical or real objects such that the two objects are distinguishable in the image formed by the optical system. Thus, optical resolution determines the shortest distance between objects that can be measured using a particular optical system. Note that if we speak about resolution of two objects of different spectral signature (stained with different fluorescent dyes), then resolution is infinitely small – we can resolve the two objects even if they coincide (zero distance).

We require not only good resolution but we often need also good optical sectioning, e.g. when we observe stained tissue samples in 3D. We need a lot of light because fluorescence signals are often very faint. And we also need to eliminate chromatic aberrations because we want to overlay individual color channels as precisely as possible (in co-localization studies).

Let us have a look at how we can optimize the imaging system performance (so that the above-mentioned features become better). We will start with specimen preparation, continue with optical system behavior (including compensation for its imperfections), and finish with setting appropriate acquisition parameters.

3.1. Specimen preparation

Already during the preparation of the specimen we should think about possible consequences of using various compounds on the quality of the image in the microscope. Here some facts are presented that not all people preparing specimens are aware of. More details can be found in [13].

3.1.1. Refractive index

Refractive index (RI) of all compounds that the light has to pass should ideally be the same. Because this requirement is not realistic, one should try to keep the refractive indices of different compounds at least close to each other. What cannot be changed is the refractive index of glass (1.518). Therefore, when using oil-immersion objectives one should use *immersion oil* with this refractive index. Moreover, oils for fluorescence microscopy should be used that are usually denoted with letter F.

Another important factor is the refractive index of the *embedding medium* that the objects are surrounded with. It ranges from 1.33 (RI of water and cells) to 1.518 (RI of glass). Most commonly glycerol-based medium is used (RI about 1.4). Unfortunately, it is usually not possible to match RI of embedding medium to both glass and observed objects. Therefore some aberrations should always be expected.

Somewhat easier is the situation in the case of water-immersion objectives. Because no coverglass is used, objective immersion medium and object embedding medium are identical (water). The RI of water is well matched to the RI of observed objects and the objective is designed to be immersed into water.

3.1.2. Coverglass thickness

Four types of coverglass thickness values exist:

- No. 0 (0.08-0.12mm)
- No. 1 (0.13-0.16mm)
- No. 1.5 (0.16-0.19mm)
- No. 2 (0.19-0.25mm)

The actual thickness of coverglasses may vary (even within one batch). Mostly the value is close to the middle of the specified interval, i.e.:

No. 0 ≈0.10mm

- No. 1 ≈0.145mm
- No. 1.5 ≈0.175mm
- No. 2 ≈0.22mm

Because the vast majority of the existing objectives is corrected for 0.17mm coverglass thickness, best imaging results are obtained using No. 1.5 coverglass.

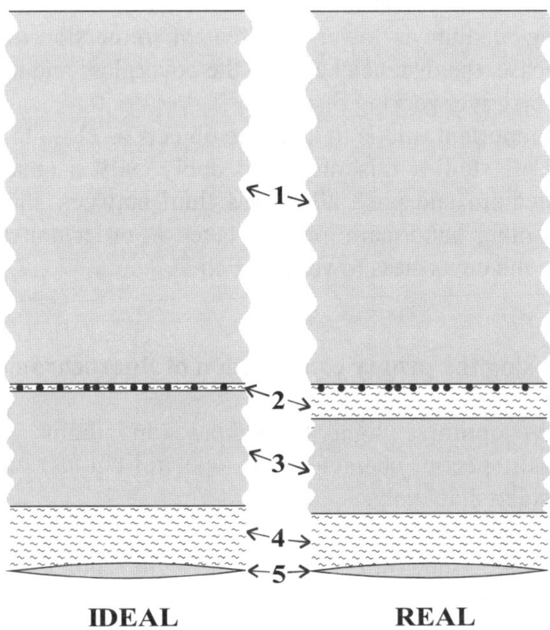

Figure 2. Schematic drawing of the optical set-up. Objects stained with fluorescent dyes (black circles) are attached to the surface of a microscope slide (1), poured with an embedding medium (2), covered with a coverglass (3) and observed through immersion oil (4) using an oil embedding objective (5). Although inverted microscope geometry is shown, the upright microscope drawing can easily be obtained by turning the image upside down. Ideal (left) and real (right) situations are compared. Reprinted from (Kozubek, 2001a).

3.1.3. Depth of objects below coverglass

The larger the depth below the coverglass is, the worse the aberrations caused by refractive index mismatch are. The effect of depth is much more severe than the effect of coverglass thickness. Good imaging properties can be observed at depths of up to several micrometers. If the depth is tens of micrometers, then the aberrations are already rather severe and the resolution is much worse. Therefore, ideally, the objects should be fixed to the coverglass surface, not to the microscope slide. Unfortunately, this is not always possible due to handling procedures during specimen preparation.

The difference between ideal and real case is illustrated in Figure 2. Ideally the coverglass thickness is 170 μm, the observed objects are located just beneath the coverglass and the refractive index of the embedding medium is equal to the refractive index of the immersion oil (1.518). In this case, the distance between the coverglass and the objective is equal to the objective working distance specified by the manufacturer (typically 90-100 μm for highest NA objectives). In real situation, the coverglass thickness is usually smaller (typically 140 μm if No. 1 coverglass is used), the observed objects are located at a certain depth beneath the coverglass (typically several micrometers or even several tens of micrometers) and the refractive index of the embedding medium is lower than that of immersion oil (typically 1.33-1.48). In this case, the distance between the coverglass and the objective is not equal to the objective working distance.

The most important rule is to keep the objects as close to the coverglass as possible. In cell studies this means to apply only a small amount of the embedding medium and push all excess fluid between slide and coverglass away. On the other hand, care must be taken in order not to distort the cells while pushing the coverglass towards the slide.

3.1.4. Choosing the proper combination of fluorochromes

In order to optimize light throughput (and light yield of imaged fluorochoromes), spectral characteristics (spectral curves) of all components should be considered, namely:

- Light source spectrum (also white light sources have significant emission peaks)
- Spectral transmission curve of the glass (lenses in the optical path) (especially if UV or IR light is used)
- Spectral transmission curves of all optical filters in the light path (excitation and emission filters, neutral density filters, hot filters, dichroic or polychroic mirrors)

- Spectral characteristics of the available fluorochromes (excitation as well as emission curves)
- Spectral curve of the quantum efficiency of the light detector

The spectra of different light sources can be found, for example, in [14]. The spectra of optical components should be obtained from the producer (e.g. `www.chroma.com` or `www.omegafilters.com`). Table of fluorochromes and their excitation/emission spectra can be found, for example, at `fluorescence.biorad.com`. Quantum efficiency curves of cameras and PMTs can be found in the detector manuals (see e.g. `www.roperscientific.com`).

3.2. Noise and artifacts

Noise and artifacts are phenomena that are very unpleasant but unfortunately occur in each imaging system. Everybody who acquires images has to continually struggle with them and try to minimize them. There are different types of noise and artifacts (see Figure 3). The three main types of noise that influence imaging in optical microscopy are:

- Readout noise
- Dark charge noise
- Photon shot noise

Based on these 3 types of noise the total noise and signal-to-noise ratio can be computed. Other artifacts can be divided into the following groups:

- Monochromatic aberrations (spherical, curvature of field, astigmatism, distortion and coma)
- Chromatic aberrations (usually much more crucial than mono-chromatic ones)
- Other optical artifacts (dust in optical path, reflections, thermal drifts, etc.)
- Specimen-induced artifacts ("biological" dust, imperfect staining, dust on slide, etc.)

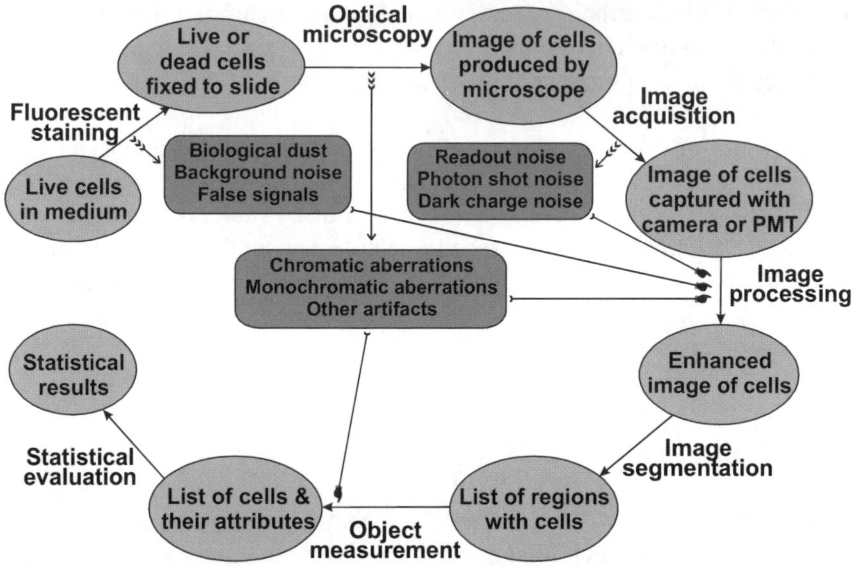

Figure 3. Noise and artifacts in fluorescence microscopy. The diagram shows in which phase individual types of noise or artifacts arise and also in which phase they are corrected for by computer. Typically they are handled during image processing phase but some types can be dealt with also after object measurement.

3.2.1. Readout noise

Readout noise is the noise caused by the uncertainty in quantifying the electronic signal. It depends mainly on pre-amplifier design and post-amplifier electronics, detector temperature (the lower the temperature, the lower the noise) and readout rate (the faster the readout rate, the higher the noise). The readout noise is specified by the producer in electrons per pixel. The value for the best cameras is 4-8 electrons/pixel.

3.2.2. Dark charge noise

Dark charge noise is the noise caused by statistical fluctuations of dark charge signal, i.e. signal generated internally in the detector even if no light is falling on it (e.g. in dark room). *Dark charge signal* is linearly proportional to the exposure time. The producer specifies dark charge generated per pixel per unit time (in electrons/pixel/s). The best cameras generate less than 0.1 electrons/pixel/s. If this constant is multiplied by the exposure (acquisition) time, the total dark charge signal is obtained. Because dark charge obeys Poisson statistics, the dark charge noise is equal to the square root of the dark charge signal. Dark charge signal (and consequently dark charge noise) is strongly dependent on the detector temperature: the lower the temperature, the lower the noise. Thus, dark charge noise can be considerably reduced by cooling.

3.2.3. Photon shot noise

Photon shot noise is the noise caused by statistical fluctuations of the acquired signal level. The signal level is the product of the photon flux (light intensity), quantum efficiency of the detector (average number of electrons generated by one photon, see Section 2.2.4) and exposure time. Because the signal also obeys Poisson statistics, the photon shot noise is equal to the square root of the signal level. Thus, increasing the signal level can decrease the percentage of the photon shot noise. The signal level can be increased by increasing the fluorescence light intensity (this is not always easy in fluorescence imaging) or by increasing the quantum efficiency of the detector (this is possible only within a certain range at the time of purchase) or by increasing the exposure time (longer integration is usually possible but limited by fading of the fluorescent dyes). In general, the longer the exposure time, the better the acquired signal. Therefore, anti-fading agents should be used during cell staining in order to facilitate long exposure times. This is especially true in 3D studies where the fluorochromes should keep their intensity for longer time than during simple 2D acquisition.

3.2.4. Total noise

The *total noise* is computed as the square root of the sum of squares of individual noise types. The total noise together with the acquired signal level determines the image quality that is measured as the ratio of the signal level and the total noise. This ratio is called *signal-to-noise ratio (SNR)*. In fluorescence imaging, we deal with low light levels and, therefore, the dominant noise is the photon shot noise. The image quality (SNR) in fluorescence imaging can be consequently improved only by choosing brighter fluorescent dyes or by increasing the exposure (integration) time for each image.

3.2.5. Monochromatic and chromatic aberrations

Modern objectives are quite well corrected for different types of monochromatic (spherical aberration, curvature of field, astigmatism, distortion and coma) as well as chromatic aberrations. Detailed description of individual aberration types can be found, for example, in [15]. In general, the more expensive the objective, the better the corrections. Nevertheless even the best objectives available on the market exhibit some residual aberrations.

In fluorescence microscopy, the most important aberrations are *chromatic aberrations* (also called *chromatic shifts*). Let us explain where they come from. According to a basic principle of optics, focal length of lens (f) changes with wavelength (λ). A straightforward consequence of this phenomenon is an axial offset between focal planes of different wavelengths. Hence, the same feature in the observed specimen changes its axial in-focus position when wavelength is changed. This mismatch is called *axial (or longitudinal) chromatic aberration*. According to another principle of optics, magnification (M) is proportional to 1/f. If we combine the two principles together we obtain the following consequence: M changes with λ. Hence, the same feature in the observed specimen also changes its lateral off-axis position when imaged at different wavelengths. This magnification error is called *lateral chromatic aberration*. Chromatic aberrations are caused by imperfect chromatic correction of the optical system (usually by imperfect correction of the objective). Although the manufacturers try to compensate for chromatic aberrations in their optical components (mostly objectives), even the best objectives (and microscopes) exhibit chromatic aberrations to some extent.

Chromatic aberrations are not negligible in some types of cell studies. A typical example of such a study is a short distance measurement between two targets visualized using two different fluorescent stains (such as a pair of heterologous genes involved in a translocation). Such measurements are often performed in the range 0-0.5 μm (nowadays mostly in 3D) and a precision of <0.1 μm (in both lateral and axial direction) is highly desirable. Obviously, such precision cannot be obtained using a fluorescence microscope with chromatic aberrations of typically 0.1-0.3 μm in lateral and 0.2-0.5 μm in axial direction.

A method for measuring and software correction of chromatic aberrations has recently been described in [16]. The method uses fluorescent beads stained with several fluorescent dyes simultaneously (such as TetraSpeck from Molecular Probes) for the determination of the aberration function (which is dependent on the lateral position in the field of view) and subsequent software correction.

3.2.6. Other optical and specimen-induced artifacts

Specimen-induced artifacts ("biological" dust, imperfect staining, dust on slide, etc.) must be minimized already during preparation of the slide (see also Section 3.1). Other optical artifacts (dust in optical path, reflections, thermal drifts, etc.) must be minimized before image acquisition. It is a good idea to perform both specimen preparation as well as image acquisition in a dust-free room (room with a good air cleaning system). Regular maintenance and alignment of the optical set-up also helps.

3.3. Acquisition parameters

For each acquisition software, the user has to select a lot of various acquisition parameters that determine how the image should be captured. Let us have a look at the most important parameters that must be set before the acquisition of fluorescence images.

3.3.1. Sampling frequency

The sampling frequency (i.e. the number of pixels per unit length in x-, y- or z-direction) should be set so that the *Nyquist criterion* is met, i.e. the sampling frequency should be at least twice the largest frequency which we want to record. In practice this means that the finer are the details that we want to resolve in the image, the higher should the sampling frequency be set. When fine details are not important (e.g. in the case of counterstain image of cell nuclei), the sampling frequency can be lower. But when small features have to be resolved (e.g. hybridization dots) higher sampling frequency is desirable.

In PMT-based systems, the sampling frequency is changed by setting a different step for the rotation of mirrors that tilt the laser beam in x- and y-directions. In camera-based systems, the sampling frequency is simply equal to the total magnification divided by the camera pixel size (or superpixel size if binning is used). As the spacing of the pixels in the chip cannot be varied, the only way of changing the sampling frequency is to change the overall magnification. This can be done either by changing the objective or by changing the intermediate zoom, if the microscope contains a zoom-unit. The axial (z-) sampling frequency is usually easily changed by setting a different step of the z-stepper motor or piezo-electric device.

3.3.2. Pixel (voxel) size

The pixel (voxel) size in x-, y- (x-, y-, z-) direction is an inverse of the x-, y-, (x-, y-, z-) sampling frequency. In PMT-based systems, the software provides the user with the possibility to set sampling frequency (pixels/μm) or pixel (voxel) size (μm/pixel). The effect of both approaches is identical. In camera-based systems, such controls are usually not available due to the constant sampling frequency for the given magnification (see above). The only exception is the use of the binning feature of the chip. This enables to group squares of $N \times N$ camera pixels into larger superpixels (i.e. to increase pixel area N^2 times). See [3] for more details.

3.3.3. Image size

Usually the user wants to acquire as large field of view as possible. Each system has got some limitation posed on the maximum number of pixels in each direction. This maximal number of pixels divided by the sampling frequency determines the maximal field of view. So mostly the maximum number of pixels is set. However, sometimes one may want to acquire only a small subregion (e.g. one cell) of the whole field of view. The majority of the modern systems allow doing this. The only constraint posed on the subregion in such systems is that it is rectangular. This is because of the easy storage and handling of rectangular matrices in computer memory.

3.3.4. Bit depth

The bit depth says how many bits are used for storing the light intensity value of one pixel in computer memory. This number is usually set so that it corresponds to the dynamic range of the detector. However, if the dynamic range of the detector (in bits) is not a multiple of 8 (e.g. if we have a 10- or 12-bit camera), then the bit depth is set to the closest higher multiple of 8 (e.g. 16 for 10- or 12-bit cameras). This means that some zero bits are appended to each pixel. It is also possible to omit 2 or 4 least significant bits and transform a 10- or 12-bit image into an 8-bit image. Some information is however lost using this approach. The multiple of 8 is required because computers work with bytes. Most software packages, however, do not provide the user with the possibility to change the bit depth.

3.3.5. Light intensity

For most light sources, the light intensity can be varied. The user changes either the power of a laser or the voltage of a lamp source. For mercury lamps, the intensity cannot be changed but it is possible to regulate the intensity using different neutral density filters. On one hand, the intensity should be as high as possible in order to obtain good image quality (SNR). On the other hand, the intensity of fluorescence fades rapidly for high intensity levels. Using low light levels (at which the fluorescence does not fade) and long exposure times one can obtain images of higher quality than using a short snap at high intensity level. Therefore, a compromise must be found experimentally.

If the signal is too faint, another fluorochrome should be considered instead of thinking about light source intensity (see Section 3.1.4).

3.3.6. Optical sectioning (degree of confocality)

An example of the optical sectioning properties of different microscope types has already been presented in Table 1. In general, confocality is primarily determined by the pinhole size in front of the detector [11, 12] in CLSM and by the pinhole size in the Nipkow disk systems. The smaller the pinhole, the better the confocality of the optical system. However, small pinholes decrease light throughput. Therefore, one must always find a compromise between confocality and light throughput. In Nipkow disk systems, the pinhole size cannot be changed (one would need several disks, each with a different pinhole size).

3.3.7. Readout speed

Some light detectors allow choosing the readout speed (e.g. 10 MHz or 20 MHz in state-of-the-art cameras). Smaller values are better because of better signal-to-noise ratio (lower readout noise). If readout time is not crucial, the smallest available value should be used.

3.3.8. Gain

Some systems allow the user to select analogue gain of the detector (PMT or CCD). The gain determines the degree of signal amplification. High gain settings improve image brightness but at the expense of image quality. The reason is that the higher the gain, the larger the readout noise. So a good idea is to keep the gain settings at some medium values (usually recommended by the producer) and amplify the signal digitally.

The only exception is *electron-multiplying back-illuminated CCDs* that use high gain setting (although at the expense of higher readout noise) in order to get the signal level sufficiently high above the readout noise level. This trick should be used only if the signal level cannot be increased by other means.

3.3.9. Exposure (integration) time

As discussed above, the longer the exposure time, the better the image quality (SNR). However, exposure time is limited by fading of fluorescent dyes or by speed of observed events. So a reasonable compromise must be found for each particular application. In older CLSMs and older CCD-based systems, the exposure time cannot be varied. Therefore, another approach is used – namely *multiple exposures*. This means that several images of the same field of view are acquired (with the same acquisition parameters) and averaged. The effect is comparable with the longer integration but the image has to be read several times, which means that the time consumption increases and additional readout noise is added.

4. FISH IMAGING AND INVESTIGATION OF 3D GENOME STRUCTURE

As an example of fluorescence imaging application, fluorescence in situ hybridization (FISH) technique and its automation will be presented in the following text. FISH is the main biological tool for the visualization of the genetic material (selected chromosomes as well as individual genes) inside cell nucleus. The FISH technique is not new but it is still widely used worldwide. Its main advantage is the possibility of mapping spatial (3D) structure of the genetic material with high specificity (one can precisely select the genome regions to be visualized). Different targets (e.g. different genes) can be stained with different colors (different spectral signatures) so that they can be distinguished in the acquired image. In this way, complex 3D genome topology studies can be performed.

4.1. The FISH technique

4.1.1. Principle of *in situ* hybridization

The *in situ* hybridization technique enables the detection of specific DNA or RNA sequences in the cytoplasm, organelles, chromosomes or cell nuclei. This ability is of great importance in the analysis of chromosome and genome structure and aberrations, the construction of physical maps of chromosomes, the determination of the spatial and temporal expression of genes, the localization of transformation sequences, oncogenes, etc.

The DNA or RNA sequences are visualized by means of DNA or RNA probes. *Probes* are DNA or RNA sequences that are complementary to the target sequence. Both the probe and the target sequence must be single-stranded and that is why the first step of the FISH technique is denaturation of both the probe and the target sequence in cells that are usually fixed to a microscope slide. Then, the probe is let to anneal to the target sequences by dropping the probe on the slide and letting the mixture hybridize for some time in a suitable environment. The probe that did not anneal to any target is then washed away in post-hybridization washing steps. Thus, the remaining probe is annealed exactly to the target sequences.

In order to visualize the probes, they must be labeled with a certain dye (so-called *direct method*) or chemically modified (so-called *indirect method*) before hybridization. The indirect method means that the probe is modified with molecules of biotin or digoxigenin and then (after hybridization) labeled anti-biotin or anti-digoxigenin molecules are applied (the same approach as in immunocytochemistry). The most frequently used labels are fluorescent labels although there exist also some non-fluorescent ones (e.g. light-absorbing or radioactive). If fluorescent labels are used, the technique is called *fluorescence in situ hybridization (FISH)*.

The ideal length of the probe is 100-300 bp. Shorter probes result in lower nucleic acid hybrid stability and larger probes have difficulties to penetrate into cell structures. However, using a lot of probes simultaneously (i.e. by visualization of many targets simultaneously), it is possible to visualize large objects such as whole chromosomes. The final targets (which we want to see in the microscope) can be *unique or low-copy sequences* such as protein-coding genes or transformation sequences, *repetitive sequences* such as satellite sequences in centromeric regions, *specific chromosomes or genomes*, *mRNA sequences* or *viral sequences*.

Different probes are suitable for different purposes. Probes for repetitive sequences produce the brightest signal and are, therefore, suitable for automatic detection and scoring. Probes for specific chromosomes are important in structural aberration scoring. Unique sequence probes are invaluable for the detection of specific translocations occurring in different malignant diseases.

It is possible to use several types of probes simultaneously. Different types have different labels that are observed at different excitation/emission wavelength (i.e. different colors) in a fluorescence microscope. This approach is called *multi-color FISH*. It is also convenient to paint total DNA (besides the probes) by a certain dye in order to see the nuclear boundary. This stain is called *counterstain* and mostly propidium iodide (PI) or 4',6-diamidine-2-phenylindole (DAPI) are used.

A typical example of FISH image data is presented in Figure 4.

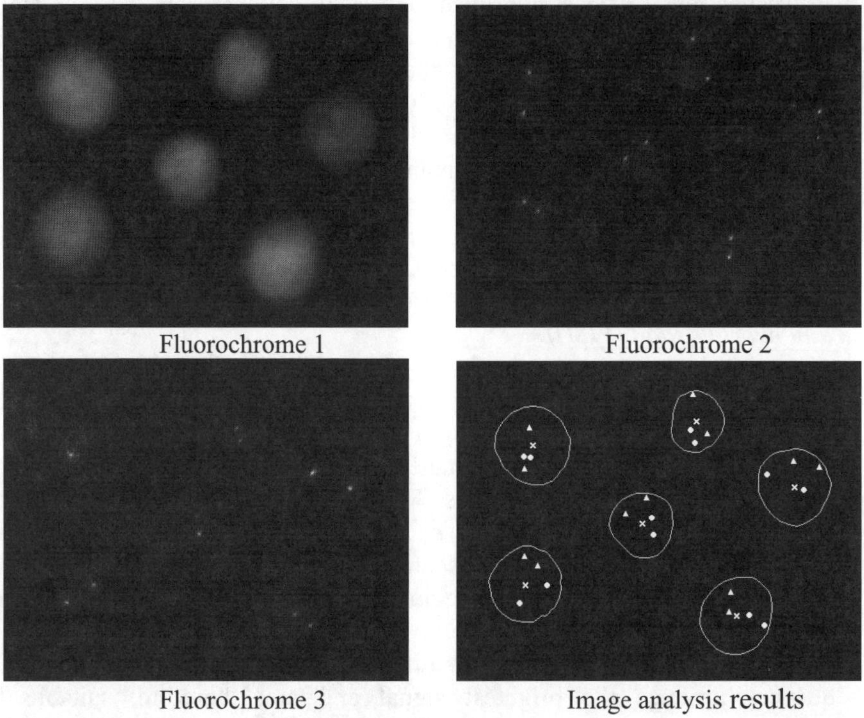

Figure 4. An example of image data and corresponding image analysis results. Cell nuclei were stained using the FISH technique. One fluorochrome was used for staining the interior of the cell nucleus and two were used for staining two different types of genes. The corresponding excitation/emission light was UV/blue, blue/green and green/red. 3D images of the three different fluorochromes were sequentially captured using a grey-scale CCD camera and appropriate optical filters. Maximum XY projection (i.e. pixel-wise maximum along z-axis) is shown for each of the three channels. Image analysis results for individual channels are visualised in one co-ordinate system (in this case a simple overlay of the three channels was sufficient for image registration). The closed curves represent boundaries of cell nuclei, the cross signs represent their weight centres and the circles and triangles represent the two types of genes. The vector graphics is projected into XY plane. Only part of the whole x-y image range is shown (about 60 x 45 μm out of 160 x 120 μm).

4.1.2. Modifications of FISH method

Two other modifications of the FISH method *are comparative genomic hybridization (CGH)* and the *primer-mediated DNA synthesis (PRINS)*. CGH provides a comprehensive analysis of imbalanced chromosomal material of the entire genome. CGH results in a ratio profile of two fluorescent signals. The relative strengths of the signals indicate chromosomal subregions that may be amplified or deleted. In research, CGH is used to examine chromosomal aberrations in several types of tumors to identify chromosomal regions containing genes that may contribute to the disease. The procedure includes the extraction of DNA from the tissue of interest, DNA labeling with fluorochrome-dUTP and hybridization of metaphase chromosomes with the 1st fluorochrome-labeled specimen DNA and the 2nd fluorochrome-labeled normal reference DNA [17, 18].

PRINS method is based on the primer-mediated DNA synthesis. It starts with the annealing of an oligonucleotide DNA primer adjacent to the DNA region of interest. This annealed oligonucleotide serves as a primer for the Taq DNA polymerase. The polymerase incorporates four nucleotides dATP, dGTP, dCTP and dUTP, of which the dUTP is labeled with a fluorochrome. The applications of the PRINS methodology are focused on the detection and localization of different genes in the cell nucleus [19, 20].

4.1.3. Preparation of probes for FISH

Probes are mostly produced by *cloning*: the DNA is inserted into a *vector*, and both the vector and the insert are amplified inside appropriate host cells. The amplified DNA is then extracted. The detection of genetic loci by means of FISH requires techniques for the efficient cloning of large DNA fragments. In the beginning, the most frequently used vectors were cosmids (derived from λ phage) and yeast artificial chromosomes (YACs). Both these clones have serious disadvantages. Cosmids accept comparatively small inserts (35-45 kb) and the clones are relatively unstable [21, 22]. YAC systems enable the cloning of inserts up to 1 Mb [23, 24]; however, the inserts are occasionally unstable and the manipulation with the long DNA is difficult. Efforts to overcome the limitations of cosmids and YACs have resulted in alternative large-insert cloning approach using a bacteriophage P and the *Escherichia coli* fertility plasmid (F-factor). The bacteriophage P1 cloning system (PAC) can accept inserts in the 70-100 kb range [25, 26]. The F-factor-based BAC vectors can accept inserts up to 300 kb [27].

DNA probes can also be prepared and amplified using the long template PCR system. The method uses the mixture of the DNA Taq polymerase and a thermostable DNA polymerase with proof-reading activity to overcome limitations in the length of amplified fragments [28]. The system can amplify fragments of 20 kb and more from isolated genomic DNA using specific primers consisting of 20-30 nucleotides suitable for the region of interest. The procedure includes denaturation of template, annealing of specific primers, primer extension by DNA polymerases using dNTP, where dUTP is labeled with a fluorochrome. The yield of labeled DNA depends on the number of PCR cycles.

Synthetic oligonucleotides can also be used as probes. They can be tailor-made to hybridize to specific sequences, but their synthesis is quite complicated and only 10-50 bp long oligonucleotides are feasible to produce.

Because of the short length, they are easily washed away during the post-hybridization washing steps, and must therefore be handled carefully.

4.1.4. Applications of FISH technique

The FISH technique is the most frequently used technique in the structural cytology of the cell nucleus. It provides a reliable means for studying the genetic composition of cells in mitosis as well as in interphase [29]. Gradually, FISH has reached a high detection sensitivity (individual genes can be detected) and a high multiplicity (several different probes can be applied to the same nucleus) [30]. It is used both in research and in clinical applications for the identification and quantification of numerical and structural chromosomal aberrations. Interphase cytogenetics is of great value for the detection of disease-related translocations such as the t(9;22) translocation observed in chronic myeloid leukemia [31], in prenatal diagnosis [32] or for studying various solid tumors [33]. In research, it can be used to study the spatial organization of the genetic material inside cell nuclei, relative positions of different genetic loci, organization of chromosomes, etc. [34, 35, 36, 37, 38, 39, 40, 41, 42, 43]. Also the influence of different carcinogenic or differentiation agents on nuclear topography [44, 45, 46] and the correlation between chromatin structure and malignant diseases [47, 48] can be studied. Not only isolated cells, but also cells in tissues can be hybridized [49] as well as living cells [50]. In order to increase the number of probes, the same slide can be re-hybridized several times with different sets of probes each time and the same objects can be re-acquired and superimposed in computer memory [51, 52].

The latest trend is to perform even many parallel hybridizations in a grid on one slide [53, 54]. The slide with the grid is called a *microarray* or *"DNA-chip"*. This technology enables the analysis of human single-nucleotide polymorphisms and the detection of disease-causing mutations in large population studies.

4.2. Automated systems for FISH imaging

In this section different types of automated systems for the acquisition of FISH-stained cells will be presented ranging from 1D flow cytometry approach up to 5D high-resolution image cytometry.

4.2.1. Flow versus image cytometry

In the research of the spatial organization of the genetic material inside cell nucleus various methods of cytometry are used. *Cytometry* means performing measurements on cells (from Greek *kytos*=cell and *metrum*=measure). Not only cells themselves are measured but also their components such as cell nuclei, proteins, ions, selected DNA or RNA sequences, etc. Such measurements can provide both quantitative and qualitative information on the measured components. Therefore, cytometry plays a key role in cell research. Mostly the word „cytometry“ is used when large quantities of cells are processed.

In order to measure certain cell components, they must be visualized by means of one of the available visualization methods. Most often, fluorescent dyes are used and the most common staining techniques are fluorescence in situ hybridization (FISH) and immuno-fluorescence.

There exist two main types of cytometry: flow cytometry and image cytometry. *Flow cytometry* means measuring cells and their components while the stained cells flow one after another in a narrow stream of liquid through a special chamber where they are illuminated with a laser beam. The detector is PMT and the acquired information is basically 1D signal (light intensity profile) and the only dimension is time. *Image cytometry* means measuring cells and their components fixed on a certain surface (typically on a glass slide) so that they do not move (or move very slowly in live cell imaging) during the acquisition process. In this case 2D or even 3D image data is recorded (hence the technique is called image cytometry).

Flow cytometry is an automated technique and enables to process large quantities of cells (thousands or even tens of thousands) in a short time (typically several minutes). Therefore, it is widely used in hospitals as well as in research. However, very little information is obtained. On the other hand, image cytometry provides a lot of information about cells and their content but few automated systems exist and, therefore, a small number of cells is usually processed (typically tens, rarely hundreds), which is not enough for statistically significant results.

Our results have shown that it is possible to increase the speed and the number of processed cells also in image cytometry using a suitable combination of state-of-the-art tools provided by molecular biology, optical microscopy and computer science.

4.2.2. Low-resolution image cytometry

If 2D imaging and low resolution are sufficient for the given task, then one can purchase a so-called laser-scanning cytometer (LSCM) developed by Kamentsky et al. [54, 55] that provides an interesting compromise between flow cytometry and conventional wide-field microscopy. *Laser scanning cytometry (LSCM)* is a tool for fast evaluation of FISH-stained interphase nuclei fixed on a microscope slide and hybridization dot counting. It is similar to confocal scanning light microscopy because it also uses a laser light source, point-by-point scanning and 2-4 photomultipliers (plus appropriate beam-splitters and filters) to detect signals from different fluorescent dyes. However, the laser beam spot is somewhat larger (2.5-3.0 μm in diameter for 40x objective) and the system uses non-confocal imaging in order to increase the depth of field. This is necessary because only 2D images are recorded. Furthermore, laser beam scanning is performed only in one lateral direction and the scanning in the other lateral direction is performed using stage movement. The stage moves from one end of the slide to the other one and, consequently, a long scan strip is acquired. Then, a neighboring strip is acquired and then the next one and so forth until the whole slide is scanned. Typically, a 40x objective, 2.5 μm beam spot size and 170 μm strip width are used. Appropriate software performs on-line image analysis and plots histograms or scattergrams similar to flow cytometry. Up to 1000 nuclei per minute can be processed using this technique.

LSCM approach is fast but deals only with low-resolution images (because of the large beam spot size). Low resolution is acceptable for nuclei segmentation (nuclei are large and we are not interested in fine details) but problems arise when hybridization dots have to be detected. Images of two dots that are close to each other merge and the dots are detected as one dot. Thus, in LSCM only large targets are detectable such as pericentromeric regions while unique sequences are not detectable.

Nevertheless, LSCM is a good compromise when large quantities of cells are to be analyzed at high speed and information provided by flow cytometry is not sufficient. An application example can be numerical aberration scoring where pericentromeric probes can be used and only the number of chromosomes within each nucleus is to be computed. LSCM is now a commercially available technique (from CompuCyte Corp., Cambridge, MA, USA) and can be a good supplement to flow cytometers in clinical cytogenetics.

4.2.3. High-resolution image cytometry: Early dot counters

The automation of high-resolution image acquisition and analysis has been performed only for very specific tasks that are not too complicated. An example of such a relatively easy task is counting of hybridization dots (FISH dots) in 2D images of interphase cell nuclei under the assumption that their density on the microscope slide is not very high (the nuclei do not form clusters). The first automated system for this task was constructed at the Delft University of Technology in the Netherlands by Netten et al. [57, 58]. The system could count hybridization dots within interphase cell nuclei using a motorized epi-fluorescence microscope and 2D image analysis. The result of the analysis was the dot frequency distribution. Another automated hybridization dot counter was reported by Ortiz de Solórzano et al. [59]. However, both systems worked only in 2D and performed only dot counting. They did not enable to perform spatial arrangement studies.

4.2.4. High-resolution image cytometry: Brno HRCM instruments

High-resolution cytometry (HRCM) was introduced in 1999 [7] and it is a technique that enables to perform automatic 2D as well as 3D image acquisition and analysis of a large number of cells with high resolution using a computer-driven microscope. In HRCM, a special set of automated microscope components are used and a special driving software. All components and all software modules are optimized for speed and quality. High resolution is obtained by means of high NA plan-apochromat objectives typically combined with a cooled high-resolution digital CCD camera.

This system is more advanced than the dot counters. It is not only designed to cope with multi-color fluorescence images using sequential acquisition of signals from individual probes through proper single-band filters but it can also perform complex 2D as well as 3D analysis of the stained objects including their spatial distribution. Not only point-like signals (dots) but also other types of signals (e.g. chromosome territories) can be analyzed. The system can automatically find cell nuclei in 2D or 3D images and FISH-stained genetic loci (genes, centromeres, chromosomes) inside cell nuclei. Targets stained using immunofluorescence (proteins) can also be acquired and analyzed. Not only object counting but also complex topological analysis can be performed including the analysis of gene position inside the corresponding chromosome domain.

A special software system for the automated 2D and 3D acquisition and analysis of fluorescence-stained objects has been developed. This software works in conventional wide-field [7] as well as in confocal mode [1] and the two modes can be combined. The speed of the acquisition and analysis depends on the mode used and ranges from 50-100 cells per minute in 2D (wide-field mode) to 0.5-2 cells per minute in 3D (confocal mode).

The system also enables the performance of *repeated acquisition*, i.e. several stainings (e.g. hybridizations) of the same microscopic slide with different sets of targets (probes). In this case, the positions of all objects (cells) acquired during the first meander scan (after the first staining) are stored onto the hard disk and exactly the same objects are re-allocated and re-acquired after the second, third, etc. staining. The analysis results of different stainings can be combined in computer memory. Repeated acquisition allows the user to visualize even more cellular targets than a single multi-color staining.

Recently HRCM technique has been extended also to time-lapse imaging [3]. In this way, it is possible to record 5D information at high resolution in an automated way.

4.2.5. High-resolution image cytometry: Commercial systems

Commercial software for automated high-resolution image acquisition and analysis is still quite rare and very expensive. For a long time only automated metaphase spread acquisition and analysis (plus semi-automatic karyotyping) has been commercially available. These systems have been offered for conventional Giemsa or band stainings and later also for FISH-stained metaphase spreads. However, automation of interphase cell studies has been out of scope of commercial vendors for a long time. Only very recently there have begun to appear also systems for automated interphase cell studies. They are far from being ideal (they acquire mostly only 2D images) but the development, optimization and improvement of these systems will surely continue in future.

The following companies offer automated high-resolution solutions for image cytometry (see also Section 2.6):

- Applied Imaging, UK (`www.aii.co.uk`)
- Imstar, France (`www.imstar.fr`)
- Loats Associates, USA (`www.loats.com`)
- MetaSystems, Germany (`www.metasystems.de`)

It should be noted that these companies usually supply complex solutions, i.e. software and hardware together including the microscope and the camera. So the installation of the software for automated imaging to current hardware setup may be impossible. Therefore, care must be taken when purchasing the microscope hardware components if automated imaging is planned in the future. Another disadvantage of the commercial systems is that one cannot easily modify them or add a new feature into the driving software. Moreover, the companies do not publish details of the implemented image analysis algorithms. Therefore, care must also be taken when selecting a particular software package from the above-mentioned vendors. It is always a good idea to request a demonstration on own specimens in order to check the behavior of the particular software package on own image data.

5. ACKNOWLEDGMENTS

I thank all the colleagues from the Laboratory of Optical Microscopy, Masaryk University, and Institute of Biophysics, Academy of Sciences, for long-term participation on the research in the field of automation of image acquisition in fluorescence microscopy. This research has been supported by the Ministry of Education of the Czech Republic (Project No. MSM-143300002).

6. REFERENCES

1. Kozubek M., Kozubek S., Lukasova E., Bartova E., Skalnikova M., Matula P., Matula P., Jirsova P., Ganova A., Koutna I. Combined confocal and wide-field high-resolution cytometry of FISH stained cells. Cytometry, 2001; 45: 1-12.
2. Kozubek M. "FISH imaging." In *Confocal and two-photon microscopy: Foundations, applications, and advances*, Diaspro, A. ed. New York:Wiley-Liss, 2001.
3. Kozubek M., Matula P., Matula P., Kozubek S. Automated acquisition and processing of multidimensional image data in confocal in vivo microscopy. Microsc Res Techniq, 2004; 64: 164-175.
4. Groen F.C.A., Young I.T., Ligthart G. Comparison of Different focus functions for use in autofocus algorithms. Cytometry, 1985; 6: 81-91.
5. Firestone L., Cook K., Culp K., Talsania N., Preston K. Comparison of autofocus methods for automated microscopy. Cytometry, 1991; 12: 195-206.
6. Boddeke F.R., Van Vliet L.J., Netten H., Young I.T. Autofocusing in microscopy based on the OTF and sampling. Bioimaging, 1994; 2: 193-203.
7. Kozubek M., Kozubek S., Lukasova E., Mareckova A., Bartova E., Skalnikova M., Jergova A. High-resolution cytometry of FISH dots in interphase cell nuclei. Cytometry, 1999; 36: 279-29.
8. Sonka M., Hlavac V., Boyle R. *Image Processing, Analysis and Machine Vision*. Pacific Grove: PWS Publishing, 1999.
9. Pratt W.K. *Digital Image Processing*. New York: John Wiley & Sons Inc., 2001.
10. Wilson T. (ed) *Confocal microscopy*. London: Academic Press, 1990.
11. Pawley J.B. *Handbook of biological confocal microscopy*. New York: Plenum Press, 1995.
12. Diaspro A. *Confocal and two-photon microscopy: Foundations, applications, and advances*. New York: Wiley-Liss, Inc., 2001.
13. Kozubek M. Theoretical versus experimental resolution in optical microscopy. Microsc. Res. Techniq., 2001; 53: 157-166.
14. Murphy D.B. *Fundamentals of light microscopy and electronic imaging*. New York: Wiley-Liss, 2001.
15. Born M., Wolf E. *Principles of optics*. Oxford: Pergamon Press, 1975.
16. Kozubek M., Matula P. An efficient algorithm for measurement and correction of chromatic aberrations in fluorescence microscopy. J. Microsc., 2000; 200: 206-217.
17. Weiss M.M., Hermsen M.A., Meijer G.A., van Grieken N.C., Baak J.P., Kuipers E.J., van Diest P.J. Comparative genomic hybridisation. Mol Pathol, 1999; 52: 243-251.
18. Wilkens L., Tchinda J., Burkhardt D., Nolte M., Werner M., and Georgii A. Analysis of hematologic diseases using conventional karyotyping, fluorescence in situ hybridisation and comparative genomic hybridisation. Hum Pathol, 1998; 29: 833-839.
19. Koch J., Kolvaa S., Peetersen K., Gregersen N., and Boland L. Oligonucleotide priming methods for the chromosome specific labeling of alpha satelite DNA *in situ*. Chromosoma, 1989; 98: 259-265.
20. Koch J. "Non-radioactive labeling and detection of nucleicacids in situ by "Primed In situ Labeling (PRINS)"". In *Non-radioactive in situ hybridization* Boehringer Mennheim Biochemica (ed). Application Manual, pp 41-33 (1992).
21. Collins J., Hohn B. Cosmids: a type of plasmid gene-cloning vector that is packagable in vitro in bacteriophage λ heads. P Natl Acad Sci USA, 1978; 75: 4242-4246.

22. Craig A.G., Nizetic D., Hoheisel J.D., Zehetner G., Lehrach H. Ordering of cosmid clones covering the herpes simplex virus type 1 (HSV-1) genome: a test case for fingerprinting by hybridisation. Nucleic Acids Res, 1990; 18: 2653-2660.
23. Burke D.T., Carle G.F., Olson M.V. Cloning of large segments of exogenous DNA into yeast by means of artificial chromosome vectors. Science, 1987; 236: 806-812.
24. Burke D.T., Olson M.V. Preparation of clone libraries in yeast artificial chromosome vectors. Method Enzymol, 1991; 194: 251-270.
25. Pierce J.C., Sternberg A., Sauer B. A mouse genomic library in the bacteriophage P1cloning system. Organization and characterisation. Mamm Genome, 1992; 3: 550-558.
26. Ioannou P.A., Amemiya C.T., Garnes J., Kroisel P.M., Shizuya H., Chen C., Batzer M.A., de Jong P.J. A new bacteriophage P1-derived vector for the propagation of large human DNA fragments. Nat Genet, 1994; 6: 84-89.
27. Hosoda F., Nishimura S., Uchida H., Ohki M. An F-factor based cloning system for large DNA fragments. Nucleic Acids Res, 1990; 11: 3863-3869.
28. Cheng S., Fockler C., Barnes W.M., Higuchi R. Effective amplification of long targets from cloned inserts and genomic DNA. P Natl Acad Sci USA, 1994; 91: 5694-5699.
29. Cremer T., Lichter .P, Borden J., Ward D.C., Manuelidis L. Detection of chromosome aberrations in metaphase and interphase tumor cells by in situ hybridization using chromosome-specific library probes. Hum Genet, 1988; 80: 235-246.
30. Tanke H.J., Florijn R.J., Wiegant J., Raap A.K., Vrolijk J. CCD microscopy and image analysis of cells and chromosomes stained by fluorescence in situ hybridization. Histochem J., 1995; 27: 4-14.
31. Tkatchuk D., Westbrook C., Andreef M., Donlon T.A., Cleary M.L., Suryanarayan K., Homge M., Redner A., Gray J.W., Pinkel D. Detection of BCR-ABL fusion in chronic myelogeneous leukemia by two color fluorescence in situ hybridization. Science, 1990; 220: 559-562.
32. Pinkel D., Landegent J., Collins C., Fuscoe J., Segraves R., Lucas J., Gray J. Fluorescence in situ hybridization with human chromosome specific libraries: detection of trisomy 21 and translocation of chromosome 4. Proc. Natl. Acad. Sci. USA, 1988; 85: 9138-9142.
33. Hopman A.H.N., Moesker O., Smeets A.W.G.B., Pauwels R.P.E., Vooijs G.P., Ramaekers F.C.S. Numerical chromosome 1, 7, 9 and 11 aberrations in bladder cancer detected by in situ hybridization. Cancer Res, 1991; 51: 644-651.
34. Cremer T., Kurz A., Zirbel R., Dietzel S., Rinke B., Schröck E., Speicher MR., Mathieu U., Jauch A., Emmerich P., Scherthan H., Ried T., Cremer C., Lichter P. Role of chromosome territories in the functional compartmentalization of the cell nucleus. Cold Spring Harbor Symp Quantitative Biology, 1993; 58: 777-792.
35. Cremer T., Kreth G., Koester H., Fink R.H.A., Heintzmann R., Cremer M., Solovei I., Zink D., Cremer C. Chromosome territories, inter chromatin domain compartment and nuclear matrix: An integrated view on the functional nuclear architecture. Crit. Rev. Eukar. Gene, 2000; 12: 179-212.
36. Eils R., Bertin E., Saracoglu K., Rinke B., Schröck E., Parazza F., Usson Y., Robert-Nicoud M., Stelzer E.H.K., Chassery J.M., Cremer T., Cremer C. Application of confocal laser microscopy and three-dimensional Voronoi diagrams for volume and surface estimates of interphase chromosomes. J Microscopy, 1995; 177: 150-161.
37. Ferreira J., Paolella G., Ramos C., Lamond AI. Spatial organization of large-scale chromatin domains in the nucleus: a magnified view of single chromosome territories. J Cell Biol, 1997; 139: 1597-1610.
38. Kozubek S., Lukasova E., Ryznar L., Kozubek M., Liskova A., Govorun R.D., Krasavin E.A., Horneck G. Distribution of ABL and BCR genes in cell nuclei of normal and irradiated lymphocytes. Blood, 1997; 89: 4537-4545.
39. Croft J.A., Bridger J.M., Boyle S., Perry P., Teague P., Bickmore W.A. Differences in the localization and morphology of chromosomes in the human nucleus. J Cell Biol, 1999; 145: 1119-1131.

40. Munkel C., Eils R., Dietzel S., Zink D., Mehring C., Wedemann G., Cremer T., Langowski J. Compartmentalization of interphase chromosomes observed in simulation and experiment. J Mol Biol, 1999; 285: 1053-1065.
41. Sachs R.K., Engh G. van den, Trask B., Yokota H., Hearst J.E. A random-walk / giant-loop model for interphase chromosomes. Proc. Natl. Acad. Sci. USA, 1995; 92: 2710-2714.
42. Sadoni N., Langer S., Fauth C., Bernardi G., Cremer T., Turner B.M., Zink D. Nuclear organization of mammalian genomes: polar chromosome territories build up functionally distinct higher order compartments. J Cell Biol, 1999; 146: 1211-1226.
43. Skalnikova M., Kozubek S., Lukasova E., Bartova E., Jirsova P., Cafourkova A., Koutna I., Kozubek M. Spatial arrangement of genes, centromeres and chromosomes in human blood cell nuclei and its changes during the cell cycle, differentiation and after irradiation. Chromosome Res, 2000; 8: 487-499.
44. Bartova E., Kozubek S., Kozubek M., Jirsova P., Lukasova E., Skalnikova M., Buchnickova K. The influence of the cell cycle, differentiation and irradiation on the nuclear location of the abl, bcr and c-myc genes in human leukemic cells. Leukemia Res, 2000; 24: 233-241.
45. Bartova E., Kozubek S., Kozubek M., Jirsova P., Lukasova E., Skalnikova M., Cafourkova A., Koutna I. Nuclear topography of the c-myc gene in human leukemic cells. Gene, 2000; 244:1-11.
46. Jirsova P., Kozubek S., Bartova E., Kozubek M., Lukasova E., Skalnikova M., Cafourkova A., Koutna I., Buchnickova K. Spatial distribution of selected genetic loci in nuclei of human leukemic cells after irradiation. Radiat Res, 2001; 155: 311-319.
47. Kozubek S., Lukasova E., Mareckova A., Skalnikova M., Kozubek M., Bartova E., Kroha V., Krahulcova E., Slotova J. The topological organization of chromosomes 9 and 22 in cell nuclei has a determinatiove role in the induction of t(9,22) translocations and in the pathogenesis of t(9,22) leukemias. Chromosoma, 1999; 108: 426-435.
48. Nikiforova M.N., Stringer J.R., Blough R., Medvedovic M., Fagin J.A., Nikiforov Y.E. Proximity of chromosomal loci that participate in radiation-induced rearrangements in human cells. Science, 2000; 290: 138-141.
49. Koutna I., Kozubek S., Zaloudik J., Kozubek M., Lukasova E., Matula P., Bartova E., Skalnikova M., Cafourkova A., Jirsova P. Topography of genetic loci in tissue samples: towards new diagnostic tool using interphase FISH and high-resolution image analysis techniques. Anal Cell Pathol, 2000; 20: 173-185.
50. Zink D., Cremer T.Cell nucleus: Chromosome dynamics in nuclei of living cells. Current Biology, 1998; 8: R321-R324.
51. Lukasova E., Kozubek S., Kozubek M., Kroha V., Mareckova A., Skalnikova M., Bartova E., Slotova J. Chromosomes participating in translocations typical of malignant haemoblastoses are also involved in exchange aberrations induced by fast neutrons. Radiat Res, 1999; 151: 375-384.
52. Cafourkova A., Lukasova E., Kozubek S., Kozubek M., Govorun R.D., Koutna I., Bartova E., Skalnikova M., Jirsova P., Pasekova R., Krasavin E.A.Exchange aberrations induced with gamma-rays among 11 chromosomes of human lymphocytes. Int J Radiat Biol, 2001 77: 419-429.
53. Bubendorf L., Kononen J., Koivisto P., Schraml P., Moch H., Gasser T.C., Willi N., Mihatsch M.J., Sauter G., Kallioniemi O.P. Survey of gene amplifications during prostate cancer progression by high throughput fluorescence in situ hybridization on tissue microarrays. Cancer Res, 1999; 59: 803-806.
54. Pinkel D., Seagraves R., Sudar D., Clark S., Poole I., Kowbel D., Collins C., Kuo W.L., Chen C., Zhai Y., Dairkee S.H., Ljung B.M., Gray J.W., Albertson D.G.High resolution analysis of DNA copy number variation using comparative genomic hybridization to microarrays. Nat Genet, 1998; 20: 207-211.
55. Kamentsky L.A., Burger D.E., Gershman R.J., Kamentsky L.D., Luther E. Slide-based laser scanning cytometry. Acta Cytologica, 1997; 41: 123-143.

56. Kamentsky L.A., Kamentsky L.D., Fletcher J.A., Kurose A., Sasaki K.Methods for automatic multiparameter analysis of fluorescence in situ hybridized specimens with a laser scanning cytometer. Cytometry, 1997; 27: 117-125.
57. Netten H., Van Vliet .LJ., Vrolijk H., Sloos W.C.R., Tanke H.J., Young I.T. Fluorescent dot counting in interphase cell nuclei. Bioimaging, 1996; 4: 93-106.
58. Netten H., Young I.T., Van Vliet L.J., Tanke H.J., Vrolijk H., Sloos W.C.R. FISH and chips: Automation of fluorescent dot counting in interphase cell nuclei. Cytometry, 1997; 28: 1-10.
59. Ortiz de Solórzano C., Santos A., Vallcorba I., García-Sagredo J.M., Pozo F. del: Automated FISH spot counting in interphase nuclei: Statistical validation and data correction. Cytometry, 1998; 31: 93-99.

STEREOLOGICAL AND DIGITAL METHODS FOR ESTIMATING GEOMETRICAL CHARACTERISTICS OF BIOLOGICAL STRUCTURES USING CONFOCAL MICROSCOPY

Lucie Kubínová[1], Jiří Janáček[1], Jana Albrechtová[2,3], Petr Karen[1]
[1]*Department of Biomathematics, Institute of Physiology, Prague, Academy of Sciences of the Czech Republic, Czech Republic*
[2]*Department of Plant Physiology, Faculty of Sciences, Charles University in Prague, Czech Republic*
[3]*Institute of Botany, Academy of Sciences of the Czech Republic, Pruhonice, Czech Republic*

Abstract: Stereological methods and automatic digital methods for the quantitative evaluation of the structure of three-dimensional (3D) objects are introduced. Classical stereological methods based on observations made on two-dimensional (2D) sections, applying point, linear or planar test probes are presented and their application for estimating geometrical characteristics such as volume, surface area, and length of microscopical structures is demonstrated together with two practical trainings presented in a "do it yourself, can do approach". Further, stereological methods based on evaluation of 3D image data obtained by confocal microscopy from thick specimens and using computer-generated virtual test probes are presented. Such methods, allowing arbitrary orientation of the thick slice, can be used for estimating number, volume, surface area, and length in a very efficient way. The methods using spatial grid of points, disector, fakir and slicer probes are described and compared with automatic digital methods. The applications of presented methods are shown on practical examples. A special attention is paid to the sampling design and problems connected with practical implementation of the methods.

Keywords: Confocal Microscopy, Digital Methods, Geometrical Characteristics, Image Analysis, Length, Number, Spatial Grid, Stereology, Surface Area, Volume.

1. INTRODUCTION

Measurements of geometrical parameters of microscopical structures, i.e. structural components of organs, tissues, or cells, are the main prerequisite for quantitative analysis in a number of studies in biological research,

V. Evangelista et al. (eds.), From Cells to Proteins: Imaging Nature across Dimensions, 271–321.

especially when the relationships between function and structure are analyzed.

Some of the biological structures are so flat that they can be regarded as planar, i.e. two-dimensional (2D) objects (Figure 1) but most biological structures are truly three-dimensional (3D), i.e. they are located (distributed) in 3D space. In contrary, classical histological microscopical techniques usually provide us only with sections (i.e. 2D data) of 3D objects (Figure 2). The fact that only parts of the structure under study, such as sections, or tissue blocks, are available for the measurement has to be taken into account in any rigorous quantitative analysis when proper sampling of such parts must be applied. When applying more advanced techniques such as confocal microscopy or electron tomography we can get 3D images of biological structures directly but even in this case it is rarely possible to measure the entire structure under study exhaustively (e.g. to count all neurons in a cerebral cortex) so for rigorous measurements we still have to sample only parts of the structure to be evaluated.

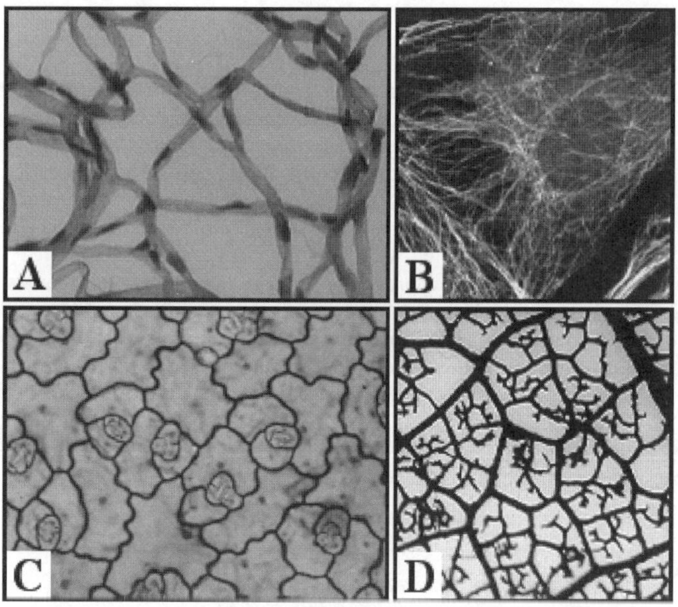

Figure 1. Examples of planar biological structures and parameters, which are usually of interest and can be measured by stereological and/or digital methods. A. Retinal digest preparation showing microvessel network – length density of microvessels in retina, number of endothelial cell nuclei in retina. B. Flat cells like fibroblasts with visualized cytoskeleton – length density of microtubules in a cell. C. Leaf peel or imprint, example of dicot epidermis - number of stomata, number of epidermal cells in a leaf, mean stoma length, mean area of epidermal cell. D. Flat dicot leaf with veins – the length of venation, venation density.

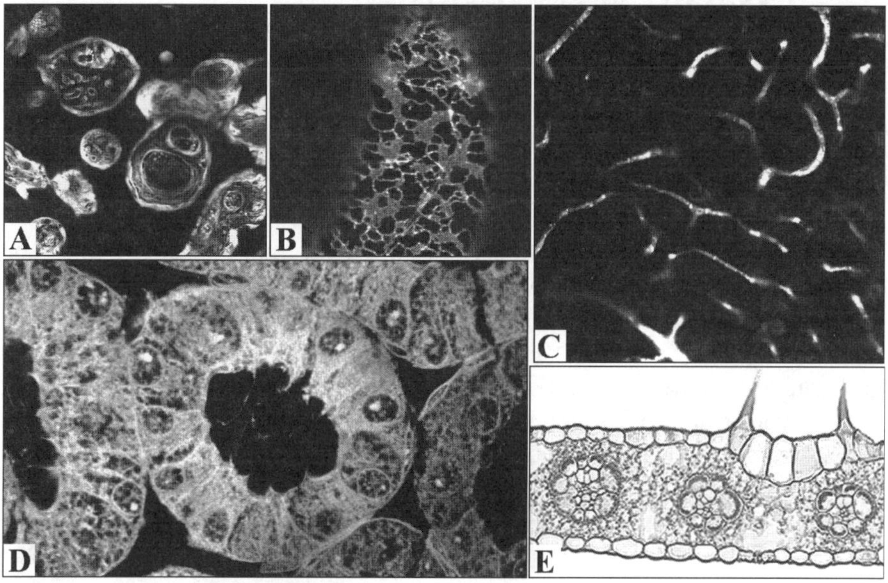

Figure 2. Planar sections of 3D structures and their geometrical characteristics of interest. A. Human placenta – surface area of capillary walls per surface area of terminal villi. B. Endoplasmic reticulum (ER) of a tobacco cell – length density of the ER bundles in a cell. C. Capillaries in a rat brain cortex – capillary length in the cortex. D. Chick embryonal kidney – mean volume of proximal tubular cells. E. Cross section of a corn leaf – volume density of mesophyll, vascular tissues, and epidermis in the leaf.

In our lecture, we present solutions for measurements of geometrical parameters of biological microscopical structures in all above situations, using stereological and digital methods. These methods, whether based on observations made on 2D sections or 3D subsamples, represent useful tools for providing quantitative information about microscopical structures under study. First, stereological and digital methods will be introduced in general. Then some of the specific methods for volume, surface area, and length estimation, based on evaluation of 2D images will be described together with two practical trainings presented in a "do it yourself, can do approach". Further, we will describe the methods using 3D image data, acquired e.g. by a confocal microscope, and demonstrate them on examples. Finally, we will discuss the possible pitfalls connected with the measurements of geometrical characteristics of biological microscopical structures.

2. GENERAL OVERVIEW OF STEREOLOGICAL AND DIGITAL METHODS

2.1. Stereological Methods

Stereological methods are precise tools for the quantitative evaluation of parameters of three-dimensional objects, such as volume, surface area, length, number, or spatial distribution. They are based mainly on observations made on 2D sections or 3D subsamples of tissue, applying test probes of different dimensions, i.e. zero-dimensional (0D, i.e. points), 1D (i.e. lines) or 2D (i.e. planes) and counting the interactions of the probes with the structures under study. For example, the number of test points falling into a given structure (Figure 3) or number of intersection points of test lines with a structure surface is counted.

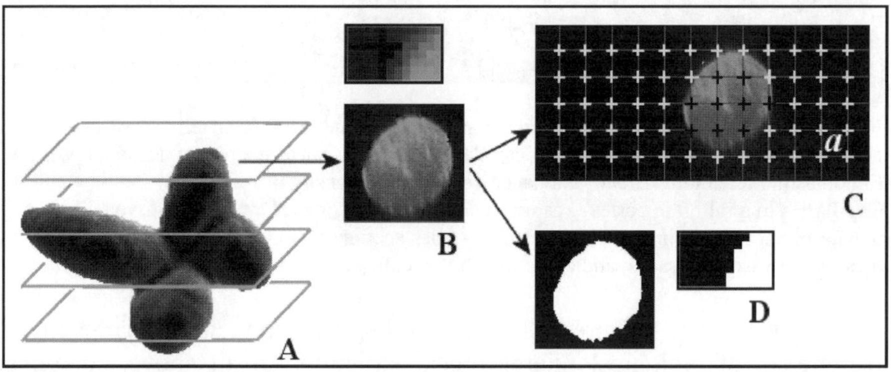

Figure 3. Estimation of the object volume by the Cavalieri principle (see 3.1.1) in combination with a stereological method or digital method. A. Systematic sections of the object are made. B. One of the sections with a detail of grayscale pixels. C. Stereological method: The section is overlaid by a point test system. The object volume is estimated by the number of test points falling in the sections multiplied by the unit area of the test system, *a*, and the distance between sections. D. Digital method: After automatic segmentation of grayscale images of sections, their binary images are obtained and the object volume is estimated by the number of pixels of the binary images of sections multiplied by the area of the pixel and the distance between sections. The binary image of the object section with a detail of binary pixels is shown. The foreground pixels are white.

The term of stereology was coined in 1961 when the International Society for Stereology was founded. The founding of stereology was motivated by the need of investigators in material and life sciences to establish a rigorous theoretical basis for the solution of problems encountered in morphometry , in order to measure objectively structure volume, surface area, length, number, etc.

One problem was to determine with reasonable accuracy the total number of neurons in the brain. The total number of neurons in the human cerebral cortex measured by different authors over the previous 100 years varied from 0.6 to 16.5 billions (Table 1, [1]). This clearly showed that something was wrong with the techniques applied. A rigorous theoretical basis led to the proper recognition of mathematical foundations of stereology, which are based on integral and stochastic geometry, probability theory, spatial and mathematical statistics. The theory as well as the applications of stereological methods to material and biomedical sciences are still evolving (for reviews see [2-7]). As far as the above problem was concerned, the total number of neurons in the human cerebral cortex was estimated by an unbiased stereological method to be 27.4 billions with an inter-individual variation coefficient of 12 % [8]. The variation in total neuron counts obtained by other authors, listed below, in Table 1, resulted from poor sampling as well as biased rules for counting. Also, the tissue shrinkage occurring during brain fixation and other tissue processing should have been taken into account.

Table 1. Total number of neurons in the human cerebral cortex in billions (10^9) identified by different authors (listed in [1]) using different methods. The two extreme results are underlined.

Author (Year)	Number in billions (10^9)	Author (Year)	Number in billions (10^9)
Meynert (1868)	<u>0,6</u>	Agduhr (1931)	5,0
Donaldson (1895)	1,2	Shariff (1953)	6,9
Thompson (1899)	9,3	Haug & Rebhan (1956)	<u>16,5</u>
Berger (1821)	5,5	Pakkenberg (1966)	2,6
Von Economo (1926)	14,0		

2.2. Digital Methods

Digital methods enable automatic measurements of geometric characteristics of 3D objects by evaluation of their digital images, i.e. the lattices of spatial elements (spels, pixels in 2D, voxels in 3D) with assigned numbers or vectors given by a specific function, $f(i)$, e.g. in a grayscale image different gray levels are assigned to spels. The digital methods can be applied directly to the binary image obtained by automatic segmentation of the grayscale image. The binary image consists of spels with values 0 or 1 that indicate if spels belong to the background or to the objects under study (foreground), respectively (see Figure 3).

In the simplest and most common situation the lattice of spels is represented by an orthogonal parallelepiped $\{B_1 \ldots E_1\} \times \{B_2 \ldots E_2\} \times \ldots \times \{B_n \ldots E_n\}$ with uniform steps $u_1, u_2, \ldots, u_n$ (calibration constants) between the spel centers in each dimension, i.e. by linear spels of equal length in 1D, rectangular pixels of the same area in 2D, and voxels of equal volume in 3D.

For example, in 1D there are only two basic textural (geometrical) characteristics: the total length of intervals (1D volume) and the number of intervals (0D topological characteristic). The total length can be calculated by multiplying the number of the foreground spels by the calibration constant. The number of intervals, χ, can be estimated by counting the rightmost spels of the intervals:

$$est\chi = \sum_{i=B}^{E-1} f(i) - f(i) \wedge f(i+1) \qquad (1)$$

where $\wedge$ is logical AND.
Alternatively, the leftmost spels can be counted, which yields:

$$est\chi = \sum_{i=B}^{E-1} f(i) \vee f(i+1) - f(i) \qquad (2)$$

where $\vee$ is logical OR. If the spels are coded by bits of integers data type, the formulas can be implemented very efficiently in current computers using a right bit shift and logical operators acting in a parallel manner on the bits. The two estimates are different if the value of just one of the boundary spels is 1, which can be corrected by subtracting value $0.5(f(B)-f(E))$ from the first formula and adding it to the second one. Similar boundary corrections can be applied to all estimators used in this section.

It is obvious that the n-dimensional volume can be calculated by summing up the volumes occupied by the individual spels, e.g. the 2D volume (i.e. area) of the 2D object of interest is estimated by summing up the areas of foreground pixels, as in Figure 3C.

For the definition of the topological characteristics of digital images, such as the connectivity number (Euler-Poincaré characteristic), reflecting the number of simple objects, the adjacency (neighboring relation) of spels must be defined. The simplest examples are the “weak” adjacency when neighboring spels are those differing by one step in one coordinate only (4-adjacency in 2D, see Figure 4A, 6-adjacency in 3D, 2n-adjacency in nD) or “strong” adjacency, when neighboring spels differ at most by one step in each coordinate (8-adjacency in 2D, see Figure 4B, 26-adjacency in 3D, (3^n-1)- adjacency in nD). First, we define the connectivity number in dimension n=0 when the image has a single binary spel and connectivity is 0 or 1, according to the spel value. Let $f_{i,\bullet}$ be the (n-1)-dimensional slice through the image at position i. Then the connectivity number under assumption of weak or strong adjacency of the n-dimensional image can be defined (similarly to [9]) by:

$$\chi_w = \sum_{i=B_n}^{E_n-1} \chi_w(f(i,\bullet)) - \chi_w(f(i,\bullet) \wedge f(i+1,\bullet)) \tag{3}$$

$$\chi_s = \sum_{i=B_n}^{E_n-1} \chi_s(f(i,\bullet) \vee f(i+1,\bullet)) - \chi_s(f(i,\bullet)) \tag{4}$$

Such definition enables direct and efficient implementation of the measurement algorithm: In the first step, formulas (1) or (2) for 1D case are used. In higher dimensions, we can calculate the estimate $est_w \chi$ from formula (3). In 2D, we obtain classical formulas (Figure 5; see [10]). By application of induction step to 2D formula for $est_w \chi$, we obtain the formula for $est_w \chi$ in 3D, which is equivalent to currently used method [11] with the lookup table of 2x2x2 voxels configurations. In the same way we can obtain the formula for $est_s \chi$ in 3D from (4). Using strong adjacency for foreground and weak adjacency for the background (or vice versa) yields consistent results (sum of characteristics of the foreground and the background is the characteristic of the whole image, equal to 1).

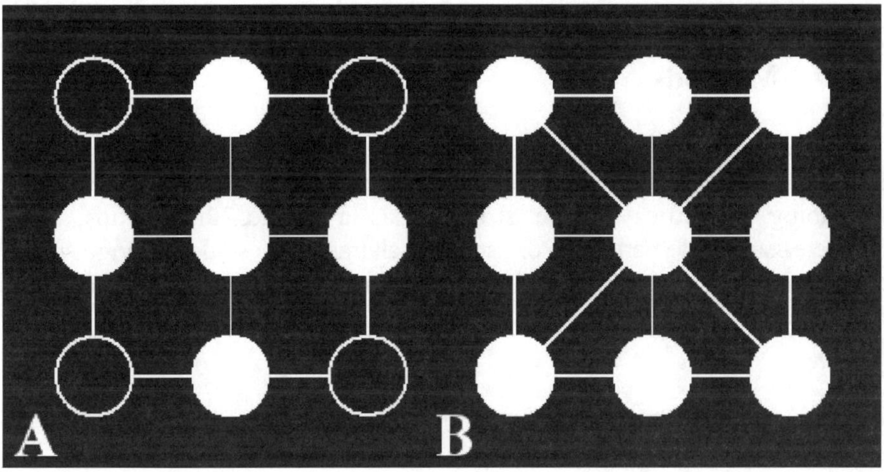

Figure 4. Weak and strong adjacency in 2D. A. In weak adjacency four (white) pixels are adjacent to the central pixel. B. In strong adjacency eight (white) pixels are adjacent to the central pixel.

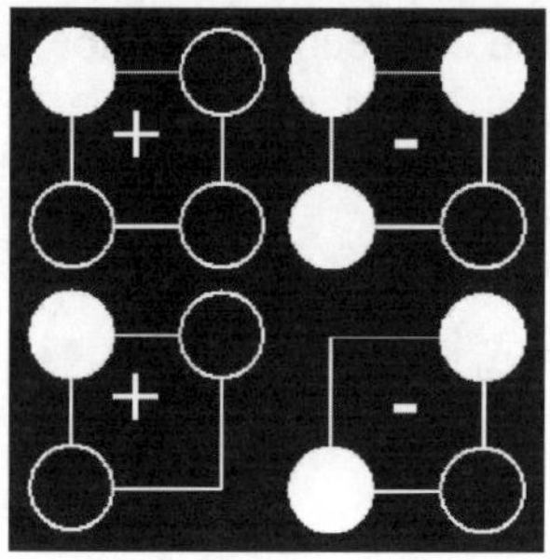

Figure 5. Configurations of pixels counted during the estimation of connectivity number in 2D. The confogurations should be counted with respective signs shown.

Other important characteristics, such as length and surface area, can be estimated by the Crofton formula [12, 13] combining the volume and connectivity estimates in lattices of lower dimensions. The perimeter in 2D can be thus estimated by calculating connectivity numbers of intersection of the set with horizontal rows, multiplying them by the step between rows, summing the results and multiplying the result by π. The result is clearly correct for circle, thus it is an unbiased estimate of a set with isotropic random orientation but its precision can be low (its maximal variation coefficient, CV, attained by a randomly oriented segment, is equal to $\sqrt{(\pi^2/8 - 1)} \cong 0.48$). Using the average of estimators for mutually perpendicular directions, CV is less than 0.098. In practice, diagonal directions are also taken into account, which in the case of $u_1 = u_2$ yields CV below 0.023 [14].

2.3. Parameters Measured by Stereological and Digital Methods

Stereological and digital methods can be applied to the estimation of different geometrical characteristics, including global characteristics describing properties of the structure as a whole, their ratios, local characteristics of particles or spatial characteristics describing spatial distribution of structures including their mutual spatial relationships. Examples are listed below:

2.3.1. Global Characteristics

- Volume (V): volume of an organ and/or its compartments, e.g. volume of embryonal kidney, volume of a brain, volume of hippocampus, volume of cell nucleus.

- Number (N): number of particles in a tissue, e.g. number of neurons in the cerebral cortex, number of glomeruli in a kidney, number of mesophyll cells in a needle.
- Length (L): length of tubular structures, e.g. total length of proximal tubules in a kidney, length of veins in a leaf.
- Surface area (S): surface area of an organ and/or its compartments, e.g. surface area of a lung, inner surface area of a leaf, surface area of inner mitochondrial membranes.

2.3.2. Ratios

- Volume density (V_V): proportion of tissues in an organ, e.g. of mesophyll in a leaf, proportion of nucleus in a cell.
- Numerical density (N_V): number of satellite cells per unit volume of a muscle.
- Length density (L_V): cortical capillary length per cortex volume.
- Length per surface area (L_S): capillary length per surface area of muscle fibers.
- Ratio of surface areas (S_S): microvillous amplification factor (surface area of enterocyte microvilli per surface area of apical membrane).

2.3.3. Characteristics of Particles

- Mean particle volume($\bar{v}_N$): mean volume of glomerulus in a kidney, mean volume of cells.
- Mean particle surface area ($\bar{s}_N$): mean surface area of a mesophyll cell in a leaf.

2.3.4. Spatial Characteristics

- Pair correlation function: pair correlation function of immunogold particles marking a specific structure in a cell nucleus, showing e.g. clustering of particles.
- Pair cross-correlation function: pair cross-correlation function of large and small immunogold particles marking two different structures in a cell nucleus, showing e.g. colocalization of the nuclear structures.

2.4. Sampling

In any quantitative analysis of biological structures, the successful realization of a study depends on a rigorous technique of sampling tissue blocks, sections, test frames, point grids, etc. Any geometrical property of an object, such as the number of cells, is compiled from the information collected from multiple small samples. In order to obtain a reliable estimate, sufficient number of samples must be taken and the technique of sampling must be unbiased, as that used in a public opinion poll, i.e. each tissue block or cell, etc., must have the same probability of being sampled. Figure 6 illustrates biased and unbiased sampling. Note that the variation in total neuron counts obtained by different authors, listed above (Table 1), resulted also from poor sampling.

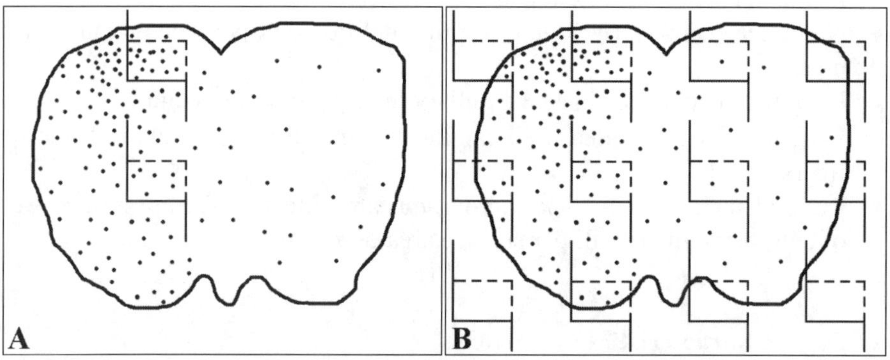

Figure 6. Biased and unbiased sampling. A. Biased sampling of counting frames: If the frames are placed into a fixed position in the object, the estimate of the density of points in the object is biased: The density of points counted from the frames is higher than their actual density in the section. B. Unbiased sampling of counting frames: The frames are distributed within the examined section in a systematic uniform random manner (see 2.4.2) such that each point has the same probability of being sampled. The density of points counted from the frames represents an unbiased estimate of their actual density in the section. (After [15] Copyright © 2003 Radiation Research Society.)

In the implementation of a stereological method, two kinds of errors must be minimized: first, a random error caused by the variance of the estimation depending on the inhomogeneity of the evaluated structure, and secondly, a systematic error, i.e. bias, which is usually caused by the false assumptions about the shape or homogeneity of the structure under study or by the inherent defect of the methodology (e.g. the method of counting cells by counting their nuclear profiles in histological sections is not correct, in general, because larger nuclei are more likely to be seen in the sections than smaller ones; see [16]). The random error can be decreased by taking a larger sample, e.g. more tissue blocks, while the systematic error cannot be eliminated: It is neccessary to choose the correct method, proper for the experimental material, directly from the beginning.

The most often used ways of unbiased sampling are described below.

2.4.1. Systematic Uniform Random Sampling of Sections

First, distance between sections, *T*, is chosen. Then the number *z* is sampled at random from {0,1,2, … , *T*-1}. The first section is then cut at the distance z from the origin, the second one at *z*+*T*, etc. (Figure 7)

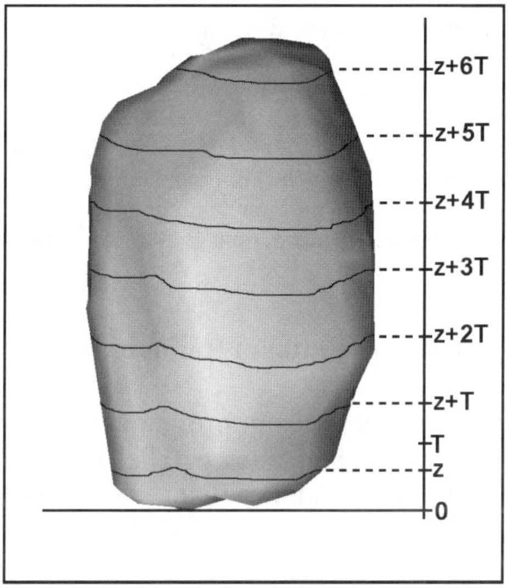

Figure 7. Example of a systematic uniform random (SUR) sampling of sections: If *T*=10 mm, we choose the number *z* from {0,1,2, … , 9}. It can happen to be *z*=6. Then the first section is cut at 6 mm from the object bottom, the second one 16 mm, the third one 26 mm, etc. (After [15] Copyright © 2003 Radiation Research Society)

2.4.2. Systematic Uniform Random Sampling of Probes, Fields of View, Tissue Blocks

The system of probes has a uniform random position if the probes are uniformly distributed within the examined area such that each position has the same probability of being sampled. It can be, e.g. for rectangular squares, achieved by placing the first frame into a point (*x,y*), with *x*, *y* being random numbers selected from {0,1,2,...,*X*-1}, or {0,1,2,...,*Y*-1}, respectively, where *X* and *Y* are the distances between the neighboring frames in x and y direction. An analogous approach can be used for uniform random superimposition of test systems. For illustrative examples see Figure 8.

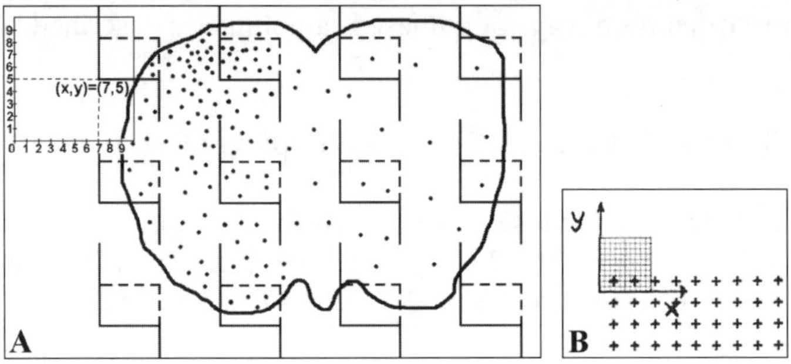

Figure 8. Systematic uniform random (SUR) sampling of probes. A. Unbiased sampling of counting frames: The system of frames has a uniform random position achieved by placing the lower left corner of the first frame into a point (*x,y*) within the unit square of the frame system; *x*, *y* are random numbers selected from {0,1,2,...,9} provided the length of the unit square side is equal to 10. In the figure, (*x,y*) = (7,5). B. Uniform random superimposition of point test system: The distance between points is 4 mm here. The point of the system is placed in the position given by (*x,y*) where *x*, *y* are random numbers selected from {0,1,2,3}. In the figure, (*x,y*) = (3,2). (Figure A after [15] Copyright © 2003 Radiation Research Society)

2.4.3. Sampling Isotropic Uniform Random Direction in 2D

The construction of an isotropic uniform random line in a planar object is shown in Figure 9. If equidistant lines parallel to this line are added, we obtain an isotropic uniform random (IUR) superposition of a linear test system on the planar object.

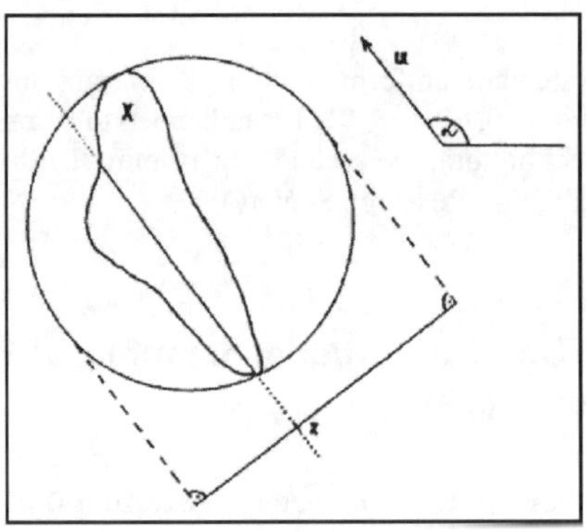

Figure 9. Construction of an isotropic uniform random line (IUR) in a planar object X. A circle circumscribing the object X is constructed. Firstly, a uniform random angle α, $0 \le \alpha < \pi$, is taken (it can be sampled as a random number from the set {0°, 10°,...,170°}). Secondly, UR point *z* from the circle projection perpendicular to the direction given by the angle α is taken (which can be done by taking a random number from the set {0 mm, 1 mm, ...,57 mm} in this case). Then the line going through the point *z* and parallel to the direction ***u***, given by the angle α, is IUR line in a planar object X, provided it is lying inside the X. (If it falls outside the object X, the above procedure is repeated so long as the line is lying inside X.)

2.5. Typical Procedure of Measuring Geometrical Characteristics of Biological Structures

The scheme of measurement of geometrical characteristics of the object under study is shown in Figure 10. Stereological or digital methods or their combination can be applied. Any of these methods must follow proper sampling (see above) and so their main difference is in the way of evaluation of 2D or 3D images sampled for the measurement. In stereology, test systems are superimposed over images and their interactions with the structures under study are counted while digital methods are usually applied to segmented images obtained by suitable image processing algorithms. This automatic approach is faster but it also has some drawbacks, e.g. it requires automatic segmentation of biological structures which is often hardly feasible (Figure 11). On the one hand, interactive evaluation using stereological methods can be speeded up by applying special stereological software generating relevant test grids over digitized images of the structure under study, on the other hand the scope of application of digital methods can be broadened by implementing possibilities of some user interaction during the segmentation process.

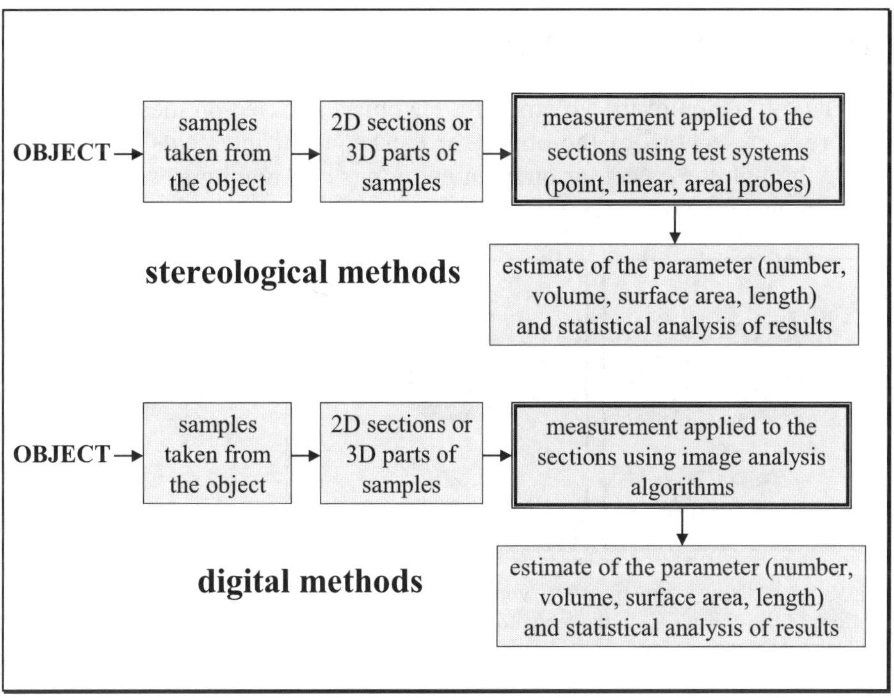

Figure 10. Scheme of measurement of geometrical characteristics of the object of interest.

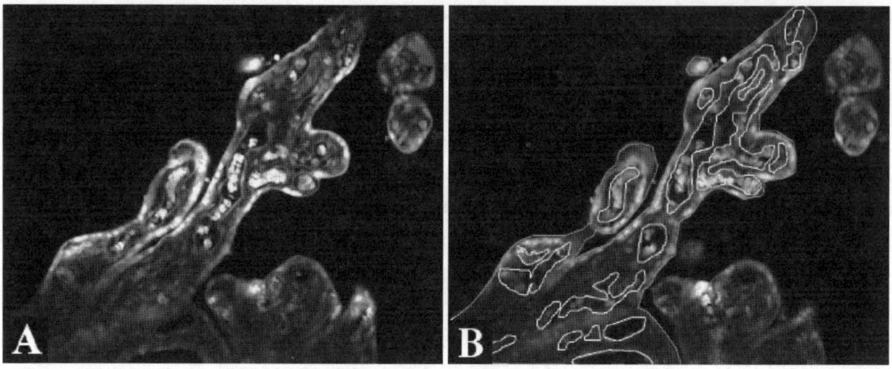

Figure 11. Example of an image where the biological structures cannot be distinguished (segmented) from the background automatically. A. Capillaries in human placental villi in this confocal image are recognizable by an experienced researcher but a fully automated algorithm for detecting capillary contours can be hardly developed. B. Manually traced contours of capillaries and villi that can represent input data for relevant 3D reconstruction. (Figure B after [17] Copyright © 2003, John Wiley & Sons, Inc.)

3. MEASUREMENTS BASED ON 2D SECTIONS

3.1. Volume Estimation

The estimation of the volume of a 3D object is based on measurement of the areas of sections of the object, or the length of intercepts of test lines going through the object, or on counting points of a point grid superimposed on sections of the object (Figure 12).

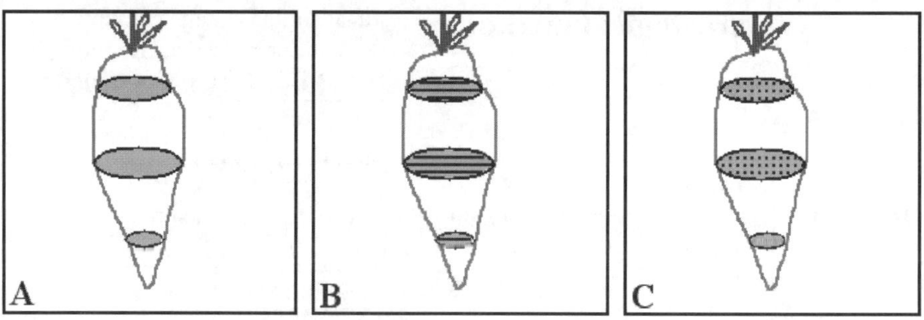

Figure 12. Volume estimation. The volume of a 3D object is proportional to the areas of sections of the object (A), to the length of the intercepts of test lines going through the object (B), or to the number of test points lying in sections of the object (B).

3.1.1. Cavalieri Principle

The volume of a 3D object can be expressed by the integral:

$$V = \int_0^h A(x)dx \tag{5}$$

where $A(x)$ is the area of the section of the object passing through the point $x \in [0,h]$ and h is the calliper diameter of the object perpendicular to section planes.

If systematic uniform random points z, $z+T$, …, $z+(n-1)T$ are selected from the interval $[0,h]$, i.e. systematic uniform random sampling of sections is applied (Figure 7), it is possible to estimate the object volume by the Cavalieri estimator (discovered by the Italian mathematician Buonaventura Cavalieri in 1635; Figure 3; [18]):

$$estV = T.\sum_{j=1}^{n} A_j \tag{6}$$

where T denotes distance between sections, n is the number of sections, and A_j is the area of j-th section (j=1,...,n), i.e. $A_j = A(z+(j-1)T)$.

3.1.1.1. Cavalieri Principle in Combination with Point Counting

In practice, the area A_j is usually estimated by the point-counting method when a point test system is superimposed on the j-th section and the number of test points (P_j) falling into the section is counted (see Figure 3B). The following relation holds:

$$estA_j = a \cdot P_j \tag{7}$$

where a is the area corresponding to a single test point. Thus, it is possible to estimate the object volume by the formula:

$$estV = T \cdot a \cdot \sum_{j=1}^{n} P_j \tag{8}$$

3.1.1.2. Cavalieri Principle in Combination with a Digital Method Using Pixel Counting

If the images of sections can be segmented so that we get their binary images (white foreground image on black background, see section 2.2), the area A_j is easily estimated by counting pixels that are lying within the j-th section (P_j). Formula (8) can be used if a denotes the size (area) of the pixel (Figure 4C,D).

3.1.1.3. Practical Training 1 – Estimation of the Carrot Volume

1. First, measure the carrot volume in a measuring glass cylinder filled with water (using Archimedes law), and record its value.
2. Cut sections of the carrot into segments according to the principle of SUR sampling (2.4.1; Figures 7, 13A):

- Identify the distance T between two subsequent sections.
- In this case we have chosen distance between two sections $T = 10$ mm, normally you can first measure the mean length of the object under study for the whole variant, and then divide it by the number of required sections, what gives you the number T. You should have at least 5 sections per object in average, but the more sections in average per object, the more precise volume estimation.
- Identify a random position of the first section
- using the table of random numbers, i.e. select a random number z from a set $\{0, 1, 2,..., T-1\}$, in our case of T = 10 mm, $z \in \{0, 1, 2,..., T-1\}$.
- Make SUR sections of the carrot (Figure 13A,B).

3. Superimpose a transparency with the point test system (Figure 13C) on the carrot sections at uniform random (Figure 8B).
4. Count test points falling into carrot sections (P_j).
5. Calculate the carrot volume using formula (9):

$$estV(carrot) = T \cdot a \cdot \sum_{j=1}^{n} P_j(carrot) \tag{9}$$

where n is the number of examined sections, T is the distance between two consecutive sections, a (mm^2) is the area unit corresponding to one test point of the point grid and $P_j(carrot)$ (j=1, ...,n) is the number of test points hitting the j-th carrot section. A suitable point grid is a square grid with the distance between neighboring points equal to 5 mm, i.e. $a = 5 \times 5$ mm^2 (Figure 13C).

For solution see section 6.1.

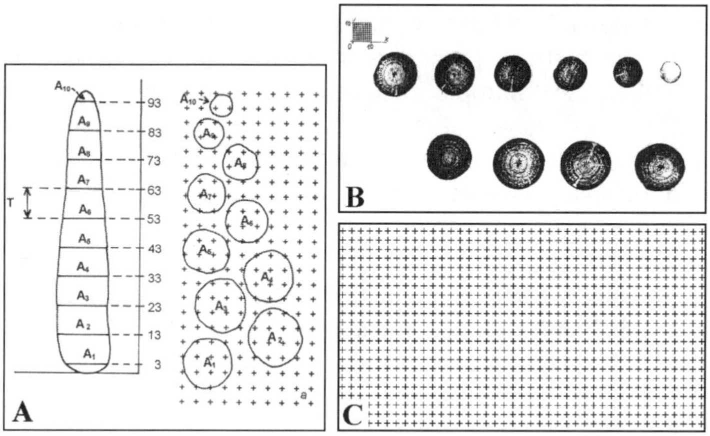

Figure 13. Practical Training 1 – Estimation of the carrot volume using the Cavalieri principle. Carrot segments in B were obtained by systematic uniform random sampling design demonstrated in A. The scale of magnification used in the picture can be derived from the square placed in the upper left corner of B. where the length of the side of the square corresponds to 10 mm. Point test system with the actual distance between points of 5 mm is demonstrated in C. Each 0-dimensional test point is represented by the intersection point between the upper edge of the cross horizontal line and the right edge of the cross vertical line.

3.2. Surface Area Estimation

The estimation of the surface area of a 3D object (e.g. cell surface area) is based on counting intersection points between the object surface and linear probes or on the measurement of the length of the intersection between the surface and test planes going through the object (Figure 14).

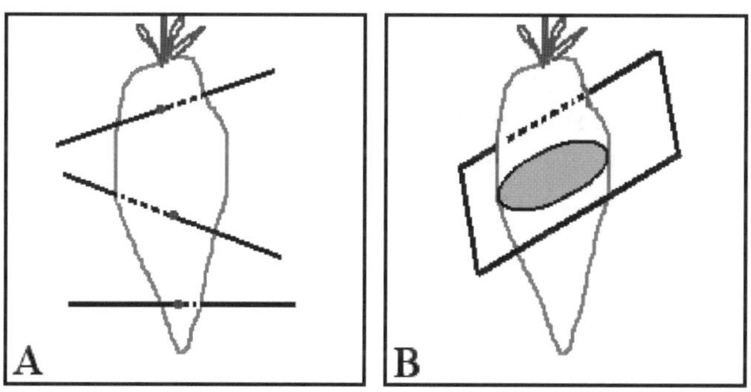

Figure 14. Surface area estimation. Surface area of a 3D object is proportional to the number of intersection points between the object surface and linear probes (A) or to the length of the intersection between the surface and test planes going through the object (B).

The orientation of test lines or planes can be arbitrary only if the structure has the same geometrical properties in all directions of space, i.e. if it is isotropic. Otherwise isotropic random (IR) direction of test probes is required. It means that every direction in 3D must have the same probability of being chosen for the measurement (Figure 15). In other words, IR direction in 3D space must be sampled at random from all directions in 3D. Each direction in 3D space can be expressed by a point on a unit sphere (in polar coordinates), i.e. it is given by two angles: θ, co-latitude, and ϕ, longitude. Isotropic uniform random (IUR) line probe is then a line with its direction sampled uniformly at random from all directions in 3D.

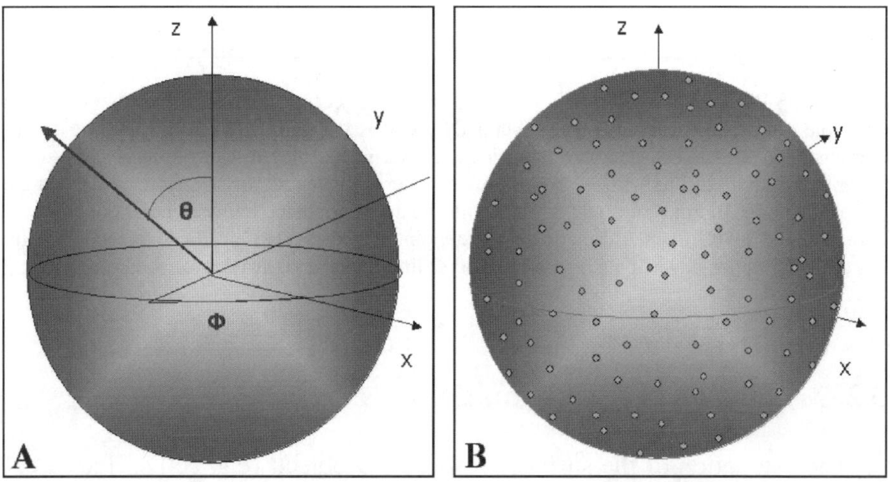

Figure 15. IR direction in 3D space. A. Each direction in 3D space can be expressed by a point on a unit sphere, given by two angles: θ, co-latitude, and ϕ, longitude. B. IR distribution of directions, represented by randomly distributed points on a unit sphere.

There is a simple relationship between the surface area per unit volume of the reference space (surface density, S_V) and

the number of intersections between the surface and the test lines per unit length of the test lines lying in the reference space (I_L) [19] provided the test lines have isotropic random orientation:

$$estS_V = 2 \cdot I_L \tag{10}$$

The reference space is usually represented by a 3D set containing the objects of interest, e.g. a specific organ or tissue.

It is usually more practical to count points than to measure lengths, thus a combination with the point-counting method is used:

$$estS_V = 2 \cdot \frac{a}{l} \cdot \frac{I}{P} \qquad (11)$$

where a/l is the ratio of the area unit of the point grid to the length of the test system, I is the number of intersections between the surface and the test lines, and P is the number of test points falling in the reference space.

In measurements based on 2D sections, two ways of generating IR line probes (Figure 15)

are most frequently used as shown below.

3.2.1. Orientator Method Using Test Line Systems or Digital Crofton Method in 2D

In the first approach, isotropic uniform random (IUR) sections are generated. The probes are then usually represented by systematically spaced straight lines or circles (Figure 16) and then formula (10) or (11) is applied. This approach is used in the orientator method [20] shown in Figure 17. Another alternative is offered by the isector method [21] where a small specimen is embedded into a resin ball, rolled at random and re-embedded into a rectangular mould for cutting.

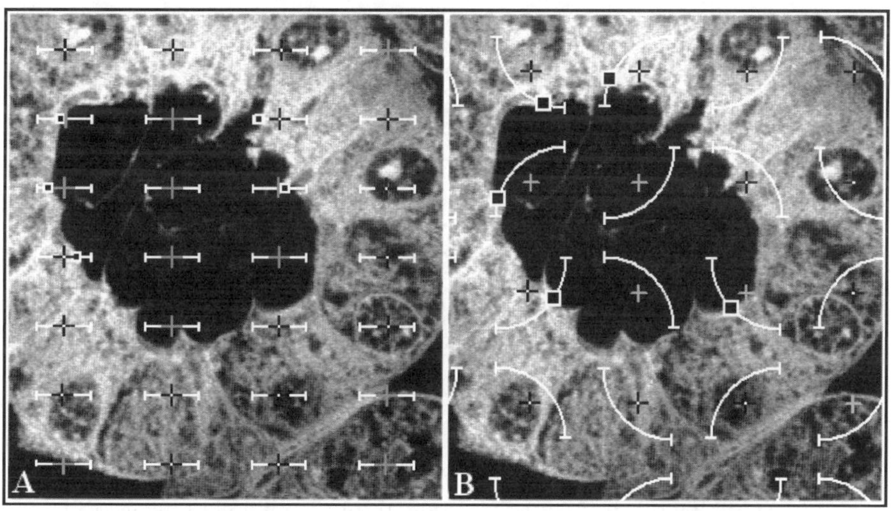

Figure 16. Line probes superimposed on IUR sections of the chick embryonal kidney. The surface density of the tubular lumina in the tubular walls is estimated by determining I, i.e. the number of intersections between the test lines and the luminal surface, and P, number of test points falling within the tubular walls (i.e. the reference space in this case). A. Test system formed by straight line segments in combination with a point grid. I=5, P=16 here. B. Test system formed by circle segments in combination with a point grid. I=5, P=11 here.

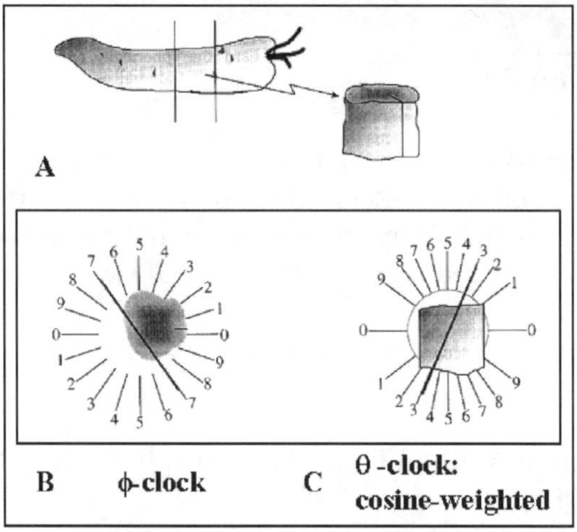

Figure 17. Generation of IR planes by the method of orientator shown on the example of carrot modified for simple practical implementation. A. A uniform random slab of the carrot is removed and a vertical direction is defined as parallel to the longitudinal axis of the carrot. B. The slab is placed with uniform random position and isotropic random orientation (see 2.4.3) on a uniformly divided ϕ-clock with the original vertical axis of the slab perpendicular to the plane of the clock. A uniform random number from {0, 1, …, 9} is sampled, i.e. 7 has been selected here. The slab is then cut perpendicular to the plane of the clock and along the random direction selected, i.e. 7-7 line. C. One of the two pieces of the slab that has just been generated is chosen at random. The newly generated flat face of this piece is then placed with uniform random position, face down to the cosine-weighted θ-clock, with the original vertical axis of the piece along the 0-0 line of the clock. Another uniform random number from {0, 1, …, 9} is sampled, i.e. 3 has been selected here. The piece is then cut perpendicular to the plane of the clock along the random direction selected, i.e. 3-3 line. This final cut face is an IUR plane section through the carrot. (From [6] Copyright © 1998 BIOS Scientific Publishers)

If we have digitized images of IUR sections available and the surface of interest, i.e. curves that form intersections of the surface with the 2D sections, can be segmented automatically, it is possible to estimate the surface density, S_V, by:

$$estS_V = \frac{4}{\pi} \cdot L_A \qquad (12)$$

where L_A is the length of the intersections of the surface with the sections per reference space area in the sections (also called boundary length density and denoted by B_A). The lengths of these curves in 2D can be estimated by the digital Crofton method (see 2.2), i.e. by calculating connectivity numbers of intersection of the curves with horizontal rows, multiplying them by the step between rows, summing the results and multiplying the result by $\pi/2$. If the same is done with vertical rows and diagonals, the average of results in these 4 directions gives a more precise estimate of curve lengths (see 2.2).

3.2.2. Method of Vertical Sections

In the second approach, vertical uniform random (VUR) sections are generated (Figures 18, 19). Vertical sections are planar sections longitudinal to a fixed (but arbitrary) axial direction, or perpendicular to a given horizontal plane (Figure 19A). The probes are then represented by systematically spaced sine-weighted curves (cycloids, see Figure 19B,D,E,F). This approach is used in the method of vertical sections (Figure 19; [22]).

Why cycloids have to be used as test lines? If IR test lines are generated on vertical sections, we *do not* get IR line probes in 3D as illustrated in Figure 18.

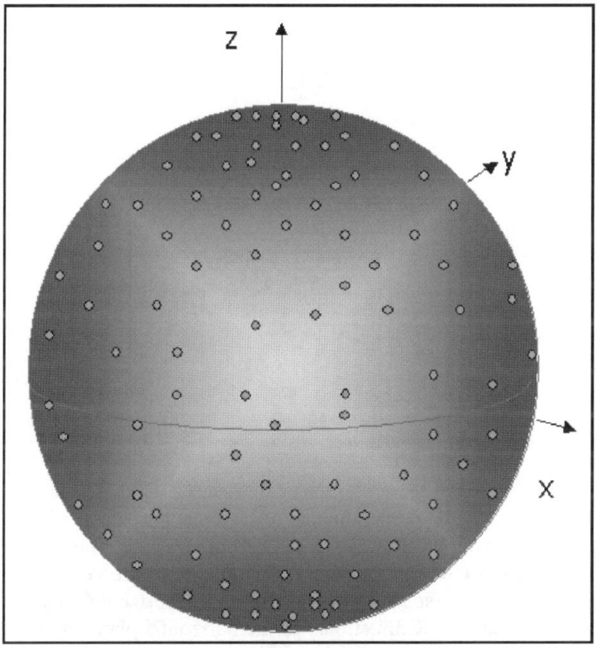

Figure 18. IR test lines generated on vertical sections *do not* result in IR line probes in 3D but the directions represented by points near the "poles" are sampled more frequently than those going near the "equator".

The total area of the surface under study can be estimated either by multiplying the surface density by the volume of the reference space or directly by the combination of the method of vertical sections with the Cavalieri principle, as shown in Practical Training 2 below.

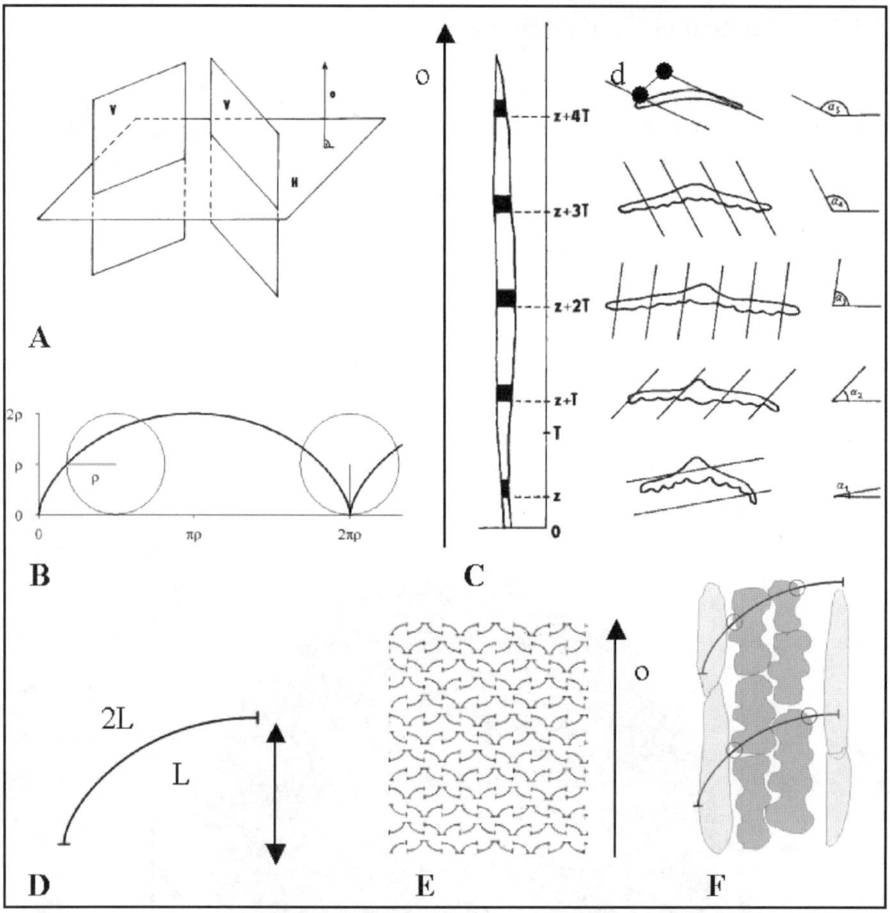

Figure 19. Method of vertical sections. Vertical planes (V), longitudinal to a fixed (but arbitrary) axial direction (o), or perpendicular to a given horizontal plane (H) (see A) are generated: First, uniform random segments are cut from an object at the distance T apart (C). Vertical uniform random (VUR) sections are cut parallel with the vertical axis (o) and along isotropic uniform random lines (see 2.4.3) generated in the horizontal plane (C). It is efficient to use a systematic sampling of IR directions: In the horizontal plane of the first segment (i.e. the nearest one to the leaf base), angle α_1 ($0° \leq \alpha_1 < 180°$) is selected uniformly at random. The systematic VUR sections of the first segment, at distance d apart, are cut in this direction. With m segments in the leaf, the orientation of vertical sections in the j-th segment is given by the angle $\alpha_j = \alpha_1 + (j-1) \times (180°/m)$ (j=1,...,m). Then we generate systematic VUR sections of the all m segments. The test probes are represented by systematically spaced sine-weighted curves (cycloids) (B,D,E). The vertical direction is indicated by the arrow (o). If L is the vertical height of the cycloid arc, the length of one cycloid arc is equal to 2L (D). When applying a cycloid test system, the vertical axis (o) has to be parallel to the shorter cycloid axis (E,F). The intersections are counted between the probes and the surface, e.g. when estimating the internal leaf surface area, intersections between test probes and mesophyll cell walls exposed to intercellular spaces are counted (denoted by circles).

3.2.2.1. Practical Training 2 – Surface Area of a Carrot

1. Generate VUR sections (Figures 19C, 20):

- First, cut the carrot into SUR segments according to the principle of systematic uniform random sampling, as described in Practical Training 1.
- Generate a random direction of the uniform random line in the first segment by selection of a random angle α from the interval [0°, 180°). Then generate vertical uniform random VUR sections at a fixed distance T apart, T=1 cm.
- Cut VUR sections from remaining SUR carrot segments: With m segments in the leaf, the direction of vertical sections in the j-th segment is given by the angle $\alpha_j = \alpha_1 + (j-1) \times (180°/m)$ ($j=1,...,m$). For example, if m=5 and α_1=10°, then α_2=10° + 1 x (180°/5) = 46°, α_3 = 82°, α_4=118°, and α_5=154°.

2. Superimpose the cycloid test system on the VUR sections at uniform random (Figure 8B). Proper o
rientation of a cycloid probe system has to be kept, i.e. shorter cycloid axis parallel to the vertical axis indicated by arrow (Figures 19 D,E,F, 20 B,C).
3. Count intersections between test probes and the carrot surface (I_j), not between the test probes and the cut surface of carrot segments (dotted in Figure 20A).
4. Calibrate the test system used, i.e. determine the ratio a/l(cyc). Remember the height of the shorter cycloid axis corresponds to one half of the arc of a cycloid (Figure 19D).
5) Calculate the carrot surface area $S(carrot)$ using formula (13):

$$estS(carrot) = 2 \cdot \frac{a}{l(cyc)} \cdot T \cdot \sum_{j=1}^{n} I_j \tag{13}$$

where n is the number of examined vertical sections, $a/l(cyc)$ (mm) is the ratio of the area corresponding to one cycloid arc (actual) length (the length of a cycloid arc equals twice its height, T is the distance between two consecutive sections and I_j is number of intersections of the exposed surface of the carrot and the cycloids.

T was chosen 10 mm in this case. In the test system shown in Figure 20C, $l(cyc)$ = 6 mm, a = 36 mm^2.

For solution see section 6.2.

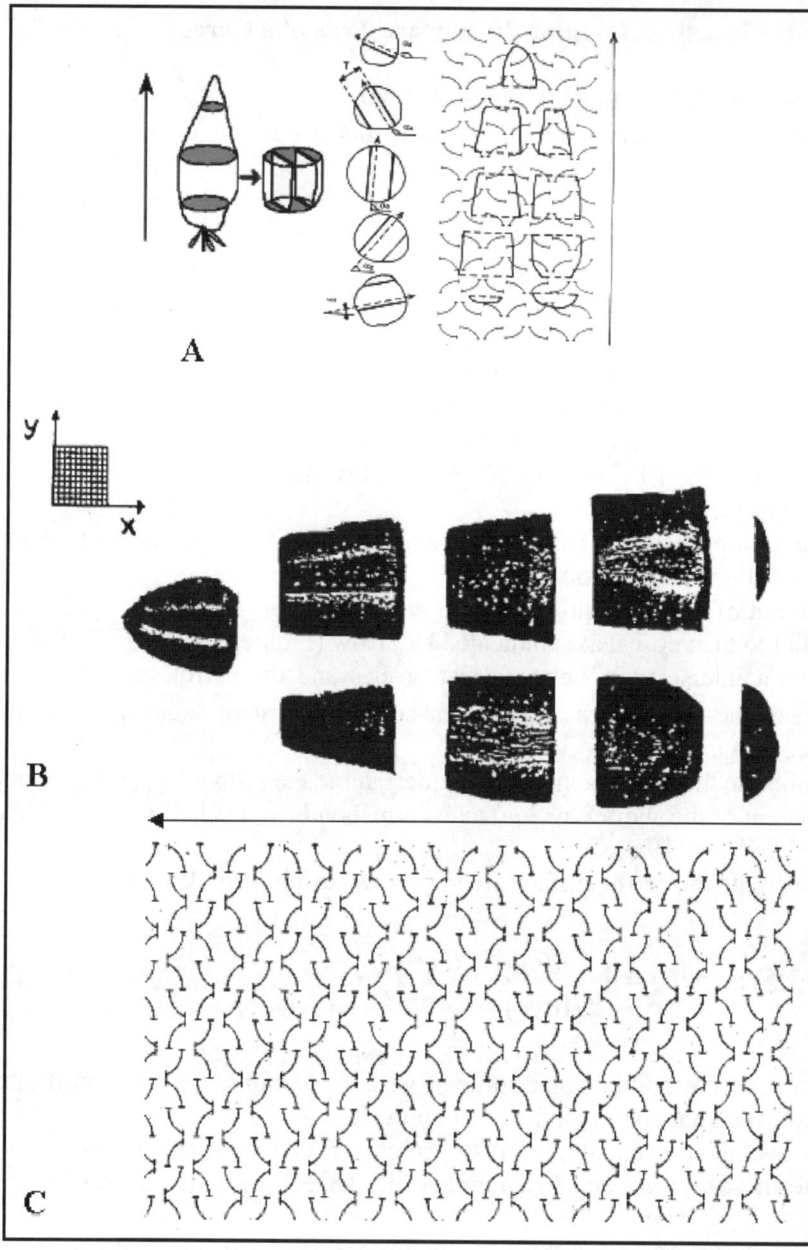

Figure 20. Practical Training 2 - Estimation of a carrot surface area using the method of vertical sections in combination with the Cavalieri principle. Carrot segments in B were obtained by systematic vertical uniform random sampling design demonstrated in A. The scale of magnification used in B and C can be derived from the square placed in the upper left corner of B where the length of the side of the square corresponds to 10 mm. Cycloid test system with *l(cyc)* = 6 mm, $a = 36\ mm^2$, is demonstrated in C. Each 1-dimensional test line is represented by the the upper edge of the drawn cycloid lines (which in practice always have a non-zero thickness).

3.3. Length Estimation

The unbiased estimation of the length of curves in 3D (e.g. tubules, capillaries) is based on counting intersection points between the curves and planar probes (Figure 21). Analogously to the surface area estimation, the direction of the planar probes can be arbitrary only if the curves are isotropic in 3D. Otherwise, in general, IR direction of the planar probes is required, i.e. every direction of the test plane must have the same probability of being chosen for the measurement.

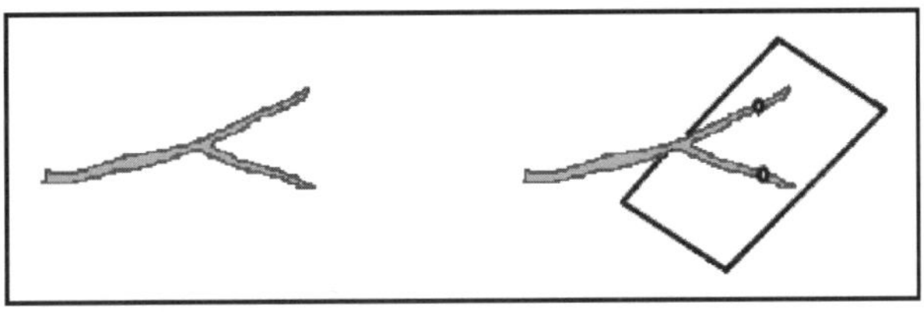

Figure 21. Length estimation. The length of curves in 3D is propotional to the number of intersections between the curves and planar probes.

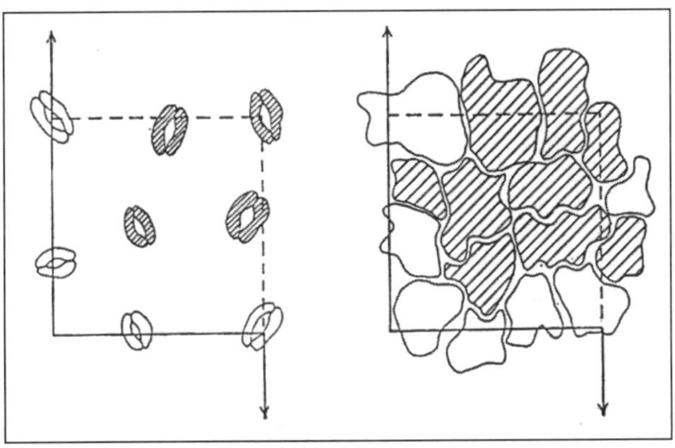

Figure 22. Unbiased estimation of particle number in 2D.
Sample just the particles that are lying at least partly in the frame and are not intersected by the (full-drawn) exclusion line. Four (hatched) stomata are sampled on the left and nine ordinary epidermal cells on the right.

In measurements of the length of curves, such as tubules, based on 2D sections, IUR sections are generated, e.g. using the orientator method [20], see Figure 17. Then the number of intersections between the tubules and these sections is counted, i.e. the number of tubular profiles lying in the sections. This counting of 2D features (profiles) must follow an unbiased counting rule, when usually unbiased sampling frames [23] are used (Figures 22, 23). Length density, L_V, is then estimated:

$$estL_V = 2 \cdot Q_A = 2 \cdot \frac{Q}{P \cdot a} \tag{14}$$

where Q_A is the number of tubular profiles per area of all sampling frames lying within the reference space, which can be easily estimated by the point counting method when one point is placed in the middle of the sampling frame (Figure 23), P is the number of test points lying in the reference space, and a is the area of the sampling frame.

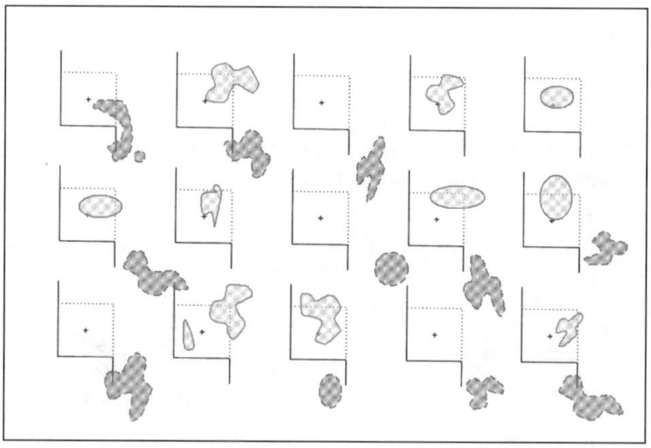

Figure 23. Counting tubular (curve) profiles. Light gray profiles with full-line boundaries lying fully or partly in unbiased sampling frames and not crossing the full-drawn exclusion line are counted. Dark gray profiles with dotted boundary are not selected by the frames.

4. MEASUREMENTS BASED ON 3D IMAGES

4.1. Acquisition of 3D Images

Three-dimensional image data can be acquired by different techniques, such as confocal microscopy, CT, MRI, and electron tomography. Confocal microscopy is especially useful as it enables to obtain perfectly registered stacks of thin serial optical sections (having thickness from approx. 350 nm) within thick specimens (Figure 24) [24] in a large variety of biological samples, including living specimens. The principle of a confocal microscope was patented by Marvin Minsky in 1957, but confocal microscopy became a useful and efficient tool not earlier than almost 30 years later, after the confocal microscope with a laser light source was introduced (confocal laser scanning microscope, CLSM, [25]). Recently, two-photon microscopy representing a new type of laser scanning microscopy providing images of thin optical sections has emerged [26], which is reported to be able to focus even deeper into the thick specimen (up to several hundred micrometers, see [27]).

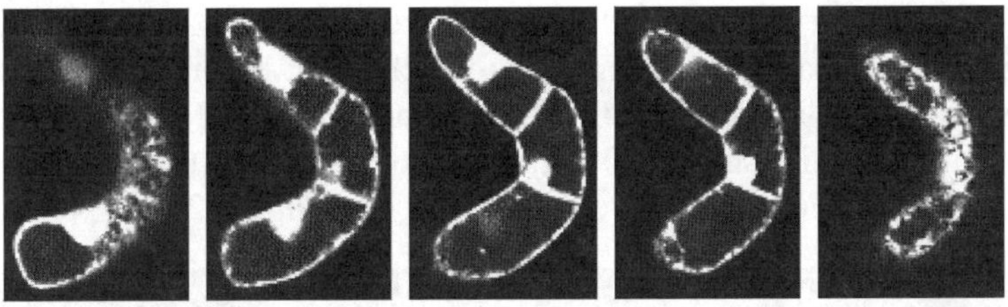

Figure 24. Confocal optical sections of tobacco cell chains 10 μm apart.

Digital images of perfectly registered stacks of optical sections from thick specimens recorded by confocal and two-photon microscopes represent suitable data for quantitative measurements as well as for computer 3D reconstructions that can be made without the necessity to solve the tedious problem of alignment of images of successive sections [24, 28]. While the role of confocal microscopy in computer-aided 3D imaging is well known, its advantages for measuring geometrical characteristics of microscopical structures are just beginning to be recognized [29 – 32]. However, especially in comparative studies, it is often more desirable to obtain quantitative information on geometry of the structure of interest than its visualization (see the example below). In such cases it is not necessary to make 3D reconstructions but it is sufficient to evaluate the optical sections captured by the confocal microscope. Deformations and possible tissue shrinkage due to the technical processing and image deformation during acquisition by the microscope should be considered and minimized in such measurements (see section 5).

4.2. Example 1: Study of Capillary Supply of the Rat Skeletal Muscle Fibers

Two functionally different types of skeletal muscle: *extensor digitorum longus* (EDL) and *soleus* (SOL) were studied [33, 34]. EDL and SOL muscles have a different level of their oxidative metabolism, as defined by histochemical studies: SOL is an exclusive slow oxidative muscle with long contraction times and EDL is a fast-contracting skeletal muscle with predominance of fast glycolytic fibers, some oxidative and oxidative glycolytic fibers [35 – 37].

In the study of differences in capillarization of the EDL and SOL muscle fibers, we can ask: 1. How does the course of capillaries along fibers look like? 2. What is the capillary supply of different types of skeletal muscle fibers?

The above questions can be answered by analysis of stacks of confocal sections within transverse thick muscle slices stained for detection of capillaries (capillary endothelial cells labeled by antibody against CD 31 and visualized by Alexa Fluor 488 in the green channel; Figure 25A) and muscle fiber walls (basal lamina present in the fiber walls labeled by the antibody against laminin and visualized by Alexa Fluor 546 in the red channel; Figure 25B), and the images from two channels merged (Figure 25C; [34]).

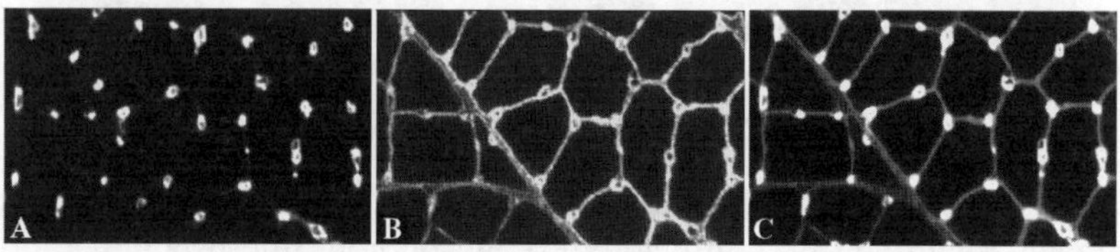

Figure 25. Confocal section of a skeletal muscle. A. Immunostaining of capillaries using CD 31 and Alexa Fluor 488. B. Immunostaining of basal lamina in fiber walls using laminin and Alexa Fluor 546. C. Merged image showing how capillaries are adjacent to fibers.

4.2.1. What Does the Course of Capillaries along Fibers Look Like?

This question can be revealed by computer 3D reconstructions of muscle fibers and adjacent capillaries, based on series of optical sections acquired by a confocal microscope (Figure 25C) as shown in Figure 26. Such visualization can give us idea on the fiber capillarization but does not provide objective, quantitative data on the structure of muscle fibers and capillaries. For example, Figure 26 suggests the length of capillaries per fiber length is larger in SOL muscle fibers than in EDL fibers but this hypothesis must be tested by objective, unbiased measurements shown below.

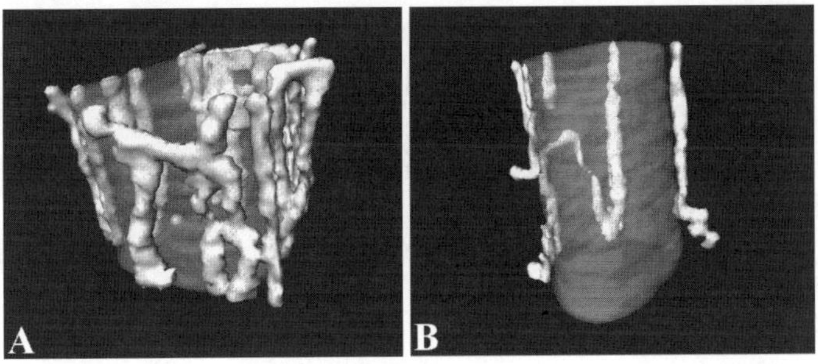

Figure 26. 3D reconstructions of muscle fibers and adjacent capillaries. A. *Soleus* muscle (slow muscle) fiber: The fiber segment shown is 70 μm long. B. *Extensor digitorum longus* muscle (fast muscle) fiber: The fiber segment shown is 80 μm long. (Reproduced by permission from [33] Copyright © 2001, Kluwer Academic Publishers.)

4.2.2. What is the Capillary Supply of Different Types of Skeletal Muscle Fibers?

This question can be answered by measuring relevant geometrical characteristics of muscle fibers and adjacent capillaries, using stereological or digital methods based on series of optical sections acquired by a confocal microscope (Figure 25):

- Mean fiber surface area within the muscle slice (*S*(fib))
- Mean fiber volume within the muscle slice (*V*(fib))
- Mean length of capillaries adjacent to fiber within the muscle slice (*L*(cap))
- Mean capillary length/adjacent fiber surface area (*L*(cap)/*S*(fib))
- Mean capillary length/adjacent fiber length (*L*(cap)/*L*(fib))
- Mean capillary length/adjacent fiber volume (*L*(cap)/*V*(fib))
- Mean number of capillary branchings/adjacent fiber length (*N*(br)/*L*(fib))

Using stereological methods shown below, according to [33], SOL has a statistically significant larger muscle fiber surface area, muscle fiber volume, number of capillary branchings per fiber, and length of capillaries per fiber length. There has not been found any significant difference between EDL and SOL in the mean capillary length/adjacent fiber surface area, the mean capillary length/adjacent fiber volume and the number of capillary branchings per capillary length.

4.3. General Overview of Stereological and Digital Methods Based on 3D Images

There is a wide variety of methods for estimating geometrical characteristics of microscopical structures based on evaluation of stacks of perfectly registered sections [30]. Below some of the relevant interactive stereological methods and automatic digital methods are described and compared.

4.3.1. Stereological Methods Applied to 3D Images

In contemporary design-based stereology, there is a trend towards spatial estimators evaluating small 3D samples of structure under study. A 3D sample of examined tissue can be analyzed if a rectangle within a field of view of a microscope is focused through. The digitized images of serial optical sections from thick specimens recorded by confocal and two-photon microscopes are especially suitable for the application of stereological spatial estimators. By using special software, it is possible to generate different virtual test probes with arbitrary pre-defined (e.g. random) position and orientation within the stack of sections and apply them directly to this image data.

The idea of estimating stereological parameters from optical sections within a thick slice was first used for counting particles in optical disector principle [38, 39] and in unbiased sampling brick rule [40] , and then in many other stereological methods, e.g., nucleator [41] and planar rotator [42] applied to a stack of optical sections and estimating the mean particle volume, methods using optical rotator [43, 44], spatial grid [45] , vertical spatial grid [46], virtual fakir probes [47] , and virtual cycloids [48] for estimating the surface area, method of vertical slices [49], of total vertical projections [50], global spatial sampling [51], and spherical probes [52] used for the length estimation.

Some of these methods still require randomizing the direction of the thick slice while other ones can be applied to slices cut in an arbitrary direction. In practice, it may be inevitable to take sections at a particular fixed direction. For example, it is not easy to orientate cells in a suspension or a cell culture at random and it can be easier to determine the mutual localization of different tissues in sections with a specific direction. Therefore, below we describe the methods without assumptions on thick slice orientation, enabling to estimate such important parameters like number, volume, surface area and length. The main idea here is to use virtual, randomly oriented spatial grids of points, lines or planes which can be generated by a computer and visualized together with the structure under study. Thus, both test grid and microscopical structure are "focused through" by browsing through the stack of optical sections and so interactions between the grid and structure can be evaluated.

4.3.2. Digital Methods Applied to 3D Images

The automatic measurements of geometric characteristics of 3D objects can be applied directly to their binary images obtained by automatic segmentation of the grayscale images (in a binary image foreground voxels belonging to the objects of interest are white, i.e. of value 1, see 2.2). The validity of the results of such automatic measurement, i.e. their unbiasedness and precision, depends on how precisely the model consisting of foreground voxels describes the object under study. Appropriate spatial resolution and high image quality are necessary for geometrical measurements. High quality contrast images enable segmentation by simply thresholding the image values while image inhomogeneities due to uneven staining or heterogeneous acquisition conditions require more advanced techniques of segmentation based on region and edge detection techniques. The images distorted by noise must be pre-processed by filtration. A heavy signal distortion may complicate or even preclude the exact segmentation. The exact spatial calibration of the image, enabling to calculate the position of individual elements, is also a prerequisite for the measurement of any dimensional characteristic. If a usual regular orthogonal grid is used, the calibration is defined by the lengths of intervals between the image elements in all orthogonal directions (see 2.2).

A 3D image formed by a dense sequence of exactly registered 2D images may contain more complete spatial information on the object under study, e.g. separate 2D sections do not contain information on gradients in the directions perpendicular to the image plane. Further, topological properties, as the continuity of objects, cannot be judged from separate sections and the counting of objects requires 3D information in general. Finally, the 3D image processing, using more spatial information, can be more effective and robust than the 2D processing of individual slices. Basic algorithms of 3D image processing can be derived from those used in 2D image processing in a straightforward way [13].

In another approach, triangulated surfaces are used for the measurement. Such surfaces can be obtained from the 3D grayscale digital images by isosurface detection using the marching cube algorithm [53] or by detecting the object contours in subsequent horizontal slices first and then connecting the contours by triangulated stripes [54]. The contours can be detected interactively or by (semi-) automatic tracking of the object boundaries in the slices [55]. Geometric characteristics of the models delimited by the triangulated surfaces can be used as estimates of the characteristics of the objects under study [56].

4.4. Number Estimation

4.4.1. Number Estimation by Stereological Methods

First, it is necessary to stress that 3D particles (e.g. cells) cannot be counted in one 2D plane (Figure 27): Larger (higher) particles are more likely to be hit by the plane, thus the number of their profiles is higher than that of smaller particles. An unbiased counting or sampling of 3D particles can be achieved by using a 3D probe, such as a disector (Figure 28; [38, 39]) or unbiased sampling brick [40]. In a physical disector only two parallel physical sections representing the top and bottom of the probe are examined which means that it is necessary to find corresponding locations in two separate sections (Figure 28B).

The optical disector [38] and also a similar unbiased sampling brick principle [40] are based on focusing through the 3D probe within a thick section with e.g. cells: We can place a counting frame on the slice and by focusing through the tissue we get the "virtual" box of tissue. During focusing up from the bottom we count only those cells within the box that disappear during focusing (i.e. they are not intersected by the upper side) and at the same time are not intersected by the left and front sides of the virtual box (i.e. their sections are not intersected by the exclusion lines of the unbiased counting frame).

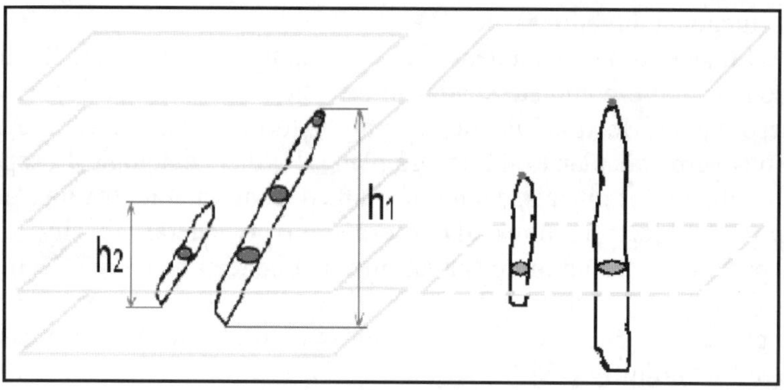

Figure 27. Counting particles in 3D. As shown on the left, particles cannot be counted by one 2D plane. In this case the higher particle is hit by three test planes while the lower one by one plane only. As shown on the right, particles can be counted by two planes or by a 3D probe when they are counted by their upmost points. Here, both particles are counted as their upmost points are lying between the two test planes delimiting the 3D probe. (reproduced by permission from [32] Copyright © 2002, John Wiley & Sons, Inc.)

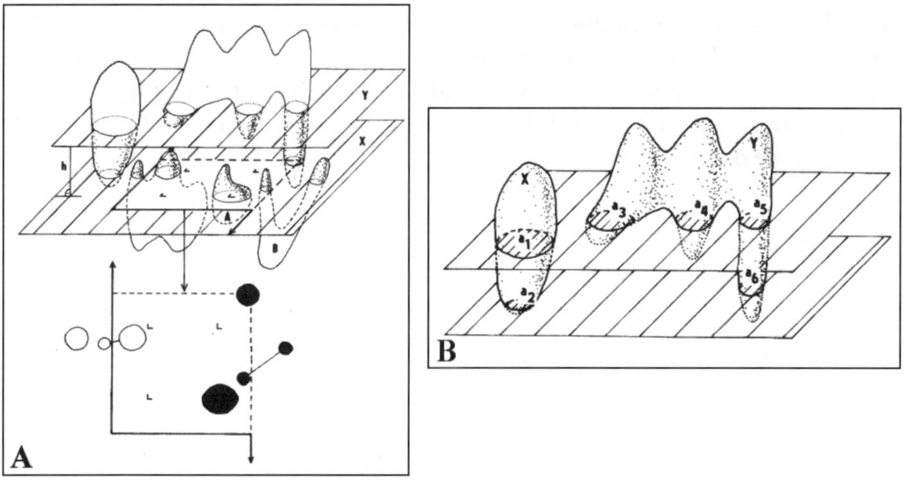

Figure 28. Estimation of number by the disector principle. A. Particles can be counted by two planes or by a 3D probe: Sample just the particles with profile sets sampled by the unbiased sampling frame in the lower plane which are not intersected by the upper, look-up plane. Particles A and B are sampled by the disector here. B. General Requirement for the physical disector:

Any particle profile from the set of particle profiles should be unambiguously identifiable as belonging to the same particle. In the situation shown here we must be able to recognize that profiles a_1 and a_2 belong to the particle X while profiles a_3, a_4, a_5, and a_6 belong to the particle Y.

If such counting of cells is made in many boxes sampled in a systematic uniform random manner (Figure 8; 2.4.2) within the reference tissue, the total number of cells, *N*(cell), can be estimated by the formula:

$$estN(cell) = \frac{Q^{-}(cell)}{P(ref)} \cdot \frac{p}{a \cdot h} \cdot V(ref) \qquad (15)$$

where $Q^{-}(cell)$ is the number of cells sampled by all disector boxes, *p* is the number of test points of the grid placed into one sampling frame, *P(ref)* is the total number of test points of this grid falling in the reference space in all sampling frames, *a* is the actual area of one sampling frame, *h* denotes the height of the disector box, and *V(ref)* is the volume of the reference space.

The second approach uses the fractionator method [38], which is based on direct counting of cells in a known predetermined fraction of the organ in which we wish to count cells. The organ is divided arbitrarily into blocks. (These can be histological blocks that will be sectioned.) Then a fraction (bsf) of blocks (e.g. every third or fifth or more) is designated to be totally cut into thick sections. Next, a fraction (ssf) of the thick sections is sampled (say every third, tenth or twentieth) and then a fraction (asf) of the area of the thick section is sampled by unbiased counting frames of the optical disector. The height of the disector (i.e. the virtual box height) is some fraction (hsf) of the mean thickness of the thick sections. Finally, the cells are counted by the disector in the way described above. Then the number of cells can be estimated by:

$$estN = \frac{1}{bsf} \cdot \frac{1}{ssf} \cdot \frac{1}{asf} \cdot \frac{1}{hsf} \cdot \sum Q^{-} \qquad (16)$$

where $\sum Q^{-}$ is the total number of cells counted.

4.4.2. Number Estimation by Automatic Methods

Number (N) and numerical density (N_V) of 3D objects (particles) can be estimated by counting the connected components in 3D binary image, which means that the automatic segmentation of analyzed objects from the original grayscale 3D image is required.
Under the assumption that the connected objects are simple (i.e. without holes or tunnels) their number can be estimated by the connectivity number (Euler-Poincaré characteristic) of the image foreground [10], see 2.2. The connectivity number of one simple object is equal to one and the correct result of counting is achieved due to the additivity of the connectivity number with respect to the disjoint union of sets. The advantage of this approach is that the connectivity number can be calculated by counting specific local configurations by a simple algorithm (see formulas 3, 4 and [11]).

4.5. Volume and Surface Area Estimation

4.5.1. Stereological Method for Volume Estimation Using Spatial Grid of Points

The method of volume estimation using spatial grid of points [57] is a modification of the Cavalieri principle (see 3.1.1). If a cubic spatial grid of points (Figure 29) is applied, the object volume can be estimated by the formula:

$$estV = u^{3} \cdot P \qquad (17)$$

where u is the grid constant (distance between two neighboring points of the grid) and P is the number of grid points falling into the object. The position of the point grid should be uniform random (see 2.4.2 with clear modification to 3D). The practical implementation of the spatial grid of points for estimating the volume of a tobacco cell chain is shown in Figure 30.

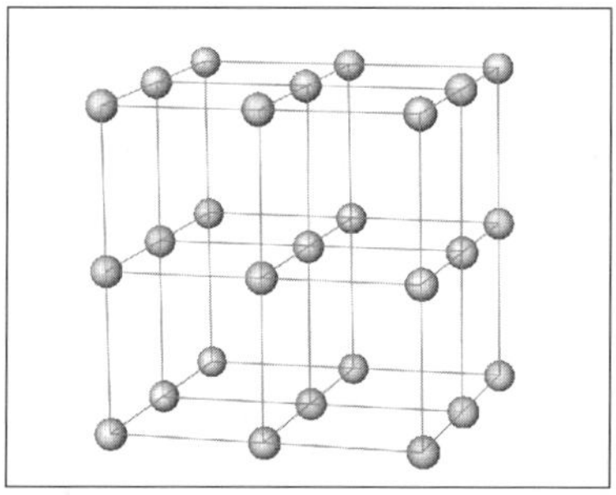

Figure 29. Cubic spatial grid of points. (Reproduced by permission from [58] Copyright © 2001, John Wiley & Sons, Inc.)

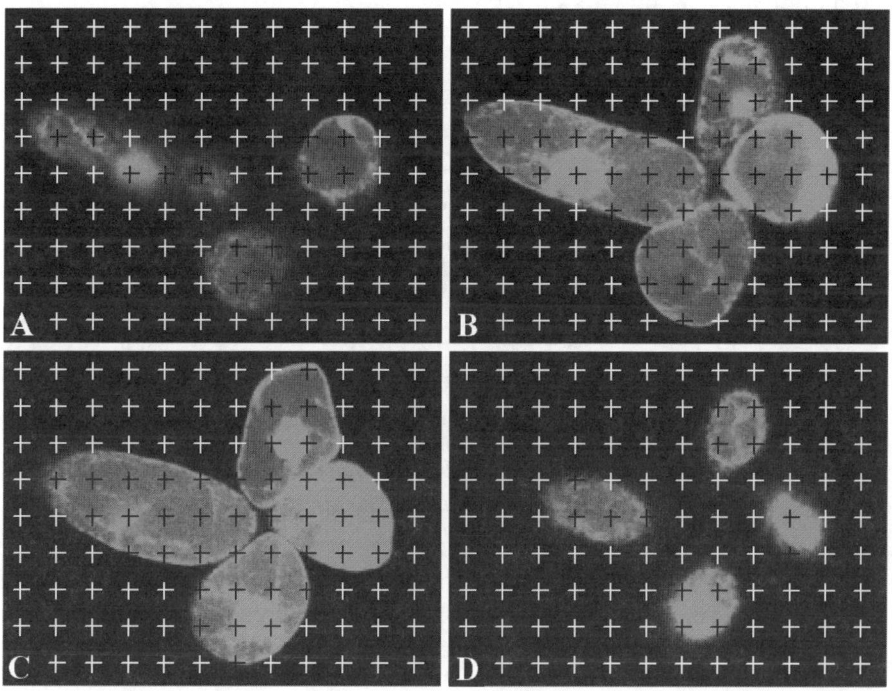

Figure 30. Estimation of the volume a tobacco cell chain by the spatial grid of points. Four serial sections of the cell chain, 20 μm apart, are shown in A-D. The test points of the cubic spatial grid (Figure 29) falling within the object are counted in each section. The grid constant $u = 20$ μm, $P = 102$ here, i.e. est $V = 20^3 . 102 = 816{,}000$ (μm^3).

4.5.2. Stereological Method for Surface Area and Volume Estimation Using the Fakir Method

The surface area of an object, e.g., cell, can be estimated using fakir probes [47]. The fakir probe (named by [59]) is a systematic probe consisting of parallel test lines (resembling nails of a fakir bed piercing the surface, see Figure 31). When estimating the cell surface area, the intersections between the cell surface and the fakir probe are counted. We can imagine that the cell is pierced through by the nails of the fakir bed and we are counting how many times the nails went into or out of the cell. The object surface area, *S,* can be estimated by the following formula:

$$estS = 2 \cdot u^2 \cdot I \quad (18)$$

where u is the distance between neighboring parallel lines of the probe, I is the number of intersections between the fakir probe and the object surface. The above estimator is unbiased if the orientation of the spatial grid is isotropic random (see Figure 15) and its position is uniform random (see 2.4.2 with clear modification to 3D).

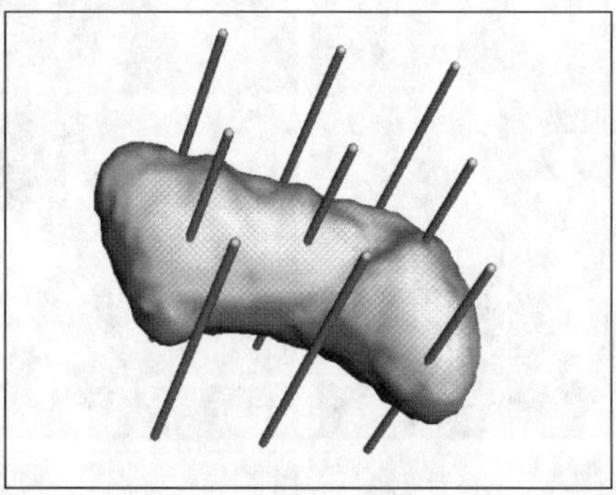

Figure 31. Fakir probes piercing the cell surface. (Reproduced by permission from [58] Copyright © 2001, John Wiley & Sons, Inc.)

A high efficiency of the measurement can be achieved if we use a cubic spatial grid consisting of three mutually perpendicular (orthogonal) fakir probes, halfway shifted with respect to each other (Figure 32B). In such case the average over the three fakir probes is considered in the surface area estimation:

$$estS = \frac{2}{3} \cdot u^2 \cdot (I_1 + I_2 + I_3) \quad (19)$$

where u, the grid scale factor, is the distance between neighboring parallel lines of the grid and I_j (j=1,2,3) is the number of intersections between the j-th fakir probe and the object surface. Figure 32 shows the efficiency of the fakir method with different arrangements of triplets of fakir probes.

The measurement can be easily performed using our interactive Fakir module running in *Ellipse* (ViDiTo, Slovakia) environment (see http://www.ellipse.sk) or FAKIR program (see [58] and http://www.biomed.cas.cz/fgu/fakir/3dtools.htm for free download). This software generates an isotropic set of virtual fakir probes and so it is not necessary to randomize the direction of the stack of sections.

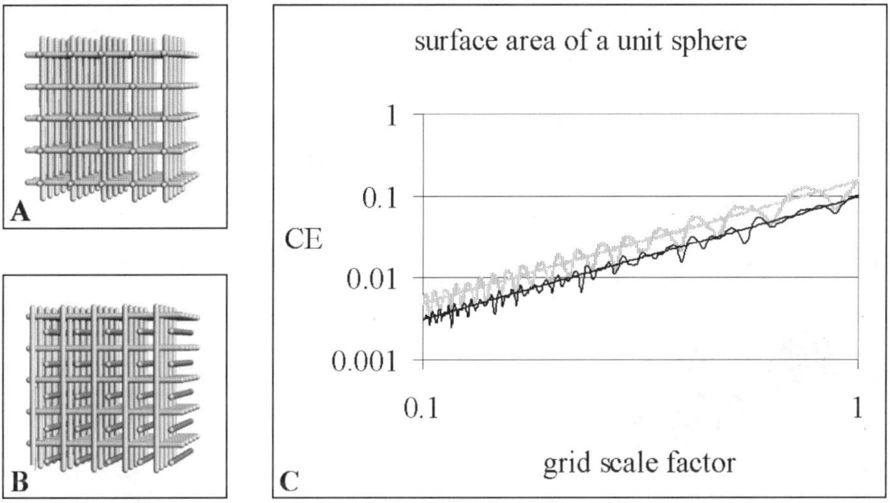

Figure 32. Efficiency of the fakir method. A. Sandau spatial grid (SG). B. Orthogonal triplet of halfway shifted fakir probes (OTSFP). C. Coefficient of error of different arrangements of three mutually perpendicular (orthogonal) fakir probes when estimating the surface area of a unit sphere: OTSFP (black) is more efficient than SG (light gray). (Figures A and B reproduced by permission from [30] Copyright © 1999, Wiley-Liss, Inc., a subsidiary of John Wiley & Sons, Inc.)

The fakir probe can be used also for estimating the object volume, when the total length of intercepts between the object and the probe is measured [57, 59]. If orthogonal triplet of shifted fakir probes (Figure 32B) is used, the object volume (V) can be estimated by the following formula:

$$est\,V = \frac{1}{3} \cdot u^2 \cdot (L_1 + L_2 + L_3) \qquad (20)$$

where L_j (j=1,2,3) is the length of intercepts between the object and the j-th probe, u is the grid constant, i.e. the distance between neighboring parallel lines of the triple grid. The position of the spatial grid should be uniform random (see 2.4.2 with clear modification to 3D).

4.5.3. Automatic Method for Volume Estimation Using Voxel Counting

As indicated in 2.2, the volume of an object can be estimated by the voxel-counting method when the number n of foreground voxels of the object digital image (Figure 33A) is multiplied by the voxel volume:

$$est\,V = u_1 \cdot u_2 \cdot u_3 \cdot n \tag{21}$$

where u_j (j=1,2,3) are the distances between centers of neighboring voxels in the directions of x, y, and z axes.

Note that the surface area cannot be estimated analogously, i.e. by the sum of areas of exposed voxel faces – such method would clearly

overestimate the surface area.

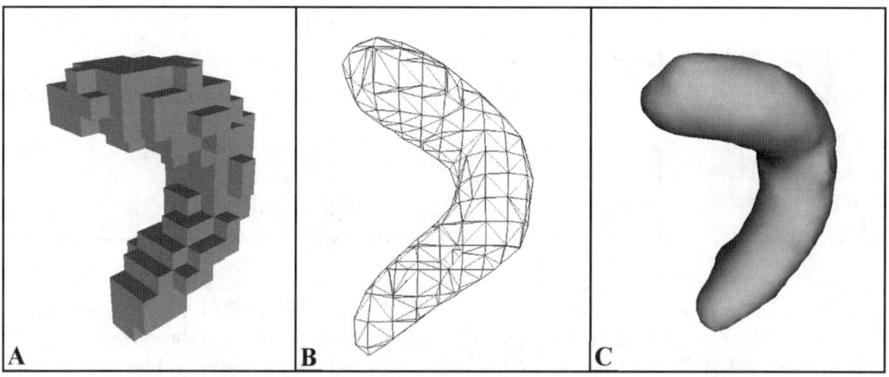

Figure 33. 3D digital representation of tobacco cell chain based on image data captured by a confocal microscope shown in Figure 24. A. Digitized image of the chain with resolution 8μm^3, where the voxels are represented as cubes. B. Chain model with triangulated surface. C. 3D reconstruction of the chain based on rendering of triangulated surface. (Figures A and C are reproduced by permission from [32] Copyright © 2002, John Wiley & Sons, Inc.)

4.5.4. Estimation of Object Surface Area by Digital Crofton Method

In automatic surface area measurements based on image processing it is advantageous to use the Crofton formula [13], as pointed out in 2.2. In the digital Crofton method, the intersections of the object surface with the rows and diagonals of an orthogonal grid of voxels, called the principal lines, are used for digital measurements of surface area. Three directions of principal lines connecting the neighbors differing in one co-ordinate are the directions of the orthogonal voxel axes (rows), six directions of principal lines connecting the neighbors differing in two co-ordinates are the directions of the voxel surface diagonals, four directions of principal lines connecting the neighbors differing in three co-ordinates are the directions of the voxel solid diagonals (Figure 34). Then the surface area can be estimated by summing up the pairs of neighbors with alternate values:

$$est\,S = 2 \cdot \frac{1}{m} \cdot \sum_{j=1}^{m} \frac{1}{L_{Vj}} \cdot n_j \tag{22}$$

where m=13 is the number of directions (Figure 34), n_j is the number of configurations of neighboring points with alternating values (i.e. the number of intersections of the object surface with the principal lines in the specified directions), and L_{Vj} is the length density of a principal line in the j-th direction, i.e. the length of a principal line per unit volume (Table 2). In a cubic lattice, where $u=u_1=u_2=u_3$, clearly $L_V = 1/u^2$ for rows, $L_V = \sqrt{2}/u^2$ for surface diagonals, and $L_V = \sqrt{3}/u^2$ for solid diagonals.

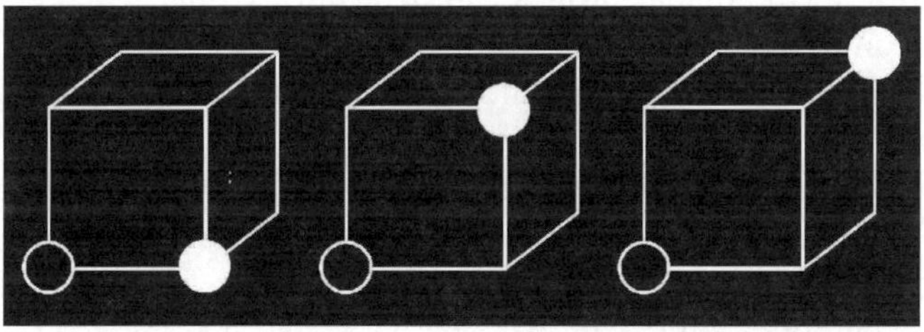

Figure 34. The digital Crofton method for surface area estimation is based on automatic counting of intersections between the surface and 13 main directions of the lattice of voxels shown here. The directions correspond to the principal lines in the direction of rows (left), voxel surface diagonals (middle) and voxel solid diagonals (right) and their orthogonal rotations. (Reproduced by permission from [32] Copyright © 2002, John Wiley & Sons, Inc.)

Table 2. The densities (L_V) of principal lines of the orthogonal grid of points with co-ordinates ($i\,u_1, j\,u_2, k\,u_3$) for integers i, j, k and calibration constants $u_1 > 0$, $u_2 > 0$ and $u_3 > 0$. The densities are listed for all three types of lines, i.e. for those going in the direction of rows, surface diagonals or solid diagonals of the cubic lattice. (Reproduced by permission from [32] Copyright © 2002, John Wiley & Sons, Inc.)

line type	L_V	
rows	$u_i / u_1u_2u_3$	$1 \le i \le 3$
surface diagonals	$\sqrt{(u_i^2+u_j^2)} / u_1u_2u_3$	$1 \le i < j \le 3$
solid diagonals	$\sqrt{(u_1^2+u_2^2+u_3^2)} / u_1u_2u_3$	

4.5.5. Estimation of Object Surface Area and Volume Based on Surface Triangulation

After the triangulated surfaces are obtained from 3D grayscale digital images by isosurface detection or by connecting the object contours by triangulated stripes (Figure 33B), the object surface area can be estimated by summing up the area of all triangles. Analogously, the object volume can be estimated by the volume delimited by triangulated surfaces.

4.6. Length Estimation

4.6.1. Stereological Method for Length Estimation Using Spatial Grid of Planes

The unbiased estimation of the length of curves in 3D (e.g. capillaries) is based on counting intersection points between the curves and planar probes. The orientation of the planar probes can be arbitrary only if the curves have the same geometrical properties in all directions of space, i.e. they are isotropic. Otherwise, in general, the isotropic random orientation of the planar probes is required (see 3.3). In addition to stereological methods for estimating the length using isotropic thin physical sections (see 3.2.1; 3.3), recently, new methods using thick sections with less or no requirements on physical section orientation have been developed [49 – 52]. We will describe the "slicer" method (called "global spatial sampling" by [51]) that uses arbitrarily oriented thick sections and spatial grid of planes generated by a computer.

The slicer probe is a systematic probe consisting of parallel test planes (Figure 35). Its direction should be isotropic random in 3D space. When estimating, e.g., the capillary length, a thick tissue slab is cut in an arbitrary direction. Then the slab is cut by the "slicer", i.e. by virtual, software generated slicer planes. The length of capillaries, L, in the given stack of sections within the thick tissue slab can be estimated by the following formula:

$$estL = 2 \cdot d \cdot Q \tag{23}$$

where d is the distance between neighboring parallel planes, Q is the number of intersections between the slicer probe and the capillaries within the given stack. This length estimator is unbiased if the orientation of the slicer probe is isotropic random, its position is uniform random and the intersections are counted properly (see [51]).

A higher efficiency of the measurement can be achieved if a cubic spatial grid consisting of three mutually perpendicular slicer probes is used (Figure 35B; [32, 33]). In such case the average over the three slicer probes is considered in the length estimation:

$$estL = \frac{2}{3} \cdot d \cdot (Q_1 + Q_2 + Q_3) \tag{24}$$

where *d* is the distance between neighboring parallel planes, Q_j (j=1,2,3) is the number of intersections between the *j*-th slicer probe and the capillaries within the given stack.

The above measurement requires the use of a specific software generating an isotropic set of virtual slicer probes that are applied to stacks of transmission, or preferably, confocal images (e.g. C.A.S.T.-GRID by Olympus Danmark A/S, SLICER module running in *Ellipse* environment by ViDiTo, Slovakia, see http://www.ellipse.sk, or SLICER program, see [32], and http://www.biomed.cas.cz/fgu/fakir/3dtools.htm for free download).

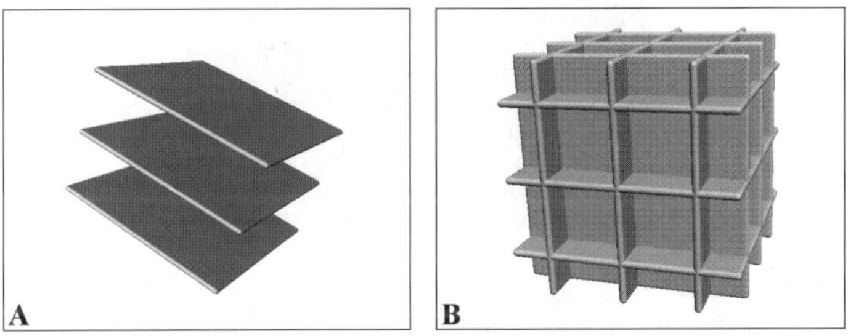

Figure 35. A. Slicer probe. B. Triplet of slicer probes. (Reproduced by permission from [32] Copyright © 2002, John Wiley & Sons, Inc.)

4.6.2. Length Estimation by Digital Crofton Method

In automatic length measurements based on image processing the Crofton formula can also be used in a similar way as for surface area measurements [13, 32] as pointed out in 2.2. The number of intersections of the object and the test plane can be calculated using the connectivity number (Euler-Poincaré characteristic; 2.2) of the intersection of the object and the test plane if the intersection consists of simple profiles, e.g. in the case of fibers. The length measured in this way is equal to the integral of the mean curvature of the surface divided by π [12].

The length of a feature set in a 3D binary image can be estimated from its intersections with the principal planes of the orthogonal grid. The principal planes of the grid are the planes intersecting the triplets of the mutually neighboring vertices; the densities of the principal planes of the grid (S_V, i.e. the ratio of the test plane areas to the unit volume; the reciprocal $1/S_V$ corresponds to the distance between successive planes) are listed in Table 3.

The 3D configurations that should be counted with respective signs during the measurement of the length using the digitized projection formula are shown in Figure 36, the rest of them can be obtained by orthogonal rotations and proper deformations of these configurations. The length of the segmented object is then estimated:

$$est\,L = 2 \cdot \frac{1}{m} \cdot \sum_{j=1}^{m} \frac{1}{S_{Vj}} \cdot n_j \tag{25}$$

where m is the number of directions, n_{ji} is the number of configurations and S_{Vj} is the surface density of principal planes (Table 3) for the j-th direction.

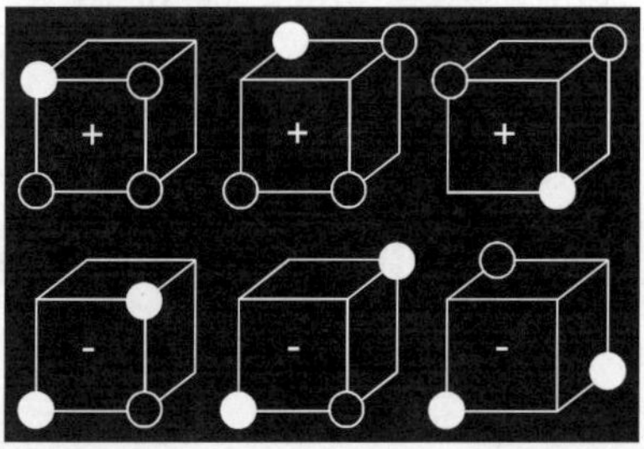

Figure 36. Examples of configurations of neighbors in the cubic lattice used for the length measurements. All relevant local configurations of vertices in the general orthogonal lattice can be obtained by orthogonal rotations and deformations of the examples. The configurations should be counted with respective signs shown. (Reproduced by permission from [32] Copyright © 2002, John Wiley & Sons, Inc.)

Table 3. The surface densities (S_V) of principal planes of the orthogonal grid of points with co-ordinates ($i\,u_1$, $j\,u_2$, $k\,u_3$) for integers i, j, k and calibration constants $u_1 > 0$, $u_2 > 0$ and $u_3 > 0$. The densities are listed for all three types of plane types, i.e. for those intersecting only rows, rows and surface diagonals or only surface diagonals of the cubic lattice. (Reproduced by permission from [32] Copyright © 2002, John Wiley & Sons, Inc.)

plane type	S_V	
rows	$1 / u_i$	$1 \le i \le 3$
rows and diagonals	$\sqrt{(u_i^{-2}+u_j^{-2})}$	$1 \le i < j \le 3$
diagonals	$\sqrt{(u_1^{-2}+u_2^{-2}+u_3^{-2})}$	

4.7. Example 2: Comparison of Different Stereological and Automatic Digital Methods Using Tobacco Cell Chains as Model Objects

The tobacco cell line BY-2 (Nicotiana tabacum L., cv. Bright Yellow) was cultured in vitro and stained without prefixation by adding the aliquot part of 0.01% solution of fluorescein diacetate (FAD) into the culture medium. Fourteen individual cells or short cell chains were selected with the aim to obtain a set of test objects having as large variability of shapes as possible (Figure 37). The images of serial optical sections of each cell chain, 1 μm apart, were recorded by a Bio-Rad MRC 600 confocal laser scanning microscope (using excitation wavelength of 488 nm), five such optical sections are shown in Figue 24. For details of specimen preparation and image acquisition see [30].

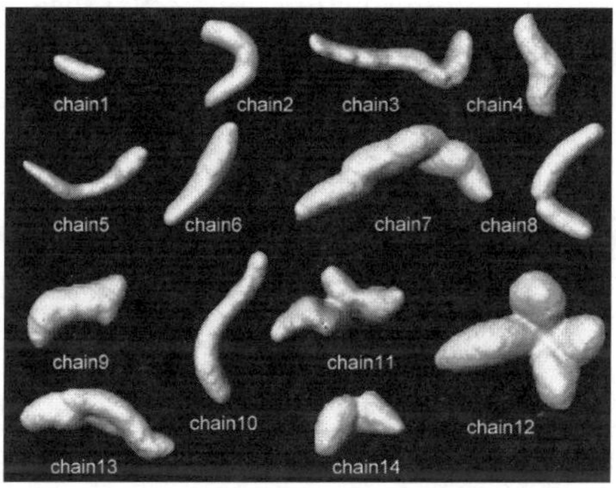

Figure 37. Three-dimensional reconstructions of fourteen tobacco cell chains used for the measurements. (Reproduced by permission from [30] Copyright © 1999, Wiley-Liss, Inc., a subsidiary of John Wiley & Sons, Inc.)

The volume data sets originating from the captured series of grayscale images were resampled to spatial resolution of 1 μm^3 , closed by 2D morphological closing [10] and smoothed by a 3D gaussian filter using a 3^3 neighborhood applied twice with standard deviation σ=1.5 μm. The binary volume data sets were obtained by thresholding, morphological closure with 2^3 structural element, filling of cavities and opening by reconstruction. The cell chain volume was estimated by the spatial grid of points (4.5.1). The method was applied to grayscale as well as binary volume data and repeated independently five times. The cell chain surface area and volume was estimated by the fakir method (4.5.2) using the OTSFP probe (Figure 32B). The fakir method was applied to grayscale as well as binary volume data while each measurement was performed independently five times. Both the digital Crofton method (4.5.4) and the method based on surface triangulation (4.5.5) were applied to four sets of image data differing in their degrees of smoothing and resolution. The voxel-counting method (4.5.3) was applied to the same four sets of binary data as used for surface area measurements by the digital Crofton method. Finally, the algorithm for volume measurement based on surface triangulation (4.5.5) was applied to the same data as used for surface area measurements by the triangulation method (4.5.5). For details see [32].

All above image processing and implementation of digital methods were performed on a SGI INDY workstation using our special custom-made modules developed in the IRIS Explorer (NAG, UK) visualization environment. The stereological methods were implemented in our special programs, recently transferred to modules running in Ellipse (ViDiTo, Slovakia) environment.

The results showed that all tested methods for volume estimation gave precise results and they were not sensitive to resolution (voxel size) and degree of smoothing. If segmentation is feasible simple voxel counting method is suitable for volume estimation, otherwise stereological method using spatial grid of points can be recommended. Digital methods for surface area estimation, especially triangulation method were more sensitive to image processing. It is necessary to be cautious especially with noisy images. Enlargement of voxel size increases while smoothing decreases surface area values. Automatic digital methods are faster than interactive stereological methods but require segmentation of analyzed objects. Before applying any of the automatic methods, it should be carefully tested for the given type of image data if it gives reasonably precise results. Stereological fakir method can be recommended for testing the precision and applicability of other, less time-consuming methods.

4.8. Example 3: Three-Dimensional Analysis of Photosynthetic Structures in Norway Spruce Needle

Within the last decades air pollution has become a potential threat to forest production world-wide. Air-borne pollutants can cause not only visible injuries such as chloroses and necroses, but their effect is recognizable at first on metabolic detection level and subsequently on anatomical one. Information on quantification of inner structural damage of needles could be useful for detection of early damage of needles of Norway spruce (Figure 38A), what is indispensable for monitoring of forest health.

Transverse SUR sections (50 μm thick) of Norway spruce needles, 2 mm apart, were prepared using a hand microtome (Figure 38B). Sections were placed in distilled water between two cover glasses. No sample pre-processing was necessary since natural autofluorescence of phenolic compounds in cell walls was used.

Needle sections were observed using a confocal microscope Bio-Rad MRC 600, with either objective 4x to acquire the whole section image (Figure 38B,C) or water immersion objective (60x, N.A.= 1.2) to capture SUR fields of view (Figure 38D). In the latter case, series of 40 optical sections (1 μm apart) were obtained (Figure 38E-H).

The following stereological methods were applied for estimation of quantitative anatomical characteristics:

1. Cavalieri principle and point counting method (see 3.1.1.1): needle and mesophyll volume, proportion of tissues in a needle.
2. Fakir method (see 4.5.2): needle inner surface area. The system of parallel fakir probes was computer generated and intersections with the exposed mesophyll cell surface and test probes were manually marked using Fakir module in *Ellipse* SW environment (Figure 38E,F).
3. Optical disector: number and mean volume of mesophyll cells in a needle. (Figure 38G,H).

In our studies of the effect of simulated acid rain (SAR) treatment on the above needle anatomical parameters, we have found that SAR causes a decrease in needle volume and increase in proportion of conductive tissues. It does not affect the total number of mesophyll cells in a needle but their mean volume is decreased.

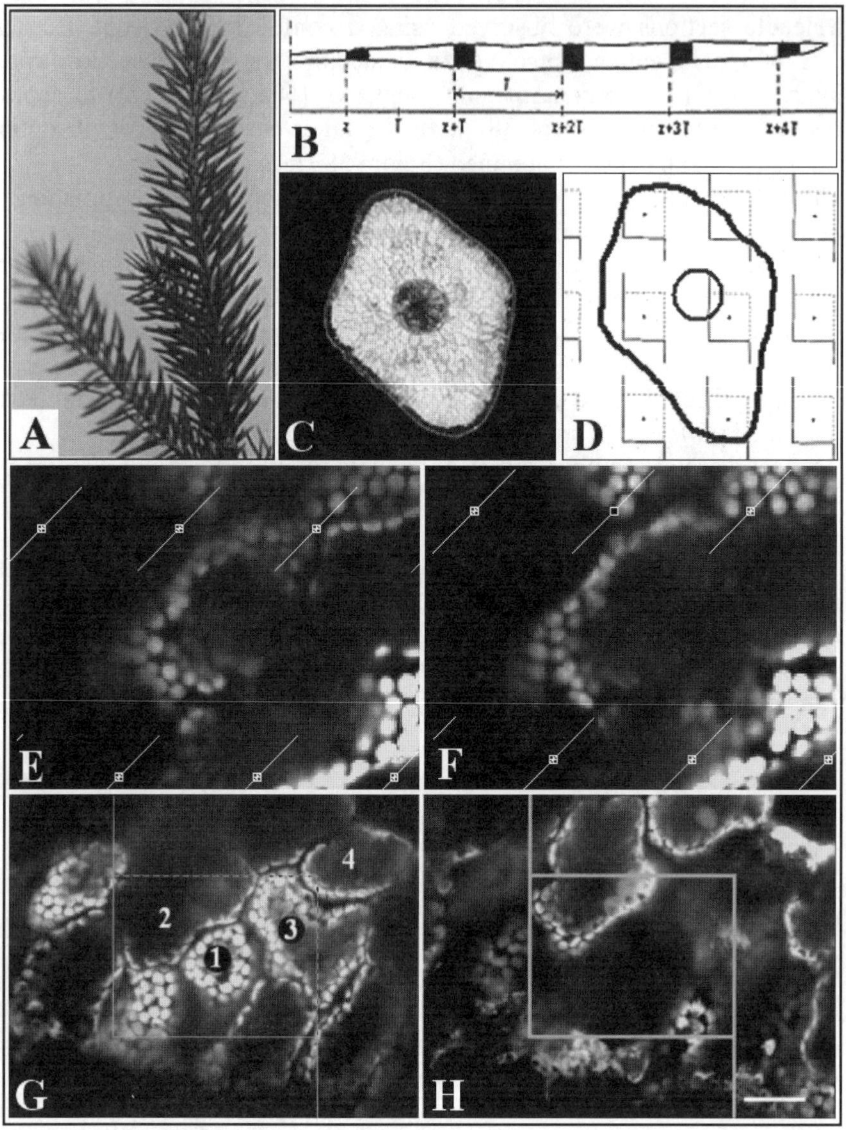

Figure 38. Stereological analysis of photosynthetic structures in Norway spruce needle. A. Photograph of a Norway spruce branch. B. SUR sampling of needle segments. C. Section of a needle captured by a confocal microscope using 4x objective. D. SUR fields of view from which series of optical sections were captured by a confocal microscope. E-F. Fakir method applied to such series. Two subsequent optical sections are shown, intersection of a fakir probe with the cell surface is marked by a black square. G-H. Optical disector method. The first, reference disector plane (G) and the last, look-up plane (H) are shown. Clearly, cells No. 1, 3, and 4 are counted, while cell No. 2 is not sampled because it is intersected by the look-up plane. (Figures G and H are reproduced by permission from [32] Copyright © 2002, John Wiley & Sons, Inc.)

5. POSSIBLE PITFALLS IN IMPLEMENTATION OF STEREOLOGICAL AND DIGITAL METHODS USING CONFOCAL MICROSCOPY

Whether interactive stereological or automatic image analysis techniques are applied, it is always necessary to follow proper sampling, i.e. the fields of view chosen for evaluation must be selected in a representative, unbiased manner. Usually, systematic sampling is a good and efficient way to select the sections and sampling frames for analysis [18, 60]. It is also necessary to define precisely conditions and aims of the study. The above methods as any other technique lead to reliable results only if the reference space and structures under study can be unambiguously identified. Further, it is necessary to take into account possible bias due to technical processing of the tissues under study, especially deformation caused by the tissue shrinkage during fixation, embedding and cutting of material [61]. Such deformations should be minimized by developing suitable processing techniques. The possible deformations should be measured and controlled in different steps of technical processing. This requires to calibrate the microtome or vibratome used for cutting the tissue and to measure lateral and axial deformations – a confocal microscope enabling to measure axial and lateral distances in different steps of tissue processing can be a very useful tool for such evaluation.

Confocal microscopy, like any other technique, has some drawbacks. Sometimes it can be difficult to find a suitable fluorescence staining of structures to be examined. It should be noted that in histological specimens a non-specific staining like eosin can often be used [62] while the cell surface can be visualized by immunofluorescence techniques when antibodies are bound to integral membrane proteins of cellular plasma membranes and labeled by a common fluorescent dye. Another drawback consists in the axial resolution (though higher than in a conventional optical microscope) being lower than the lateral resolution. The shape of the point spread function of a confocal microscope is elongated in the direction of z-axis [63] which causes defocusing that can possibly result in an overestimation of the surface area and volume of the examined objects. This can be eliminated by applying special deconvolution algorithms to perform 3-D deblurring of the images before the measurements (for comparison of volume and surface area before and after applying deconvolution see [64]). In two-photon excitation fluorescence microscopy, the axial resolution is higher than in confocal microscopy [26, 65, 66] and it is also possible to penetrate more deeply into the specimen with decreased bleaching of fluorescence dyes. However, possible deblurring of 3-D images should be considered even here. Aberrations, especially the axial displacement due to the unmatched refractive indices [67], should also be taken into account in the measurements of microscopical structures.

6. SOLUTIONS TO PRACTICAL TRAININGS

6.1. Solution to Training 1

The value of the carrot volume measured by the Archimedes law is usually no more than 5-10% different from the value estimated by the Cavalieri principle. The volume of the sample carrot is 40.5 cm^3 .

6.2. Solution to Training 2

The surface area of the sample carrot is 44 cm^2 .

7. ACKNOWLEDGEMENTS

We wish to thank Dr.J.O.Archambeau (Department of Radiation Medicine, Loma Linda University, CA, USA) for Figure 1A, Dr.M.Jirkovská (Institute of Histology and Embryology, First Medical Faculty, Charles University, Prague, Czech Republic) for providing us with placenta specimen shown in Figures 2A and 11, Dr.K.Schwarzerová (Department of Plant Physiology, Faculty of Sciences, Charles University, Prague, Czech Republic) and Dr.B.Radochová (Department of Biomathematics, Institute of Physiology, ASCR, Prague, Czech Republic) for Figure 2B, Dr.X.W.Mao (Department of Radiation Medicine, Loma Linda University, CA, USA) for Figure 2C, Dr.Z.Zemanová (Department of Epithelial Physiology, Institute of Physiology, ASCR, Prague, Czech Republic) for preparing the specimen of chick embryonal kidney shown in Figures 2D and 16, Prof.Z.Opatrný (Department of Plant Physiology, Faculty of Sciences, Charles University, Prague, Czech Republic) for preparing the specimen of tobacco cells shown in Figures 3, 24, 30, 31, 33, and 37, Dr.I.Eržen and Dr.V.Čebašek (Institute of Anatomy, Medical Faculty, University of Ljubljana, Slovenia) for preparing the specimen of skeletal muscle shown in Figures 25 and 26.

The study was supported by the Academy of Sciences of the Czech Republic (Grants KJB6011309, A600110507, A100110502, and Grant AVOZ 5011922), by the Ministry of Education of the Slovak Republic together with the Ministry of Education, Youth and Sport of the Czech Republic (KONTAKT grant No.139), and by the Ministry of Development (G.S.R.T.) of the Hellenic Republic together with the Ministry of Education, Youth and Sport of the Czech Republic (KONTAKT grant No. ME688 and RC_3_19).

8. REFERENCES

1 Haug H. History of Neuromorphometry. J Neurosci Meth 1986; 18: 1-17.
2 Weibel E.R. *Stereological Methods, Vol.1: Practical Methods for Biological Morphometry*. London: Academic Press, 1979.
3 Gundersen H.J.G., Bendtsen T.F., Korbo L., Marcussen N., Møller A., Nielsen K., Nyengaard J.R., Pakkenberg B., Sørensen F.B., West M. J. Some new, simple and efficient stereological methods and their use in pathological research and diagnosis. APMIS 1988; 96: 379-94.
4 Gundersen H.J.G., Bagger P., Bendtsen T.F., Evans S.M., Korbo L., Marcussen N., Møller A., Nielsen K., Nyengaard J.R., West M.J. The new stereological tools: disector, fractionator, nucleator and point sampled intercepts and their use in pathological research and diagnosis. APMIS 1988; 96: 857-81.
5 Cruz-Orive L.M., Weibel E.R. Recent stereological methods for cell biology: a brief survey. Am J Physiol 1990; 258: L148-56.
6 Howard C.V., Reed M.G. *Unbiased Stereology: Three-Dimensional Measurement in Microscopy. Microscopy Handbooks. Vol. 41.* New York: Springer-Verlag, 1998.
7 Mouton P.R., Gokhale A.M., Ward N.L., West M.J. Stereological length estimation using spherical probes. J Microsc 2002; 206: 54-64.
8 Braendgaard H., Evans S.M., Howard C.V., Gundersen H.J.G. The total number of neurons in the human neocortex unbiasedly estimated using optical disectors. J Microsc 1990; 157: 285-304.
9 Nagel W., Ohser J., Pischang K. An integral-geometric approach for the Euler-Poincaré characteristic of spatial images, J Microsc 2000; 198: 54-62.
10 Serra J. *Image Analysis and Mathematical Morphology*. London: Academic Press, 1982.
11 Lee C.N., Poston T., Rosenfeld A. Winding and Euler numbers for 2D and 3D digital images. Computer Vision, Graphics and Image Processing: Graphical Models and Image Processing 1991; 53: 522-37.
12 Santaló L.A. *Integral Geometry and Geometric Probability*. Reading, Massachusetts: Addison-Wesley, 1976.
13 Meyer F. Mathematical morphology: from two dimensions to three dimensions. J Microsc 1992; 165: 5-28.
14 Moran P.A.P. Measuring the length of a curve. Biometrika 1966; 53: 359-64.
15 Kubínová L., Mao X.W., Janáček J., Archambeau J.O. Stereology techniques in radiation biology. Radiat Res 2003; 160: 110-9.
16 Coggeshall R.E., Lekan H.A. Methods for determining numbers of cells and synapses: A case for more uniform standards of review. J Comp Neurol 1996; 364: 6-15.
17 Karen P., Jirkovská M., Tomori Z., Demjénová E., Janáček J., Kubínová L. Three-dimensional computer reconstruction of large tissue volumes based on composing series of high-resolution confocal images by GlueMRC and LinkMRC software. Microsc Res Tech 2003; 62: 415-22.
18 Gundersen H.J.G., Jensen E.B. The efficiency of systematic sampling in stereology and its prediction. J Microsc 1987; 147: 229-263.
19 Saltykov S.A. The method of intersections in metallography (In Russian). Zavodskaja laboratorija 1946; 12: 816-25.
20 Mattfeldt T., Mall G., Gharehbaghi H., Moller, P. Estimation of surface area and length with the orientator. J Microsc 1990; 159: 301-317.
21 Nyengaard J.R., Gundersen H.J.G. The Isector: a simple and direct method for generating isotropic, uniform random sections from small specimens. J Microsc 1992; 150: 1-20.
22 Baddeley A.J., Gundersen H.J.G., Cruz-Orive L.M. Estimation of surface area from vertical sections. J Microsc 1986; 142: 259-76.
23 Gundersen H.J.G. Notes on the estimation of the numerical density of arbitrary profiles: the edge effect. J Microsc 1977; 111: 219-23.

24 Pawley J.B., ed. *Handbook of Biological Confocal Microscopy. Second Edition.* New York: Plenum Press, 1995.
25 Åslund N., Carlsson K., Liljeborg A., Majlof L. PHOIBOS, a microscope scanner designed for micro-fluorometric applications, using laser induced fluorescence. Proceedings of the Third Scandinavian Conference on Image Analysis, Studentliteratur; 1983; Lund: 338.
26 Denk W., Strickler J.H., Webb W.W. Two-photon laser scanning fluorescence microscopy. Science 1990; 248: 73-6.
27 Svoboda K., Denk W., Kleinfeld D., Tank D.W. In vivo dendritic calcium dynamics in neocortical pyramidal neurons. Nature 1997; 385: 161-5.
28 Russ J.C. *Computer Microscopy: The Measurement and Analysis of Images*. New York: Plenum Press, 1990.
29 Rigaut J.P., Carvajal-Gonzales S., Vassy, J. "3-D Image Cytometry". In: *Visualization in Biomedical Microscopies*, Kriete, A., ed., Weinheim - New York: VCH, 1992.
30 Kubínová L., Janáček J., Guilak F., Opatrný Z. Comparison of several digital and stereological methods for estimating surface area and volume of cells studied by confocal microscopy. Cytometry 1999; 36: 85-95.
31 Peterson, D.A. Quantitative histology using confocal microscopy: Implementation of unbiased stereology procedures. Methods 1999; 18: 493-507.
32 Kubínová L., Janáček J., Krekule I. "Stereological Methods for Estimating Geometrical Parameters of Microscopical Structure Studied by Three-dimensional Microscopical Techniques." In *Confocal and Two-photon Microscopy*, A. Diaspro, ed. New York: Wiley-Liss, 2002.
33 Kubínová L., Janáček J., Ribarič S., Čebašek V., Eržen I. Three-dimensional study of the capillary supply of skeletal muscle fibers using confocal microscopy. J Musc Res Cell Motil 2001; 22: 217-27.
34 Čebašek V., Kubínová L., Ribarič S., Eržen I. A novel staining method for quantification and 3D visualisation of capillaries and muscle fibres. Eur J Histochem 2004; 48: 151-8.
35 Romanul F.C.A. Capillary supply and metabolism of muscle fibres. Arch Neurol 1965; 12: 497-509.
36 Gray S.D., Renkin E.M. Microvascular supply in relation to fibre metabolic type in mixed skeletal muscles of rabbits. Microvasc Res 1978; 16: 406-25.
37 Punkt K., Unger A., Welt K., Hilbig H., Schaffranietz. Hypoxia-dependent changes of enzyme activities in different fibre types of rat soleus and extensor digitorum longus muscles. A cytophotometrical study. Acta Histochem (Jena) 1996; 98: R1154-62.
38 Gundersen H.J.G. Stereology of arbitrary particles. A review of unbiased number and size estimators and the presentation of some new ones, in memory of William R. Thompson. J. Microsc. 1986; 143: 3-45.
39 Sterio D.C. The unbiased estimation of number and sizes of arbitrary particles using the disector. J Microsc 1984; 134: 127-36.
40 Howard C.V., Reid S., Baddeley A., Boyde A. Unbiased estimation of particle density in the tandem scanning reflected light microscope. J Microsc 1985; 138: 203-12.
41 Gundersen H.J.G. The nucleator. J Microsc 1988; 151: 3-21.
42 Jensen E.B.V., Gundersen H.J.G. The rotator. J Microsc 1993; 170: 35-44.
43 Kiêu K., Jensen E.B.V. Stereological estimation based on isotropic slices through fixed points. J Microsc 1993; 170: 45-51.
44 Tandrup T., Gundersen H.J.G., Jensen, E.B.V. The optical rotator. J Microsc 1997; 186: 108-20.
45 Sandau K. How to estimate the area of a surface using a spatial grid. Acta Stereol 1987; 6: 31-6.
46 Cruz-Orive L.M., Howard C.V. Estimation of individual feature surface area with the vertical spatial grid. J Microsc 1995; 178: 146-51.
47 Kubínová L., Janáček J. Estimating surface area by isotropic fakir method from thick slices cut in arbitrary direction. J Microsc 1998; 191: 201-11.

48 Gokhale A.M., Evans R.A., Mackes J.L., Mouton, P.R. Design-based estimation of surface area in thick tissue sections of arbitrary orientation using virtual cycloids. J Microsc 2004; 219: 25-31.
49 Gokhale A.M. Unbiased estimation of curve length in 3D using vertical slices. J Microsc 1990; 159: 133-41.
50 Cruz-Orive L.M., Howard C.V. Estimating the length of a bounded curve in three dimensions using total vertical projections. J Microsc 1991; 163: 101-13.
51 Larsen J.O., Gundersen H.J.G., Nielsen, J. Global spatial sampling with isotropic virtual planes: estimators of length density and total length in thick, arbitrarily orientated sections. J Microsc 1998; 191: 238-48.
52 Mouton P.R. *Principles and Practices of Unbiased Stereology. An Introduction for Bioscientists*. Baltimore: The Johns Hopkins University Press, 2002.
53 Lorensen W.E., Cline H.E. Marching cubes: a high resolution 3D surface construction algorithm. Computer Graphics 1987; 21:163-9.
54 Oliva J-M., Perrin M., Coquillart S. "3D Reconstruction of Complex Polyhedral Shapes from Contours Using a Simplified Generalized Voronoi Diagram". In *Eurographics 96*, Rossignac J., Sillion, F. , eds. Blackwell Publishers, 1996.
55 Baba, N. "Computer Aided Three-dimensional Reconstruction from Serial Section Images." In *Image Analysis in Biology*, Donat-P. Häder, ed. Boca Raton, Ann Arbor, London: CRC Press, 1992.
56 Guilak F. Volume and surface area measurement of viable chondrocytes in situ using geometric modelling of serial confocal sections. J Microsc 1994; 173: 245-56.
57 Cruz-Orive L.M. Stereology of single objects. J Microsc 1997; 186: 93-107.
58 Kubínová L., Janáček J. Confocal microscopy and stereology: Estimating volume, number, surface area and length by virtual test probes applied to three-dimensional images. Microsc Res Tech 2001; 53: 425-35.
59 Cruz-Orive L.M. Systematic sampling in stereology. Bull Intern Statis Inst Proceedings 49th Session, 1993; Florence: 55(2): 451-68.
60 Gundersen H.J.G., Jensen E.B.V., Kiêu K., Nielsen J. The efficiency of systematic sampling in stereology - reconsidered. J Microsc 1999; 193: 199-211.
61 Dorph-Petersen K.A., Nyengaard J.R., Gundersen H. J. G. Tissue shrinkage and unbiased stereological estimation of particle number and size. J Microsc 2001; 204: 232-46.
62 Jirkovská M., Kubínová L., Krekule I., Hach P. Spatial arrangement of fetal placental capillaries in terminal villi: a study using confocal microscopy. Anat Embryol 1998; 197: 263-72.
63 Shaw P. Deconvolution in 3-D optical microscopy. Histochem J 1994; 26: 687-94.
64 Difato F., Mazzone F., Scaglione S., Fato M., Beltrame F., Kubínová L., Janáček J., Ramoino P., Vicidomini G., Diaspro A. Improvement in volume estimation from confocal sections after image deconvolution. Microsc Res Tech 2004; 64: 151-5.
65 Nakamura O. Fundamental of two-photon microscopy. Microsc Res Tech 1999; 47: 165-71.
66 Diaspro A., ed. *Confocal and Two-photon Microscopy*. New York: Wiley-Liss, 2.
67 Sheppard C.J.R., Török, P. Effects of specimen refractive index on confocal imaging. J Microsc 1997; 185: 366-74.

WAVELENGTH AS THE FOURTH DIMENSION IN LIGHT MICROSCOPY

Valtere Evangelista, Laura Barsanti, Anna Maria Frassanito, Vincenzo Passarelli and Paolo Gualtieri
Istituto di Biofisica C.N.R., Via G. Moruzzi,1 56124 Pisa Italy

Abstract: This paper describes a microspectroscopic instrumentation suitable for performing absorption and fluorescence spectra of cellular compartments inside their functional environment. The system consists of an optical microscope, measuring diaphragms, diffraction gratings, and photodetectors. A personal computer acquires the signals from the photodetectors and handles the diffraction gratings. In this chapter the absorption and fluorescent spectra of *Euglena gracilis* photoreceptor and chloroplast are presented.

Keywords: Absorption Microspectroscopy, Fluorescence Microspectroscopy, *Euglena gracilis*, Polychromator

1. INTRODUCTION

Light microscopes enable us to explore biological matter from cells to proteins, allowing biological matter to be imaged in its functional environment. Light Microscopes directly produce three dimensional images in the space (x, y, and z), but they can also produce absorption or fluorescence spectra of cells or sub-cellular components; in this case the wavelength (λ) is the fourth dimension of the microscope output. In this chapter we will discuss how light microscopes can produce absorption and fluorescence spectra for the investigation of cellular compartments.

2. ABSORPTION MICROSPECTROSCOPY

Absorption Microspectroscopy has several advantages:

- the integrity of the organism or of the subcellular component is not disturbed, and it is possible to examine the uninjured system with its physiological functions intact (*in vivo*);
- cellular morphology can be easily correlated with pigment type by direct observation;

V. Evangelista et al. (eds.), From Cells to Proteins: Imaging Nature across Dimensions, 323–333.

- quantitative determinations of various reactions at the time of their occurrence in the sample are possible.

2.1. Hardware set-up

The hardware set-up is shown in Figure 1.

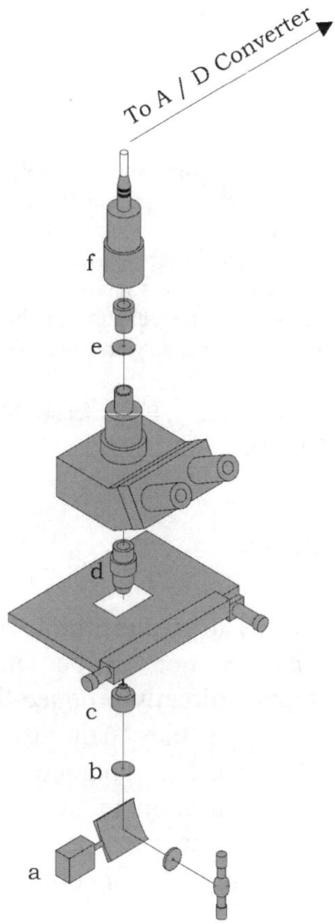

Figure 1: Set-up of the absorption microspectrophotometer

The absorbing structure is visualized in a high-magnification microscope placed in the optical path of a monochromator (a, Figure 1). A submillimeter light beam defined in a pinhole (b, Figure 1) is demagnificated by an inverted microscope (c, Figure 1) on the specimen. The wavelength is changed by the computer controlled stepping motor (RS Electronics 440-420, USA) inserted on the axis of the knurled knob that moves the filter housing. The light beam is collected by a objective (d, in Figure 1) and detected through a pinhole (e, Figure 1) by a photomultiplier (Hamamatsu R928, Japan) (f, Figure 1), mounted on the binocular phototube by means of a 40 mm diameter adapter ring. The output signal of the photomultiplier is amplified by a current amplifier (Model 427, Keithley Instruments, U.S.A.) and transmitted to the A/D card (Controller Pontech STP 100, USA) plugged in the computer bus. The computer controls the position of the monochromator, acquires the data, performs the calculation of optical density, produces the graphic output and stores the data. All the absorption spectra are recorded from 400 nm and 700 nm, with a step size of 0.5 nm and scan speed of 100 nm $*$ sec^{-1}. For each wavelength 10,000 values of optical density are averaged [1, 2].

2.2. How to measure

The beam can be superimposed alternatively on the pigmented structure and on an area free of structures or cells, i.e. the reference area.

The ratio of the intensities of these two beams (I_i, intensity incident on the sample; I_t, intensity transmitted by the sample) is then calculated to give the absorbance of the specimen.

Using the Lambert-Beer's law we can consider the absorbance of a specimen (A) as derived from the measures as follows:

$$A=log(I_i)-log(I_t)$$

Lambert-Beer's law is valid for non-scattering substances, and for dilute solutions; knowing A, the path of light in the absorbing material, and the extinction coefficient, we can calculate the concentration of the absorbing structure.

The light impinging on the surface of the photocathode (I) is equal to:

$$I = IP * SB * OV * TR * IS$$

where,

IP is the initial power of the lamp, called spectral concentration of radiance that is the radiant flux per unit area, unit solid angle, and unit spectral bandwidth ($W * cm^{-2} * sr^{-1} * nm^{-1}$);

SB is the spectral bandwidth in nanometers (nm);

OV is the volume of the optical system through which the beam has passed, named optical flux ($cm^2 * sr$). In turn OV is equal to $\boldsymbol{F * NA^2}$ where $\boldsymbol{F}$ is the area of aperture diaphragm (or of the image of the diaphragm), and ***NA*** is the effective numerical aperture;

TR is the total transmittance of the optical system, which is a measure of the fraction of light that remains after losses (by absorption, reflection, scattering and diffraction);

IS is the interactions of the light with the sample consisting in absorption, reflection, fluorescence, interference, refraction and diffraction; these interactions obviously can influence each other.

2.3. Experimental cases

The biological structures we have selected for demonstrating the applications of microspectroscopy are the photoreceptor and the chloroplast of the photosynthetic alga *Euglena gracilis*.

2.3.1. The Photoreceptor

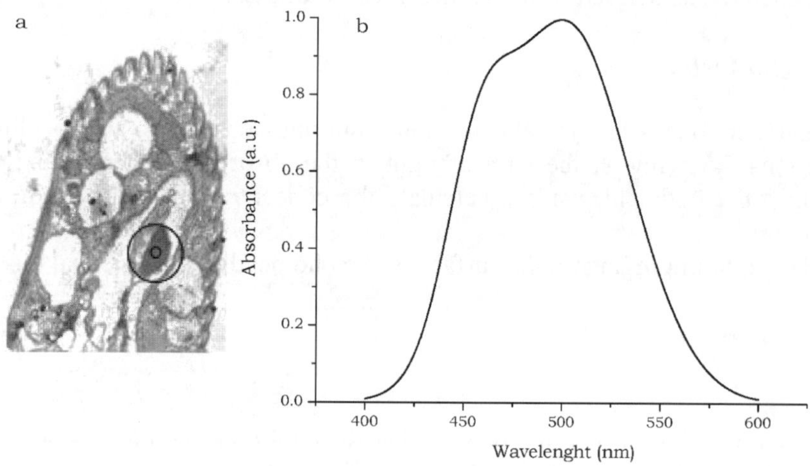

Figure 2: TEM image of the photoreceptor and its absorption spectrum

Figure 2a shows a TEM image of the apical zone of *Euglena*, where the photoreceptor is located. The size of the photoreceptor is 2 μ and the two measuring pinholes are superimposed on the structure in order to show their relative position. Figure 2b shows the absorption spectrum recorded on the photoreceptor. It is centered at 503 nm and shows an asymmetrical structure that fits very well with an absorption spectrum of a rhodopsin-like protein [3, 4].

2.3.2. The Chloroplast

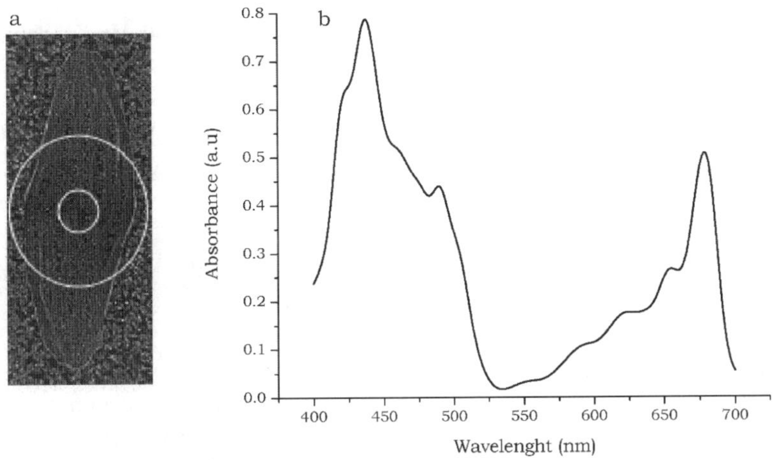

Figure 3: TEM image of a chloroplast and its absorption spectrum

Figure 3a shows a freeze etching image of a chloroplast. The size of the chloroplast is 10 μ and the two measuring pinholes are superimposed on the structure in order to show their relative position. Figure 3b shows the absorption spectrum recorded on the chloroplast.

It clearly shows that *Euglena gracilis* belongs to the green lineage of eukaryotic algae since only chlorophylls *a* and *b* and carotenoids are present in this spectrum [5].

3. FLUORESCENCE MICROSPECTROSCOPY

Fluorescence Microspectroscopy has several advantages:

- the background is much reduced, and therefore fluorescence spectroscopy of the living cell achieves more reliable results compared to absorption spectroscopy, where very small change in absorption often have to be measured in the presence of a strong total signal. (Photon noise is proportional to the square root of the intensity of the incident light; background provides a high level of noise, which makes the evaluation of small quantities of a substance very difficult in absorption microspectroscopy);

- the sensitivity of detection is not limited by the signal-to-noise ratio, but rather by the presence, virtually unavoidable, of fluorescent contaminants;
- phosensitive structures are created to collect light. Therefore, if these structures emit radiation, the emitted radiation has a very low quantum yield. Several pigments absorb light, but do not emit;
- the energy of fluorescent light is much lower than that of normal transmitted or reflected light; therefore, the photocathode should be very sensitive and chilled to reduce shot noise;
- the measured radiant flux emanating from a self-luminous object is proportional to the optical flux of both the imaging and illuminating systems, so that it is advisable to use an objective with numerical aperture as high as possible;
- in order to avoid the bleaching phenomena due to the high light intensity, it is advisable to use a very fast method of measure, or an acousto-optic modulator to control the power of the excitation beam;
- the majority of emission spectra of fluorophores consists of a single band that can be described as a skew Gaussian, without fine structure; often it can be difficult to understand which is the fluorophore which emits, i.e. fluorescence has a relatively low degree of specificity.

3.1. Hardware set-up

The hardware set-up is show in Figure 4. The emission fluorescent spectra are measured using a high-magnification microscope. A high pressure Mercury 100 W lamp (a, Figure 4) is used to excite the sample. A filter cube, placed in the epi-illumination optical path (b, Figure 4), selects the incident monochromatic light that reaches the sample. The fluorescent light emitted by the sample, collected by an objective (c, Figure 4), reaches the high quality inspection probe that acts as a pinhole (d, Figure 4).

The high quality fluorescence probe is mounted in the back focal plane of the ocular in the TV-microscope paths. The exit pupil of this probe is connected to a flat field imaging concave grating polychromator (e, Figure 4) and its dispersion image is focused on a digital slow scan cooled CCD camera (f, Figure 4), connected to a Frame Grabber plugged in a computer bus. Each spectrum represents the average of 100 measurements [6].

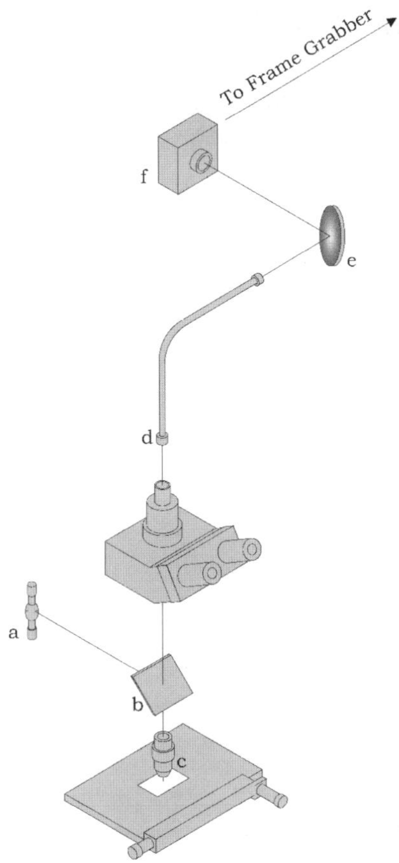

Figure 4: Set-up of the fluorescence microspectrophotometer

3.2. How to measure

The dispersion image of the sample focused on a digital slow scan cooled CCD camera is the fluorescence emission spectrum of the sample.
The fluorescence intensity is given by:

$$F_{\lambda}=I_A \Phi_F Z$$

The population in the excited state depends on the amount of light absorbed (I_A) and the quantum yield (Φ_F) that is proportional to the cross section ($\Phi_F \propto \varepsilon_{\lambda}$) and to Z, which represents the fraction of fluorescence collected by the objective.

3.3. Experimental cases

As for the absorption spectra the fluorescent spectra, are performed on *Euglena* photoreceptor and chloroplast.

3.3.1. The Photoreceptor

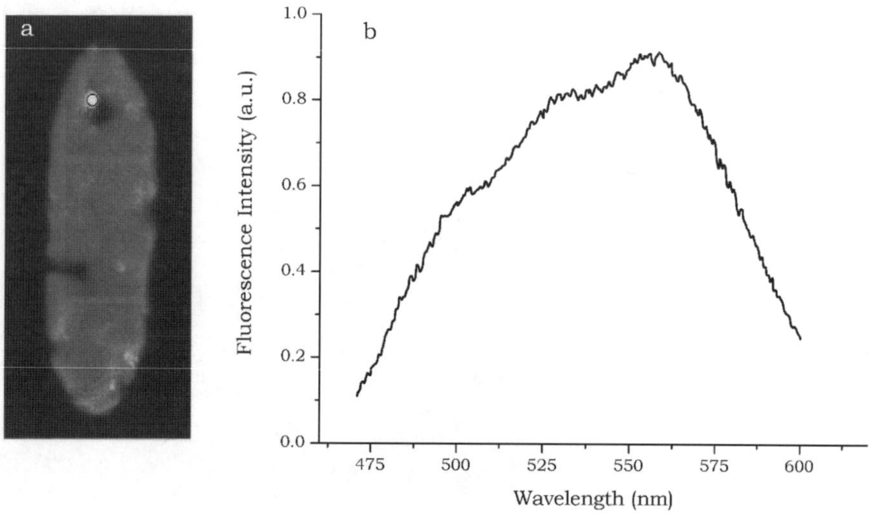

Figure 5: Fluorescent image of the cell and fluorescent spectrum of the photoreceptor

Figure 5a shows a fluorescent image of Euglena. The photoreceptor is recognizable as a the bright spot located in the apical portion of the cell. The measuring diaphragm of the fluorescent probe is superimposed on the structure in order to show its position. Figure 5b shows the fluorescence spectrum of the photoreceptor under 436 nm excitation light, after 10 seconds of excitation with the 365 nm light. Gaussian bands decomposition of the emission spectrum reveals 3 bands with different intensity centered at about 500 nm, 525 nm, and 556 nm (not shown).

The bands obtained by the Gaussian decomposition indicate the presence of a mixture of very similar conformers in the photoreceptor, each one capable of photocycling between a non-fluorescent parent species and a fluorescent excited species [6,7].

3.3.2. The Chloroplast

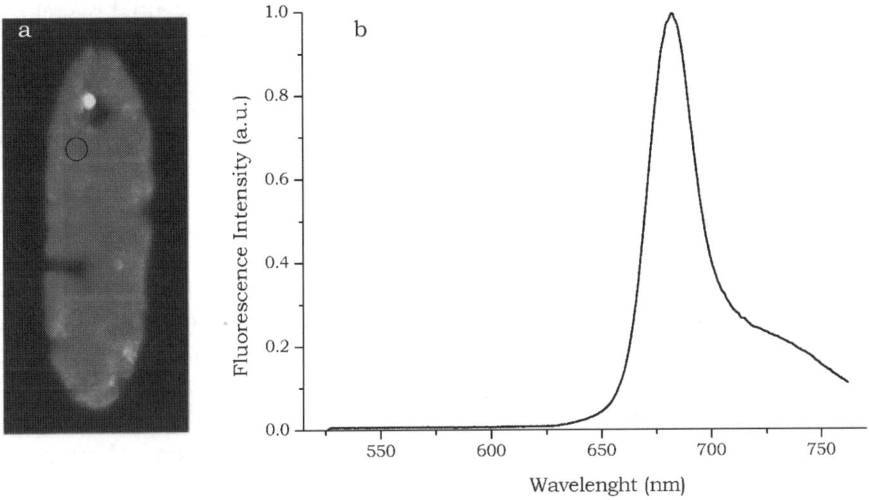

Figure 6: Fluorescent image of the cell and the fluorescent spectrum of the chloroplast

Figure 6a shows a fluorescent image of Euglena. The cell is filled with red emitting chloroplasts. The measuring diaphragm of the fluorescent probe is superimposed on a chloroplast in order to show its position. Figure 6b shows the typical emission spectrum of a chloroplast that contains only chlorophyll *a* and *b* [8].

4. CONCLUSION

To conclude, let us detail and compare the characteristics of the microspectroscopic techniques we have described with those of the other techniques used to define the properties of pigments. Each technique presents its own assets and liabilities, some of which have been summarized in Figure 7.

4.1. Spectroscopy of Extracts

Extraction by detergents could be unique in providing large quantities of the pigments present in the photoreceptive structures, in an accessible "in vitro" form upon which detailed chemical analyses can be pursued. Such massive quantities of a single pigment allow a very accurate determination of spectroscopic parameters. Spectroscopy of solubilized pigment may be complicated by the simultaneous extraction of several other pigments present in the cell, which cause distortion of absolute spectra and necessitate special procedure of correction, such as bleaching spectra and differential bleaching. Finally, since the pigment extraction forever removes the identifying link that tells in what cell structure the pigment originates, this detrimental factor would suggest a careful evaluation of the results obtained with this method.

4.2. Action Spectroscopy

By means of action spectroscopy we can measure the photosensitivity of a cell at different wavelengths, thus gathering information on the chemical nature of the pigments involved in photoreception. The direct measure of an action spectrum is much more difficult than that of an absorption spectrum; moreover, action spectra may be not directly correlated with absorption peaks of the pigments involved in photoreception, because several errors can be caused by light scattering. Action spectra have been determined by measuring the system response at the same photon or light fluence rate, or by measuring the light fluence rate which induces the same response at different wavelength. A stable monochromatic radiation is obviously essential, and this can be obtained by using a continuous spectrum source and a monochromator; the energy must be accurately measured with a calibrated radiometer. Action spectra have poor wavelength resolution and sensibility, i.e. they do not discriminate between the many candidate pigments with similar absorption characteristics in the visible range. Mass-movement methods or computerized single-cell methods can be used for the determination of action spectroscopy. The utilization of action spectroscopy should be limited to the study of the change, better of the increase, of photosensitivity of a mutant cell, which is defective of the presumed photoreceptor pigment, after the exogenous addition of that pigment, [1].

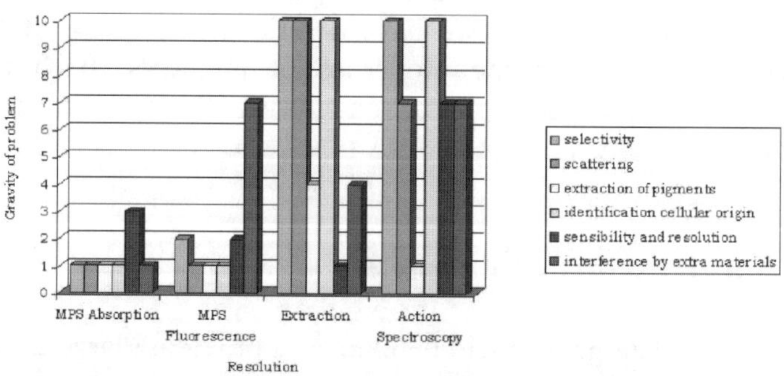

Figure 7: Characteristics of the techniques used to identify cell pigments.

5. REFERENCES

1. Gualtieri P. Microspectroscopy of photoreceptor pigments in flagellated algae. Critical Rev. Plant Sci. 1991; 9 (6): 474-495.
2. Liebman, P. A.. "Microspectrophotometry of photoreceptors." In, *Handbook of sensory physiology VII/B*, Dartnall, H. J. A. ed. (pp. 481-528). Springer Verlag, 1972.
3. Gualtieri P., Barsanti L. and Passarelli V. Absorption spectrum of a single isolated paraflagellar swelling of *Euglena gracilis*. Biochim. Biophys. Acta 1989;993: 293-296.
4. Passarelli V., Barsanti L., Evangelista V., Frassanito A. and Gualtieri P. *Euglena gracilis* photoreception interpreted by microspectroscopy. Eur. J. Protistol. 2003; 39:404-408.
5. MIllie D., Evens TJ. Using absorbance and fluorescence spectra to discriminate microalgae. Eur. J. Phycol. 2002; 37: 313-322.
6. Evangelista V., Passarelli V., Barsanti L. and Gualtieri P. Fluorescence behavior of *Euglena* photoreceptor. Photochem. Photobiol. 2003; 78: 93-97.
7. Barsanti L., Passarelli V., Walne P.L.,, and Gualtieri P. In vivo photo-cycle of the *Euglena gracilis* photoreceptor, Biophys. J. 1997; 72: 545-553
8. Govijndjee Sixty-three years since Kautsky: chlorophyll *a* fluorescence. Aust. J. Plant Physiol. 1995; 22: 131-160

TIME AS THE FIFTH DIMENSION IN MICROSCOPY

Primo Coltelli[1], Laura Barsanti[2], Valtere Evangelista[2], Anna Maria Frassanito[2] and Paolo Gualtieri[2]
[1] *Istituto di Tecnologia dell'Informazione C.N.R., Via G. Moruzzi,1 56124 Pisa Italy*
[2] *Istituto di Biofisica C.N.R., Via G. Moruzzi,1 56124 Pisa Italy*

Abstract. This paper describes an image processing system, suitable for real time image understanding and the analysis of moving objects under the optical microscope, such as microorganisms orienting toward a source of light. The system consists of an optical microscope, a desktop computer, a commercially available image-processing hardware module plugged in the computer bus, a b/w TV-camera, video monitors, and special software developed for the use. The structure and the capability of this system are explained.

Key words: Real Time Image Processing, Detection of Moving Objects, Motion Interpretation

1. INTRODUCTION

Time is an essential dimension in the study of biological processes. If image-processing techniques are applied to the analysis of time sequences of microscope images, we can obtain a deterministic interpretation of the biological processes varying during time. A very intriguing case of the latter is the behaviour of microorganisms under different external stimuli. If the analysis is performed in real time we get a reliable statistics of the output data, and a very powerful tool to study cellular integrated responses, such as movement [1,2,3, 4].

The real time image processing of microscope images substitutes the traditional routine qualitative analysis prone to human errors, delayed outputs, and ambiguous interpretations.

The motion-analysis workstation we set up and describe in this chapter, links a very fast image acquisition with precise and accurate determination of cell features in the image frames film, achieving the exact determination of space and time parameters of swimming microorganisms, and the complete and automatic analysis of the cell trajectories under light or chemical stimuli.

V. Evangelista et al. (eds.), From Cells to Proteins: Imaging Nature across Dimensions, 335–348.

The time resolution of our system, which is 40 msec, (CCIR standard), is well adequate to determine motion parameters of moving microorganisms, since the study of physiological aspects of motility is usually based on microscope observation of these phenomena and human visual system, due to its integration time, has a time resolution of 250 msec, which is 6-times greater than that of our system.

2. HARDWARE CONFIGURATION

Our system consists of a Pulnix TM860 (Pulnix, USA) CCD video camera mounted onto a Zeiss Axioplan microscope (Zeiss, Germany) equipped with 16x and 60x objectives and a 100 W halogen lamp as light source. Cells were placed in a small chamber obtained by fixing a PVC ring onto a microscope slide. The chamber was closed by means of a cover slip so as to avoid sample drying-out. The signal of the camera was the input of a Frame Grabber (FG100 AT, Imaging Technology USA) plugged into a Pentium III personal computer 750 MHz clock, (Figure 1).

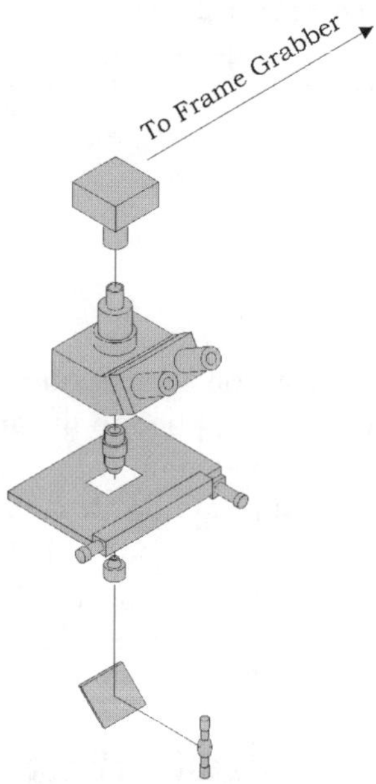

Figure 1. The hardware set-up of the tracking microscope

The Frame Grabber board consists of a frame memory (64 MB x 12-bit), a multiplexer, a 4096x12-bit Feedback Input Look-up-Table (LUT) and an A/D converter. The multiplexer has 24 input bits: twelve from frame memory, eight from the A/D converter and four from the LUT control register. The multiplexer can be programmed in order to optimize different types of image processing. In any given mode, the multiplexer passes twelve bits to the LUT which processes them before the storage in the frame memory. The A/D converter samples the analog video signal of the TV-camera at 12.5 Mhz. The acquired image is reduced, and stored in the frame memory, by means of the Zoom, Pan and Scroll hardware capabilities of the board. These capabilities act during the frame acquisition, (Figure 2).

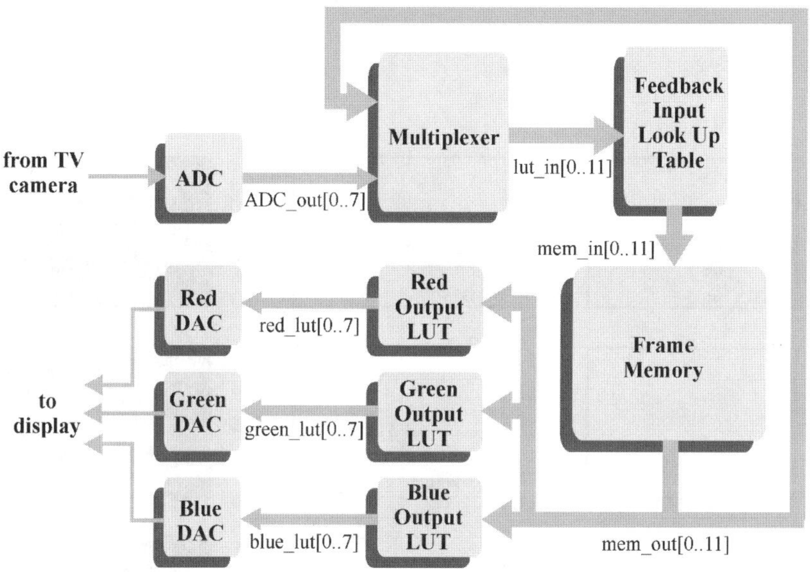

Figure 2. Configuration of the programmable frame grabber board

Real time detection of moving objects is performed by subtracting continuously each frame of the video image from a previous frame, with an experimentally determined delay. This operation is performed by programming the 12-bit input Look-up- table (LUT) of the board. This LUT, which is located between the digitization circuit and the frame memory, transforms the image before it is stored into the frame memory. Thanks to a feed-back circuit between the frame memory and the input LUT, operations are made on combinations of stored and newly-acquired data. The input LUT is programmed in order to move the six most significant bits from the A/D converter (these bits represent the newly acquired data) to the six most significant bits of the frame memory; then the LUT subtracts the newly acquired data from the data previously stored in the six most significant bits of the frame memory (these bits represent the previously acquired data). The resulting six bits are then stored in the least significant bits of the frame memory. The difference-image produced by these operations is always available in the lower six 1-bit planes of the frame memory, while the upper six 1-bit planes contain the most recent data of the A/D converter, which are used as input for the next frame subtraction. In the case that the images of a moving cell in two successive frames are partly overlapping, the subtraction operation gives a zero value for the overlapping region of the cell and for the background, a positive value for that part of the cell image which is present only in the newly acquired frame, and a negative value for that part of the cell image which is present only in the previously acquired frame, (Figure 3).

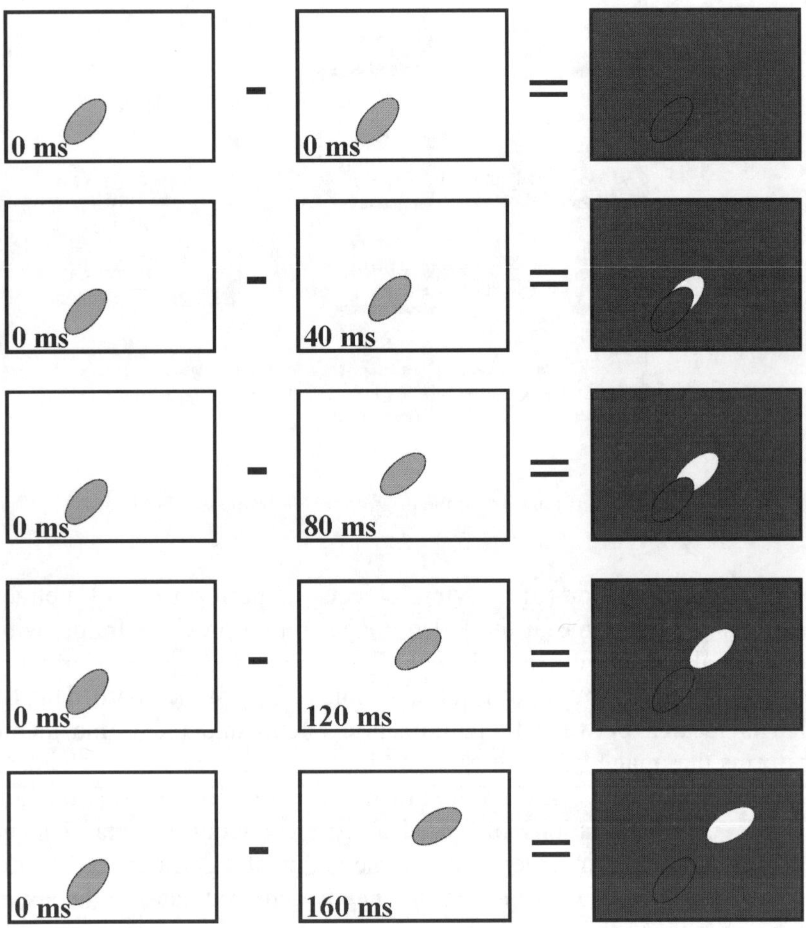

Figure 3. Subtraction operation

The input LUT is programmed to clip to zero the negative values, in order to discard them (Figure 4). However, for a further simplification of this problem, we choose the delay between the two frames so that the two images do not overlap. This delay is chosen by testing the object area variation in the difference-images.

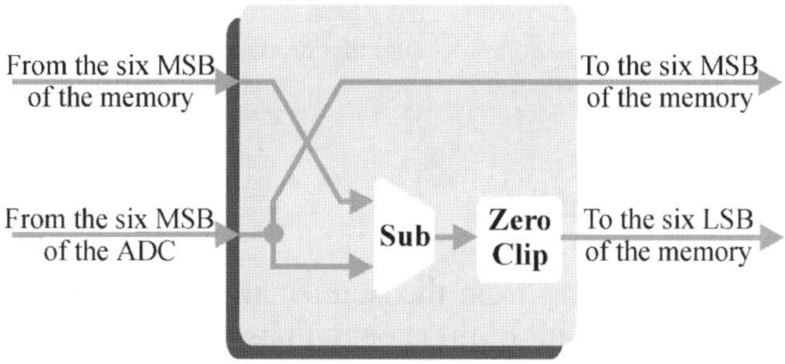

Figure 4. Feedback Input Look up Table configuration

A black and white monitor (Philips BM 7542) is used to display the signal output of the TV-camera, while RGB high persistence monitor (Mitsubishi Model C-3479 Japan) is used to display the digital image in false colour mode. A Windows based application software with graphical objects such as menus, windows, etc.., using predefined Windows API (Application Program Interface) classes was developed for tracking microscope control.

3. PROCESSING

Processing moving objects may be subdivided in:

- detection of moving objects
- memory storage of the detected objects
- identification of moving objects in successive frames
- tracking moving objects
- motion interpretation.

3.1. Detection of moving objects

The segmentation method we used to discriminate the moving objects in the difference-images is based on threshold finding in the bimodal difference-images grey level histogram. The sharp dip between the peak corresponding to the objects and the peak corresponding to the background is the threshold, (Figure 5).

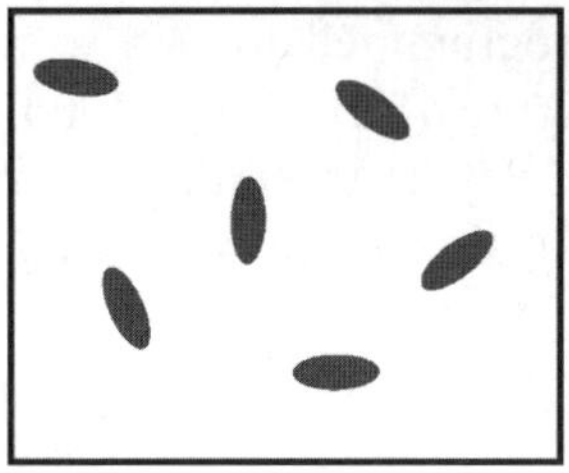

Figure 5. An example of segmentation procedure that produces binary image: black objects, white background

3.2. Memory storage

The binary images resulting from the segmentation procedure are stored in the frame memory. The result of these operations is a film of time sequences. In order to store several difference-images in the frame memory (in Figure 6 the difference-images are 12) we reduce the spatial resolution of the image being acquired by means of the hardware Zoom. The result of this operation is a 320x256 pixel image, whose size corresponds to 1/2 of the original image size. In order to store the reduced image into its proper position of the frame memory, the X and Y coordinates of its origin are shifted by means of hardware Pan and Scroll operations. In this way we can store 12 images, by moving the coordinates of their origin toward the right and downward. At the end of this procedure, the frame memory is displayed as a patchwork of reduced images. The time necessary for the 12-image reduction depends only on the delay between the two frames on which the temporal difference has been performed.

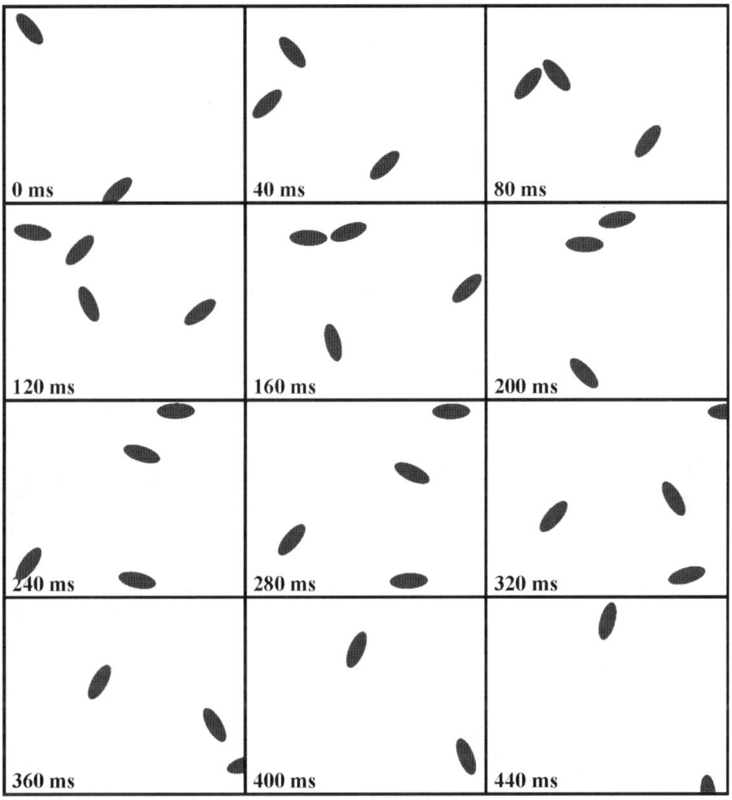

Figure 6. Result of the storing procedure

3.3. Identification of moving objects in successive frames

After threshold calculation, a labelization procedure is applied to each difference-image in order to extract object features such as baricenter coordinates and areas, (Figure 7). Because the labeling process of connected regions is a typical recursive procedure, which requires backtracks at each point when regions are found to recombine, it is necessary a second pass analysis through the entire image to complete the process. The algorithm we use combines in a single pass the labeling of connected regions with the generation of a data structure containing global features of all labeled regions. Backtrack from the connected points to previously labeled domains is avoided. The speed of this algorithm depends on the handling of the random memory of a personal computer as if it were an associative memory, thus allowing parallel computation, since a single instruction can change the content of all the cells having the same content, [5].

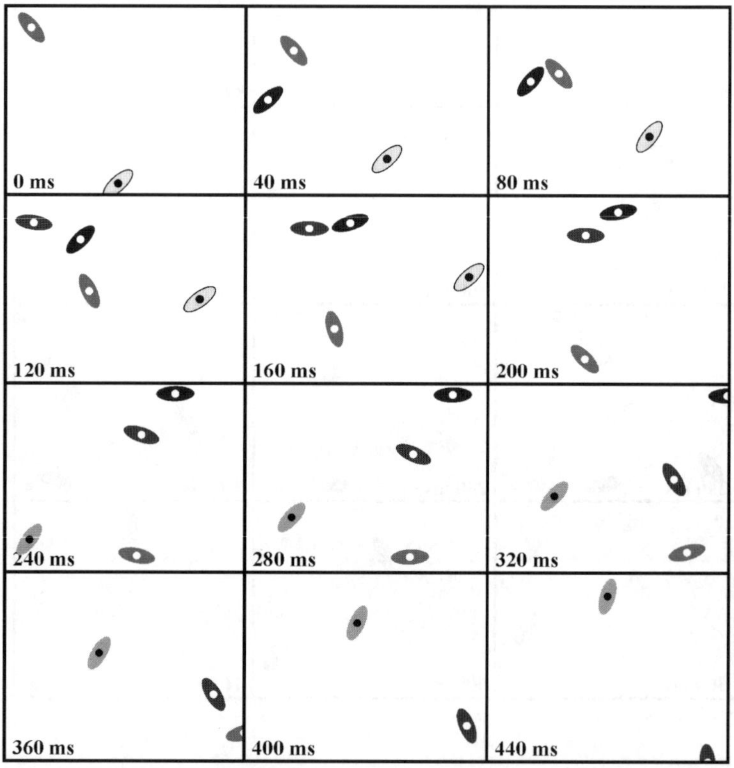

Figure 7. An example of labelling procedure: the identified microorganisms are labelled with a dot in their baricenter

3.4. Tracking of the moving objects

The data structure generated by the labelization algorithm has a list organization; for each difference-image, the data structure is composed by a pointer, which points to a list of elements, and this list of elements. The number of the elements of each list corresponds to the number of labelized cells in the corresponding difference-image. The lists of the data structure have not equal length, because of the different number of cells present in each difference-image. A list element consists of the cell baricenter coordinates, of the cell area value and of a pointer to the successive element of the list.

The algorithm of track reconstruction consists of two procedures which generate a new data structure, whose lists represent the reconstructed tracks. The first procedure of the reconstruction algorithm selects and removes the first element which is pointed by the first valid pointer in the data structure generated by the labelization algorithm. This element represents the starting element of the first track list, i.e. the first point of the track, and it is pointed by a header. Each header consists of the number of the starting difference-image, of the number of elements of the track, of a pointer to the first element of the list, and of a pointer to the header of the successive track. Starting from the element selected by the first procedure, the second procedure scans the successive list of the data structure, in order to find and remove the element whose baricenter lies in a preestablished distance range, and whose calculated motion vector has a pre-established angle value respect to the forecast motion vector. The distance range can vary depending on the speed of the microorganisms. The last calculated motion vector is assumed as the forecast motion vector. If this element does not exist, the track ends; if it exists, it represents the second element of the track list, i.e. the second point of the track. In the latter case, the second procedure repeats its operations on the successive lists iteratively, until all the elements of the track being reconstructed are removed from the data structure generated by the labelization procedure. It can happen that the second procedure encounters more than one element whose baricenter satisfy the selection criteria, i.e. more than one motion vector could be constructed; in this case we choose the baricenter which allows to construct the vector with the smallest deviation from the forecast motion vector. For this purpose, the difference between each of the possible vectors and the last motion vector is calculated. The element whose baricenter has the smallest deviation will then be selected. When a track is completed, the first procedure selects a new starting element, and the second procedure uses this starting element to reconstruct another track. These operations are performed until all the elements are removed from the data structure generated by the labelization procedure. Instantaneous and average velocity, acceleration, the space covered in the sampled time and the angles of the trajectory are calculated for each cell.

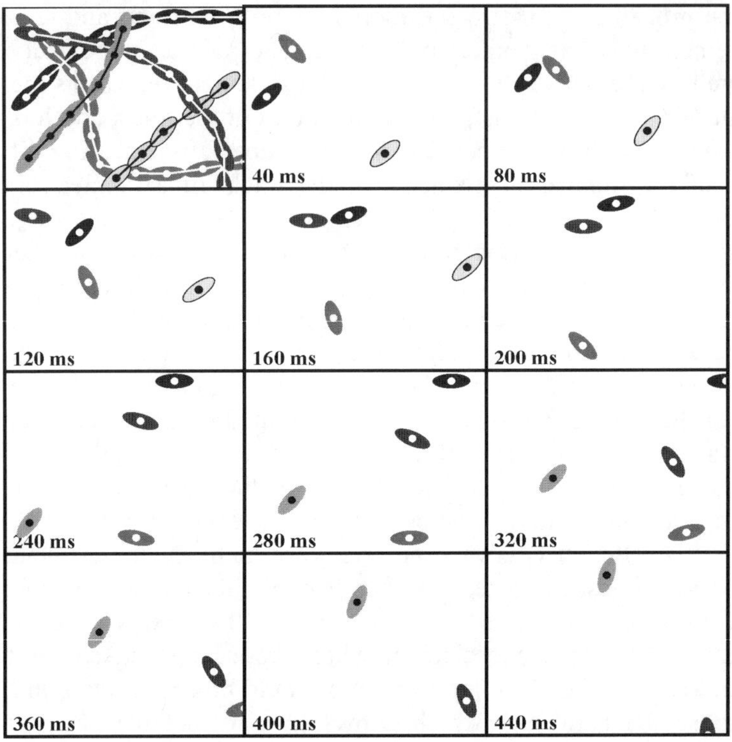

Figure 8. An example of reconstruction: the tracks are shown superimposed on the first framed image

The reconstruction of the tracks of microorganisms freely swimming in a microscope slide can present the problem of overlapping cells, i.e. intersecting tracks. However, the peculiar characteristics of our experimental apparatus allow minimizing this problem. The microscope has a low depth of focus, and the microorganisms swim within a narrow layer of medium between a slide and a cover slip. In this situation the cells, which swim turning along their major axis in the focused layer, usually are not occluded by the cells that swim in the out-of-focus layers, because the microscope do not focus them. However, it can happen that two cells hide each other, but these rare events are left unsolved in the routine work, since for this kind of analysis is very important to reconstruct more than five hundred tracks. In these cases the labelization algorithm detects a cell with an area value greater than the average area value, but only one baricenter. Therefore, this baricenter is inserted in the track of one cell, while the other cell will have two tracks: the first track stops just before the intersection, the second track starts just after the intersection. For the statistical analysis our samples demand, it has no importance to link the two sections of the incomplete track. Therefore, in order to save time, in the routine work, we do not implement any procedure for a second-step analysis of incomplete tracks.

However, for a more precise analysis, we apply a procedure to solve these misinterpreted tracks. Because the data structure contains the positions of the cells, we can calculate the motion vectors in each difference-image and on the basis of this calculus we can forecast the behavior of the cells. As the choice of the delay avoids the occurrence of the occlusion phenomena in more than one difference-image, the procedure scans the headers of the tracks in order to find a track which ends in a difference-image and a track which starts two difference-images later. If this couple of tracks is found, the remaining tracks are then scanned in order to find in the intermediate difference-image a track element which could join the two selected tracks. Therefore, its baricenter coordinates has to satisfy the distance and angle criteria, and its area value has to be greater than the average area value. When a track element with these characteristics is found, the two sections of the incomplete track can be joined.

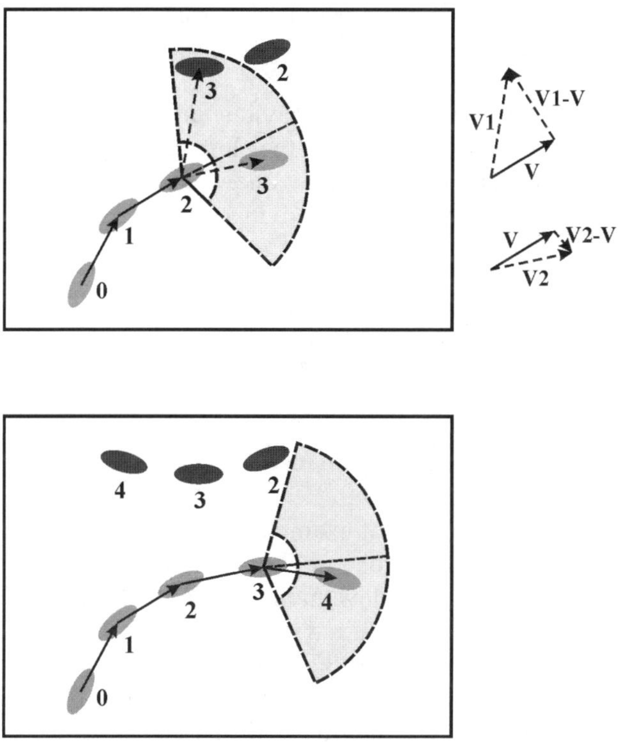

Figure 9. How to find the successive position of the cell in its track. The difference between each of the possible vectors is calculated, and the smallest deviation is selected.

3.5. Motion interpretation

An example of motion interpretation is the production of polarograms. Polarograms (circular histograms) represent the vectorial path distributions of cells with isotropic or anisotropic external stimuli.

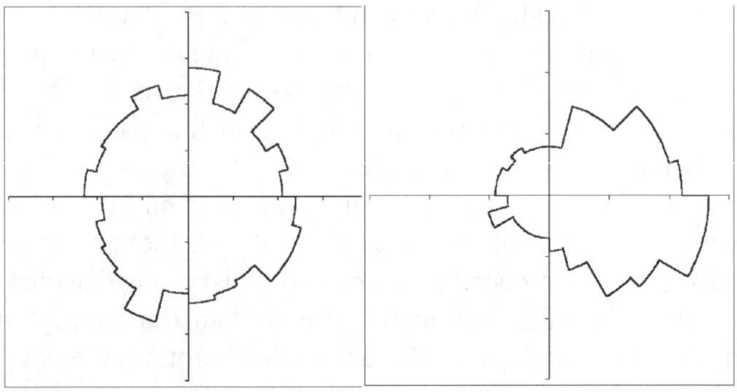

Figure 10. Polarograms

Figure 10 shows the different behaviour of dark-adapted cells with and without unilateral light stimulus: the left polarogram shows uniform distribution (r = 0.05), while the right one, with high degree of orientation (r = 0.51), shows that the cells are moving away from a light source placed on the left.

4. EXPERIMENTAL CASE:

4.1. Microorganisms motion

Figures 11 and 12 are a real example of the digital procedure that stores the elapsed frames of microscope fields acquired in transmitted light into the computer memory. A very slow motile mutant of *Dunaliella salina* was selected, [6]. Frames were acquired every 200 ms. Six selected images are shown. The total elapsed time is 37 min. Each cell is numbered and can be followed in the successive frames where its trajectory is represented with a black line starting from the previous cell baricenter. Since cell trajectories are randomly directed, we can say that the cells move and are not transported by a flow in the medium. The calculated average velocity of the mutant is 0.2 mm$*$s $^{-1}$. Tracking experiment was performed also on wild type cells

(Figure 7). In this case frames were acquired very 40 ms, for a total elapsed time of 400 ms. Six selected images are shown. Cell trajectories are not oriented, and indicate an active, straight swimming of the cells. The calculated average velocity of the wild type is 95 mm$*$s^{-1}.

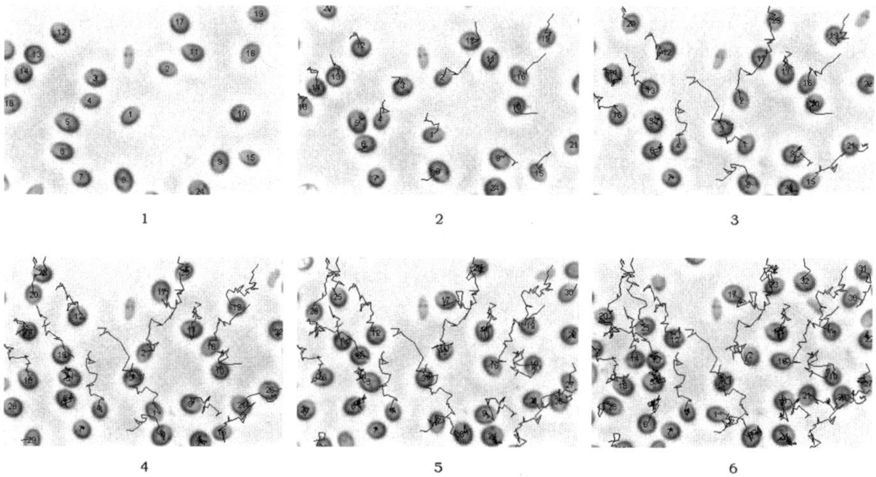

Figure 11. Result of the track reconstruction experiment for the mutant. The black lines represent the cell trajectories. The total elapsed time between the first and the last image is 37 min.

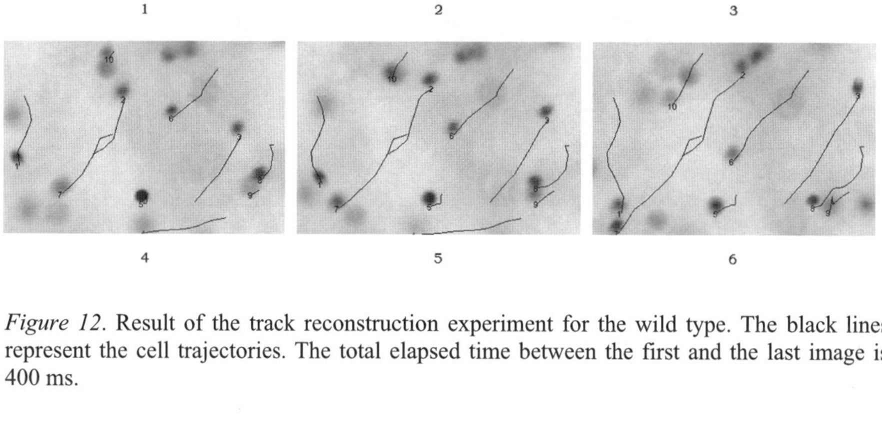

Figure 12. Result of the track reconstruction experiment for the wild type. The black lines represent the cell trajectories. The total elapsed time between the first and the last image is 400 ms.

5. REFERENCES

1 Allen R.D. "New direction and refinement in videoenhanced microscopy, applied to problems in cell motility." In *Advanced in Microscopy*, Cowden, R.R. and Harrison F.W., eds., New York: Alan Riss, pp.3-11,1983

2 Gualtiei P., "Image analysis techniques for studying photomovements." In *Image Analysis in Biology*, Hader ,ed., Boca Raton: CRC Press, pp. 55-75, 1991

3 Gualtieri P., and Coltelli P. A real time Automate d system for the analisis of Moving images. J of Computer Assisted Microscopy 1991; 3,1:15-21,

4 David R. Soll and Deborah Wessels (eds) *Motion analysis of living Cells*, New York:Wiley 1997

5 Coltelli P and Gualtieri P. A procedure for the extraction of object features in microscope images. J. Biomed. Comput. 1990 ;25, 3: 69-76.

6 Vismara R., Verni F., Barsanti L. Evangelista V. Gualtieri P. A short flagella mutant of Dunaliella salina. Micron 2004;35: 337-344

IMAGE DECONVOLUTION

M. Bertero and P. Boccacci
DISI, Universita' di Genova,
Via Dodecaneso 35, I-16146 Genova, Italy
bertero@disi.unige.it, boccacci@disi.unige.it

Abstract Image deconvolution is a basic problem in the processing of microscopic images. It is ill-posed, as the vast majority of inverse problems, and therefore it requires an accurate modeling taking into account all the known properties of the process of image formation and acquisition. In this chapter, after a brief discussion of the ill-posedness of image deconvolution in a continuous setting, we develop a detailed statistical model which applies to the case of fluorescence microscopy. Two approximate models, denoted as the Gaussian case and the Poisson case, are also introduced. The statistical model is the starting point for a maximum likelihood approach. In the Gaussian case one re-obtains the standard least squares problem which is also ill-posed. The constraint of non-negativity is introduced and two iterative algorithms converging to nonnegative least-squares solutions are presented. The need of early stopping of the iterations to produce sensible results is discussed. Moreover, an iterative method converging to the maximum-likelihood estimates of the Poisson case is presented: it is the classical RL (or EM) method which also requires early stopping of the iterations. Finally Bayesian methods, based on the use of *a priori* statistical information on the object to be restored, are introduced and their relationship with the standard regularization theory of inverse problems is discussed.

Keywords: Image deconvolution, maximum likelihood, regularization, iterative methods

Introduction

In several domains of applied science images are degraded by blurring and noise. Blurring is a perturbation due to the imaging system (caused, for instance, by diffraction, aberrations etc.) while noise is intrinsic to the detection process. Therefore image deconvolution is basically a post-processing of the detected images aimed to reduce the disturbing effects of blurring and noise.

When formulated in a naive way, image deconvolution implies the solution of a linear equation (a linear system of algebraic equations in the discrete case), but this problem turns out to be ill-posed: the solution may not exist or may not be unique. Moreover, even if a unique solution can be found (as in the case

V. Evangelista et al. (eds.), From Cells to Proteins: Imaging Nature across Dimensions, 349–370.

of discrete problems), this solution is strongly perturbed by noise propagation. The standard approach to the treatment of ill-posed problems is provided by the so-called *regularization theory* (for an introduction see, for instance, [1]).

In general, the ill-posedness implies that there are too many approximate solutions of the problem which are able to reproduce the data within a given noise level. To get a sensible solution one must reformulate the problem of image deconvolution by taking into account, as far as possible, all the known properties of image formation and acquisition as well as *a priori* information about properties of the object to be restored.

In the case of space invariance the process of image formation can be described by a *Point Spread Function* (PSF), which must be computed or measured. On the other hand, in applications such as Astronomy and Fluorescence Microscopy, images are detected by means of CCD cameras and therefore the process of image acquisition is affected by a combination of Poisson (photon counting) and Gaussian noise (read-out noise due to the amplifier). Finally *a priori* information about the unknown image, such as, for instance, non-negativity, is available.

The need of taking into account all the properties mentioned above, and, in particular, the statistical properties of the noise, implies that statistical approaches can be used for reformulating the problem of image deconvolution. *Maximum likelihood* looks as the most natural one and, in such a way, the image deconvolution problem can be reduced to the minimization of a suitable functional. Since the new problem is still ill-posed, two approaches are possible: the first is to use iterative minimization algorithms with *regularization properties*, in the sense that it is possible to control noise propagation by a suitable stopping of the iterations (such a property is called *semiconvergence*); the second is to use *a priori* information about statistical properties of the unknown object in the framework of Bayesian estimation.

In Section 2 we discuss the ill-posedness of the problem in a continuous setting and we prove the need of additional information as a way to reduce the uncertainty of the approximate solutions. Section 3 is devoted to a detailed statistical model which applies to the case of fluorescence microscopy; moreover two approximate models, denoted respectively as the Gaussian and the Poisson case, are introduced. In Section 4 the maximum likelihood approach is described and applied to the two approximate models of the previous section. In particular, in the Gaussian case we show that one re-obtains the classical least squares problem which is also ill-posed; next the additional constraint of non-negativity is introduced and two iterative methods for approximating the non-negative least squares solutions are discussed. Moreover the classical iterative method for approximating the maximum likelihood solutions in the Poisson case, the so-called Richardson-Lucy (RL) or Expectation-Maximization (EM) method, is presented and its main properties are discussed. Finally Section 5 is devoted to the Bayesian methods and their relationship with the methods provided by regularization theory.

1. Basic Properties of Image Deconvolution

The main feature of image deconvolution is ill-posedness. The practical implication of this property is that the solution of a (discrete) convolution equation is completely corrupted by an excess of noise propagation. However, the most simple way for understanding the difficulties generated by ill-posedness is to discuss the problem in a continuous setting.

Continuous Model

We assume that images are described by functions of 2 or 3 variables, $\mathbf{x} \in \mathbf{R}^n$ (n=2 in the 2D case and n=3 in the 3D case). We denote by $f_0(\mathbf{x})$ the intensity, at the point $\mathbf{x}$, of the object to be imaged, by $g_0(\mathbf{x})$ its image produced by the optical instrument before detection (also called the noiseless image) and by $g(\mathbf{x})$ the detected image.

In most imaging systems the noiseless image is approximately a linear function of the object. Therefore the imaging system defines a linear operator A such that:

$$g_0 = A\, f_0 \ . \tag{1}$$

If the system is space invariant, then the operator A is a *convolution operator*, i. e. there exists a function $h(\mathbf{x})$ such that:

$$(A\, f_0)(\mathbf{x}) = \int h(\mathbf{x}-\mathbf{x}')f_0(\mathbf{x}')\, d(\mathbf{x}' \ . \tag{2}$$

The function $h(\mathbf{x})$ is the PSF of the imaging system. Models of 3D PSFs for different kinds of microscopes are discussed in the accompanying chapter *Image Formation in Fluorescence Microscopy* by G. Vicidomini.

If we denote the Fourier Transform (FT) of a function by the corresponding capital letter, then from the *convolution theorem* and Eqs. (1)- (2), we get:

$$G_0(\omega) = H(\omega)\, F_0(\omega) \ , \tag{3}$$

$\omega \in \mathbf{R}^n$ being the coordinates in Fourier space, also called *space frequencies*.

The function $H(\omega)$ is the *transfer function* (TF), which describes the behaviour of the imaging system in the frequency domain. If the TF is zero outside a bounded domain $\mathbf{\Omega}$, the PSF is said to be *band-limited* and the set $\mathbf{\Omega}$ is called the *band* of the imaging system. It is obvious that the noiseless image g_0 is also band-limited.

In most cases the PSF has the following properties:

$$i)\ \ h(\mathbf{x}) \geq 0 \ \ ; \ \ ii)\ \ \int h(\mathbf{x})\, d\mathbf{x} < +\infty \ . \tag{4}$$

The first property implies that the noiseless image g_0 is also non-negative, while the second property implies that the imaging system is a low-pass filter. In fact, the Riemann-Lebesgue theorem implies that the TF $H(\omega)$ is bounded and continuous and that it tends to zero when $|\omega| \rightarrow +\infty$.

Ill-Posedness of Image Deconvolution

The detected images are corrupted by several kinds of noise (this point will be discussed in the next section). In the continuous setting this situation is modeled by writing the relation between the detected image g and the noiseless image g_0 in the following form:

$$g(\mathbf{x}) = g_0(\mathbf{x}) + w(\mathbf{x}) = (A\, f_0)(\mathbf{x}) + w(\mathbf{x}) \ . \tag{5}$$

This expression is not related to specific assumptions about the noise; in particular it does not mean that we are assuming additive noise. The term $w(\mathbf{x})$ is just the difference between the detected and the noiseless image.

In terms of the FTs we have:

$$G(\omega) = H(\omega)\, F_0(\omega) + W(\omega) \ , \tag{6}$$

and therefore, even if the noiseless image is band-limited, the detected image, in general, is not, because $W(\omega)$ may not be zero where $H(\omega) = 0$ (*out-of-band noise*).

Given the detected image g and the PSF h, the problem of *image deconvolution* is to determine a sound estimate f of f_0. If the noise term w is small, then the most natural approach is to look for a solution of the linear equation:

$$Af = g \ , \tag{7}$$

which, in Fourier space, becomes:

$$H(\omega)\, F(\omega) = G(\omega) \ . \tag{8}$$

This elementary equation clarifies the difficulties of image deconvolution. In fact, if the PSF is band-limited while the noise is not, the equation is inconsistent outside the band of the instrument. In other words, *no solution exists*! Moreover, again in the case of a band-limited system, even if a solution exists, *the solution is not unique*: one can add to it an object whose FT is zero over the band and takes arbitrary values outside the band. Such a pathological object is sometimes called an *invisible object*, because its image is zero even if it is not zero. Finally, even if a solution exists and is unique, a small variation of the noise in points of the band where the TF is small can modify completely the solution, as we can understand by substituting Eq. (6) into Eq. (8) and solving for F; the result is given by:

$$F(\omega) = F_0(\omega) + \frac{W(\omega)}{H(\omega)} \ . \tag{9}$$

All these remarks imply that the problem is *ill-posed*, because the solution may not exist, may not be unique and may not depend continuously on the data.

However, it is obvious that the true object f_0 is an approximate solution of Eq. (7), in the sense that Af_0 is not exactly g but is close to g. This remark suggests to investigate the set of the approximate solutions of Eq. (7), i. e. the set of the objects whose noiseless images approximate g within a given error level. The difficulty is that this set is too broad: it contains both sensible and crazy objects. To find methods for extracting the sensible ones one must reformulate the problem by taking into account all the available information both on the process of image acquisition (noise) and on the object itself (*a priori* information, such as non-negativity).

2. A Statistical Model

The first step in the reformulation of image deconvolution is to model the noise corrupting the data. We restrict the analysis to the case of fluorescence microscopy where the noise is a random process intimately related to photon emission and acquisition, so that the best framework is provided by a discrete setting. For the sake of generality we consider the 3D case; the reduction to the 2D case is obvious.

If we assume magnification one, then the object and the image are defined in the same volume which can be partitioned into the same number of voxels with the same size. The latter is defined by the acquisition process: it is given by the size of the elements of the CCD camera in the lateral directions, and by the scanning distance in the axial direction. The voxels can be characterized by a multi-index $\mathbf{n}$, which is the triple of indexes $\{n_1, n_2, n_3\}$, defining the position of the voxel within the image volume and taking respectively N_1, N_2, and N_3 values. This triple will be denoted by $\mathbf{N}$. Moreover, given a physical quantity $h(\mathbf{x})$, its value at the voxel $\mathbf{n}$, also called *voxel value*, is denoted by $h(\mathbf{n})$. It is given, for instance, by the integral of $h(\mathbf{x})$ over the voxel volume.

Photons, emitted in the object volume, are collected by the microscope and detected in the image volume. This is a statistical process. We denote by $\xi(\mathbf{n}')$ the Random Variable (RV) describing the statistical distribution of the number of photons emitted at voxel $\mathbf{n}'$ and collected by the microscope during the integration time T. Then the *first basic assumption* is the following:

- $\xi(\mathbf{n}')$ is a Poisson RV, with expected value $f(\mathbf{n}')$, i. e.

$$P_{\xi(\mathbf{n}')}(m) = \frac{e^{-f(\mathbf{n}')} f(\mathbf{n}')^m}{m!} \tag{10}$$

  is the probability of the emission of m photons at voxel $\mathbf{n}'$;

- the RVs $\xi(\mathbf{n}')$ and $\xi(\mathbf{n}'')$, corresponding to different voxels, are statistically independent.

Next we must provide a statistical interpretation of the sampled PSF, whose voxel values are denoted by $h(\mathbf{n})$. If it is normalized in such a way that:

$$\sum_{\mathbf{n}} h(\mathbf{n}) = 1 , \quad (11)$$

and if we assume space invariance, then $h(\mathbf{n}-\mathbf{n}')$ *is the probability that a photon emitted at voxel* $\mathbf{n}'$ *of the object volume is collected at voxel* $\mathbf{n}$ *of the image volume*. Moreover, if we assume, as usual, a periodic extension of the voxel values of f and h, so that their cyclic convolution can be computed by means of FFT, we denote by A the block-circulant matrix defined by:

$$(Af)(\mathbf{n}) = \sum_{\mathbf{n}'} h(\mathbf{n}-\mathbf{n}')f(\mathbf{n}') . \quad (12)$$

Let $\xi(\mathbf{n},\mathbf{n}')$ be the RV describing the statistical distribution of the number of photons emitted at voxel $\mathbf{n}'$ and collected at voxel $\mathbf{n}$, then the *second basic assumption* is the following:

- $\xi(\mathbf{n},\mathbf{n}')$ is a Poisson RV with expected value given by:

$$E\{\xi(\mathbf{n},\mathbf{n}')\} = h(\mathbf{n}-\mathbf{n}')f(\mathbf{n}') ; \quad (13)$$

- for any fixed $\mathbf{n}''$, and $\mathbf{n} \neq \mathbf{n}'$, $\xi(\mathbf{n},\mathbf{n}'')$ and $\xi(\mathbf{n}',\mathbf{n}'')$ are statistically independent.

We remark that the previous assumptions are consistent. In fact, it is obvious that the following relationship holds true between the RVs we have introduced:

$$\xi(\mathbf{n}') = \sum_{\mathbf{n}} \xi(\mathbf{n},\mathbf{n}') . \quad (14)$$

Since, for fixed $\mathbf{n}'$, the RVs $\xi(\mathbf{n},\mathbf{n}')$ are independent and Poisson distributed, it follows that $\xi(\mathbf{n}')$ is also Poisson distributed and that its expected value is the sum of the expected values:

$$E\{\xi(\mathbf{n}')\} = \sum_{\mathbf{n}} E\{\xi(\mathbf{n},\mathbf{n}')\} = \sum_{\mathbf{n}} h(\mathbf{n}-\mathbf{n}')f(\mathbf{n}') = f(\mathbf{n}') , \quad (15)$$

where the normalization condition of Eq. (11) has been used. This result is in agreement with the first assumption.

The previous analysis concerns the emission process. Let us consider now the detection process in the image volume, where photons are detected by means of a CCD camera. *If we assume efficiency one*, then all the photons collected by the microscope are detected by the CCD camera and therefore the statistical properties of the detected photons can be deduced from the statistical properties of the emitted ones. If we denote by $\eta_{obj}(\mathbf{n})$ the RV describing the photons emitted by the object and detected at voxel $\mathbf{n}$, we have;

$$\eta_{obj}(\mathbf{n}) = \sum_{\mathbf{n}'} \xi(\mathbf{n}, \mathbf{n}') \ . \tag{16}$$

Thanks to the previous assumptions, this RV is Poisson distributed with an expected value given by:

$$E\{\eta_{obj}(\mathbf{n})\} = \sum_{\mathbf{n}'} h(\mathbf{n} - \mathbf{n}')f(\mathbf{n}') = (Af)(\mathbf{n}) \ . \tag{17}$$

Moreover, the RVs $\eta_{obj}(\mathbf{n})$ associated to different voxels are statistically independent. We will denote by $g_{obj}(\mathbf{n})$ the number of photons which are actually detected at voxel $\mathbf{n}$, i. e. the realization of the RV $\eta_{obj}(\mathbf{n})$.

Now we can consider the following simplified version of the model developed in [2] for data acquired by a CCD camera. If we denote by $g(\mathbf{n})$ the value at voxel $\mathbf{n}$) of the detected image, then this is the realization of a RV $\eta(\mathbf{n})$ which is given by:

$$\eta(\mathbf{n}) = \eta_{obj}(\mathbf{n}) + \eta_{bkg}(\mathbf{n}) + \zeta_{ron}(\mathbf{n}) \ , \tag{18}$$

where:

- $\eta_{obj}(\mathbf{n})$ is the number of object photons given in Eq. (16), with expected value given in Eq. (17);
- $\eta_{bkg}(\mathbf{n})$ is the number of photons due to background emission; it is an independent Poisson RV with expected value $b(\mathbf{n})$, which, in most cases, can be assumed to be constant; in general it can be estimated and, with the detected image $g(\mathbf{n})$ and the PSF $h(\mathbf{n})$, is one of the data of the problem;
- $\zeta_{ron}(\mathbf{n})$ is the additive Read-Out Noise (RON); it is an independent Gaussian process, whose expected value and variance can also be estimated.

We denote by η the set of the independent RVs $\eta(\mathbf{n})$, by g the set of their realizations $g(\mathbf{n})$ and by f the set of the object expected values $f(\mathbf{n})$. Since the RVs $\eta(\mathbf{n})$ are statistically independent and each of them is the sum of a Poisson RV with expected value $(Af)(\mathbf{n}) + b(\mathbf{n})$ and of an independent Gaussian process, the joint probability density of the RVs η for a given f is given by:

$$P_\eta(g|f) = \tag{19}$$

$$= \prod_{\mathbf{n}} \sum_{m=0}^{+\infty} e^{-[(Af)(\mathbf{n})+b(\mathbf{n})]} \ \frac{[(Af)(\mathbf{n}) + b(\mathbf{n})]^m}{m!} \ P_{ron}(g(\mathbf{n}) - m) \ ,$$

where $P_{ron}(u)$ is the probability density of the RON. In the frequent case of white Gaussian noise, with expected value r and standard deviation σ, we have:

$$P_{ron}(u) = \frac{1}{\sqrt{2\pi}\sigma} e^{-\frac{(u-r)^2}{2\sigma^2}} . \tag{20}$$

A useful, even if very rough, approximation is obtained by neglecting the photon noise, so that the randomness of the image is only due to additive white noise. In such a case, if we assume, for simplicity, a RON with zero expected value, then we have:

$$P_\eta(g|f) = \left(\frac{1}{\sqrt{2\pi}\sigma}\right)^{|N|} exp\left\{-\frac{1}{2\sigma^2}||g-(Af+b)||_2^2\right\} , \tag{21}$$

where $|N| = N_1N_2N_3$ and $||.||_2$ denotes the Euclidean norm:

$$||h||_2 = \left(\sum_{\mathbf{n}} |h(\mathbf{n})|^2\right)^{1/2} . \tag{22}$$

Another approximation, in general more accurate than the previous one and frequently used both in Microscopy and Astronomy, is obtained by neglecting the RON with respect to the photon noise. Since in such an approximation each detected value $g(\mathbf{n})$ should be an integer number, we can write:

$$P_\eta(g|f) = \prod_{\mathbf{n}} e^{-[(Af)(\mathbf{n})+b(\mathbf{n})]} \frac{[(Af)(\mathbf{n})+b(\mathbf{n})]^{g(\mathbf{n})}}{g(\mathbf{n})!} . \tag{23}$$

In the following, the approximation, where only the RON is considered, will be called the **Gaussian case**, while that, where only the photon noise is considered, will be called the **Poisson case**.

3. Maximum Likelihood Methods

In Statistics the Maximum Likelihood (ML) approach provides the most frequently used methods for parameter estimation. Its application to image deconvolution is based on the knowledge of the random properties of the detected image, i. e. on the knowledge of the probability density $P_\eta(g|f)$. If the detected image g, the PSF h, and the background b are given, then $P_\eta(g|f)$ is a function only of f and the problem of image deconvolution becomes the problem of estimating these unknown parameters. In such a situation the ML estimator answers to the following question:

Which object f is most likely to produce the detected image g ?

Definition 1 - For a given detected image g the *likelihood function* is the function of the object f defined by:

$$L_g(f) \;=\; P_\eta(g|f) \;. \tag{24}$$

Definition 2 - A ML-estimate of the object f is any object f_{LM} which maximizes the likelihood function:

$$f_{ML} \;=\; arg\, max_f\; L_g(f) \;. \tag{25}$$

Remark - In the case of image deconvolution this problem is, in general, ill-posed. We will justify this statement in a particular case.

In all practical applications to imaging the probability density of η for a given f is the product of a very large number of factors, so that it is useful to introduce the following log-likelihood function:

$$l_g(f) \;=\; log\, L_g(f) \;. \tag{26}$$

Since the logarithm is an increasing and concave function, the set of the maximum points of $l_g(f)$ coincides with the set of the ML-estimates. However, to clarify the relationship with the standard approach to Inverse Problems provided by regularization theory, it is convenient to consider functionals of the following type:

$$J_g(f) \;=\; -\; log\, L_g(f) \;+\; (terms\ depending\ only\ on\ g) \;. \tag{27}$$

The terms depending on g are added or subtracted, in some cases, to obtain well-known discrepancy functionals, as we will show in the following.

In conclusion, the ML-estimates are solutions of a variational problem of the following type:

$$f_{ML} \;=\; arg\, min_f\; J_g(f) \;. \tag{28}$$

In such an approach it is obvious that it is possible to introduce additional constraints such as non-negativity. This case is quite important and it deserves additional comments.

The non-negative ML-estimates are solutions of the problem:

$$f^+_{ML} \;=\; arg\, min_{f\geq 0}\; J_g(f) \;. \tag{29}$$

If the functional $J_g(f)$ is convex, then necessary and sufficient conditions for a minimum point are provided by the *Kuhn-Tucker (KT) conditions* (see, for instance, [3]):

$$f_{ML}(\mathbf{n})\nabla_f\, J_g(f)\,|_{f=f_{ML}}(\mathbf{n}) \;=0 \;\;, \tag{30}$$

$$f_{ML}(\mathbf{n}) \geq\, 0 \;\;\;,\;\;\; \nabla_f\, J_g(f)\,|_{f=f_{ML}}(\mathbf{n}) \;\geq\, 0 \;. \tag{31}$$

These conditions imply that, in the voxels where f^+_{ML} is zero, the gradient of the functional is non-negative and that f^+_{ML} is non-negative in the voxels where the gradient is zero. In general these solutions are zero in a large part of the voxels and therefore they are not reliable from the physical point of view.

The Gaussian Case

In the Gaussian case the likelihood function is given by Eq. (21) and therefore the functional $J_g(f)$ is given by:

$$J_g(f) = ||g - (Af + b)||_2^2 = ||Af - g_s||_2^2 \ , \tag{32}$$

where g_s is the subtracted image defined by:

$$g_s = g - b \ . \tag{33}$$

It follows that the ML-method is equivalent to the classical Least-Squares (LS) method, which is just the starting point of regularization theory [1]. Any minimum point of this functional is called a *LS-solution* and is denoted by f_{LS}; it is also a solution of the *Euler equation*:

$$A^T A f_{LS} = A^T g \ , \tag{34}$$

which can be easily obtained by zeroing the gradient of the functional. Here A^T denotes the transposed of the matrix A.

Since A is a block-circulant matrix, Eq. (34) is diagonalized by the Discrete Fourier Transform (DFT) and we obtain:

$$|H(\mathbf{k})|^2 \, F_{LS}(\mathbf{k}) = H^*(\mathbf{k}) \, G_s(\mathbf{k}) \ , \tag{35}$$

where $H(\mathbf{k})$ is the DFT of $h(\mathbf{n})$, etc., and the * denotes complex conjugation.

This equation is always consistent, and therefore a solution always exists. But in the Fourier voxels $\mathbf{k}$ where $H(\mathbf{k}) = 0$, the value of the DFT of the LS-solution is not determined, and therefore, in such a case, the solution is not unique; on the other hand, in the voxels $\mathbf{k}$ where $H(\mathbf{k}) \neq 0$, the value of the DFT of the LS-solution is given by:

$$F_{LS}(\mathbf{k}) = \frac{G_s(\mathbf{k})}{H(\mathbf{k})} \ . \tag{36}$$

These values coincide with those provided by the linear equation $Af = g_s$.

We conclude that the LS-problem is ill-posed, first because the solution may not be unique, as indicated above, and second because the LS-solutions are strongly perturbed by noise. Uniqueness can be restored by looking for the LS-solution with minimal Euclidean norm, also called *generalized solution*; it

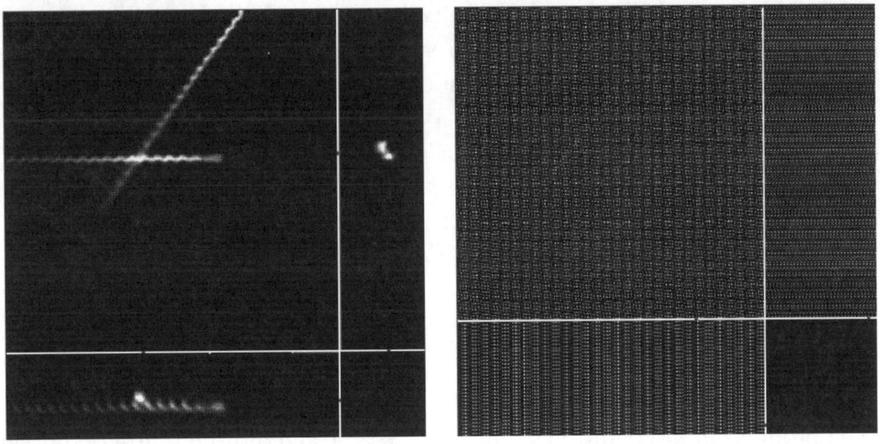

Figure 1. Illustration of the numerical instability of the generalized solution. Left panel: three orthogonal sections of a two photon 3D image of the mature sperm head of the Octopus Eledone Cyrrhosa (details are given in the text) - Right panel: the "reconstruction" provided by the generalized solution.

is obtained by setting $F_{LS}(\mathbf{k}) = 0$ in the Fourier voxels where $H(\mathbf{k}) = 0$ and therefore is given by:

$$f_{LS}(\mathbf{n}) = \frac{1}{|N|} \sum_{H(\mathbf{k})\neq 0} \frac{G_s(\mathbf{k})}{H(\mathbf{k})}\, e^{i2\pi \mathbf{n}\cdot\frac{\mathbf{k}}{\mathbf{N}}} . \tag{37}$$

In Figure 1 we give the result obtained by means of this solution in the case of an image recorded with a two-photon microscope. The object is the mature sperm head of the Octopus Eledone Cirrhoses. The image is obtained with a $100X/1.4NA$ oil immersion lens, "open pinhole" condition, excitation at $720nm$, filter emission at $450nm$; it is a cube of $256 \times 256 \times 64$ 8bits voxel values. The pixel dimension is $137nm$ while the plane distance is $200nm$. In the left panel of Figure 1 a transversal section and two orthogonal axial sections of the image are shown. The PSF used for deconvolution is that modeled in the chapter on *Image Formation*. In the right panel the result provided by the generalized solution is given. The image has been completely destroyed by noise propagation.

As it was already observed in Section 2, the numerical instability of the solution given in Eq. (37) is due to the small value of $H(\mathbf{k})$ which amplify the noise corrupting the corresponding components of the subtracted image. This effect is quantified by the *condition number* of the problem which is given by:

$$\alpha = \frac{H_{max}}{H_{min}} , \tag{38}$$

where H_{max}, and H_{min} are respectively the largest and smallest non-zero value of $|H(\mathbf{k})|$. In fact, if δg_s is a small variation of the subtracted image and δf_{LS} is the corresponding variation of the generalized solution, the following inequality holds true:

$$\frac{\|\delta f_{LS}\|_2}{\|f_{LS}\|_2} \leq \alpha \frac{\|\delta g_s\|_2}{\|g_s\|_2} , \tag{39}$$

which easily follows from the inequalities (obtained by means of the Parseval equality for the DFT):

$$\|\delta f_{LS}\|_2 \leq \frac{1}{H_{min}} \|\delta g_s\|_2 \ , \ \|g_s\|_2 \leq H_{max} \|f_{LS}\|_2 \ . \tag{40}$$

The problem is ill-conditioned when $\alpha >> 1$ and a discrete ill-conditioned problem is typically the result of the discretization of an ill-posed problem [1].

The condition number gives also a measure of the *uncertainty* characterizing the approximate LS-solutions. In fact, the set of all the objects f compatible with the data within a given noise level ϵ is given by:

$$S_{LS}^{(\epsilon)} = \{ f \mid \|Af - g_s\|_2 \leq \epsilon \} , \tag{41}$$

and, by means of Parseval equality, we find the following equation for its boundary in Fourier space:

$$\frac{1}{|N|} \sum_{\mathbf{k}} |H(\mathbf{k}) F(\mathbf{k}) - G_s(\mathbf{k})|^2 = \epsilon^2 , \tag{42}$$

or also:

$$\sum_{\mathbf{k}} \frac{\epsilon^2 |H(\mathbf{k})|^2}{|N|} |F(\mathbf{k}) - \frac{G_s(\mathbf{k})}{H(\mathbf{k})}|^2 = 1 . \tag{43}$$

Therefore the boundary is an ellipsoid with center the generalized solution and half-axis inversely proportional to $|H(\mathbf{k})|$; since the condition number is just the ratio between the largest and smallest half-axis, it provides a measure of the extent of the set of the approximate solutions.

Since this set is too broad, one can try to identify a subset of interest by means of additional constraints. If we use non-negativity, then we can define non-negative LS-solutions as follows:

$$f_{LS}^{+} = arg\, min_{f \geq 0} \|Af - g_s\|_2 . \tag{44}$$

The KT conditions of Eqs. (30)-(31) become now:

$$f_{LS}(\mathbf{n}) \left\{ (A^T A f_{LS})(\mathbf{n}) - (A^T g_s)(\mathbf{n}) \right\} = 0 , \tag{45}$$

$$f_{LS}(\mathbf{n}) \geq 0 , \ (A^T A f_{LS})(\mathbf{n}) - (A^T g_s)(\mathbf{n}) \geq 0 . \tag{46}$$

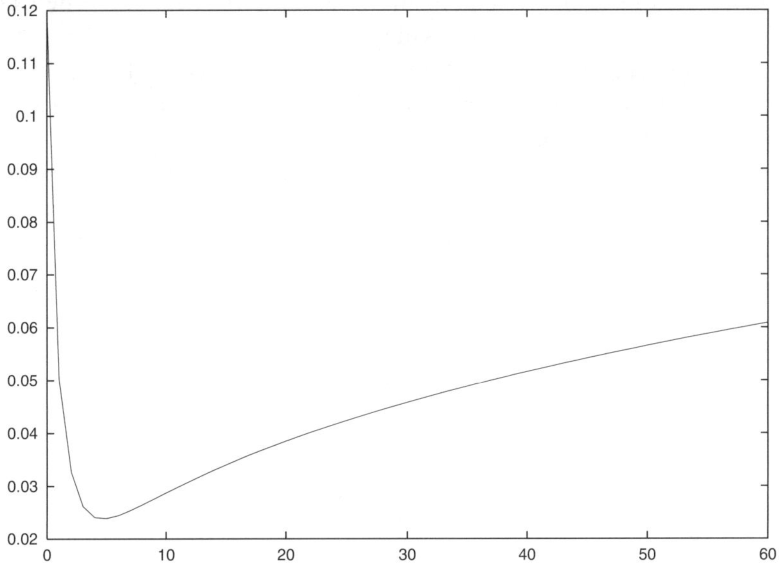

Figure 2. Example of the behaviour of the reconstruction error as a function of the number of iterations in the case of an iterative method with the semiconvergence property. The optimal number of iterations is defined by the minimum of the curve.

The solution of this problem is also affected by numerical instability and may not be unique. However it is possible to introduce iterative algorithms converging to the non-negative LS-solutions and having the following *semiconvergence* property (see [1] for a discussion): *first the iterates approach a sensible solution and then go away.*

In Figure 2 we plot a typical example of the behaviour of the reconstruction error, as a function of the number of iterations, for one of these iterative algorithms. If $f^{(k)}$ is the result of the $k-th$ iteration and f_0 is the unknown object, then the relative reconstruction error is defined by:

$$\rho^{(k)} = \frac{||f^{(k)} - f_0||_2}{||f_0||_2} \quad . \tag{47}$$

As follows from Figure 2, the reconstruction error has a minimum after a suitable number of iterations. Therefore the algorithm must not be pushed to convergence and an early stopping is required. It is obvious that the optimal number of iterations, corresponding to the minimum of the reconstruction error, can be computed only in the case of numerical simulations. Its estimation in the case of real data is a difficult problem and must be investigated for any particular application of image deconvolution. As a general rule one can say that, for a given imaging system, the optimal number of iterations depends on the Signal-to-Noise Ratio (SNR) (which depends on the integration time) and decreases for decreasing values of the SNR. It is obvious that the quality of the restoration is also decreasing; however, lower quality reconstructions can be obtained at a lower computational cost.

A first example of these iterative methods is provided by the **projected Landweber method**, which is basically a gradient method with a projection, at each iteration, on the closed cone of the non-negative objects. If we denote by P_+ this projection, which is defined by:

$$(P_+ f)(\mathbf{n}) = \begin{cases} f(\mathbf{n}) & \text{if } f(\mathbf{n}) > 0 \\ 0 & \text{if } f(\mathbf{n}) \leq 0, \end{cases} \tag{48}$$

then the iterative algorithm is as follows:

$$\begin{aligned} &i) \quad give \;\; f^{(0)} \geq 0; \\ &ii) \quad given \;\; f^{(k)}, \; compute: \\ &f^{(k+1)} = P_+ \left\{ f^{(k)} + \tau \left(A^T g_s - A^T A f^{(k)} \right) \right\} , \end{aligned} \tag{49}$$

where τ is a relaxation parameter which must be selected by the user within the following range of values:

$$0 < \tau < \frac{2}{H_{max}^2} . \tag{50}$$

The convergence of this algorithm is proved in [4]. Moreover, by taking the limit in Eq. (49), it is easy to verify that the limit satisfies the KT conditions and therefore is a solution of the constrained LS-problem. The semiconvergence of the algorithm is discussed in [1]: it can be proved in the non-projected case (namely when the projection P_+ is replaced by the identity) and is an experimental result in the projected case, namely it can be verified by means of numerical experiments.

In the practical applications, a frequent choice for the initial guess is $f^{(0)} = 0$ because, in the case of non-uniqueness, this choice presumably provides the minimal norm solution (this result holds true in the non-projected case). Moreover, the computational burden of the algorithm can be estimated by remarking that each iteration requires the computation of two FFTs.

A second example is provided by the **Iterative Space Reconstruction Algorithm (ISRA)**:

$$\begin{aligned} &i) \quad give \;\; f^{(0)} \geq 0; \\ &ii) \quad given \;\; f^{(k)}, \; compute: \\ &f^{(k+1)} = f^{(k)} \frac{A^T g_s}{A^T A f^{(k)}} , \end{aligned} \tag{51}$$

where the quotient of two images is defined voxel by voxel.

The convergence of this algorithm is proved in [5]. By taking the limit in Eq. (51), it is easy to verify that it satisfies the first of the KT conditions. The semiconvergence of the algorithm is an experimental result derived from numerical simulations.

In the practical applications the iterations are, in general, initialized with a constant image. Moreover the computational burden of each iteration coincides with that of the projected Landweber method.

The Poisson Case

In the Poisson case the likelihood function is given by Eq. (23), and one can obtain the following expression of the function $J_g(f)$:

$$J_g(f) \; = \; \sum_{\mathbf{n}} \left\{ g(\mathbf{n}) \; ln \; \frac{g(\mathbf{n})}{(Af+b)(\mathbf{n})} \; + \; [(Af+b)(\mathbf{n}) - g(\mathbf{n})] \right\} \; . \tag{52}$$

This function is a generalization of the Kullback-Leibler divergence and is also called the Csiszar I-divergence [6]; it provides a measure of the discrepancy between the detected image g and the computed image $Af+b$, associated with f. It is defined for nonnegative values of g and positive values of $Af+b$. If the PSF is nonnegative, then the natural domain of the functional is the closed cone of the nonnegative objects.

The functional $J_g(f)$ is convex, as it can be shown by verifying that the Hessian matrix is positive semi-definite (it is positive definite if and only if the equation $Af = 0$ implies $f = 0$, i. e. in the case of uniqueness of the solution of the deconvolution equation). Moreover the functional takes nonnegative values (as follows from the elementary inequality $a \, lna - a \, lnx + x - a \geq 0$, which holds true for fixed $a > 0$ and any $x > 0$) and therefore is bounded from below. These two properties imply that all the minima, namely the ML-solutions f_{ML}, are global and that they satisfy the KT conditions which, as follows from the computation of the gradient, take the following form:

$$f_{ML}(\mathbf{n}) \left\{ 1 - \left(A^T \frac{g}{Af_{ML}+b} \right) (\mathbf{n}) \right\} \; = 0 \; , \tag{53}$$

$$f_{ML}(\mathbf{n}) \geq 0 \; , \; 1 - \left(A^T \frac{g}{Af_{ML}+b} \right) (\mathbf{n}) \geq 0 \; , \tag{54}$$

where the normalization condition of Eq. (11) has been used and, again, the quotient of two images is defined voxel by voxel.

An iterative method for the computation of the ML-solutions was proposed by several authors, in particular by Richardson [7] and Lucy [8] for the deconvolution of astronomical images, and by Shepp and Vardi [9] for emission tomography. As shown in [9], this method is related to a general approach for the solution of ML problems, known as *Expectation-Maximization* (EM). For these reasons, the algorithm is known as *Richardson-Lucy* (RL) method in Astronomy and as EM method in tomography. We will use the first name; it is as follows:

$$\begin{aligned} i) \quad & give\ f^{(0)} \geq 0; \\ ii) \quad & given\ f^{(k)},\ compute: \\ & f^{(k+1)} = f^{(k)} A^T \left(\frac{g}{Af^{(k)} + b} \right) . \end{aligned} \tag{55}$$

It is evident that, *if the image g and the PSF h are nonnegative, then each iterate is also nonnegative.*

The convergence of this algorithm is proved in [10] and [11] (an incomplete proof is also given in [9]). By taking the limit in Eq. (55), it is easy to verify that it satisfies the first of the KT conditions. The semiconvergence of the algorithm is an experimental result derived from numerical simulations [1].

The computation of one iterate of this method is more expensive than the computation of one iterate of ISRA or of the projected Landweber method; in fact it requires the computation of four FFTs: two for the denominator and two for applying the transposed matrix to the quotient. Moreover the convergence is slow, so that, in general, a large number of iterations is required for reaching the optimal solution.

For these reasons it is important to find methods for accelerating the convergence. One approach is proposed in [12] where an *acceleration exponent* is introduced and Eq. (55) is modified as follows:

$$f^{(k+1)} = f^{(k)} \left(A^T \frac{g}{Af^{(k)} + b} \right)^{\omega} , \tag{56}$$

with $\omega > 1$; the authors show that a reduction in the number of iterations by a factor ω should be expected. Numerical simulations indicate that in the case $\omega = 2$ one has (semi)convergence with a reduction in the number of iterations by a factor of 2 (remark that the computational cost of one iteration is not significantly increased with respect to the standard case $\omega = 1$).

However there is an important point which must be taken into account. A nice property of the RL method is that, if $b = 0$ and the PSF satisfies the normalization condition of Eq. (11), then all the iterates have the following property:

$$\sum_{\mathbf{n}} f^{(k)}(\mathbf{n}) = \sum_{\mathbf{n}} g(\mathbf{n}) . \tag{57}$$

The physical meaning of this relation is obvious: the photon content of each iterate $f^{(k)}$ coincides with the total number of detected photons.

If the background is not zero, then this condition is not automatically satisfied and therefore it can be introduced as a constraint in the minimization problem. In the case of the accelerated algorithm of Eq. (56) this constraint is necessary in order to guarantee convergence. As shown in [13] this is equivalent to normalize each iterate in such a way that:

$$\sum_{\mathbf{n}} \left\{ f^{(k)} + b \right\} (\mathbf{n}) \; = \; \sum_{\mathbf{n}} g(\mathbf{n}) \; . \tag{58}$$

Therefore the iterative algorithm must include also this normalization step.

4. Bayesian Methods and Regularization

The ill-posedness (ill-conditioning) of ML-problems is generated by a lack of information on the object f, in particular by a lack of information at the frequencies corresponding to small values of the TF $H(\mathbf{k})$. A remedy can be the use of additional deterministic information (the classical example of non-negativity has been already discussed) or the use of additional statistical properties of f.

In a complete probabilistic approach it is assumed that both the object f and the detected image g are realizations of RVs denoted respectively by ξ and η, and that the problem is solved if we are able to estimate their joint probability density $P_{\xi\eta}(f, g)$.

As we discussed in Section 2, the conditional probability density of η given f can be deduced from known statistical properties of the noise. However, the marginal probability density of ξ, $P_\xi(f)$, in general is not known. One can guess this probability density using his knowledge, or ignorance, about f. The model used is usually called a *prior*.

Once the marginal distribution of ξ is given, the joint probability density can be obtained from *Bayes formula*:

$$P_{\xi\eta}(f, g) \; = \; P_\eta(g|f) \, P_\xi(f) \; . \tag{59}$$

Then, from the other Bayes formula, one obtains the conditional probability density of ξ given g:

$$P_\xi(f|g) \; = \; \frac{P_\eta(g|f) \, P_\xi(f)}{P_\eta(g)} \; , \tag{60}$$

which is just the solution of the image deconvolution problem in the Bayesian approaches. In fact from this probability density one can compute all the desired quantities concerning the restored image. More precisely, instead of computing a unique restored image one can compute the probability of any possible restored image.

However, in all applications, it is necessary to show at least one restored image and this can be provided by the *Maximum A Posteriori* (MAP) estimate, which is the object maximizing the *a posteriori* conditional probability :

$$f_{MAP} \; = \; arg \, max_f \; P_\xi(f|g) \; . \tag{61}$$

Introducing also in this case the log-function and neglecting the term independent of f, one finds:

$$f_{MAP} \;=\; arg\,max_f\;\{l_g(f)\;+\;ln\;P_\xi(f)\}\quad, \tag{62}$$

or, in terms of the functional $J_g(f)$ introduced in Section 2:

$$f_{MAP} \;=\; arg\,min_f\;\{J_g(f)\;-\;ln\;P_\xi(f)\}\quad. \tag{63}$$

Therefore the term $-\;ln\;P_\xi(f)$ plays the role of a regularization functional.

The most frequently used priors are of the Gibbs type, i. e.:

$$P_\xi(f)\;=\;C\;exp\{-\mu\Omega(f)\}\quad, \tag{64}$$

where μ is a parameter, controlling the amount of regularization (in regularization theory it is called the *regularization parameter*), and $\Omega(f)$ is a functional describing prior information about the object to be estimated [14]. In such a case the MAP problem takes the following form:

$$f_{MAP} \;=\; arg\,min_f\;\{J_g(f)\;+\;\mu\,\Omega(f)\}\quad. \tag{65}$$

Examples of Gibbs priors are given by the following functionals:

$$\begin{aligned}
i)\quad & \Omega(f)\;=\;\|f\|_2^2\;(white\;noise\;prior)\;; && (66)\\
ii)\quad & \Omega(f)\;=\;\|\Delta f\|_2^2\;(smoothing\;prior)\;; && (67)\\
iii)\quad & \Omega(f)\;=\;\|f\|_1\;\;(impulse\;noise\;prior)\;; && (68)\\
iv)\quad & \Omega(f)\;=\;\||\nabla f|\|\;\;(total\;variation\;prior)\;. && (69)
\end{aligned}$$

Moreover, in the Gaussian case, the functional of Eq. (65) becomes:

$$J_{g,\mu}(f)\;=\|Af\;-\;g_s\|_2^2\;+\;\mu\Omega(f)\quad, \tag{70}$$

and we obtain the basic functional of the classical regularization theory of inverse problems. Therefore all functional analytic methods developed for this theory (see, for instance, [15]) apply also to the investigation of the MAP solutions in the Gaussian case.

In the particular case of the prior $i)$ we get the classical *Tikhonov regularization method*, which, in our case, consists in minimizing, for each value of the regularization parameter μ, the following functional:

$$\begin{aligned}
J_{g,\mu}(f)\;&=\|Af\;-\;g_s\|_2^2\;+\;\mu\|f\|_2^2\;\;= && (71)\\
=\;\frac{1}{|N|}\sum_{\mathbf{k}}&\left\{|H(\mathbf{k})F(\mathbf{k})-G_s(\mathbf{k})|^2\;+\;\mu\;|F(\mathbf{k})|^2\right\}\quad.
\end{aligned}$$

The unique object minimizing this functional is called *regularized solution*, it is denoted by $f^{(\mu)}$ and is also the unique solution of the Euler equation:

$$(A^TA\;+\;\mu I)f^{(\mu)}\;=\;A^Tg_s\quad, \tag{72}$$

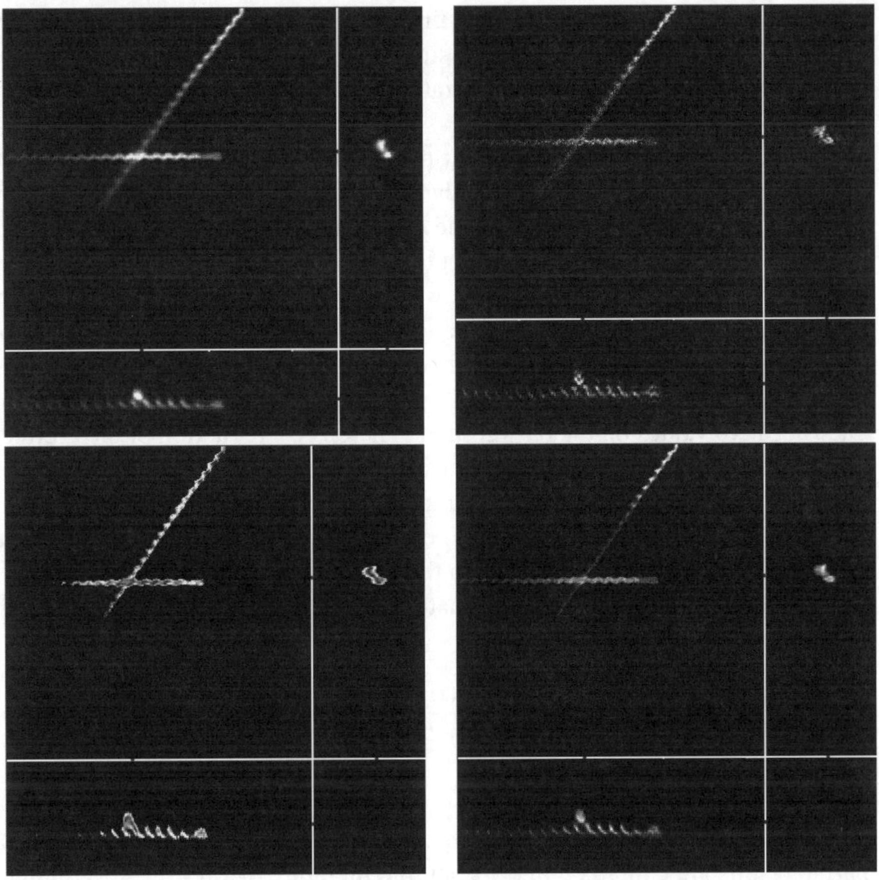

Figure 3. Comparison of the results provided by three different deconvolution methods in the case of the image of Figure 1 - Top-left panel: the original image - Top right panel: the reconstruction provided by the Tikhonov method with $\mu = 10^{-3}$ - Down-left panel: the reconstruction provided by the projected Landweber method with 50 iterations and relaxation parameter $\tau = 1.8$ - Down-right panel: the reconstruction provided by the RL (EM) method with 22 iterations.

which must be compared with Eq. (34). The DFT of the regularized solution is given by:

$$F^{(\mu)}(\mathbf{k}) = \frac{H^*(\mathbf{k})}{|H(\mathbf{k})|^2 + \mu} G_s(\mathbf{k}) \ . \tag{73}$$

For $\mu = 0$ we re-obtain the generalized solution while we obtain 0 in the limit $\mu \to \infty$. The problem is to estimate the optimal value of μ, a problem similar to that of the optimal number of iterations in the iterative methods described above. If we define a relative restoration error as in Eq. (47), then it can be proved that, as a function of μ, $\rho^{(\mu)}$ first decreases, goes through a minimum and then increases, with a behaviour similar to that shown in Figure 2. Several criteria for the choice of the regularization parameter have been proposed (see again [1] or [15]).

In Figure 3 we compare the result provided by the Tikhonov method with those provided by the projected Landweber and RL methods in the case of the image of Figure 1. The values of the parameters used in the reconstructions are given in the caption. We point out that Tikhonov reconstruction is affected by artifacts generated by the well-known Gibbs effect, related to the truncation of Fourier series in the case of functions with sharp edges. In fact, as follows from Eq. (73), Tikhonov regularization is equivalent to a linear low-pass filter. The Gibbs artifacts appear as "oscillations" around the edges and introduce negative values on the background. Therefore a reduction of these effects should be provided by the constraint of non-negativity. The reconstructions obtained with the iterative methods implementing this constraint confirms this statement. The best reconstruction is obtained with the RL method which, in this case, is a method consistent with the type of noise corrupting the image.

In the Poisson case the MAP estimates are obtained by minimizing the functional:

$$
\begin{aligned}
J_{g,\mu}(f) &= \sum_{\mathbf{n}} \left\{ g(\mathbf{n}) ln \frac{g(\mathbf{n})}{(Af+b)(\mathbf{n})} + [(Af+b)(\mathbf{n}) - g(\mathbf{n})] \right\} + \quad (74) \\
&+ \ \mu \, \Omega(f) \ .
\end{aligned}
$$

A complete theory has not yet been developed in this case, even if several partial results are contained in the scientific literature. Moreover, several iterative methods have been proposed for the minimization of these functionals. A unified approach is proposed in [13].

5. Concluding Remarks

As outlined in this Chapter, image deconvolution is a difficult problem. First, it requires an accurate modeling of image formation, including both a model of the PSF and a model of the noise. Second, accurate reconstruction algorithms must be designed, coherent with the model of image formation. In this Chapter we have only discussed the two approximate models we called the Gaussian case and the Poisson case. But in some circumstances (faint objects), the more accurate model of Eq. 19 should be used. Methods applying to this case have been proposed [16]; their utility in Microscopy must still be demonstrated.

Software packages containing tools for image deconvolution are available but, in general, their use is not easy; moreover, in the 3D case the computational burden and the storage requirements can be excessive. In such a situation, it may be difficult for a user to improve the images of his/her microscope by means of deconvolution methods. From this point of view, the project “Power Up Your Microscope” looks interesting. It provides a free service based on web tools (see www.powermicroscope.com): the microscopist uploads the images and the characteristics of the microscope; the latter are used by the system for the computation of the PSF which is required for the deconvolution of the images; finally, after notification, he/she can downloads the deconvolved images. In this way he/she has easy access to refined methods for image deconvolution.

Acknowledgments

We thank G. Vicidomini for kindly providing the images and the reconstructions of the figures of this Chapter.

References

[1] Bertero M., and Boccacci P., (1998) *Introduction to Inverse Problems in Imaging*, Bristol, UK: IOP Publishing

[2] Snyder D. L.,Hammoud A. M., and White R. L., (1993) Image recovery from data acquired with a charge-coupled-device camera, *J. Opt. Soc. Am.*, **A 10**, 1014-1023

[3] Avriel M., (1976) *Nonlinear Programming: Analysis and Methods*, Englewood Cliff (NY): Prentice Hall

[4] Eicke B., (1992) Iteration methods for convexly constrained ill-posed problems in Hilbert space, *Num. Funct. Anal. Optim.*, bf 13, 413-429

[5] De Pierro A. R., (1987) On the convergence of the iterative space reconstruction algorithm for volume ECT, *IEEE Trans. Med. Imaging*, **6**, 124-125

[6] Csiszar I. (1991) Why least squares and maximum entropy? An axiomatic approach to inference for linear inverse problems, *Annals of Stat.*, **19**, 2032-2066

[7] Richardson W. H. (1972) Bayesian-based iterative method of image restoration, *J. Opt. Soc. Am.*, **62**, 55-59

[8] Lucy L., (1974) An iterative technique for the rectification of observed distribution, *Astron. J.*, **79**, 745-754

[9] Shepp L. A. and Vardi Y., (1982) Maximum likelihood reconstruction for emission and transmission tomography, *IEEE Trans. Med. Imaging*, **1**, 113-122

[10] Lange K., and Carson R. (1984) EM reconstruction algorithm for emission and transmission tomography *J. Comp. Assisted Tomography*, **8**, 306-316

[11] Vardi Y., Shepp L. A., and Kaufman L. (1985) A statistical model for Positron Emission Tomography, *J. Am. Stat. Association*, **80**, 8-20

[12] Lanteri H., Roche M., Cuevas O., and Aime C. (2001) A general method to devise maximum-likelihood signal restoration multiplicative algorithms with non-negativity constraints, *Signal Processing*, **81**, 945-974

[13] Lanteri H., Roche M., and Aime C. (2002) Penalized maximum likelihood image restoration with positivity constraints: multiplicative algorithms, *Inverse Problems*, **18**, 1397-1419

[14] Kaipio J. P., and Somersalo E. (2004) *Computational and Statistical Methods for Inverse Problems*, Berlin: Springer

[15] Engl W. H., Hanke M., and Neubauer A. (1996) *Regularization of Inverse Problems*, Dordrecht: Kluwer

[16] Lanteri H., and Theys C. (2004) Restoration of Poisson images with Gaussian noise - Application th astrophysical data, *Signal Processing*, in press

IMAGE FORMATION IN FLUORESCENCE MICROSCOPY

Three-dimensional mathematical model

Giuseppe Vicidomini
INFM, LAMBS-DIFI University of Genoa,16146, Genoa, Italy.

Abstract: In this chapter we will discuss the properties of three-dimensional image formation in an optical fluorescence microscope. An image is blurred and distorted by noise during its formation and acquisition. This distortion hides fine details in the image hampering both the visual and the quantitative analysis of the image. In order to appreciate the fundamental limitations of the resolution and image formation properties of an optical fluorescence microscope, it is necessary to begin by discussing the foundations of diffraction theory. Using this theory, the wave description and the quantum nature of light it is possible to derive a model for the image formation in a general epi-fluorescence microscope and discuss the conditions under which a fluorescence microscope can be modeled as a linear translation invariant system. This means that the optical system is completely described by its Point Spread Function (PSF). The image formations in a Confocal Laser Scanning Microscope (CLSM), in a Wide-Field Microscope and in a Two-Photon Excitation (TPE) Microscope are derived as limited case of the general epi-fluorescence microscope.

Keywords: Point Spread Function, Resolution, Fluorescence, Diffraction, Optical-Sectioning

1. INTRODUCTION

Within the frame of 3D methods for the representation of an object from a set of 2D images, there is one technique that is becoming rather popular because it allows analyzing biological samples *in vivo* without damaging either their structure or their functionality, and, hence, turns out to be of particular interest for the study of cells and cellular organs. This method, know as *optical sectioning*, is based on the acquisition of data at different planes of focus by means of appropriate device. Its modern form is established in confocal and two-photon microscopy techniques [1], as well as in computational methods of dataset [2] coming from wide-field microscopy operating under defocus conditions.

V. Evangelista et al. (eds.), From Cells to Proteins: Imaging Nature across Dimensions, 371–393.

When the microscope is in focus on a specific plane of a given sample, the image being acquired is obtained from the joined contribution of the plane that is in focus as well as the other planes. The resulting output of the system encapsulates the relevant information about 3D structure of the sample as a whole, but this information is affected by noise, background and interference from the plane that were not in focus. Since the last two components depend on the acquisition modality and are of a deterministic nature, they can be mathematically modeled and "subtracted" by *deconvolution* [3]from the result, thus correcting the image and transforming it into a form that better approximates the original object.

Among all degrading factors, the geometry of the system is the major cause for the blurring effect which most affects the quality of the final output, and for this reason a good model of the acquisition device is fundamental.

2. WAVE DESCRIPTION OF LIGHT AND DIFFRACTION THEORY

2.1. Scalar theory of light

Before embarking on a treatment of diffraction, we first consider some mathematical preliminaries that form the basis of the latter diffraction-theory derivations. These initial discussions will also serve to introduce the notation to be used.

Let the *light disturbance* at position P and time t be represented by the scalar function $u(P,t)$; for the case of linearity polarized waves, we may regard this function as representing either the electric or the magnetic field strength.

For a monochromatic wave, the filed may be written explicitly as

$$u(P,t) = U(P)\cos\left[2\pi\nu t + \theta(P)\right] \tag{2.1}$$

where $U(P)$ and $\theta(P)$ are the amplitude and phase, respectively, of the wave at position P, while ν is the optical frequency. Usually Equation (2.1) is written using complex notation

$$u(P,t) = \Re\left[\mathbf{U}(P)e^{(-j2\pi\nu t)}\right] \tag{2.2}$$

where $\mathbf{U}(P)$ is the following complex function of position called *complex amplitude*

$$\mathbf{U}(P) = U(P)e^{[-j\theta(P)]} \tag{2.3}$$

and $\Re$ is shorthand notation meaning *the real part of.*

If the real disturbance $u(P,t)$ is to represent an optical wave, it must satisfy the scalar wave equation [4], it follows that the complex amplitude $\mathbf{U}$ must obey the time-independent equation

$$(\nabla^2 + k^2)\mathbf{U} = 0 \tag{2.4}$$

where k is termed the *wave number,* given by $k = 2\pi/\lambda$, and λ is the wavelength. The Equation (2.4) is known as the *Helmholtz equation.*

However no sensor exists for directly measuring the amplitude and phase of optical radiation. Light detector and sensor are sensitive to the *intensity* of the incident radiation, which is proportional to the square of the modulus of the complex amplitude

$$I(P) = |\mathbf{U}(P)|^2 \tag{2.5}$$

2.2. The Huygens-Fresnel Principle

Huygens expressed an intuitive conviction that if each point on the wavelength of a light disturbance were considered to be a new source of a "secondary" spherical disturbance, then the wave front at any later instant could be found constructing the "envelope" of the secondary wavelets. Fresnel was able to account for diffraction by supplementing Huygens' construction with the postulate that the secondary wavelets mutually interfere. This combination of Huygens' construction with the principle of interference is called the *Huygens-Fresnel Principle.*

2.3. Kirchhoff diffraction

The idea of Huygens and Fresnel were put on a firmer mathematical foundation (*Green's theorem*) by Kirchhoff. He showed that the *Huygens-Fresnel Principle* may be regarded as an approximate form of a certain integral theorem. This theorem is knows as *Kirchhoff's theorem* [4, 5], it expresses the solution of the homogeneous wave equation, at an arbitrary point P in the field, in terms of the values of the solution and its first derivatives at all points on an arbitrary closed surface S surrounding P.

$$\mathbf{U}(P)=\frac{1}{4\pi}\iint_S\left\{\mathbf{U}\frac{\partial}{\partial n}\left(\frac{e^{iks}}{s}\right)-\frac{e^{iks}}{s}\frac{\partial \mathbf{U}}{\partial n}\right\}dS\,. \tag{2.6}$$

Consider now the problem of diffraction. As illustrate in Figure 1, a monochromatic single spherical wave,

$$\mathbf{U}(r)=\frac{Ae^{(ikr)}}{r} \tag{2.7}$$

where A is the amplitude at unit distance from the source P_0, propagates through an opening in an opaque screen, and let be the point at which the light disturbance is to be determined. We assume that the linear dimensions of the opening although large compared to the wavelength, are small compared to the distance of both P_0 and P from the screen.

To find disturbance at P we use *Kirchhoff's theorem* over the closed surface S formed by: the opening $\mathcal{A}$, a portion $\mathcal{B}$ of the non-illuminated side of the screen, and a portion C of a large sphere centered at P.

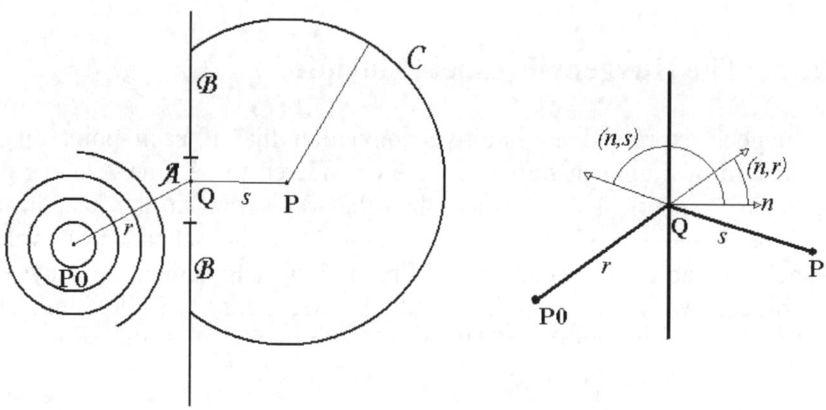

Figure 1. Diffraction of a monochromatic spherical wave

The difficulty is to find the value of $\mathbf{U}$ and $\partial\mathbf{U}/n$ on $\mathcal{A}$, $\mathcal{B}$ and C which should be substituted in(2.6). Kirchhoff formulated two assumptions, called *Kirchhoff's boundary condition*:

1. across the surface , the $\mathcal{A}$, the filed distribution $\mathbf{U}$ and its derivative $\partial\mathbf{U}/n$ are exactly the same as they would be in the absence of the screen;
2. over the portion $\mathcal{B}$ the field distribution $\mathbf{U}$ and its derivative $\partial\mathbf{U}/n$ are identically zero.

Under these conditions and noting that the distance r and s are usually many optical wavelengths we obtain:

$$\mathbf{U}(P)=-\frac{iA}{2\lambda}\iint_{\mathcal{A}}\frac{e^{ik(r+s)}}{rs}\left[\cos(n,r)-\cos(n,s)\right]dS \tag{2.8}$$

which is known as the *Fresnel-Kirchhoff diffraction formula* [4, 5].

2.4. The Rayleigh-Sommerfeld formulation of diffraction by a plane screen

The Kirchhoff theory has been found experimentally to yield remarkably accurate result and is widely used in practice. However Kirchhoff based his function formulation upon two assumptions, which were later proved inconsistent by Sommerfeld. The difficulties of the Kirchhoff theory come from the fact that boundary conditions must be imposed on both the field strength and its derivatives. Sommerfeld eliminated this inconsistence choosing an alternative function to use in the *Green's theorem.*

He obtained the so called *Rayleigh-Sommerfeld diffraction formula* [5].

$$\mathbf{U}(P)=-\frac{iA}{\lambda}\iint_{\mathcal{A}}\frac{e^{ik(r+s)}}{rs}\left[-\cos(n,s)\right]dS\,. \tag{2.9}$$

As a consequence of this criticism, Kirchhoff's formulation of the so-called *Huygens-Fresnel principle* must be regarded as a first approximation, although under most conditions it yields results that agree amazingly well with experiment, thus in the next paragraph we will use Kirchhoff's formulation.

2.5. The 3D light distribution near focus of a well-corrected lens

Consider a spherical monochromatic wave emerging from a circular aperture with radius a and converging towards the axial focal point O.

We shall consider the disturbance $\mathbf{U}(P)$ at a typical point P in the neighborhood of O.

The point P will be specified by a position vector $\mathbf{R}$ relative to O, and it will be assumed that the distance $R = OP$ as well the radius a a (□ λ) are small compared to $f = CO$ of the wave-front that momentarily fills the aperture.

Compare Figure 1 and Figure 2, we can see that $\mathcal{W}$ is different from $\mathcal{A}$. $\mathcal{W}$ is the wave front that momentarily fills the aperture; it is evident that in place of $\mathcal{A}$ any other surface could be used in Equation(2.8). If the radius of curvature of the wave is sufficiently large the contribution from $\mathcal{D}$ (see Figure 2) may be neglected.

We note also that r is replaced by f, the distance of the focal plane, and that the wave-fronts incident upon the aperture are concave to the direction of propagation, i.e. O is a centre of convergence and not of divergence of the incident wave as P_0 in Fig. 1. Also, on $\mathcal{W}$, $\cos(n, f) = 1$.Using this assumptions Equation (2.8) become

$$\mathbf{U}(P) = -\frac{Ai}{2\lambda}\frac{e^{-ikf}}{f}\iint_{w}\frac{e^{iks}}{s}\left[1-\cos(n,s)\right]ds \tag{2.10}$$

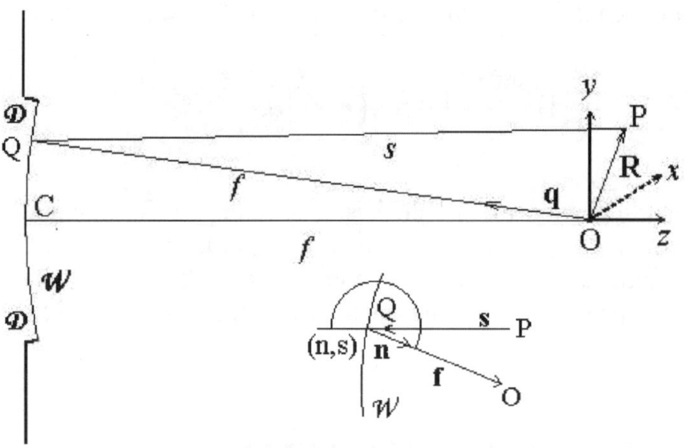

Figure 2. Diffraction of a well-corrected lens

As we assumed that f is much greater than a and R we have, to a good approximation

$$\cos(n,s) \cong -1 \tag{2.11}$$

hence

$$\mathbf{U}(P) = -\frac{Ai}{\lambda}\frac{e^{-ikf}}{f}\iint_{\mathcal{W}} \frac{e^{iks}}{s} dS . \tag{2.12}$$

Now we will reduce Equation (2.12) to more simple mathematical manipulations. If $\mathbf{q}$ denotes the unit vector in the direction OQ, we have

$$s - f = -\mathbf{q} \square \mathbf{R} \tag{2.13}$$

Also, the element dS of the wave-front is given by

$$dS = f^2 d\Omega , \tag{2.14}$$

where $d\Omega$ is the element of the solid angle that dS subtends at O. Moreover, we may replace s by f in the denominator of the integrand without introducing an appreciable error. Equation (2.12) becomes

$$\mathbf{U}(P) = -\frac{Ai}{\lambda}\iint_{\Omega} e^{-ik\mathbf{q} \square \mathbf{R}} d\Omega \tag{2.15}$$

the integration now extending over the solid angle Ω which the aperture subtends at the focus. Let (x, y, z) be the coordinates of P and (ξ, η, ζ) those of Q. We set

$$\begin{aligned} \xi &= a\rho \sin\theta \quad & x &= a\rho \sin\psi \\ \eta &= a\rho \cos\theta \quad & y &= r \cos\psi \end{aligned} \tag{2.16}$$

where $r = \sqrt{x^2 + y^2}$. Since Q lies on the spherical wave-front $\mathcal{W}$,

$$\zeta = -\sqrt{f^2 - a^2\rho^2} = -f\left[1 - \frac{1}{2}\frac{a^2\rho^2}{f^2} + \ldots\right].$$

Then

$$\mathbf{q}\square\mathbf{R} = \frac{x\xi + y\eta + z\zeta}{f} = \frac{a\rho r\cos(\theta - \psi)}{f} - z\left[1 - \frac{1}{2}\frac{a^2\rho^2}{f^2} + \ldots\right]. \quad (2.17)$$

It is useful at this stage to introduce dimensionless variables u and v, which together with ψ specify the position of P:

$$u = k\left(\frac{a}{f}\right)^2 z, \quad v = k\left(\frac{a}{f}\right) r \quad (2.18)$$

From (2.17) and (2.18) it follow that if the terms above the second power of $a\rho / f$ are neglected in comparison to unity,

$$k\mathbf{q}\square\mathbf{R} = v\rho\cos(\theta - \psi) - \left(\frac{f}{a}\right)^2 u + \frac{1}{2}u\rho^2. \quad (2.19)$$

Further, the element of the solid angle is

$$d\Omega = \frac{dS}{f^2} = \frac{a^2\rho d\rho d\theta}{f^2} \quad (2.20)$$

Hence (2.15) becomes

$$\mathbf{U}(P) = -\frac{Ai}{\lambda}\frac{a^2}{f^2} e^{i\left(\frac{f}{a}\right)^2 u} \int_1^0 \int_0^{2\pi} e^{-i\left[v\rho\cos(\theta-\psi) + \frac{1}{2}u\rho^2\right]} \rho d\rho d\theta \quad (2.21)$$

Now we have the well-know integral representation of *Bessel* functions $J_n(z)$:

$$\frac{i^{-n}}{2\pi}\int_0^{2\pi} e^{ix\cos\alpha} e^{in\alpha} d\alpha = J_n(x) \quad (2.22)$$

Hence (2.21) becomes

$$\mathbf{U}(P) = -\frac{2\pi i a^2 A}{\lambda f^2} e^{i\left(\frac{f}{a}\right)^2 u} \int_0^1 J_0(v\rho) e^{-\frac{1}{2}iu\rho^2} \rho d\rho \quad (2.23)$$

For a calculation of this integral and in particular for its intensity see [4]. At the end using the definition of *numerical aperture* (NA)

$$NA = n\sin\theta \tag{2.24}$$

(where n is the refractive index of the medium in which the lens is working and θ is one half of the angular aperture of the lens) and using the *paraxial approximation*

$$a/f = \tan\theta \cong \sin\theta \tag{2.25}$$

we obtain

$$\mathbf{U}(u,v) = -i\frac{2\pi NA^2 A}{\lambda}e^{iu/\sin^2\theta}\int_0^1 J_0(v\rho)e^{-\frac{1}{2}iu\rho^2}\rho d\rho$$
$$v = \frac{2\pi NAr}{\lambda} \quad u = \frac{2\pi NA^2 z}{n\lambda} \tag{2.26}$$

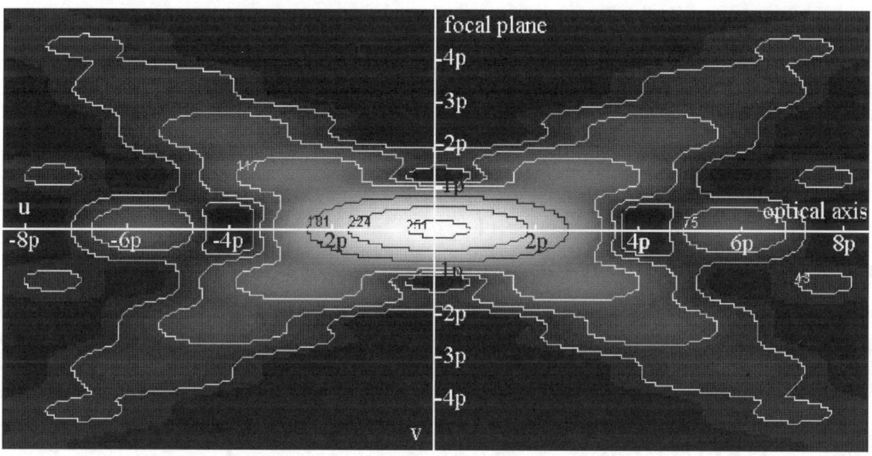

Figure 3. Intensity in a meridional plane near focus of a converging spherical wave diffracted at a circular aperture. Contour lines are calculated for intensity of 0.9, 0.5, 0.2, 0.05, 0.02 and 0.01. The image is displayed using a logarithmic contrast stretch.

3. IMAGE FORMATION AND LTI SYSTEM

The *resolution* of an optical microscope is defined as the shortest distance between two points on a specimen that can still be "sufficiently" distinguished by the observer or camera system as separate entities. The way the resolution of the microscope influences the image can be modeled by the *Point Spread Function* (PSF) of the microscope, which determines how a single point in the sample is being imaged.

3.1. Image formation

Representing the sample as a collection of points, and supposed the microscope as a *Linear Translation Invariant System* (LTI System), the image of the sample can be composed by replacing each point in the sample with the PSF weighted by the intensity of each point. This operation is defined as *convolution product* (denoting by $\otimes$) between the sample and the PSF of the microscope

$$\begin{aligned} g_0(x,y,z) &= h(x,y,z) \otimes_{3D} f_0(x,y,z) = \\ &= \iiint h(x-\xi, y-\eta, z-\zeta) f_0(x,y,z)\, d\xi\, d\eta\, d\zeta \end{aligned} \tag{3.1}$$

Besides *blurring* by the PSF, two other factors influence the image formation of a fluorescence microscope: *background* and *noise.* The Background consists of the amount of intensity that arrives to then sensor due to the auto-fluorescence of the sample. The Noise in a fluorescence microscope is due to various sources: *photon statistics* and *read-out* are the crucial ones, but we have also *quantization* and *thermal* noise [6].

$$g(x,y,z) = g_0(x,y,z) + b(x,y,z) + w(x,y,z) \tag{3.2}$$

The term w is just the differenced between the detected image and the ideal image summed to the background; it doesn't want to indicate an additive noise.

3.2. LTI System

To use linear system theory we have to show that an optical fluorescence microscope behaves (to a good approximation) as a *Linear Translation Invariant System.*

An optical system is *translation-invariant* (or *isoplanatic*) if the image of a point source changes only in position, not in form, when the point source is translated over the object field. To give the mathematical formulation of this sentence we have to introduce the *Dirac function* that describe the unit impulse at position $(0,0,0)$

$$\delta(x,y,z)=\delta(x)\delta(y)\delta(z) \quad \delta(x)=0 \Leftrightarrow \forall x\neq 0$$
$$\int_{-\infty}^{+\infty}\delta(x)dx=1 \tag{3.3}$$

This function has the following important property:

$$\int_{-\infty}^{+\infty}\delta(x)f(x)dx=f(0) \tag{3.4}$$

If $h_{\xi,\eta,\zeta}(x,y,z)$ denote the response of $\delta(x-\xi,y-\eta,z-\zeta)$, that is the unit impulse translated in position (ξ,η,ζ), the condition becomes

$$h_{\xi,\eta,\zeta}(x,y,z)=h_{0,0,0}(x-\xi,y-\eta,z-\zeta) \tag{3.5}$$

In practice, optical imaging systems are seldom isoplanatic over the whole object field, but it is usually possible to divide the object field into regions within which the system is approximately translation-invariant. For example in microscopy the PSF is axial shift variance: it changes in function of the position along the optical axis. This dependences arises from the fact that a lens is designed to work without aberration in some particulars conditions, but in the study of living biological cells and tissue, large deviation from the designed condition are introduced [7]. Finally we assume that the magnification factor of the microscope has been reduced to one by rescaling of the space variables of the image plane.

A system is said to be *linear* if weighted combination of inputs yields a weighted combination of outputs. In fluorescence microscopy, where the input is the intensity, this means that the distribution of the intensity of two light point sources must be equal to the sum of the distributions of the intensity of each point source

$$|aU(P_1)+bU(P_2)|^2=|a|^2 I(P_1)+|b|^2 I(P_2) \tag{3.6}$$

this requirements is true only in the case of *incoherent light* (the phases of all points are varying in statistically independent fashion), thus the PSF of a system in the incoherent case is (see Equation(2.5))

$$h_{incoherent}(P)=I(P)=|U(P)|^2 \tag{3.7}$$

In the case of *coherent light* (the relative phase of two object point is constant in time, i.e. even if the two phases can vary randomly in time, they vary in identical fashion) the system is linear in the complex amplitude, not respect to the light intensity, thus we have the following relation [8]

$$h_{incoherent}(P) = \left|h_{coherent}(P)\right|^2 \tag{3.8}$$

Since no fixed phase relationship exists among the fluorescence from molecules that compose the object, the fluorescence light is incoherent, and our microscope system is linear.

4. FLUORESCENCE MICROCOPY

In the first paragraph we found the *diffraction PSF* of a well-corrected lens, now by using it and the schematic setup of the different kind of fluorescence microscopes we will determine the PSFs of Confocal, Wide-Field and Two Photon Excitation microscope.

4.1. General epi-fluorescence microscope

In fluorescence microscopy the fluorescence molecules in the sample are excited by incident light of wavelength λ_{ex}. After near 10 ns (10 x 10^{-9}s) they return to the ground state, during this period they emit light of higher wavelength λ_{em} [9].

In this work we will assume that both the excitation and emission light are monochromatic. This is not true because most fluorescence molecules have a broad emission and excitation spectra, but placing narrow band pass filters both in the illumination and excitation light path we can fulfill this assumption.

In this condition of monochromatic light we can use Equation (2.26) and Equation (3.7) to find the diffraction PSF of the objective. Note also that in a epi-fluorescence microscopy the illumination and detection light path share the same objective to reduce the amount of illumination light penetrating the detection light path, thus

$$h_{ex}(x) = h(x, \lambda_{ex}) \qquad h_{em}(x) = h(x, \lambda_{em}) \tag{4.1}$$

Now using the "flow of data" of the Figure 4 we will derive the PSF of a general epi-fluorescence microscope. In Equation (4.1) and in all the following derivations we will use a single lateral coordinate, but in the next section we will reformulate the most important results for the 3D case.

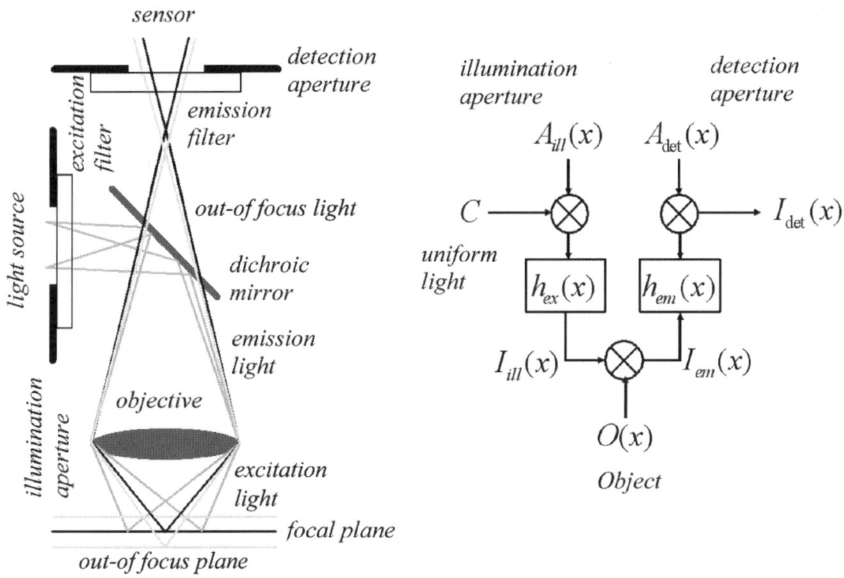

Figure 4. Model and system diagram of a general epi-fluorescence microscope

Illumination intensity is the amount of light projected onto the sample; it determines the probability that an excitation photon "hits" a fluorescence molecule at a certain point in the object. As the light source can be modeled as a collection of point sources, with constant intensity C, which passes through the aperture, the illumination intensity can be written as the convolution between the illumination aperture and the diffraction PSF of the objective at the excitation wavelength

$$I_{ill}(x) = h_{ex}(x) \otimes CA_{ill}(x) \tag{4.2}$$

For a low concentration of fluorophores and for low illumination intensity I_{ill}, the emitted light I_{em} is proportional to the illumination intensity, to the fluorophore concentration on the sample, represented by the function $O(x)$, and to the quantum efficiency Q of the fluorophore:

$$I_{em}(x) = O(x) I_{ill}(x) Q \tag{4.3}$$

With this relation we don't consider the different factors that can influence fluorescence: *photobleaching*, *quenching*, *FRET*, *self-quenching* and *solvent effect* [9]. But in this work we can neglect them contributes without introducing appreciable errors.

The emitted light is imaged by the same objective onto a photo-sensitive sensor and a detection aperture limits the emission light with respect the dimension of the sensor. Detection intensity quantifies the probability that an excitation photon emitted from a point is able to propagate to the detector.

$$I_{\det}(x) = A_{\det}(x)\left[h_{em}(x) \otimes I_{em}(x)\right] \tag{4.4}$$

Replacing Equation (4.3) in Equation (4.4) and $O(x)$ with the *Dirac Function* $\delta(x)$, and using property (3.4) we obtain

$$\begin{aligned} h_{GFM}(x) &= A_{\det}(x)\left[\int_{-\infty}^{+\infty} h_{em}(x-\xi) I_{ill}(\xi)\delta(\xi)\,d\xi\right] = \\ &= A_{\det}(x)h_{em}(x)I_{ill}(0) = A_{\det}(x)h_{em}(x)(A_{ill} \otimes h_{ex})(0) = \\ &= A_{\det}(x)h_{em}(x)\int_{-\infty}^{+\infty} A_{ill}(\xi)h_{ex}(-\xi)\,d\xi \end{aligned} \tag{4.5}$$

which describe the PSF of a general epi-fluorescence microscope and where we eliminated the constant terms.

In this model we assumed that the illumination and the detection are limited by finite sized aperture, to model them we use a block function representing the size and the shape of the aperture

$$\begin{aligned} A_{ill}(x) &= \begin{cases} 1 & x \text{ inside the aperture} \\ 0 & x \text{ outside the aperture} \end{cases} \\ A_{\det}(x) &= \begin{cases} 1 & x \text{ inside the aperture} \\ 0 & x \text{ outside the aperture} \end{cases} \end{aligned} \tag{4.6}$$

In wide-field microscope the illumination aperture is determined by the size of the diaphragm, in a confocal microscope by the illumination pinhole. The detection aperture depends on the dimensions of the CCD in the case of wide-filed microscope and by the dimension of the detection pinhole in the case of confocal microscope. Table 1 show how changing the functions that represent these apertures we can obtain the different kind of microscopes. In the next sections we will analyze separately these situations.

Table 1. Different kinds of microscopes as limiting cases of the general epi-fluorescence microscope

Microscope	*Illumination aperture*	*Detection aperture*
Wide-Field	$\forall x \; A_{ill}(x)=1$	$\forall x \; A_{\text{det}}(x)=1$
Ideal CLS	$\delta(x)$	$\delta(x)$
Two-Photon & One-Photon Scanning	$\delta(x)$	$\forall x \; A_{\text{det}}(x)=1$
Confocal TPE	$\delta(x)$	$\delta(x)$

4.2. Wide-Field microscope

In a wide-field microscope aligning and focusing the arc lamp (mercury or xenon) it is possible to obtain a homogeneous incoherent light on the entire field-of-view of the sample (Kohler illumination). In this situation focused and out of focus emission light are recorded by a CCD.

Using our model of a general epi-fluorescence microscope, we enlarge the illumination aperture to an infinite size (see Table 1).

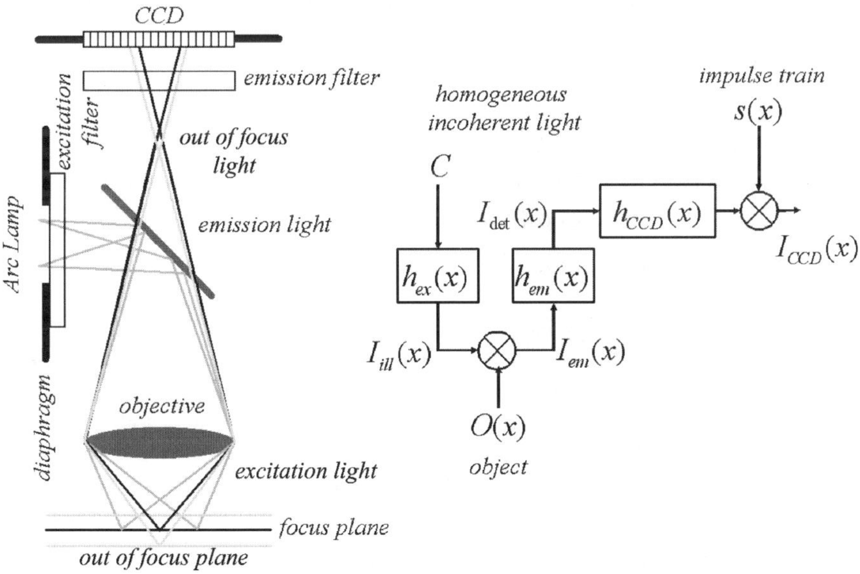

Figure 5. Model and system diagram of a wide-field microscope

From Equation (4.2) and using the definition of convolution the illumination intensity becomes a constant

$$I_{ill}(x) = \int_{-\infty}^{+\infty} A(x-\xi)h_{ex}(\xi)d\xi = \int_{-\infty}^{+\infty} h_{ex}(\xi)d\xi = C_{ill} \quad (4.7)$$

In the same wave we enlarge the detection aperture (see again Table 1). From Equation (4.4) and (4.3)

$$I_{\det}(x) = A_{\det}(x)\left[h_{em}(x)\otimes\left(O(x)C_{ill}\right)\right] = h_{em}(x)\otimes\left(O(x)C_{ill}\right) \quad (4.8)$$

The image produced by the objective is then sampled by the pixels of the CCD. This can be described by first a convolution with a block function $h_{CCD}(x,y)$, representing the size and the shape of the pixels, following by a multiplication with a square grid of unit impulse functions $s(x,y)$, representing the position of the pixel. In case of square pixel we have:

$$h_{CCD}(x,y) = \begin{cases} 1 & \text{if } |x| < p_{size}/M \text{ and } |y| < p_{size}/M \\ 0 & \text{else where} \end{cases} \quad (4.9)$$

$$s(x,y) = \sum_{n,m} \delta\left(x - mp_{spacing}/M\right)\delta\left(y - np_{spacing}/M\right) \quad n,m \in N$$
(4.10)

where p_{size} is the size of the pixel $p_{spacing}$ is the distance between the centers of two pixels and M is *the magnification factor* of the microscope. Note that p_{size} and $p_{spacing}$ can be of different size [6].
From Equation (4.8) we obtain

$$\begin{aligned} I_{CCD}(x) &= \left(I_{\det}(x)\otimes h_{CCD}(x)\right)s(x) \\ &= \left[\left(h_{em}(x)\otimes O(x)\right)\otimes h_{CCD}(x)\right]s(x) \end{aligned} \quad (4.11)$$

therefore the discrete PSF of a wide-field microscope is

$$h_{wide}[n] = h_{em}\left(np_{spacing}/M\right)\otimes h_{ccd}\left(np_{spacing}/M\right) \quad (4.12)$$

Since the distance p_{size} is 6-23 micron and a typical magnification M is 40-100 times, usually we can neglect the CCD's contribution [6].

4.3. Confocal Laser Scanning microscope

In a confocal fluorescence microscope thanks to the *illumination pinhole* the sample is illuminated by a point source. In Confocal Laser Scanning microscope (CLSM) this is implemented by using a focused laser beam and a beam expander designed to adjust the diameter of the laser beam to the entrance pupil of the objective. When an objective is illuminated by a laser beam whose diameter is too small, the optical resolution is reduced. On the other hand, overfilling the objective can lead to a loss of intensity. In wide-field microscope focused and out-of-focus light are detected. The confocal principle suppresses all structures that fall outside the focal plane of the microscope objective. This is achieved by means of an aperture (*detection pinhole*), which is placed in the beam path in an optical conjugated plane to the point light source. At the end the emission light is collected with a light detector (typically a photomultiplier tube (PMT)). To record an image in the focal plane, the sample is scanned with the point laser bean in the x- and y-direction. Spatial image is possible by moving the sample along the optical axis (z-direction). In confocal laser microscope the light that passes through the detection pinhole is collected by a single detector.

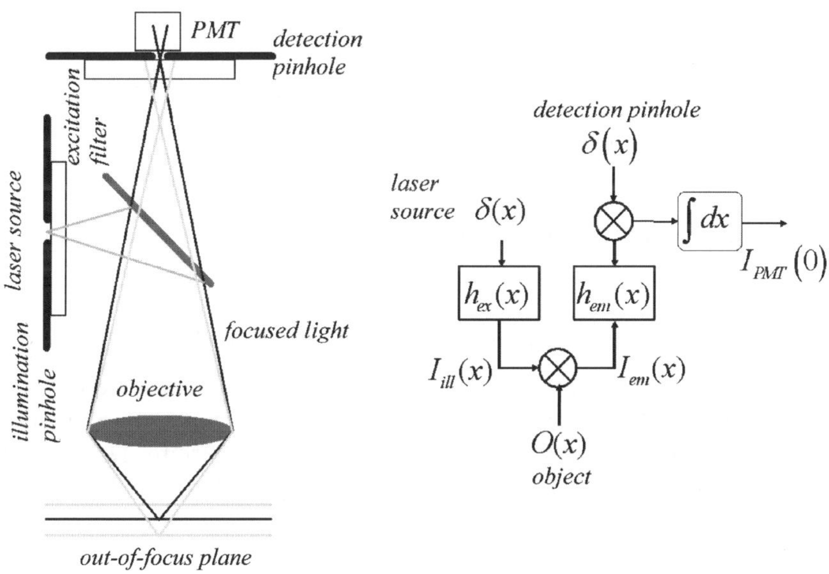

Figure 6. Model and system diagram of a confocal laser microscope

We use the Dirac function to model both the illumination point source and the detection pinhole, as we suppose that they are infinitely small

$$I_{PMT}(0) = \int_{+\infty}^{-\infty} A_{\det}(-x)\left[h_{em}(x)\otimes\left[\left[h_{ex}(x)\otimes A_{ill}(x)\right]O(x)\right]\right]dx =$$

$$= \int_{+\infty}^{-\infty} \delta(-x)\left[h_{em}(x)\otimes\left[\left[h_{ex}(x)\otimes \delta(x)\right]O(x)\right]\right]dx =$$

$$= \int_{+\infty}^{-\infty} \delta(-x)\left[h_{em}(x)\otimes\left[h_{ex}(x)O(x)\right]\right]dx = \tag{4.13}$$

$$= \left(h_{em}\otimes\left(h_{ex}\bullet O\right)\right)(0) = \int_{-\infty}^{+\infty} h_{em}(-\xi)h_{ex}(\xi)O(\xi)d\xi$$

As we already said to obtain a 3D image we have to scan the object in three dimensions. The intensity collected by the PMT for each point of the object can be derived from by shifting both the illumination and detection over the object

$$I_{PMT}(x) = \int_{+\infty}^{-\infty} \delta(x-x')\left[h_{em}(x')\otimes\left[\left[h_{ex}(x')\otimes \delta(x'-x)\right]O(x')\right]\right]dx' =$$

$$= \int_{+\infty}^{-\infty} \delta(x-x')\left[h_{em}(x')\otimes\left[h_{ex}(x'-x)O(x)\right]\right]dx' =$$

$$= \int_{+\infty}^{-\infty} \delta(x-x')\left[\int_{-\infty}^{+\infty} h_{em}(x'-\xi)h_{ex}(\xi-x)O(\xi)d\xi\right]dx' = \tag{4.14}$$

$$= \int_{+\infty}^{-\infty} h_{em}(x-\xi)h_{ex}(\xi-x)O(\xi)d\xi = \left[h_{em}(x)h_{ex}(-x)\right]\otimes O(x)$$

In practice the detection pinhole is not infinitely small, but its size (r_{phys}) is a compromise between resolution and the Signal to Noise Ratio of the detected signal [10]. Therefore we can't use the *Dirac function* to model the detection aperture, but we have to use a finite size aperture. Note that the size of the block function that models the pinhole detection is not the physics size of the pinhole but the size of its projection on the object plane (see Table 2)

$$
\begin{aligned}
I_{PMT}(x) &= \int_{+\infty}^{-\infty} A_{\det}(x-x')\left[h_{em}(x')\otimes\left[\left[h_{ex}(x')\otimes\delta(x'-x)\right]O(x')\right]\right]dx' = \\
&= \int_{+\infty}^{-\infty} A_{\det}(x-x')\left[h_{em}(x')\otimes\left[h_{ex}(x'-x)O(x)\right]\right]dx' \\
&= \int_{+\infty}^{-\infty} A_{\det}(x-x')\left[\int_{-\infty}^{+\infty} h_{em}(x'-\xi)h_{ex}(\xi-x)O(\xi)d\xi\right]dx' \\
&= \int_{-\infty}^{+\infty}\left[\int_{-\infty}^{+\infty} A_{\det}(x-x')h_{em}(x'-\xi)dx'\right]h_{ex}(\xi-x)O(\xi)d\xi \\
&= \int_{-\infty}^{+\infty}(A_{\det}\otimes h_{em})(x-\xi)h_{ex}(\xi-x)O(\xi)d\xi = \\
&= \left[\left[A_{\det}(x)\otimes h_{em}(x)\right]h_{ex}(-x)\right]\otimes O(x)
\end{aligned}
\tag{4.15}
$$

Table 2. Detection pinhole with round shape for a real CLSM microscope

Microscope	*Illumination aperture*	*Detection aperture*
Real CLS	$\delta(x)\delta(y)$	$A_{\det}(x,y)=\begin{cases}1 & \sqrt{x^2+y^2}\leq\frac{r_{phys}}{M}\\ 0 & \text{elsewhere}\end{cases}$

4.4. One-Photon and Two-Photon excitation microscope

We start from Equation (4.15) and we enlarge to infinite size the detection pinhole (see again Table 1), therefore the convolution between detection pinhole and diffraction emission PSF become a constant.

$$
\begin{aligned}
I_{PMT}(x) &= \left[\left[A_{\det}(x)\otimes h_{em}(x)\right]h_{ex}(-x)\right]\otimes O(x) = \\
&= Ch_{ex}(-x)\otimes O(x)
\end{aligned}
\tag{4.16}
$$

In practice single-photon scanning microscope is rarely used. But the same arrangement is used with multi-photon excitation. As we already say we can see the diffraction excitation PSF as the probability that a photon "hits" a fluorescence molecule in a certain position. In two-photon microscope the fluorescence emission is proportional to the probability that two low-energy photons "hit" simultaneously a fluorescence molecule. Since two independent events have to occur, this probability is proportional to the square of the excitation intensity [1].

$$I_{PMT}(x) = (h_{ex}(-x))^2 \otimes O(x) \tag{4.17}$$

Adding a pinhole in the detection light path of the two-photon microscope we obtain a two photon confocal microscope. Its model can be found by combining the model of confocal (Equation (4.14)) and two-photon microscope (Equation (4.17)):

$$I_{PMT}(x) = (h_{em}(x)(h_{ex}(-x))^2) \otimes O(x) \tag{4.18}$$

5. PSF AND RESOLUTION

In the previous section we have used a single lateral coordinate, in Table 3 we reformulate the results for the 3D case and replacing the concentration function $O(x, y, z)$ whit the unit impulse function $\delta(x, y, z)$ we found the PSFs for the different microscopes that we have analyzed.

We already introduce the definition of resolution (Paragraph 2), and we said that it is related to the PSF of the system. Now that we know how to evaluate the PSF for a fluorescence microscope, we can resolve the ambiguity as to what constitutes "sufficiently" in the resolution definition.

There are different criterions to measure the resolution of a microscope; we will use the *Full-Width-Half-Maximum* (FWHM) of the central bright region of the PSF. Another criterion that is usually utilized is the well-know Rayleigh criterion (for a comparison of them see ([1]).

To summarize, the components of the calculated intensity PSFs in the lateral ($z = 0$) and axial ($r = 0$) directions are plotted in Figure 8 for various microscope arrangements. Taking the FWHM as criterion for resolution, we can conclude (see Table 4) that the confocal microscope provides improved resolution, both axially and laterally, over the wide-field and two-photon excitation microscope.

Table 3. PSF of the different kinds of microscopes

Microscope	*Illumination aperture*
Ideal Wide-Field	$h_{em}\left(x,y,z\right)$
Real Wide-Field	$h_{CCD}\left(x,y\right)\otimes_{2D} h_{em}\left(x,y,z\right)$
Ideal CLS	$h_{ex}\left(-x,-y,-z\right)h_{em}\left(x,y,z\right)$
Real CLS	$\left(A_{Pinhole}(x,y)\otimes_{2D} h_{em}(x,y,z)\right)h_{ex}(-x,-y,-z)$
One-Photon Scanning	$h_{ex}(-x,-y,-z)$
Two-Photons Scanning	$h_{ex}\left(-x,-y,-z\right)^{2}$
Ideal Confocal-TPE	$(h_{ex}(-x,-y,-z))^{2}h_{em}(x,y,z)$
Real Confocal-TPE	$\left(A_{Pinhole}(x,y)\otimes_{2D} h_{em}(x,y,z)\right)h_{ex}(-x,-y,-z)^{2}$

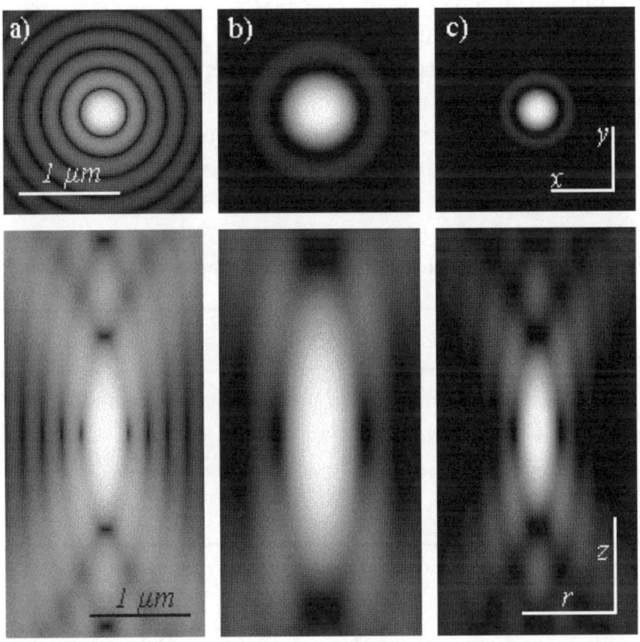

Figure 7. Computed PSFs for fluorescence microscope arrangements. *(a)* Wide-Field *(b)* Two-Photon *(c)* Confocal. All PSFs are computed using NA=1.3, n=1.518, single-photon excitation wavelength 488nm, two-photon excitation wavelength 900nm, emission wavelength 520nm. The images are displayed using a logarithmic contrast stretch.

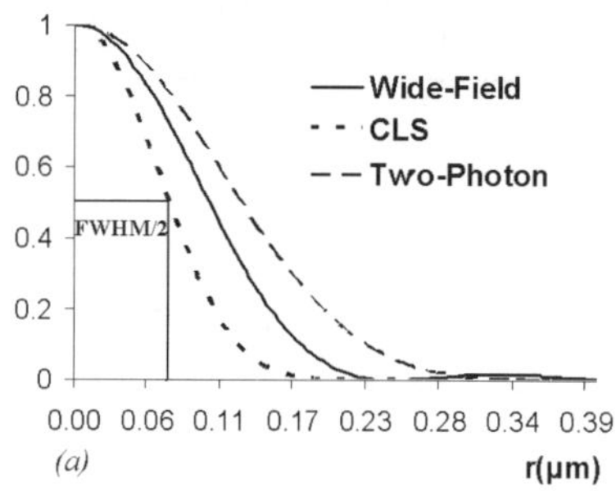

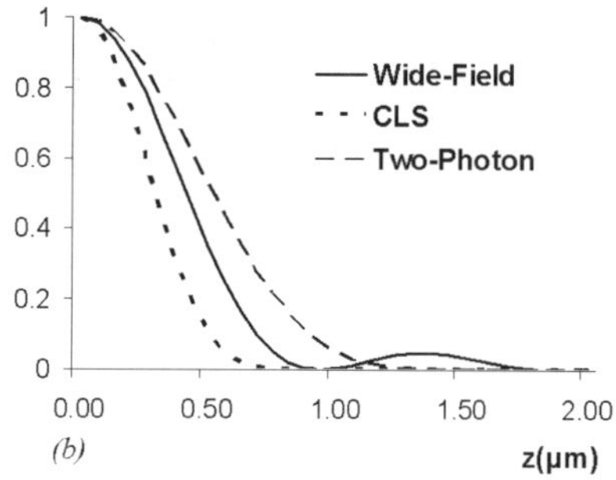

Figure 8. Orthogonal extents of the Wide-Field, Confocal and Two-Photon Excitation microscope PSFs calculated under exactly the same conditions as Figure 7. *(a)* lateral and *(b)* axial

In the lateral direction, these conclusions are valid under ideal imaging condition, but these conditions can only be achieved for certain specimens and with considerable effort. Many factors contribute to reduce the contrast and, therefore, the resolution of microscope images [1]. In the axial direction, the true advantage of confocal and two-photon microscopy over wide-field microscopy is that they provide an optical sectioning property in thick fluorescent specimens [1].

We can also see as the resolution in wide-field microscope results primarily from the lens in emission (objective). The lens in excitation (condenser) plays a much less important role. On the other hand in scanning microscope the lens in excitation (objective) is the crucial one, and the lens in emission (collector) doesn't contribute significantly to the resolution [8].

Table 4. FWHM extents of calculated PSFs. The parameters are the same condition of Figure 7 and Figure 8.

Microscope	*Axial Resolution (μm)*	*Lateral Resolution (μm)*
Ideal Wide-Field	0.83	0.2
Ideal CLS	0.58	0.14
Two-Photons Scanning	1.03	0.25

6. ACKNOWLEDGEMENTS

I thank Alberto Diaspro (Department of Physics, University of Genoa) for a number of useful comment and suggestion. I am also grateful to all the staff of the LAMBS (Laboratory for Advanced Microscopy, Bioimaging and

Spectroscopy) for their help. At the end I should like to thank Paolo Gualtieri for NATO school experience.

7. REFERENCES

1. Diaspro, A., ed. *Confocal and Two-photon Microscopy*. 2002, John Wiley & sons.: New York.
2. Bertero, M. and P. Boccacci, *Introduction to inverse problems in imaging*. 1998, Bristol, UK: IOP Publishing.
3. Bonetto, P. and G. Vicidomini, *Three-Dimensional Microscopy Migrates to the Web with "PowerUp Your Microscope"*. Microscopy Research and Technique, 2004. **64**(2): p. 196-203.
4. Born, M. and E. Wolf, *Principle of Optics*. Sixth ed. 1980, Oxford: Cambridge University Press.
5. Goodman, J.W., *Introduction to Fourier optics*. 1968, San Francisco: McGraw Hill.
6. Vliet, L.J.v., et al., *Image Detectors for Digital Image Microscopy*, in *Imaging, Morphometry, Fluorometry and Motility Techniques and Applications, Modern Microbiological Methods*, J.W. Sons, Editor. 1998: Chichester (UK). p. 37-64.
7. Gibson, S.F. and F. Lanni, *Experiment test of an analytical model of aberration in an oil-immersion objective lens used in three-dimensional light microscopy*. Journal of the optical Society of America A, 1991. **8**(10): p. 1601-1613.
8. Wilson, T. and C. Sheppard, *Theory and Practice of Scanning Optical Microscopy*. Second Edition ed. 1985, London: Academic Press.
9. Lakowicz, J.R., *Principles of Fluorescence Spectroscopy*. Second Edition ed. 1999, New York: Kluwer Academic / Pleum Publisher.
10. Gauderon, R. and C.J.R. Shepard, *Effect of a finite-size pinhole on noise performance in single-, two-, and three photon confocal fluorescence microscopy*. Applied Optics, 1999. **38**(16): p. 3562-3565.

ATOMIC FORCE MICROSCOPY STUDY ON THE PELLICLE OF THE ALGA *Euglena gracilis*

Francesca Sbrana[1], Laura Barsanti[2], Vincenzo Passarelli[2] and Paolo Gualtieri[2]
[1]*Dipartimento di Fisica, Università di Trento, I-38050 Povo (TN), Italy*
[2]*Istituto di Biofisica C.N.R., Via G. Moruzzi,1 56124 Pisa, Italy*

Abstract: AFM technique was used to investigate the organization of the pellicle complex of *Euglena gracilis*. These results mainly validate data obtained by our group using SEM and TEM investigations. In this chapter particular attention has been given to the surface characteristic of *Euglena* such as the muciferous canals and pores.

Key word: AFM, Muciferous Bodies

1. INTRODUCTION

Since the invention of scanning probe microscopes, in the early 1980s, atomic force microscopy (AFM) has gained a continue growing significant role also in biological research, opening new exciting perspectives in the investigation of biological system at the sub-molecular level [1,2,3,4]. The AFM contributes to the knowledge of the biological system, by scanning the surface sample in the x, y plane and in z direction with a very sharp tip mounted at the end of a small cantilever, and taken closer to the sample surface. Monitoring the deflection of the cantilever due to the roughness of the surface during the scan, the surface morphology can be to reconstruct in a 3D image from the x, y, and z scan data with high lateral and, overall, vertical resolution down the nanometre scale. A high resolution tip-sample positioning system, encompassing an extremely accurate scanner and an electronic feedback circuits and a very sharp tip with small radius of curvature are the key role of the AFM. AFM can reach a very high magnification with high resolution, overcoming the diffraction limits of traditional optic microscopes, it is able to investigate sample in ambient or in liquids environments, maintaining the sample in its physiologic conditions, and it requires little sample preparation, and no metal coating is required.

In the last years the number of publications of AFM application in biology study have grown rapidly, and spans smallest bio-systems as DNA,

V. Evangelista et al. (eds.), From Cells to Proteins: Imaging Nature across Dimensions, 395–403.

RNA, lipid and protein-nucleic acid complexes to biggest one as chromosomes, cellular membranes, virus; bacteria and so on. Its ability to explore the surface morphology in non-invasive way at nanometre scale makes the AFM a powerful imaging tool of investigation of the cell surface and its physical properties. The AFM is complementary to the other established techniques such as optical and electron microscopy, nuclear magnetic resonance and X-ray crystallography, but at the same time it offers several advantages. For an introduction to AFM for biologists see Morris et al., 1999. [5]

Since AFM can attain the molecular resolution on biological surfaces [6], it can be useful to analyze the peculiarities of surface structure in various taxonomic groups of algae. Imaging in liquid environment opens vast perspectives for *in situ* monitoring of dynamic processes such as cell growth, division and movement. The main goal of our work was to introduce AFM for comparative studying of the cell surface and in particular the mucus coating of the photosynthetic flagellate *Euglena gracilis*, already investigated by SEM and TEM microscopy.

2. METHODS

AFM is based on raster scanning the surface of an object with a sharp tip (probe) mounted on a flexible cantilever whose deflection caused by the interaction with the surface is detected by a high-precision optical method, (Figure 1), [1].

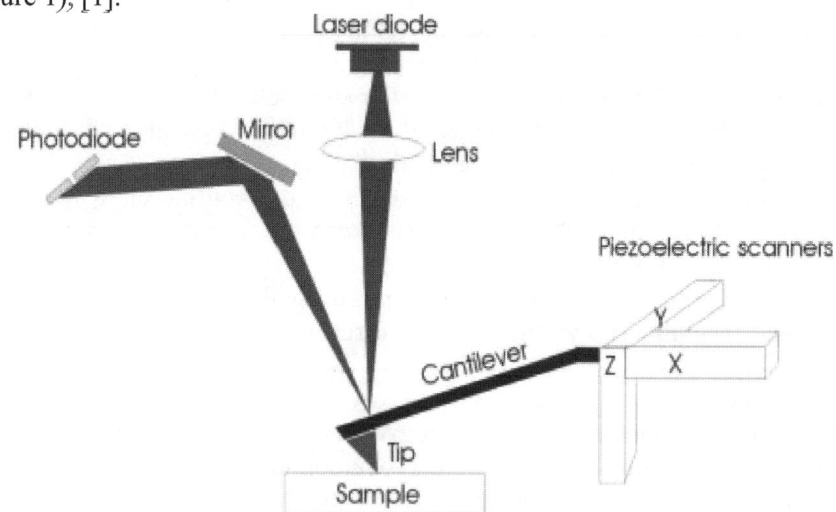

Figure 1: Generalized diagram of an atomic force microscope.

Tip movement with respect to the surface is accomplished by a piezoelectric scanner. The atomic force microscope measures the interaction force between the tip and the surface. A very sharp tip (cantilever) is scanned across the sample surface using piezoelectric scanners. The tip may be dragged across the surface or may vibrate as it moves. The measurement is monitored using the so-called optical beam reflection detection system, in which a laser beam is reflected from the reverse side of the cantilever onto a four quadrant position-sensitive photo sensor. This detector measures the bending of cantilever while the tip is scanning over the sample. The measured cantilever deflections are used to generate a map of the surface topography).

Four imaging modes can be used to produce topographic images of sample surfaces:

- In contact mode, interatomic van der Waals forces become repulsive as the AFM tip comes in close contact with the sample surface. The force exerted between the tip and the sample in contact mode is on the order of about 0.1-1000 nN. Under ambient conditions the capillary force from a thin layer of water in the atmosphere, and the mechanical force from the cantilever itself are also generally present. The capillary force is due to the fact that water can wick its way around the tip, causing the AFM tip to stick to the sample surface. The magnitude of the capillary force should vary with the tip-sample distance. The mechanical force resulting from the cantilever is similar to the force of a compressed spring and its magnitude and sign (repulsive or attractive) is dependent on the cantilever deflection and the cantilever's spring constant. Consequently, in contact mode AFM, the repulsive van der Waals forces arising for the AFM tip to sample interaction must balance the sum of the forces arising from the capillary force plus the mechanical force from the cantilever. The thin layer of water present on many surfaces in air exerts an attractive capillary force and holds the tip in contact with the surface. Thus, when scanner pulls the tip away from the surface, the cantilever bends strongly towards the surface and the scanner has to retract further so that the tip can snap off of the surface. The cantilever returns to its original unbent status as the scanner moves the tip away from the surface beyond the snap-out point. Contact mode imaging can be performed within a liquid environment; when scanning in fluids, the overall forces in contact mode are lower than in ambient air: the fluid layer/meniscus, which is formed in air, is not present furthermore, electrostatic forces can be dissipated or screened.
- In non-contact mode, the cantilever is deliberately vibrated by a piezo-electric modulator with a small amplitude at a fixed frequency that is near the intrinsic resonance frequency of the cantilever. As the tip approaches the sample, the van der Waals attractive force between the

tip and the sample acts upon the cantilever and causes changes in both the amplitude and the phase of the cantilever vibration. These changes are used by the z-servo system to control the tip-sample distance. It is important to maintain the tip-sample distance at a certain constant value (typically tens to hundreds of angstroms) and to prevent the tip from contacting the sample surface. Like contact mode, the motion of the scanner is used to generate the topographic image.

- Intermittent contact mode (tapping mode) overcomes the limitations of the conventional scanning modes by alternately placing the tip in contact with the surface, to avoid dragging the tip across the surface. The cantilever is oscillated at frequencies around 300 kHz; at these frequencies, viscoelastic surfaces become stiff and can more easily resist forces from the probe tip. This property further reduces the possibility of sample damage for extremely soft samples and causes less distortion of the sample due to tip forces. Another advantage of the tapping mode technique is that the vertical feedback system is highly stable, allowing routine reproducible sample measurements. In tapping mode AFM, the cantilever is excited into resonance oscillation with a piezoelectric driver. The oscillation amplitude is used as a feedback signal to measure topographic variations of the sample.
- In Phase Imaging, an AFM cantilever is oscillated vertically near its mechanical resonance frequency while it is in close proximity to a sample. As the AFM tip comes in very close proximity to the sample surface, the amplitude of the cantilever's oscillation is reduced. The change in amplitude is measured and is used to track changes in the surface topography and roughness of the sample. Simultaneously, as the AFM tip encounters regions of different composition, a change in phase, relative to the phase of the drive signal, is measured and recorded. This change in phase is very sensitive to variations in material properties, including surface stiffness, elasticity and adhesion. The phase shifts are measured and displayed in a very straightforward manner that facilitates quantitative analysis and interpretation. Phase Imaging can reveal material properties that can not be observed in surface topography and it can identify properties that might otherwise be obscured by topography. It is a sensitive, quantitative, high lateral resolution AFM method that is often more convenient and gentler than other surface property methods that are based on contact mode operation

The AFM investigation on *Euglena* pellicle was performed using a Solver P-47 AFM, made by NT-MDT Co., equipped with a 60 x 60 mm^2 piezoelectric-tube (PZT) scanner. Topography images of 512x512 pixel were acquired in intermittent contact mode using rectangular Si non-contact mode cantilevers, by Mikro Masch, with resonant frequency of 310 khz, in ambient environments, and scanned at a scan rate of 0.70 Hz. Prior to use, the PZT scanner was calibrated in x, y and z using reference standard grating.

3. RESULTS AND DISCUSSION

One cytoskeletal feature distinguishes the euglenids from all other protists: a pellicle composed of longitudinal articulated strips that arise in the reservoir and extends the length of the cell immediately beneath the membrane, (Figure 2), [7 - 10].

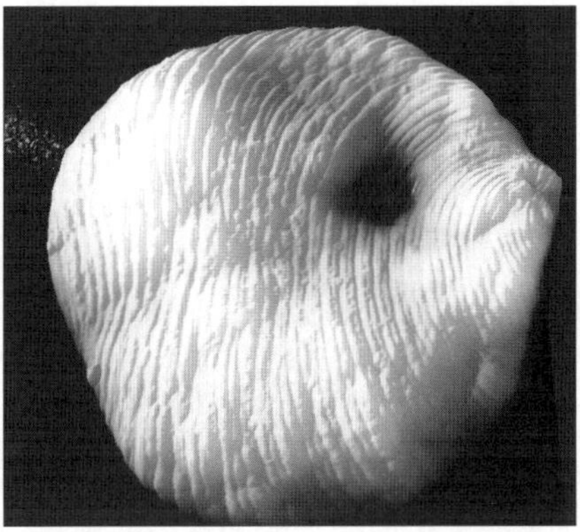

Figure 2: AF
AFM image of *Euglena gracilis.*

This very solid structure gives rigidity to the cell and in contractile species, relational sliding of the strips occurs along the articulations. The pellicle is believed to play a main role in metaboly, the peculiar squirming movement that is found in all euglenoids not restricted by a rigid pellicle. In *Euglena* the strips are formed as more or less pronounced ridges, with thinner more membranous grooves between and connected to the ridges, (Figure 3).

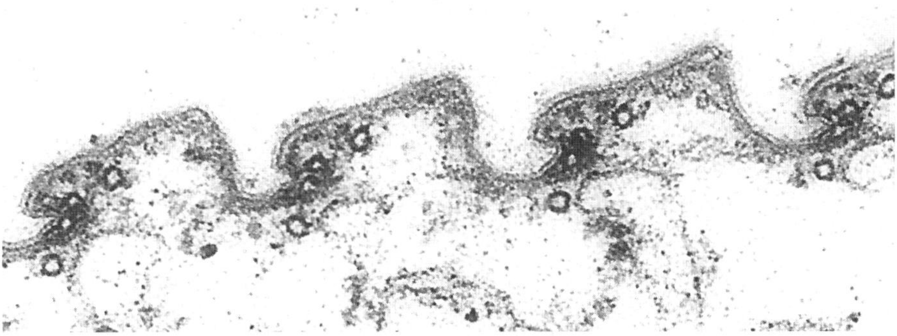

Figure 3: Transverse section of the pellicle of *Euglena*

These pellicular striations leave the reservoir opening and proceed posteriorly in the form of a left-handed or β-helix. Near the posterior vortex, these strips fuse once and then once again, reducing the number. The membrane layer is continuous and covers the ridges and grooves on the whole cell. In many species it is covered by a dense irregular material believed to be mucilage. Mucilage secretion occurs in all euglenoid flagellates, and living cells are permanently coated with a thin slime layer.

Figure 4 shows the mucus ghost left by an *Euglena* cell treated with nicotine.

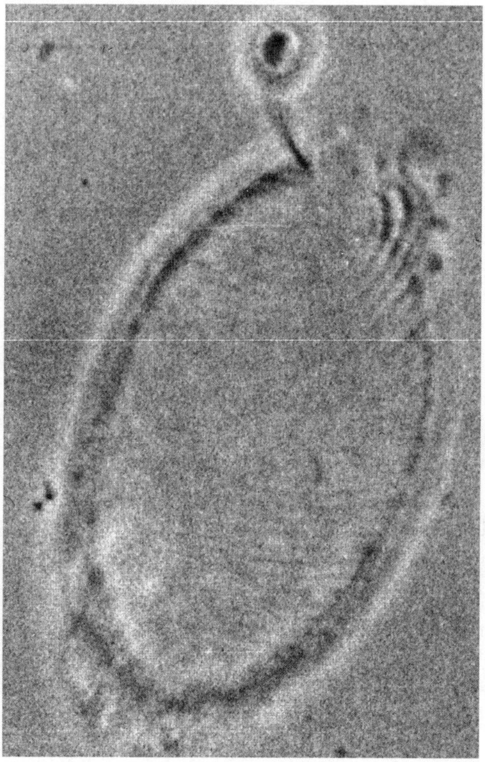

Figure 4: Mucus ghost of *Euglena* cell

The shape of the ghost tells us that the cell, not elongated as under normal conditions, has first reacted with a rounding-up movement, whereas its striated pattern indicates that the mucus is secreted as filaments paralleling the grooves of the pellicle.

Figure 5 shows the AFM image of the external surface of an *Euglena* cell and the mucus trail left by the cell upon dehydration; the coating is present both over the ridges and inside the grooves.

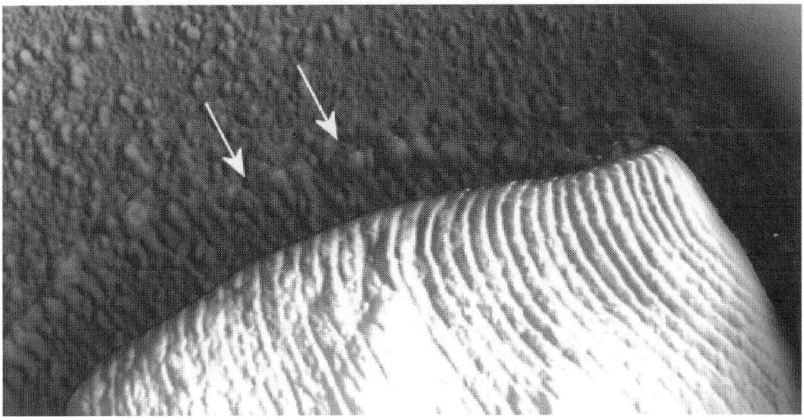

Figure 5: AFM image of an *Euglena* cell and its mucus trail (arrows)

Just beneath each ridge of the pellicle, topographically associated with the notch, there is one tubular cistern of the endoplasmic reticulum, together with closely packed rows of muciferous bodies responsible of the secretion of the mucus coating. Narrow canals from the muciferous bodies are often present, which connect them to the groove and then to the outside of the cell through pores of regular disposition. Figure 6 is a deep etching preparation of *Euglena* pellicle that shows the opening of these pores (arrows) along the ridges. The fracture has left traces of membrane from tubular cisterns around the rim of the pores, thus indicating the presence of a secreting activity in these districts of the pellicular complex.

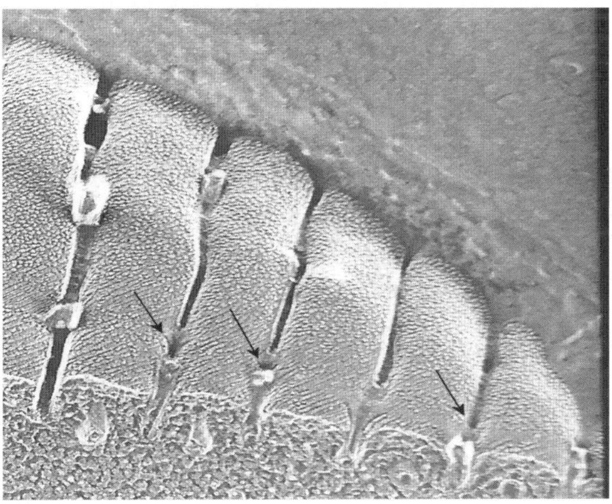

Figure 6: Pellicle pores of the muciferous canals along the ridges

Figure 7 is an AFM image of the *Euglena* pellicle showing a pore opening (arrow) between two strips.

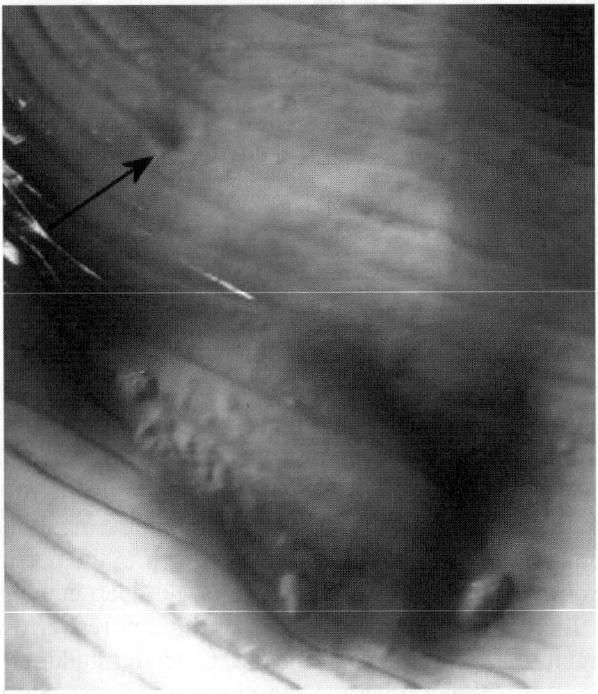

Figure7: Pellicle pore as seen by AFM.

Figure 8 show the depth profile along a line perpendicular to the pore. The left arrowhead gives the height of the strip located in proximity of the pore, while the right arrowhead gives the depth of the pore. The difference between these values gives the length of the pore canal.

Figure 8: Depth profile by AFM

4. REFERENCES

1. Binnig G., Quate C., Gerber C. Atomic force microscope. Phys Rev Lett 1986; 56: 930-33.
2. Muller D.J., Amrein M., Engel A. Adsorption of biological molecules to a solid support for scanning probe microscopy. J Struct Biol 1997; 119: 172-88.
3. Bustamante C., Keller D. Scanning force microscopy in biology. Physics Today 1995; December: 32-38.
4. Engel A., Schoenenberger C.A., Muller D.J. High resolution imaging of native biological sample surfaces using scanning probe microscopy. Curr Opin Struct Biol 1997; 7(2): 279-84.
5. Morris V.J., Kirby A.R. and Gunning A.P. *Atomic Force Microscopy for Biologists.* World Scientific Publishing Company, 1999
6. Kiselyova O. and Yaminsky I. "Atomic force microscopy of protein complexes." In *Atomic Force Microscopy: Biomedical Methods and Applications (Methods in Molecular Biology, vol. 242)*, P.C. Braga, D. Ricci eds, Humana Press, 2003.
7. Vismara R., Barsanti L., Lupetti P., Passarelli, V., Mercati, D., Dallai R., and Gualtieri, P. Ultrastrutture of the pellicle of *Euglena gracilis.* Tissue and Cell 2000; 32 (6): 451-56.
8. Leander B.S., Witek R.P., and. Farmer M.A. Trends in the evolution of the euglenid pellicle. Evolution 2001; 55(11): 2215–35
9. Leander B.S. and Farmer M.A. Comparative morphology of the euglenid pellicle. I. Patterns of strips and pores. J. Eukaryot. Microbiol. 2000; 47:469-79
10. Leander B.S. and Farmer M.A. Comparative morphology of the euglenid pellicle. II. Diversity of strip substructure. J. Eukaryot. Microbiol. 2001; 48: 202-17.

ATOMIC FORCE MICROSCOPY STUDY OF PILI IN THE CYANOBACTERIUM *SYNECHOCYSTIS* SP. PCC 6803

Evgeniy V. Dubrovin[1], Inessa A. Kirik[2], Michael M. Babykin[3], Igor V. Yaminsky[1]
[1]*Department of Physics, Moscow State University, Moscow 119899, Russia*
[2]*Department of Genetics, Moscow State University, Moscow 119899, Russia*
[3]*International Biotechnological Center, Moscow State University, Moscow 119899, Russia*

Abstract: Atomic force microscopy (AFM) is now widely used method in life sciences, specifically for investigation of the microbial surfaces. Continuing the AFM research of various bacteria we studied pilus-like appendages of the unicellular cyanobacterium *Synechocystis* sp. PCC 6803. As known the *Synechocystis* 6803 wild type cells produce pili of two morphotypes distinguished by diameter, length, morphology and relative abundance. Thick pili (one morphotype) are identified with well-known bacterial type IV pili responsible for cell gliding motility; the function of thin pili (another morphotype) is still unknown. AFM has revealed differences in piliation of the motile (wild type) and non-motile (spontaneous mutant) strains of the cyanobacterium and allowed estimating real dimensions and quantity of pili. According to the data obtained non-motile mutant cells possess thick pili increased in number (to 3-5 times) and length (more than to 10 times) relative to the wild type cells, however, both strains do not differ in thin pili. The advantages of AFM over conventional electron microscopic techniques in comparative morphology of the bacterial cells are discussed.

Key words: Atomic Force Microscopy; Cyanobacteria; Pili, Genetic Mutants.

1. INTRODUCTION

Atomic force microscopy (AFM) appeared in 1986 has already become a conventional investigation tool for studying the biological objects [1-3] including bacteria [4]. AFM is based on raster scanning the surface of object with sharp tip (probe) mounted on a flexible cantilever whose deflection caused by the interaction with the surface is detected by a high-precision optical method [5]. Tip movement with respect to the surface is accomplished by a piezoelectric scanner. In general, AFM offers significant advantages over conventional microscopic techniques, such as scanning and transmission electron microscopy. The main benefits of AFM are the

V. Evangelista et al. (eds.), From Cells to Proteins: Imaging Nature across Dimensions, 405–414.

relatively simple procedures for sample preparation, the possibility to scan samples in ambient air (as well as in liquid environment) and the possibility of quantitative data acquisition in three dimensions.

Since AFM can attain the molecular resolution on biological surfaces [6], it is very perspective method allowing to analyze the peculiarities of surface structure in various taxonomic groups of bacteria and therefore allowing to solve fundamental and applied problems of biology and medicine [4, 7]. Comparison of the AFM images of the cell surface of the *Escherichia coli* parent strain K12 J62 and its transductant, acquiring the capacity for synthesizing O-antigen of *Shigella flexnery*, revealed essential differences in the cell surface patterns [8]. Transductant cells can be used as a noninfectious living vaccine expressing certain O-antigen of *S. flexnery*. Surface alterations in strains of *Staphylococcus aureus* [9], *Heliocobacter pylory, E.coli, Bacillus cereus* and *Streptococcus pyogenes* [10] were observed after different types of antibiotic treatment that is very important for medical diagnostics.

Imaging in liquid environment opens vast perspectives for *in situ* monitoring of dynamic processes such as cell growth, division and movement. For example the germination of the *Bacillus* sp. spores [11] and the formation of 30 nm micropores in the bacterial cell wall treated with calcium ions [12] were successfully documented by AFM.

To date AFM examination of the bacteria of a great many species has been realized, e.g. *E. coli*, *Klebsiella pneumoniae*, *H. pylori*, *Lactobacillus fermentum*, *Bifidobacterium longum*, *Arthrobacter globoformis, B. sphaericus* [13], *S. aureus* [9], *B. cereus, S. pyogenes* [10] etc. Nevertheless, the AFM data on cell wall morphology of cyanobacteria considered as ancient precursors of chloroplasts and widely used as model objects in molecular biology of oxygenic photosynthesis were not yet been obtained. The entire genome of unicellular cyanobacterium *Synechocystis sp.* PCC 6803 has been sequenced one of the first [14] and systemic functional genomics of this cyanobacterium is now actively developed. Molecular genetic analysis of the mutants with altered phototactic motility has revealed a set of genes controlling biogenesis of pilus-like cellular appendages in *Synechocystis* 6803 [15-17]. According to electron microscopic data the wild type cyanobacterial cells produce pili of two morphotypes, denoted thick and thin, distinguished by diameter, length, morphology and relative abundance. The thin pilus morphotype has a diameter of 2-3 nm and a length of 500-1000 nm. These pili are profuse and appear to be distributed about the entire surface of the cell. The thick pili (second morphotype) have a diameter of 6-8 nm and a length of at least 1000-2000 nm. At higher magnification, the thick pili often appear twisted or knotted; this is a morphological detail never associated with the thin pili and may be an indication of flexibility. There are considerably fewer thick than thin pili and it is not clear whether these thick pili are more concentrated at a particular pole of the cell [16]. Together the genetic and morphologic data unambiguously indicate that phototaxis is mediated by thick pili [16, 17] clearly identified with type IV pilus structures that play an important role in twitching and gliding motility in a number of Gram-negative non-photosynthetic bacteria [18]. But it is not known whether the thin pili play a role in motility.

The main goal of our work was to introduce AFM for comparative studying the cell wall appendages in the photosynthetic cyanobacterium *Synechocystis* 6803. We investigated the cells of two different *Synechocystis* 6803 strains, motile wild type strain and non-motile mutant strain, and have revealed substantial differences in their piliation.

2. METHODS AND MATERIALS

2.1. Atomic force microscopy

All measurements were performed with a Nanoscope IIIa multimode scanning probe microscope (Digital Instruments, USA) in contact and tapping modes. For tapping mode measurements, we used commercial silicon 125 μm cantilevers (NanoProbe) with a spring constant of 42 N/m and 280-310 kHz resonant frequencies and 100 μm non-contact cantilevers with 200–300 kHz resonant frequencies (State Research Institute for Problems in Physics, Russia). For contact mode measurements we used silicon nitride cantilevers (Nanoprobe) with a spring constant of 0.06 and 0.32 N/m. FemtoScanOnline software was used for the image processing [19].

2.2. Strains and Growth Conditions

Two strains of *Synechocystis* 6803 were used: the motile wild type strain, wtR (from the Pasteur Culture Collection), and spontaneous mutant lacking motility, wtM (from the collection of Dept. of Genetics of Moscow State University). Cells were grown in BG-11 medium [20] in moderate light (50 microeinsteins (μE) m^{-2} s^{-1}) at 30°C.

2.3. Motility Assays

Liquid samples containing approximately 100-500 cells were spread on the surface of BG11 agar (0.8 %) supplemented with 10 mM glucose. The cells were incubated under unidirectional light of 20 μE m^{-2} s^{-1} for 5-8 days until colony movement was detected.

2.4. Sample preparation for AFM

The *Synechocystis* 6803 cells cultivated on BG-11 solid medium for 3-5 days were transferred into 20 μl of deionized water. The optical density (A_{750}) of final cell suspension was equal to 0.9-1.5 that was the optimum value for adsorption of single cells onto mica. A droplet of bacterial suspension was placed onto freshly cleaved mica, left for 5 min for adsorption and dried in airflow.

3. RESULTS AND DISCUSSIONS

We have acquired and analyzed both cell motility data (fig. 1) and AFM images of the unicellular cyanobacterium *Synechocystis* 6803 of two different wtR (motile) and wtM (non-motile) strains. As shown, phototactic movement of the *Synechocystis* 6803 cells is mediated by type IV thick pili [16, 17]. Two types of non-motile *Synechocystis* 6803 mutants have been described. One of them had no thick pili [15] while another was hyperpiliated [16].

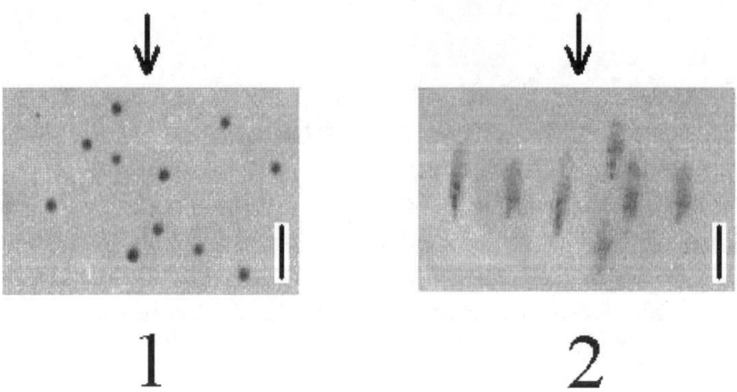

Figure 1. Directional motility assays. 1 - single colonies of the *Synechocystis* 6803 non-motile wtM strain; 2 - single colonies of the *Synechocystis* 6803 motile wtR strain, which move toward the light source (phototaxis). The arrows point to the light direction. The bar represents 2 mm.

Contact mode of AFM, at which the flexible cantilever with a probe scans across the sample surface, being in a physical contact with the sample, is frequently used for measuring the elasticity or Young's modulus of a studied object [21]; as the contact force between the tip and the surface increases, the height of imaged object diminishes. The minimization of effective cantilever force to the surface to zero is sometimes rather problematic due to capillary forces. Therefore, on images recorded in contact mode, the heights of soft samples, such as bacteria and their pili, can often be underestimated. Taking this into account, for height measurements we used tapping mode of AFM when the presence of capillary forces does not play a significant role. We used contact mode only for obtaining the general view of cyanobacteria and for the length measurements of pili.

Interestingly, that almost all *Synechocystis* PCC6803 cells deposited onto mica that we have imaged with AFM looked like dividing cells during cytokinesis (in the stage of separation of two daughter cells) (fig. 2). This simultaneous division of cells can be explained by using log-phase growing cells for AFM assay.

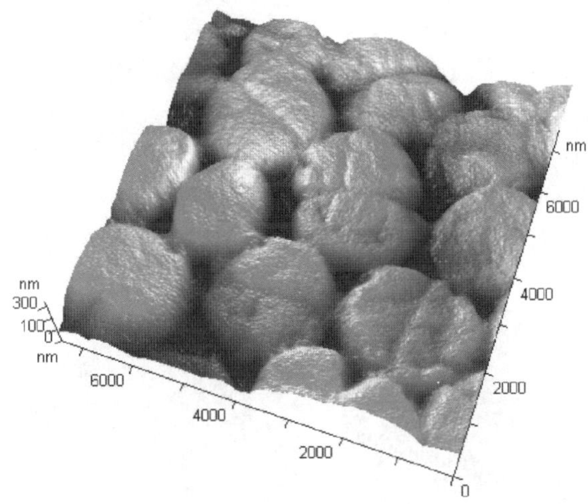

Figure 2. 3-D image of the *Synechocystis* 6803 cells (motile wtR strain) acquired with AFM

AFM images of cells with pili of motile wtR strain are shown in fig. 3. Single bacterium deposited on mica in air looks like round cell 1.5 – 2 microns in diameter and 1 micron in height. Height distribution of pili (fig. 4) reveals two clearly defined peaks at approximately 2 and 4 nm, which obviously correspond to thin and thick pili. The fact that the value of height of thick pili (4 nm) is smaller than the diameter computed from electron microscopy data (6-8 nm, see [16]) can be explained by the effect of drying on mica. This effect was previously reported for several proteins immobilized on mica surface [22, 23]. Thin pili tend to gather with each other and to form structures that look like braids. In one such braid it can be distinguished up to twenty single threads (fig. 3C) but sometimes they could be hardly counted. Thus, we are able to roughly estimate the number of thin pili, which are seen on AFM images; this estimation gives the value of 150-200 pili per one cell.

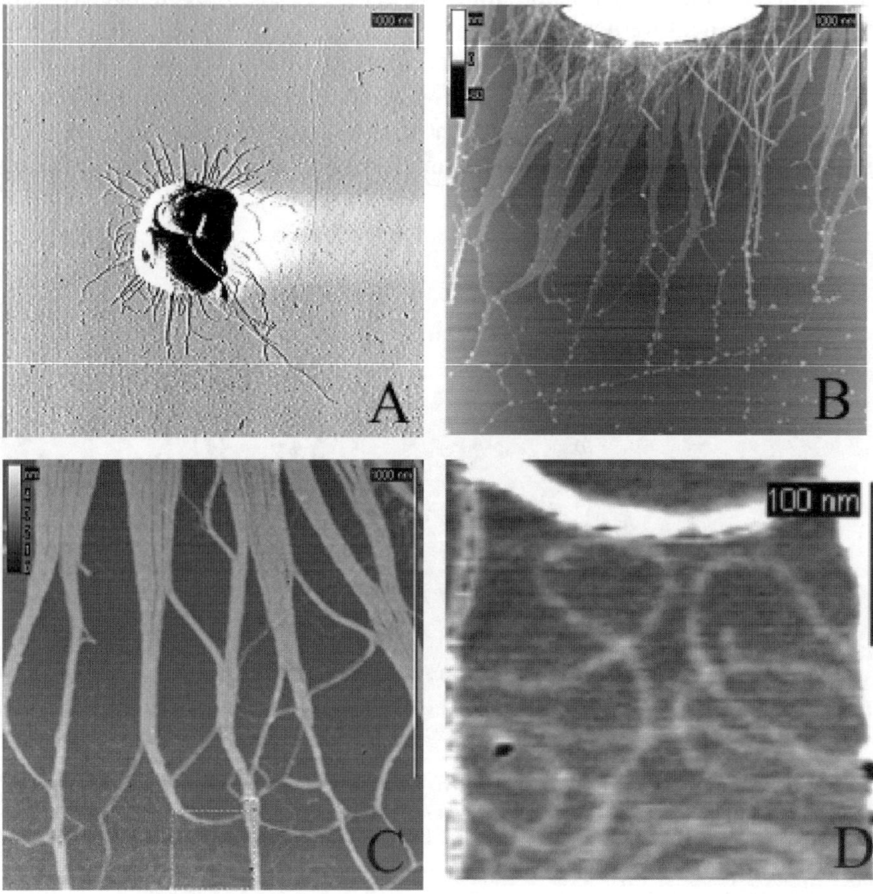

Figure 3. AFM images of the motile *Synechocystis* 6803 wtR cells: (A) general view of single cell (contact mode deflection image); (B) thick and thin pili; (C) only thin pili; (D) thin filaments coming out from thin pili (zoomed from (C)). (B), (C) and (D) are tapping mode height images.

When we say about the quantity of pili we mean a number of AFM visualized pili around the cell adsorbed onto mica. The real number of pili belonged to the cell should be higher because they can be everywhere on the cell surface (particularly thin pili, see Introduction) but we can image only those laying on the substrate surface. Probably (from the simple geometrical considerations) after drying of the sample in airflow (see Sample preparation) the greater part of pili lays onto the surface of immobilized cells and only small part – onto the substrate.

Small maximum under 1 nm in the distribution in fig. 4 appeared due to the presence of very thin filaments that come out from the pili (fig. 3D). It may indicate the presence of inner structure of pili which consist of thin filaments, interwoven with each other. The length of both thin and thick pili of the motile wtR cells is under 2 microns showing that in this case thick pili are not longer than thin ones.

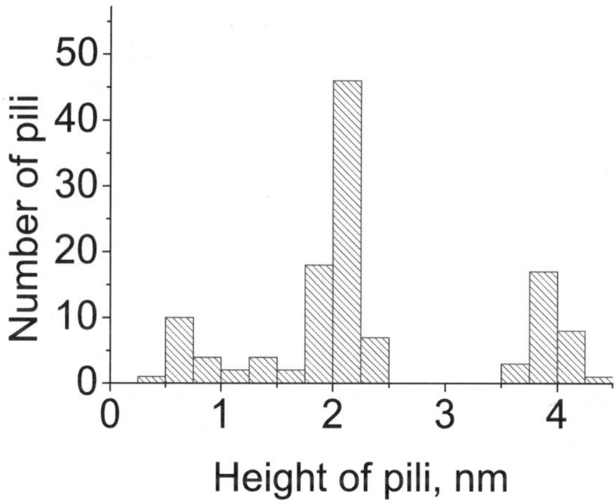

Figure 4. Histogram of the height distribution of pili in cells of the motile *Synechocystis* 6803 wtR strain.

The length of thick pili of non-motile wtM cells is, on the contrary, much higher than of thin pili and may exceed 10 microns (fig. 5). Moreover, we can see approximately 3-5-fold increasing in the number of thick pili on the cell surface relative to the wtR cells. The length and number of thin pili in the wtM cells are the same as in the wtR cells, however. Mean height values (nm) of thick and thin pili on the surface of wtM and wtR cells are summarized in the table.

Thin pili of both studied strains of *Synechocystis* 6803 are distributed evenly around the cell while thick pili – quite heterogeneously, e.g. they can concentrate on only one pole of the cell as can be seen in the fig. 5A.

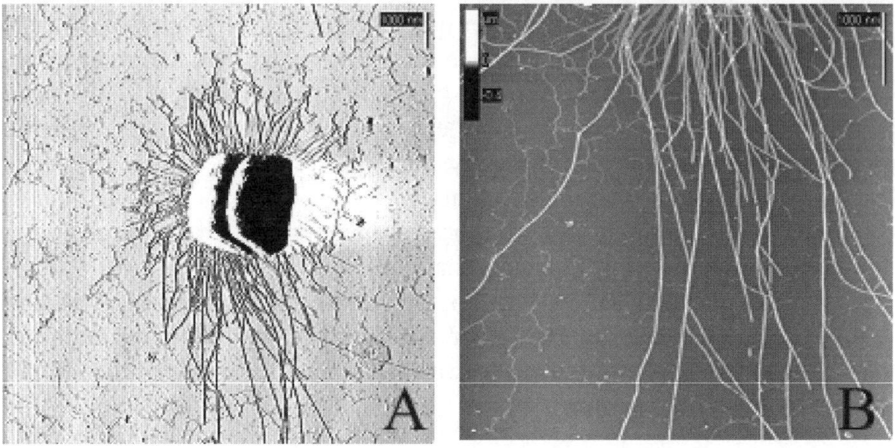

Figure 5. AFM images of the non-motile *Synechocystis* 6803 wtM cells: (A) general view of single cell (contact mode deflection image); (B) thick and thin pili (tapping mode height image).

The hyperpiliation phenotype of non-motile wtM cells is the similar to that observed in the non-motile *taxAY1* and *pilT1* mutants of *Synechocystis* 6803 [24]. These mutants also had the increased number of thick but not of thin pili. Thus, increased density of thick pili may serve as a reason for violated motility in the wtM strain. Non-motile phenotype of the wtM strain is caused by spontaneous mutation in a gene controlling pilus biogenesis. To identify this gene, further molecular genetic analysis of the wtM mutant is required. The data obtained represent the first example of using AFM analysis for genetic study of the cyanobacteria.

Table 1. Mean height values (nm) of thick and thin pili of the *Synechocystis* 6803 wtM and wtR strains calculated from the AFM images.

Type of pili/strain	Motile type (wtR)	Non-motile type (wtM)
Thin pili	2.1±0.1	2.1±0.1
Thick pili	3.9±0.2	3.5±0.2

4. CONCLUSION

For investigation of *Synechocystis* 6803 cell surface and the morphology of pilus-like appendages we used AFM because it offers substantial benefits over conventional microscopic techniques, such as scanning and transmission electron microscopy. Among the advantages are relatively simple procedures for the sample preparation (e.g. samples do not need contrasting with heavy metal atoms), the possibility to scan samples in ambient air (as well as in liquid environment) and the possibility of quantitative data acquisition in three dimensions. The surface of conventional AFM support – mica – is negatively charged in water therefore preventing from immobilizing on itself many biological objects, e.g. nucleic acids [25].

The use of AFM for study of piliation of *Synechocystis* 6803 was shown to be very promising because of both conventional advantages of AFM technique and good adhesion of these cells onto mica substrate. AFM images have revealed hyperpiliation and the increase in length of thick pili of non-motile strain in comparison with motile strain of *Synechocystis* 6803. There is no substantial difference in thin pili, which in both strains tend to aggregate in groups. These results are of great importance in genetics showing the relation of cell morphology with cell function.

5. ACKNOWLEDGEMENTS

This work was supported by the Russian Foundation for Basic Research (projects N 04-03-32472-a and 02-04-48398-a), Russian Academy of Sciences (Program "Creation and Study of Macromolecules and New Generation Macromolecular Complexes) and NATO Science Program (Grant N LST.CLG.980194).

6. REFERENCES

1. Muller, D.J., Amrein, M., Engel, A. Adsorption of biological molecules to a solid support for scanning probe microscopy. J Struct Biol 1997; 119:172-88.
2. Bustamante, C., Keller, D. Scanning force microscopy in biology. Physics Today 1995; December:32-38.
3. Engel, A., Schoenenberger, C.A., Muller, D.J. High resolution imaging of native biological sample surfaces using scanning probe microscopy. Curr Opin Struct Biol 1997; 7(2):279-84.
4. Bolshakova, A.V., Kiselyova, O.I., Yaminsky, I.V. Microbial surfaces investigated using atomic force microscopy. Biotechnology progress 2004; 20(6):1615-22.
5. Binnig, G., Quate, C., Gerber, C. Atomic force microscope. Phys Rev Lett 1986; 56:930-33.
6. Kiselyova, Olga and Yaminsky, Igor. “Atomic force microscopy of protein complexes.” In Atomic Force Microscopy: Biomedical Methods and Applications (Methods in Molecular Biology, vol. 242), eds P.C. Braga, D. Ricci: Humana Press, 2004.
7. Bolshakova, A.V., Kiselyova, O.I., Filonov, A.S., Frolova, O.Yu., Lyubchenko, Yu.L., Yaminsky, I.V. Comparative studies of bacteria with atomic force microscopy operating in different modes. Ultramicroscopy 2001; 86(1-2):121-28.
8. Yaminsky, I.V., Demin, V.V., Bondarenko, V.M. The differences in cellular surface of hybrid bacteria Escherichia coli K12, inheriting rfb-p3,4 gene of Shigella flexneri as revealed by atomic force microscopy. J Microbiol Epidemiol Immunol (in russian) 1997; 6:15-18.

9. Boyle-Vavra, S., Hahm, J., Sibener, S.J., Daum, R.S. Structural and topological differences between a glycopeptide-intermediate clinical strain and glycopeptide-susceptible strains of *Staphylococcus aureus* revealed by atomic force microscopy. Antimicrob Agents Chemother 2000; 44:3456-60.
10. Braga, Pier Carlo and Ricci, Davide. Imaging Bacterial Shape, Surface, and Appendages Before and After Treatments With Antibiotics. In Atomic Force Microscopy: Biomedical Methods and Applications (Methods in Molecular Biology, vol. 242), eds P.C. Braga, D. Ricci: Humana Press, 2004.
11. Bolshakova, A.V., Vorobyova, E.A.; Yaminsky, I.V. Indication of living bacterial cells in native soil and permafrost. Phys Low-Dim Struct 2003; 3/4:105-12.
12. Oberleithner, H., Giebisch, G., Geibel, J. Imaging the lamellipodium of migrating epithelial cells in vivo by atomic force microscopy. European Journal of Phisiology 1993; 425:506–10.
13. Wahl, R., Raff, J., Selenska-Pobell, S., Mertig, M., Pompe, W. A Fast Screening Method for Surface Layers on Gram-Positive Bacteria. Biotechnol Lett 2001; 23:1485-90.
14. Kaneko, T., Sato, S., Kotani, H., Tanaka, A., Asamizu, E., Nakamura, T., Miyajima, N., Hirosawa, M., Sugiura, M., Sasamoto, S. Nakamura, T., et al. Sequence analysis of the genome of the unicellular cyanobacterium *Synechocystis* sp. strain PCC 6803. II. Sequence determination of the entire genome and assignment of potential protein-coding regions. DNA Res 1996; 3:109–36.
15. Bhaya, D., Watanabe, N., Ogawa, T., Grossman, A.R. The role of an alternate sigma factor in motility and pili formation in the cyanobacterium *Synechocystis* sp. strain PCC 6803. Proc Natl Acad Sci USA 1999; 96:3188-93.
16. Bhaya, D., Bianco, N.R., Bryant, D., Grossman, A. Type IV pilus biogenesis and motility in the cyanobacterium *Synechocystis* sp. PCC6803. Mol Microbiol 2000; 37(4):941-51.
17. Yoshihara, S., Geng, X., Okamoto, S., Yura, K., Murata, T., Go, M., Ohmori, M., Ikeuchi, M. Mutational analysis of genes involved in pilus structure, motility and transformation competency in the unicellular motile cyanobacterium *Synechocystis* sp. PCC 6803. Plant Cell Physiol 2001; 42:63–73.
18. Wall, D., Kaiser, D. Type IV pili and cell motility. Mol Microbiol 1999; 32:1–10.
19. Filonov, A.S., Gavrilko, D.Yu., Yaminsky, I.V. "FemtoScan" Software for Three-Dimensional Image Processing. Moscow: Advanced Technologies Center, 2001.
20. Rippka, R., Deruelles, D.E., Waterbury, J.B., Generic Assignments, Strain Histories and Properties of Pure Cultures of Cyanobacteria. J Gen Microbiol 1979; 111:1–61.
21. Kiselyova, O.I., Nasikan, N.S., Yaminsky, I.V., Novikov, V.K. AFM imaging of PVX particles and PVX RNA. Phys Low-Dim Struct 2001; 3/4:167-74.
22. Lyubchenko, Y.L., Shlyakhtenko, L.S. Visualization of supercoiled DNA with atomic force microscopy in situ. Proc Natl Acad Sci USA 1997; 94:496–501.
23. Skabkin, M.A., Kiselyova, O.I., Chernov, K.G., Sorokin, A.V., Dubrovin, E.V., Yaminsky, I.V., Vasiliev, V.D., Ovchinnikov, L.P. Structural organization of mRNA complexes with major core mRNP protein YB-1. Nucleic Acids Res 2004; 32:5621–35.
24. Bhaya, D., Takahashi, A., Grossman A.R. Light regulation of type IV pilus-dependent motility by chemosensor-like elements in *Synechocystis* PCC6803. Proc Natl Acad Sci USA 2001; 98:7540–45.
25. Klinov, D.V., Dubrovin, E.V., Yaminsky, I.V. Scanning probe microscopy of DNA on mica and graphite. AIP Conference Proceedings 2003; 696:452-56.

CHANGES OF ALGAE PROTEIN COMPLEX UNDER pH EFFECT

Tatyana V. Parshikova
Kiev National University named T.Shevchenko, Vladimirskaya st., 60, 01017, Kiev, UKRAINE

Abstract: It was established that “blue liquid” formed in the decomposition of blue-green algae (*Cyanophyta, Cyanobacteria*) results from isolation of water-soluble bilichromoproteins from the cells, mainly phycocyanine and its derivatives, with the zone of principal absorption maximum within the limits of 610-630 □m. The most stable color and spectral peculiarities the liquid preserves at pH 6.8-7.2. It was shown that alkalization up to pH 9-12 and acidification of the cultural medium up to 1.5-4.5 cause almost complete destruction of “blue liquid” proportional to the degree of pH change.

In case of “blue liquid” formation the nitrogen and protein content in the cells decreases in connection with their isolation from the cells, the content of chlorophyll *a*, carotenoides and phycobiline pigments also decreases.

It was discussed that estimation of microcystine concentration in “blue liquid” (under presence of phycobiline complex) with using of ELISA requires elaboration of new methodical receptions.

Keywords: Algae, Water “Blooming”, Toxicity, Phycobilines, ph

1. INTRODUCTION

Algae protein complex attracts widespread attention of scientists in different aspects: evolutionary, structural, from the point of view for biological fullness of feed and food using, stability to influence of different factors and peculiarities of survival in extreme conditions [1, 2]. Problem of toxicity plays an important role too, at first in water environment. In this connection the main attention in the world is concerned to agents, which cause water “blooming”. At first, it is organism-cosmopolite – *Microcystis aeruginosa* Kutz. emend Elenk. [3, 4]. This organism obtained the fame as producer of strong hepatotoxin – microcystine [5]. As well known, the limit permissible concentration of microcystine in

V. Evangelista et al. (eds.), From Cells to Proteins: Imaging Nature across Dimensions, 415–421.

drinking water was established by World Organization of Public Health on level from 0.5 to 1.0 μg/l [6].

It is well known that *Microcystis aeruginosa* develop more intensively at summer. In consequence of it natural water in bays, gulfs, dams of hydroelectric power stations, many lakes and reservoirs are covered by unbroken layer of blue-green algae. After destruction of algae cells was rived light – blue liquid in reflected light have red fluorescence. Formation of "blue liquid" is connected with water-soluble phycobiliprotein – phycobiline pigments (C-phycocyanine, allophycocyanine).

It has been known that O'h Eocha [7] for a first isolated from C-phycocianine of *Microcystis aeruginosa* Kutz. emend. Elenk 3 different water-soluble pigment-protein complexes in 1963. It were complexes with corresponding maximum of absorption: phycobiline 655, phycobiline 608 and phycobiline 630. Under appropriated conditions these complexes were readily interconvertible.

As a rule, "blue liquid" is formed under cell destruction of blue-green algae at sacrifice of different factors: mechanical (destruction of cells by grinding, water hammer); chemical (influence of acid, alkali, different buffer solutions); factor of low temperature (freezing). In this liquid was found the microcystine. Consequently water gains toxicity for as well for fishes, invertebrates as for warm-blooded animals.

The aims of our experiments were estimation toxicity of "blue liquid, which formed on the base of hydrophilic complex of bilichromoproteids. For estimation of toxicity had been used standard method ELISA (Enzyme-Linked Immunosorbent Assay). Experiments had been provided as well on the cultures of *Cyanophyta* (*Cyanobacteria*) as natural algae population.

2. RESULTS AND DISCUSSION

Spectrophotometric investigations show the close association of "blue liquid" with phycobiline pigment complex of *Cyanophyta* (*Cyanobacteria*). On Figure 1 is presented spectral characteristic of "blue liquid", which is received from natural population of blue-green algae with dominating of *Microcystis aeruginosa.*

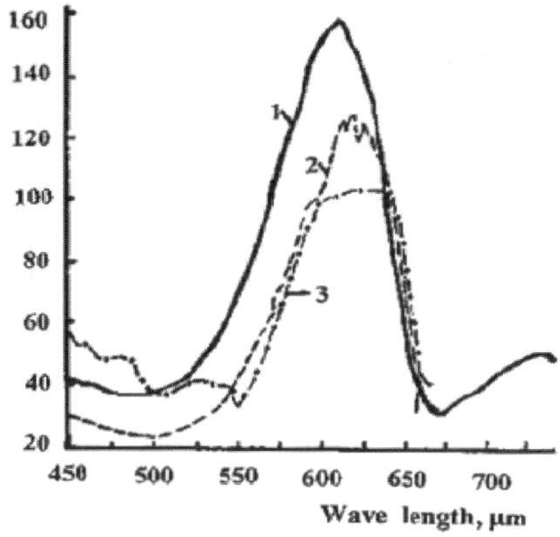

Figure 1. Spectral characteristic of "blue liquid", which was obtained from natural material of blue-green algae: 1 – initial sample, 05.08; 2 – initial sample, 28.09; 3 – initial sample, 28.10.

As indicates the Figure 1, samples of natural water, which were selected in different months, are differed by its spectral characteristics. Line 1 – sample was selected at the beginning of August. It was the height of intensive water "blooming". Line 2 – sample from the end of September, when intensity of water "blooming" decreases as the result of temperature fall. Line 3 – sample was selected in October. In all cases the spectral zone of maximum absorption the light by liquid (on wave length) was similar and typical for *Microcystis aeruginosa*. Biochemical characteristic of algae and "blue liquid" as its formation storage is presented in Table 1.

Table 1. Biochemical characteristic of algae and "blue liquid" as its formation and storage (in % to dry matter)

Index of accounting and object of research	Initial content	After 3 days of storage	After 16 days of storage	After 40 days of storage
Algae + "blue liquid"				
Total nitrogen content	12.8 ± 0.4	12.2 ± 0.2	12.8 ± 0.3	12.7 ± 0.5
Protein nitrogen	11.8 ± 0.3	9.1 ± 0.1	9.9 ± 0.2	7.8 ± 0.3
Non-protein nitrogen	1.0 ± 0.1	3.1 ± 0.2	2.9 ± 0.1	4.9 ± 0.2
Rest of algae after liquid formation				
Protein nitrogen	4.1 ± 0.01	4.9 ± 0.2	6.5 ± 0.4	5,7 ± 0.3
Non-protein nitrogen	0,6 ± 0.03	0.5 ± 0.05	0.8 ± 0.06	0.6 ± 0.02
"Blue liquid"				
Protein nitrogen	-	4,2 ± 0.05	3.4 ± 0.02	2.0 ± 0.05
Non-protein nitrogen	-	2,5 ± 0.03	2.4 ± 0.03	4.1 ± 0.06
Pigments content in algae rest				
Phycobilines	0.45 ± 0.02	-	0.22 ± 0.01	0.21 ± 0.01
Chlorophyll a	1.00 ± 0.03	0.61 ± 0.03	0.71 ± 0.02	0.62 ± 0.03
Sum of carotenoides	0.50 ± 0.02	0.48 ± 0.02	0.38 ± 0.01	0.38 ± 0.02

Referring to Table 1 at “blue liquid” formation is observed the essential changes in nitrogen balance of cell. It were registered the changes in content of fat-soluble and phycobiline pigments. It was shown that content of chlorophyll *a* and carotenoids in algae cells are decreased. Content of phycobiline pigments decreased too. Main part of it passes into resulting “blue liquid”. Only insignificant amount preserved in algae cells. The content of nitrogen and protein in algae cells decreased as more “blue liquid” are developed, as more “blue liquid” are developed and its storage. Its content in cultural medium was increased. Storage of “blue liquid” under temperature of 20-22°C is accompanied of its decoloration, decreasing of fluorescence and changes of spectral peculiarities.

Other results we received under controlled changes of pH for natural algae population. As may be seen from Figure 2, the initial samples without of pH changes kept their spectral characteristic (Lines 1, 3). Under change of pH as well to acidic as basic side the spectral peculiarities of “blue liquid” changed. Under addition of HCl (up to pH 5-3) the experimental liquid turns green. Under addition of NaOH (up to pH 10-12) to solution, on the contrary it grows brown, turns yellow and fades. Under addition of acid the height of main absorption maximum is descended, at the beginning lowly and then sharply. Under addition of same quantity of alkali solution the absorption maximum at the beginning is shifted into long-wave range and then disappears totally. It denotes on substance destruction.

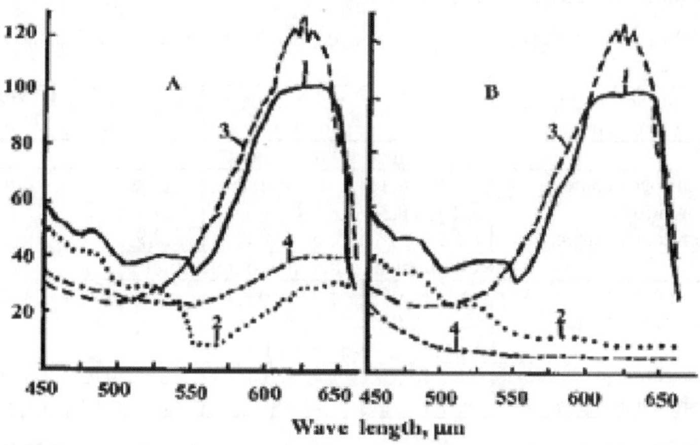

Figure 2. Influence of environmental pH on spectral properties of “blue liquid”. A – experiments with acidification: 1- initial sample, 28.10; 2 – HCl, 28.10; 3 – initial sample, 28.09; 4 – HCl, 28.09. B – experiments with alkalization: 1 – initial sample, 28.10; 2 – NaOH, 28.10; 3 – initial sample, 28.09; 4 – NaOH, 28.09.

It was shown the significant changes for algae cultures on Figures 3, 4. As may be seen, the alkalization of cultural medium from pH 6.9 to 11.9 sharply changed the spectral characteristic of "blue liquid". Under effect of pH change it is happened the "blue liquid" destruction and decreasing of main maximum of absorption, more sharply in alkaline environment. Alkalization of environment at the beginning increases slightly the blue coloration, but then fades it and causes yellowing (Figure 3). Under acidification (Figure 4) at the beginning liquid turns blue (pH 6.1-6.5) and then became green. Consequently, under pH change for algae cultures image of "blue liquid" destruction was similar to the same for natural populations.

Unfortunately, we didn't register the clear changes of toxicity under regulation of cultural medium pH. Level of water toxicity increased or decreased, but lethal effect for warm-blooded animal (white mice) kept stably though variation of time of its occurrence. Using of ELISA methods didn't help for us. The point is that the intensive blue color of water is barrier for estimation of its toxicity with help of standard method ELISA. This standard test is worked on the base of different intensive of blue tints, which were formed under addition of reagents.

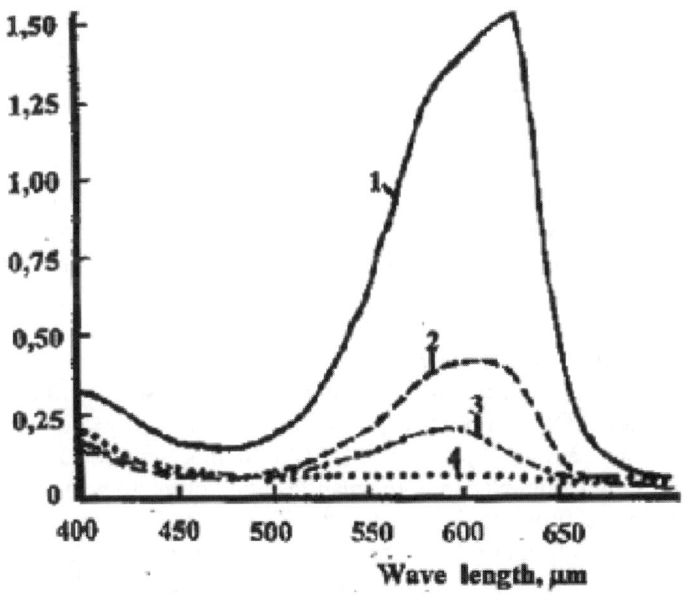

Figure 3. Influence of alkalization on spectral properties of "blue liquid" from *Microcystis*: 1 – pH 6.9, initial sample; 2 – pH 8.9; 3 – pH 10.3; 4 – pH 11.9 .

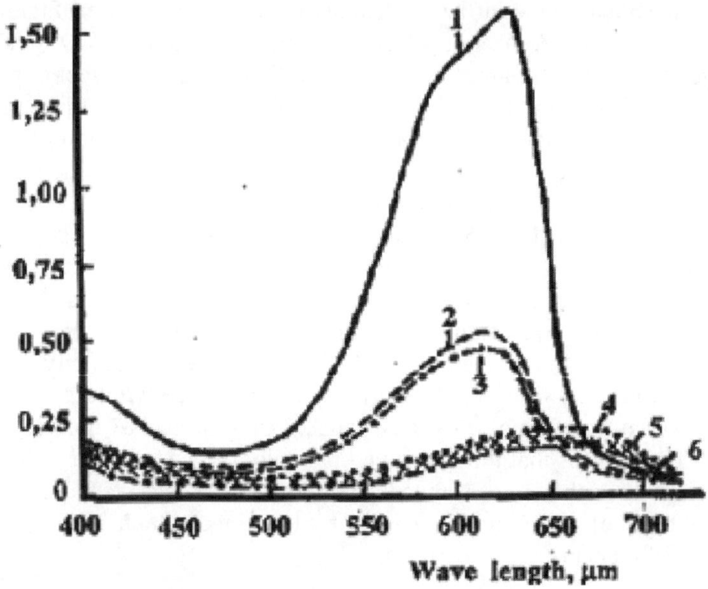

Figure 4. Influence of acidification on spectral properties of "blue liquid" from *Microcystis*: 1 – pH 6.9, initial sample; 2 – pH 4.2; 3 – pH 3.8; 4 – pH 2.4; 5 – pH 2.3; 6 – pH 1.5.

3. CONCLUSIONS

1. More stable coloring and its spectral peculiarities preserved the "blue liquid", which is formed as the result of leaving from blue-green algae cells – agents of "water blooming" the hydrophilic complex of phycobiline pigments. At parallel with it under cells destruction is increased the secretion of microcystine into water, which has high toxicity.
2. More stable color and spectral peculiarities the hydrophilic complex of bilichromeproteides is preserved under neutral pH – 6.8-7.2. Alkalization up to pH 9-12 and acidification of cultural medium cause the full destruction of blue color of phycobilines pro proportionally to pH, decreasing the absorption maximum on level of 610-630 nm, sedimentation of protein flakes, but without excepting the manifestation of microcystine toxicity.
3. Formation of "blue liquid" is caused with deep changes in nitrogen balance of cells. As its formation content of nitrogen and protein in cells is decreased in connection with its withdrawal into water, which is quick on initial stages and gradual later under joint storage of algae and liquid.

4. Toxicity of this suspension preserved during long term because of microcystine presence.
5. Estimation of microcystine concentration in “blue liquid” (under presence of phycobiline complex) with using of ELISA requires the elaboration of new methodical receptions.

4. ACKNOWLEDGEMENTS

Author is grateful to Prof. L.Sirenko for help and useful suggestions. This study was partially supported by the Grant of Ukrainian Ministry on Science and Education N 06.07 / 82.

5. REFERENCES:

1. Bryant, Donald, *The Molecular Biology of Cyanobacteria.* Dordrecht-Boston-London: Kluwer Acad.Publ, 1995.
2. Parshikova, Tatyana, *The structure-functional markers of adaptation for microalgae under surfactant effect,* Thesis of dissertation for obtaining the scientific degree (doctor of biological sciences). Specialty 03.00.12 – Physiology of plants. Kiev: Scientific World, 2003.
3. Sirenko, Lidiya and Gavrilenko, Maria, *Water “blooming” and eutrophycation.* Kiev: Naukova Dumka, 1978.
4. Canter-Lund, Hilda and Lund, John, *Freshwater algae, their microscopic world explored.* Bristol: Biopress Ltd, 1995.
5. Kirpenko, Yuriy et al., *Toxins of blue-green algae and animal organism.* Kiev: Naukova Dumka, 1977.
6. Carmichael, W.W. The toxins of Cyanobacteria. Scientific American 1994; 270 (1): 78-86.
7. Eocha, C.O. Spectral properties of the phycobilins. 1. Phycocyanobilin. Biochemistry, 1963; 2 (2): 375-382.

LEAF FLUORESCENCE AS DIAGNOSTIC TOOL FOR MONITORING VEGETATION

Simona Apostol
Physics Department, Faculty of Sciences, Valahia University, 24 Bd. Unirii, 0200 Targoviste, Romania

Abstract In this paper we present the potential offered for obtaining, in a non-invasive and non-destructive way, information about photosynthesis and physiological state of vegetation, by using laser induced fluorescence. Fluorescence is a very sensitive and versatile tool, allowing to extract information about photosynthetic apparatus architecture and functioning, from molecular level up to integrated level in leaves, plants or even for monitoring vegetation at large scale (canopy or ecosystem). We will present here results obtained by using fluorescence emitted by leaves, for discriminating nitrogen need in a corn filed. We have reconstruct from the collected samples the map of nitrogen availability in the corn field.
Some fluorescence parameters issued from UV and VIS excitation of chlorophyll, especially the ratio FRFexUV/FRFexVIS measuring the epidermis UV transmittance, seemed to be more specifically related to the nitrogen content of leaves and precluded an ambiguous response as the case of the more complex BGF/ChlF ratio.

Keywords: Photosynthesis, Chloroplast, Laser-Induced Fluorescence

1. INTRODUCTION

Remote sensing system based on laser-induced fluorescence (LIF) has a great potential in terrestrial vegetation mapping for detecting crop stress like nitrogen (N) and water deficiency that are the most yield-limiting factors for the plant growing. A precision farming for crop management may lead to more efficient use of agronomic inputs by treating specific areas in an agricultural field.

Fluorescence is well known as a very sensitive tool for giving information about molecules, structures and interactions in biological systems. But fluorescence may be used also *in vivo* for integrated systems, if the difficulties to assign the origin of signal in a very complex assembly are overpass. UV-induced fluorescence from leaves is a potential tool to monitoring vegetation since it probe the plant response to stress before

V. Evangelista et al. (eds.), From Cells to Proteins: Imaging Nature across Dimensions, 423–430.

visible symptoms occur: it is non-invasive and it can be remotely and rapidly recorded [1, 2, 3].

In figure 1 it can be seen the spectral characteristics of the fluorescence emission from a leaf under UV excitation.

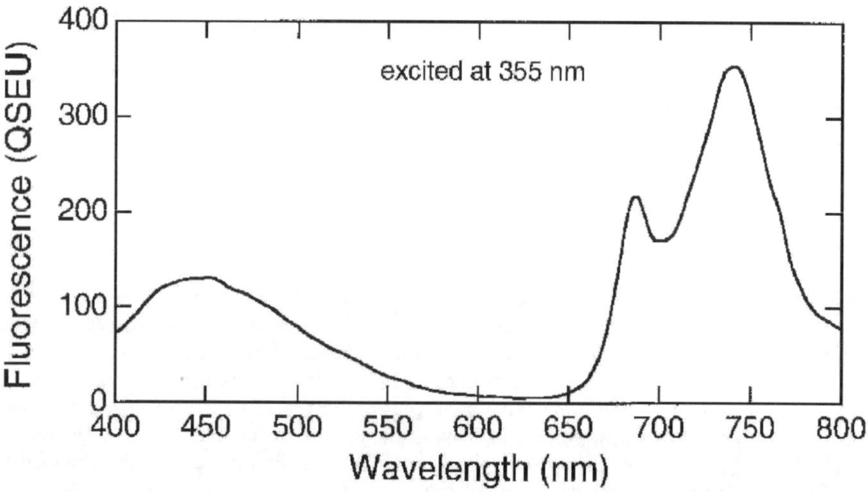

Figure 1. Fluorescence emission spectrum of a wheat leaf under 355 nm excitation.

Green plants emit a red fluorescence (RF, 630-700nm), a far-red fluorescence (FRF, 700-800nm), both from chlorophyll, and a blue-green fluorescence (BGF, 400-630 nm) coming mainly from hydroxycinnamic acids (for reviews see [3, 5]. Among the different groups of plants, the Poaceae which comprises the cereals, emitted the brightest BGF because of their large content of ferulic acid bound to cell walls [6]. The information on the anatomical and chemical origins of BGF comes from UV-fluorescence microscopy [7, 8, 9], and from *in vivo* fluorescence spectroscopy [10, 11, 12, 13]. Ferulic acid linked to cell wall was the major fluorophore identified in most plant species, taking into account the shape of spectrum and the time-resolved analysis. Moreover, it has been shown that the BGF from an intact leaf emanates mainly from the cutinized epidermis [8, 14]. BGF varies dependent on the anatomy and during leaf development [15].

On the other hand, UV-induced RF or FRF emanates from mesophyll chloroplasts and depends not only on the Chl content but also on the flavonoid content of the epidermis vacuoles that screen UV-excitation of the mesophyll [16, 17, 18].

Figure 2 show the microscopic image of the fluorescence emission from a leaf. It can be observed the main two emission of fluorescence from a leaf under UV excitation: blue-green fluorescence (BGF) and red fluorescence coming from chlorophyll.

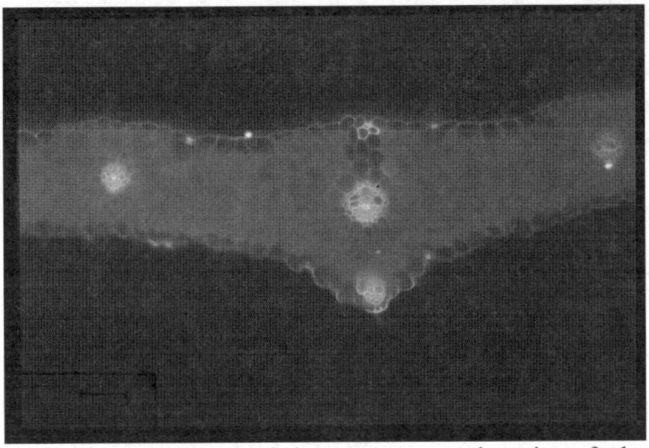

Figure 2. Color autofluorescence micrograph from a transversal section of a barley leaf.

Leaf fluorescence can detect nutrient stress on vegetation by blue-green (BGF) or chlorophyll fluorescence (ChlF) analysis, and is a non-destructive and non-intrusive probe of plant status (for a review see [5]). However, historically, it was mainly used as a close contact technique. An important limitation of the fluorescence LIDARs in detecting vegetation is the measurement of fluorescence intensity, an extensive parameter depending on distance, canopy structure and atmospheric transmission. Alternatives may be the use of fluorescence ratio from different spectral bands, similar to the reflectance issued parameters, or that of light-induced variable fluorescence, as presented in this paper.

The ratio of BGF to RF has been used for remote sensing purposes as a signature of plant health status under a variable availability of nutrients [19] and under different light condition affecting leaf development [8]. Also the ratio between RF and FRF was widely used related to different stress affecting chlorophyll concentrations in leaves. German group from Karlsruhe developed instruments for imaging these ratios at leaf level and intensively studied the spectral change and distribution of fluorescence on leaf surface depending on the physiological state of plant [19. 20, 21].

2. MATERIAL AND METHODS

Corn plants (*Zea mays* L., DeKalb 389Bt variety) were grown under field conditions. 16 experimental plots of 20 m x 20 m were prepared, to which four nitrogen fertilizer treatments, A(0+0), B(0+144), C(0+165-180), D(100+150) kg, were randomly applied at seeding and respectively topdressing in order to produce a gradient between low and high N treatments.

Four plants from each plot area were sampled and brought into the laboratory for subsequent spectral measurements. Plants were stored in a dark cool place and acclimated to the room temperature (24°C) prior to spectral measurements. Fluorescence measurements were done on the youngest full-expanded leaves of corn plants, 70 days aged (full development).

Estimation of Chl content from the same leaves was done using a SPAD chlorophyll meter (SPAD-502 Minolta, Osaka Co Ltd., Japan) measuring leaf transmittance.

Fluorescence emission spectra of dark adapted leaves following laser excitation at different wavelength 308, 360, 440, 480 and 630 nm were recorded using a compact multi-wavelength Fluorescent Lidar System model PL (FLS-PL) prototype [22]. The laser source was an excimer (XeCl) laser emitting nanosecond pulses at 308 nm, that may pump a system of four alternative dyes in order to provide additional excitation wavelengths at 360, 440, 480 and 630 nm. Fluorescence emission was remotely (5 m) sensed by a CCD array that enables the simultaneous measurements of emitted fluorescence between 380 and 850 nm, via a grating polychromator. Fluorescence emission spectra were normalized to the intensity of excitation wavelengths, then fluorescence ratios combining BGF, RF and FRF emission following different excitations were computed. Results from the most relevant of these parameters are shown. Four samples were collected from each of the 16 plots. Maps were constructed using MapInfo software using the measurements from those 64 samples uniformly distributed into the field .

After fluorescence measurements were taken, plant shoots were dried during 48 hours at 70ºC. The determination of total N in plant tissue was determined by colorimetry using an autoanalyser Traacs 800.

3. RESULTS AND DISCUSSION

The four nitrogen treatments applied on field plots produced significant differences in N content of corn plants, as well as in leaf Chl content and biomass production (see Figure 3A, B and C).

We observed that Chl fluorescence emission excited by UV radiation increased in high N-treated plants comparing to the N-deficient, where ChlF emission remained always very low. The low ChlF emission under UV compared to VIS excitation is attributable to an UV-screening effect of the epidermis on field-cultivated plants, preventing UV radiation to reach mesophyll cells in order to excite Chl. It is well known that UV radiation induces the synthesis and the epidermal accumulation of UV-absorbing compound, as flavonoids, which are involved in protective photochemical mechanisms [23, 24]. On the other hand, the relative emission of red and far-red fluorescence RF/FRF ratio differed between N-treated and N-deficient plants, depending also on the excitation wavelength (data not shown). With increasing N availability, coupled with increased chlorophyll content in the leaf, the relative red fluorescence at 690 nm becomes smaller. This is attributable to the reabsorption of the emitted shorter-wavelength red fluorescence by *in vivo* chlorophyll, since Chl absorption and fluorescence emission bands overlap in this spectral region. The FRF emission of Chl fluorescence in the region of 735 nm is very little or not affected by reabsorption [25] and develops with the increasing of Chl content.

As a results, the ratio RF/FRF630 (Figure 3C) exhibited lower values for fully green plants from high-N plots, compared to low content Chl plants, as already reported by many authors under different environmental conditions [26, 27, 28]. The (RF/FRF630) ratio was very sensitive at low Chl content. Still it reached saturation quickly at medium Chl content, which makes it unsuitable to discriminate N-stress at early stages. The ratio between far-red ChlF excited under 360 and 440 nm (FRF360/FRF440) showed an important N-discriminating potential. A significant decrease in that ratio was observed in N-deficient plants, correlated with decreases in nitrogen and chlorophyll content of leaves.

The (FRFexUV)/(FRFexVIS) ratio was shown to measure the epidermal UV transmittance [1, 22] confirmed that such ratio was proportional to the epidermal UV transmittance. The decrease of the (FRF360/FRF440) we observed in N-deficient plant (Figure 3F) is correlated with a decrease of the Chl content, but is mainly due to the increases of the epidermal UV-screening. The (FRF360/FRF440) parameter is significantly correlated with the leaf N-content (R^2=0.55).

4. CONCLUSIONS

In this study, remote sensing of plant fluorescence produced accurate signatures of nitrogen stress. The technique is rapid, non-invasive and non-destructive compared to conventional contact measurements, offering a high potential for its further use in Precision Agriculture. The LIF spectral analysis brings specific information on the nitrogen status by the (ChlFexUV)/(ChlFexVIS) ratio in addition to the more widely used (RF/FRF) or (BGF/ChlF) ratios.

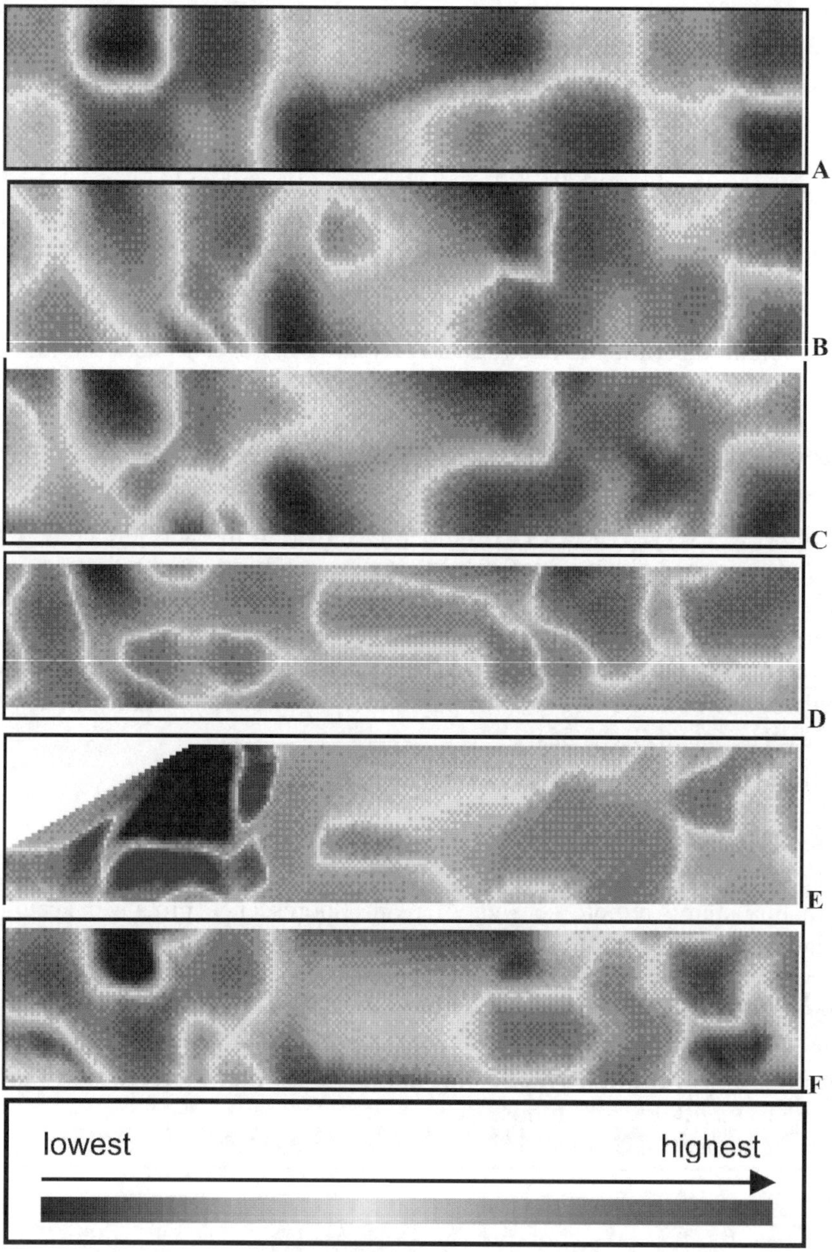

Figure 3. 2D Map of different biochemical or fluorescence computed parameters measured on corn plants at 70 days plant age from the experimental fields.
N treated plots distribution (A); Chlorophyll content (B); nitrogen content (C); FarRed/Red fluorescence ratio under 630 nm excitation FRF/RF630 (D); Blue-green over far-red emission ratio following UV excitation at 308 nm FRF/BGF308 (E); Far-red emission fluorescence ratio under UV 360 nm and VIS 440 nm excitation,FRF360/FRF440 (F) .

5. ACKNOWLEDGEMENTS

We acknowledge the contribution of Nicolas Tremblay research team from Agriculture and Agri Food Canada for their assistance in plot layout and for performing measurements in the N-stress experiment. Thanks are due also to Nicolae Moise from LURE Orsay, France for the gift of microscope fluorescence image.

6. REFERENCES

1. Chappelle, E. W., F. M. Wood, J. E. McMurtrey, and W. W. Newcomb. Laser-induced fluorescence of green plants. 1: A technique for the remote detection of plant stress and species differentiation. Appl. Opt. 1984; 23: 134-138.
2. Chappelle, E. W. "Fluorescence Measurements of Vegetation". Special issue of Remote Sensing of Environment. *In* Remote Sensing of Environment. Elsevier, New York. 105 1994.
3. Ounis, A., Z. G. Cerovic, J.-M. Briantais, and I. Moya. Dual excitation FLIDAR for the estimation of epidermal UV absorption in leaves and canopies. Remote Sens. Environ. 2001 (b); 76: 33-48.
4. Buschmann, C., and H. K. Lichtenthaler. Principles and characteristics of multi-colour fluorescence imaging of plants. J. Plant Physiol. 1998; 152: 297-314.
5. Cerovic, Z. G., G. Samson, F. Morales, N. Tremblay, and I. Moya. Ultraviolet-induced fluorescence for plant monitoring: present state and prospects. Agronomie: Agriculture and Environment. 1999; 19: 543-578.
6. Harris, P. J., and R. D. Hartley. Phenolic constituents of the cell walls of monocotyledons. Biochem. Syst. Ecol. 1980; 8: 153-160.
7. Harris, P. J., and R. D. Hartley. Detection of bound ferulic acid in the cell walls of the Gramineae by ultraviolet fluorescence microscopy. Nature 1976; 259: 508-510.
8. Stober, F., and H. K. Lichtenthaler. Studies on the localization and spectral characteristics of the fluorescence emission of differently pigmented wheat leaves. Bot. Acta. 1993 (b); 106: 365-370.
9. Hutzler, P., R. Fischbach, W. Heller, T. P. Jungblut, S. Reuber, R. Schmitz, M. Veit, G. Weissenböck, and J.-P. Schnitzler. Tissue localization of phenolic compounds in plants by confocal laser scanning microscopy. J. Exp. Bot. 1998; 49: 953-965.
10. Cerovic, Z. G., E. Langrand, G. Latouche, F. Morales, and I. Moya. Spectral characterization of NAD(P)H fluorescence in intact isolated chloroplasts and leaves: Effect of chlorophyll concentration on reabsorption of blue-green fluorescence. Photosynth. Res. 1998; 56: 291-301.
11. Johnson, G. A., S. V. Mantha, and T. A. Day. A Spectrofluorometric Survey of UV-Induced Blue-Green Fluorescence in Foliage of 35 Species. J. Plant Physiol. 2000; 156: 242-252.
12. Morales, F., Z. G. Cerovic, and I. Moya. Time-resolved blue-green fluorescence of sugar beet (*Beta vulgaris* L.) leaves: Spectroscopic evidence for the presence of ferulic acid as the main fluorophore in the epidermis. Biochim. Biophys. Acta. 1996; 1273: 251-262.
13. Lichtenthaler, H. K., and J. Schweiger. Cell wall bound ferulic acid, the major substance of the blue-green fluorescence emission of plants. J. Plant Physiol. 1998; 152: 272-282.
14. Bongi, G., A. Palliotti, P. Rocchi, I. Moya, and Y. Goulas. Blue-green fluorescence excited by UV laser on leaves of different species originates from cutin and is sensitive to leaf temperature. Plant Cell Environ. 1994; 17: 777-780.

15. Meyer, S., A. Cartelat, I. Moya, and Z. G. Cerovic. UV-induced blue-green and far-red fluorescence along wheat leaves: a potential signature of leaf ageing. J. Exp. Bot. 2003; 54: 757-769.
16. Ounis, A., Z. G. Cerovic, J.-M. Briantais, and I. Moya. DE-FLIDAR: A new remote sensing instrument for the estimation of epidermal UV absorption in leaves and canopies. In EARSeL Proceedings. R. Reuter, editor. EARSeL, Dresden/FRG. 2001 (a); 196-204.
17. Day, T. A., B. W. Howells, and W. J. Rice. Ultraviolet absorption and epidermal-transmittance spectra in foliage. Physiol. Plant. 1994; 92: 207-218.
18. Bilger, W., M. Veit, L. Schreiber, and U. Schreiber. Measurement of leaf epidermal transmittance of UV radiation by chlorophyll fluorescence. Physiol. Plant. 1997; 101: 754-763.
19. Heisel, F., M. Sowinska, J. A. Miehe, M. Lang, and H. K. Lichtenthaler. Detection of nutrient deficiencies of maize by laser induced fluorescence imaging. J. Plant Physiol. 1996; 148: 622-631.
20. Lichtenthaler, H., and J. Miehé. Fluorescence imaging as a diagnostic tool for plant stress. Trends Plant Sci. 1997; 2: 316-320.
21. Langsdorf, G., C. Buschmann, F. Babani, M. Sowinska, M. Mokry, F. Timmermann, and H. K. Lichtenthaler. Multicolour fluorescence imaging of sugar beet leaves with different N-status by flash lamp UV-excitation. Photosynthetica. 2000; 38: 539-551.
22. Samson, G., N. Tremblay, A. E. Dudelzak, S. M. Babichenko, L. Dextraze, and J. Wollring. Nutrient stress of corn plants: early detection and discrimination using a compact multiwavelength fluorescent lidar. In EARSeL Proceedings. R. Reuter, editor. EARSeL, Dresden, Germany. 2001; 214-223.
23. Stober, F., and H. K. Lichtenthaler. Characterization of the Laser-Induced Blue, Green and Red Fluorescence Signatures of Leaves of Wheat and Soybean Grown Under Different Irradiance. Physiol. Plant. 1993 (a); 88: 696-704.
24. Schweiger, J., M. Lang, and H. K. Lichtenthaler. Differences in fluorescence excitation spectra of leaves between stressed and non-stressed plants. J. Plant Physiol. 1996; 148: 536-547.
25. Gitelson, A. A., C. Buschmann, and H. K. Lichtenthaler. Leaf chlorophyll fluorescence corrected for re-absorption by means of absorption and reflectance measurements. *J.* Plant Physiol. 1998; 152: 283-296.
26. Subhash, N., O. Wenzel, and H. K. Lichtenthaler. Changes in blue-green and chlorophyll fluorescence emission and fluorescence ratios during senescence of tobacco plants. Remote Sens. Environ. 1999; 69: 215-223.
27. Hak, R., H. K. Lichtenthaler, and U. Rinderle. Decrease of the chlorophyll fluorescence ratio F690/F730 during greening and development of leaves. Rad. Environ. Biophys. 1990; 29: 329-336.
28. Lichtenthaler, H. K., O. Wenzel, C. Buschmann, and A. Gitelson. Plant stress detection by reflectance and fluorescence. Ann. N.Y. Acad. Sci. 1998; 851: 271-285.

DYNAMIC HOLOGRAPHY FOR STUDY OF NONLINEAR OPTICAL PROCESSES IN BIOLOGICAL PHOTORECEPTOR MOLECULE

E. Korchemskaya[1], D. Stepanchikov[2], S. Bugaychuk[1] and N. Burykin[3]
[1] *Institute of Physics, National Academy of Sciences, 46 Prospect Nauki, Kiev 03028 Ukraine;*
[2] *Zhytomir State University, Zhytomir 10008 Ukraine;*
[3] *Institute of Applied Optics, National Academy of Sciences, Kiev 01054 Ukraine.*

Abstract It is proposed to apply dynamic holography method for exploring real-time phototransformations in the biological photoreceptor molecule. At dynamic holography grating recording, the interference of two coherent laser beams forms a modulation of the total light intensity at the sample plane. This leads to the space modulation of both absorption and refractive index in the sample and a dynamic holography grating is recorded. The dynamic holography is a zero background technique and it is possible to vary a special grating period from 0.2 to 100 micrometers in real time. The advantages of dynamic holography method for an investigation of functionally important processes in photoreceptor molecule are demonstrated with bacteriorhodopsin sample.

Keywords: Dynamic Holography Grating Recording, Real-Time Phototransformations, Light Intensity Dependence, Nonlinear Optical Processes.

1. INTRODUCTION

The biological function of photoreceptor protein molecules is to sense and transmit information concerning the color, intensity, and pattern of light in the environment [1]. Upon absorbing visual light, a photoreceptor molecule undergoes a photochemical cycle of several distinct intermediates accompanied by a spectral shift. Light-induced processes in biological photoreceptors have the nonlinear dependence on the light intensity of actinic light [2, 3]. It can be connected with both the multistep light energy transduction and presence of a spontaneous decay of the photocycle intermediates in parallel with their photoinduced formation.

Insight into the biological photoreceptor mechanism depends heavily on a capacity of the research method to reflect the natural biological functions. We proposed to apply dynamic holography method for exploring real-time phototransformations in the biological photoreceptor molecule [4, 5]. Dynamic holography is a simple but very sensitive method for detecting

V. Evangelista et al. (eds.), From Cells to Proteins: Imaging Nature across Dimensions, 431–450.

even very small changes in the phototransformations. At dynamic holography grating recording, the interference of two coherent laser beams forms a modulation of the total light intensity at the sample plane. This leads to the space modulation of both absorption and refractive index in the sample and a dynamic holography grating is recorded. By these means, the holography grating kinetics is very sensitive to the light intensity dependence of photoreceptor phototransformations.

Biological photoreceptor and transmembrane photosensitive protein bacteriorhodopsin is unique biological holographic material [6, 7]. The advantages of dynamic holography method for an investigation of functionally important processes in photoreceptor molecule are demonstrated with bacteriorhodopsin sample.

2. POLYMER FILMS WITH BACTERIORHODOPSIN

Bacteriorhodopsin (BR) is the key protein in the light-energy transducing system of the archaea bacterium *Halobacterium salinarum*. At the same time, a structure of BR molecule is similar to the visual human pigment rhodopsin. Upon absorbing visual light, the BR molecule undergoes a well-known photochemical cycle of several distinct intermediates accompanied by conformational changes. The intermediates can be spectroscopically distinguished, which are represented in Figure 1 by their common letter abbreviations. Their absorption maxima are given as subscripts. The initial form of BR, bR570 has an absorption peak at 570 nm, and the longest-lived shortwavelength-absorbing intermediate M412 has an absorption peak at 412 nm. All photocycle intermediates are photoactive.

BR is contained in a special part of the cell membrane of the bacterium, the so-called purple membrane. The optical and functional properties of the BR molecules are preserved in isolated purple membranes and even in dry films. This enables one to fabricate samples with purple membrane fragments both to be used in real-time optical information processing and in studying phototransformations in BR by means of dynamic holography

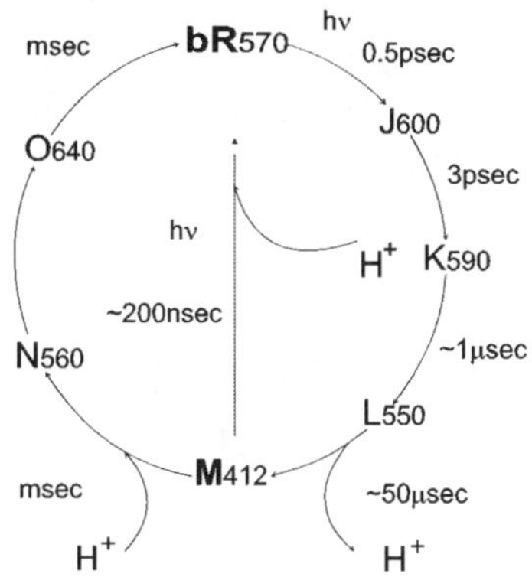

Figure 1. Simplified photocycle of bacteriorhodopsin.

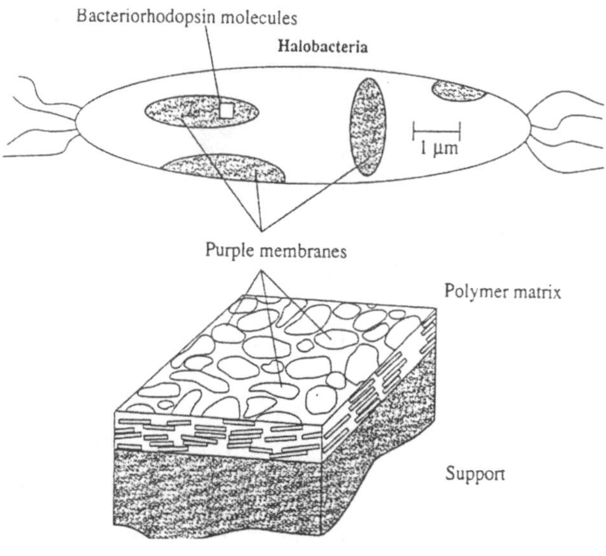

Figure 2. A schematic of a polymer film with photoreceptor bacteriorhodopsin.

methods. Figure 2 illustrates the fabrication of BR polymer films. This film comprises a dried layer of a mixture of purple membrane fragment suspension with water soluble polymer (we use gelatin) and may be produced using routine gel-casting procedure [8, 9]. During the drying process, the purple membrane fragments are self-oriented in the plane of the glass support; side orientation of purple membrane fragments is random.

Thus gelatin film with BR biological photoreceptor is a real biological material. A recording in BR films does not require a chemical treatment after an exposition. A real-time optical recording occurs. A clear example of the bacteriorhodopsin film will be useful for applying dynamic holography approaches to other biological photoreceptors.

3. ADVANTAGES OF DYNAMIC HOLOGRAPHY METHOD FOR STUDY OF REAL TIME LIGHT-INDUCED PROCESSES

3.1. Dynamic Holography Grating Recording

The interference of two coherent laser beams I_1, I_2 of equal intensity, $I_1=I_2$ (Figure 3) forms a modulation of the total light intensity I at the sample plane x (Figure 4). This leads to the space inhomogeneous generation of

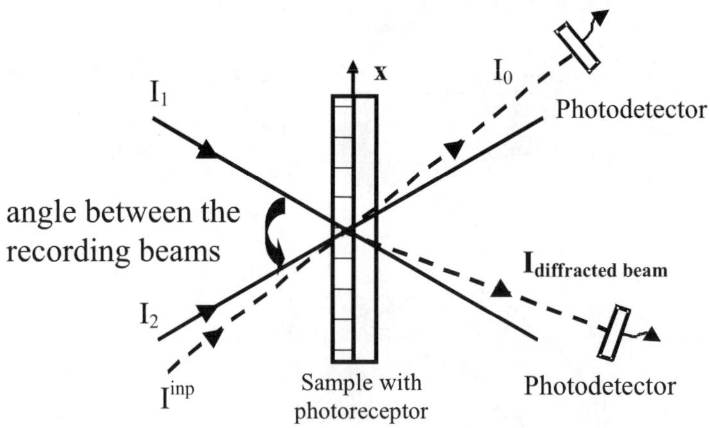

Figure 3. The scheme of dynamic holography grating recording and diffraction of independent probe beam on this grating.

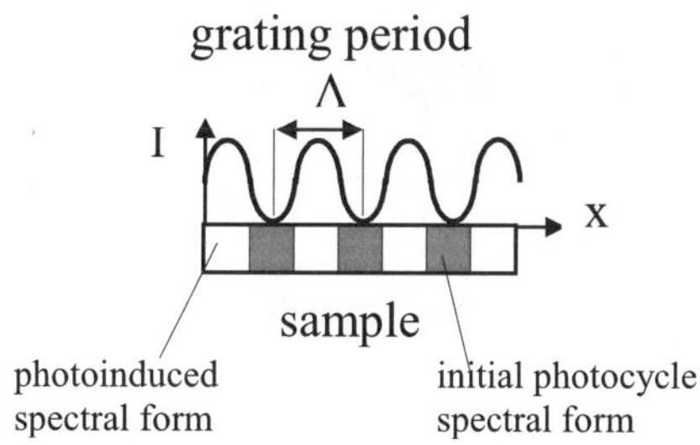

Figure 4. Schematic of space-inhomogeneous generation of initial photocycle form and photoinduced form at holographic grating recording in the photoreceptor sample. The interference of two coherent laser beams I_1, I_2 forms a modulation of the total intensity I at the film plane x.

photoproducts and protons, which results in the modulation of both the absorption and refractive index of the photoreceptor film. At the maximum of interference pattern the molecules transform to the photoinduced form, whereas at the minimum of interference pattern they leave in the initial photocycle form. By these means, a dynamic holography grating is recorded in the photosensitive sample. Once independent incoherent beam I^{inp} (Figure 3) passes throw this grating, it undergoes diffraction and new diffracted beam appear. Thus dynamic holography method is a zero background technique. A ratio between the intensities of diffracted and input beams $I^{diffracted\ beam}/I^{inp}$ is a diffraction efficiency of holographic grating. A diffraction efficiency of dynamic holography grating depends on a difference between the absorption and refractive index in the bright and dark fringes on the film plane. It enables the high sensitivity of dynamic holography method for detecting even small changes in the photoreceptor phototransformations. Phase grating based on the refractive index changes has considerably higher diffraction efficiency than grating based on absorption changes.

Another important advantage of this technique is a possibility to vary a special grating period (Figure 4) from 0.2 to 100 micrometers in real time. It is achieved by a variation of the angle between the recording beams I_1, I_2 (Figure 3). It enables the real time study of light induced processes both in the separate protein molecule and molecule cluster.

It is possible also the self-diffraction of recording beams in the thin (Raman-Nath) transmission holographic grating (Figure 5). Under self-diffraction conditions, the beams at the same time record holographic grating and diffract from the grating themselves. The He-Ne laser with λ=633 nm is

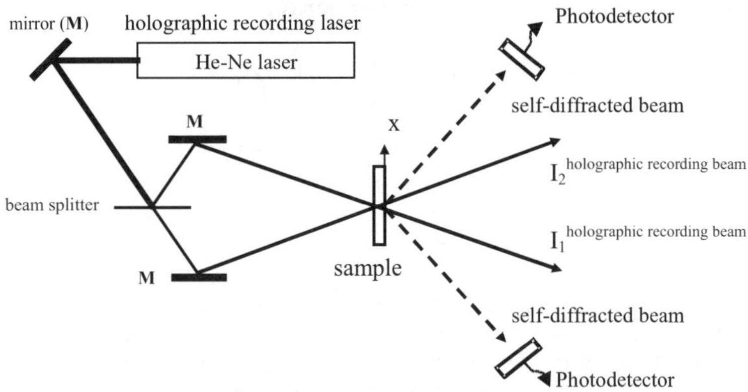

Figure 5. Experimental setup of the self-diffraction of recording beams in transmission thin holographic grating (Raman-Nath).

used for holography grating recording in BR film. On the wavelength λ=633 nm, the phase holographic grating is recorded in BR film. One of the key novel components of our approach is simultaneous observation of the kinetics of both self-diffraction of recording beams and diffraction of an independent probe beam in transmission thin grating (Raman-Nath). It enables a presence of any nonlocal process (for example, the light-induced aggregation or diffusion) to be indicated. For nonlocal processes, the kinetics of first-order self-diffracted and diffracted probe beams vary widely. For local processes, they have a similarly character.

3.2. Holographic Grating Recording Kinetics and Spatial Grating Profiles

Figure 6 shows a typical example of the diffraction efficiency kinetics of the first-order self-diffracted and diffracted probe beams for light-adapted BR film. The diffraction efficiency kinetics in BR films consists of two stages: an initial peak followed by a decrease to a steady-state. It is usual

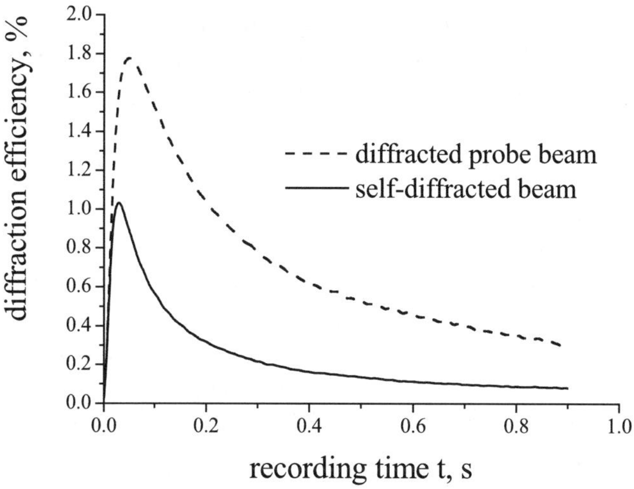

Figure 6. The typical diffraction efficiency kinetics of the first-order self-diffracted and diffracted probe beams for low-saturated optical material.

diffraction efficiency kinetics for low-saturated optical material. Biological photoreceptors have very small saturation intensity due to the high photosensitivity optimized evolutionary. A saturation intensity of all types of BR films is nearly 10 mW/cm^2. It leads to the distortion of the sinusoidal modulated intensity profile I of interference pattern (Figure 4) during the recording in BR film.

For a calculation of the spatial grating profile recorded in BR film by the sinusoidal modulated intensity I of interference pattern in Figure 4, the two level mechanism (bR570↔M412, Figure 1) of nonlinear absorption in low-light intensity saturating material was considered. The balance equation for bR570↔M412 transitions is:

$$\frac{dN_2}{dt} = \chi_1 A_1 \frac{I}{h\nu} N_1 - \frac{1}{\tau} N_2, \tag{1}$$

where N_1 is the bR570 population density, N_2 is the M412 population density, χ_1 is the absorption cross-section for the bR570, A_1 is quantum yield of bR570→M412 transition, I/hν is a photon flow density exciting initial form bR570, τ is the decay time of M412, $N_1 + N_2 = N_0$, where N_0 is the total population density of BR molecules. For equal intensities of the recording beams, I_1=I_2, the spatial-intensity modulation I(x) (Figure 4) is:

$$I(x) = I[1 + \cos(Kx)], \tag{2}$$

where $\Lambda_x = 2\pi / K$ is a grating space.

The dependence of the average decay time of M412 on the light intensity of actinic light which does not excite M $\tau(I)$ and hence $\tau(x)$ [4] is taken into account. The solution of the balance equation (1) is:

$$\frac{N_1(I_0,x,t)}{N_0} = \frac{1}{1+\beta(x)\cdot I(x)}\left[1+\beta(x)\cdot I(x)\cdot \exp\left(-\frac{t}{\tau(x)}(1+\beta(x)\cdot I(x))\right)\right] \tag{3}$$

where $\beta(x) = (\chi_1 A_1 / h\nu)\tau(x)$. The equation (3) describes the spatial modulation of light-induced absorption, or the spatial grating profile recorded in dynamic material. The spatial grating profiles at the different values of recording time t in the holographic recording kinetics of Figure 6 are shown in Figure 7. As is seen, at a steady-state a majority of BR molecules are in the photoinduced bleached form.

The grating amplitudes at the different spatial frequencies K_m can be found from the Fourier series:

$$\Delta n_m \sim \frac{K}{2\pi}\int_{-\pi/K}^{\pi/K} \frac{N_1(I_0,x,t)}{N_0}(x,I_0,t)\cdot \cos(mKx)dx\,\cdot \tag{4}$$

Thus a value of the diffraction efficiency of dynamic holography grating is determined by the spatial grating profile of Equation (3). The nonlinear optical processes are responsible for the spatial grating profile. For example, even the slight changes in the M412 decay time τ depending on the environment humidity reflect in the spatial grating profile (Figure 8). The value of total light intensity I(x) (Figure 4) was the same for humidity values 55 and 99%.

Also the light-dark adaptation of a photoreceptor molecule reflects in the spatial grating profile. In the dark BR relaxes in the mixture with equal fractions of all-*trans*- and 13-*cis*-retinal isomers. They have absorption maxima around 570 and 550 nm, correspondingly. Upon exposure BR becomes light-adapted and all the BR molecules convert to the initial photocycle form bR570 with all-*trans*-retinal. Only the molecules with all-

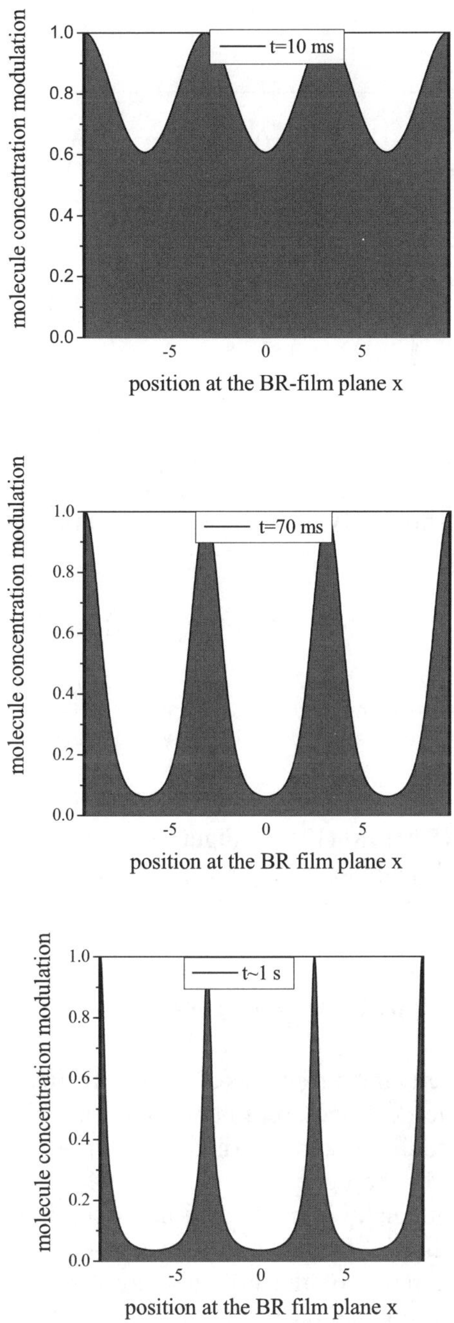

Figure 7. The spatial grating profiles recorded in BR film by the spatial-intensity modulation I(x) (Figure 4) at the different values of recording time t: t=10 ms in the recording beginning, t=70 ms in the diffraction efficiency peak, t=1 s in the diffraction efficiency steady-state (Figure 6).

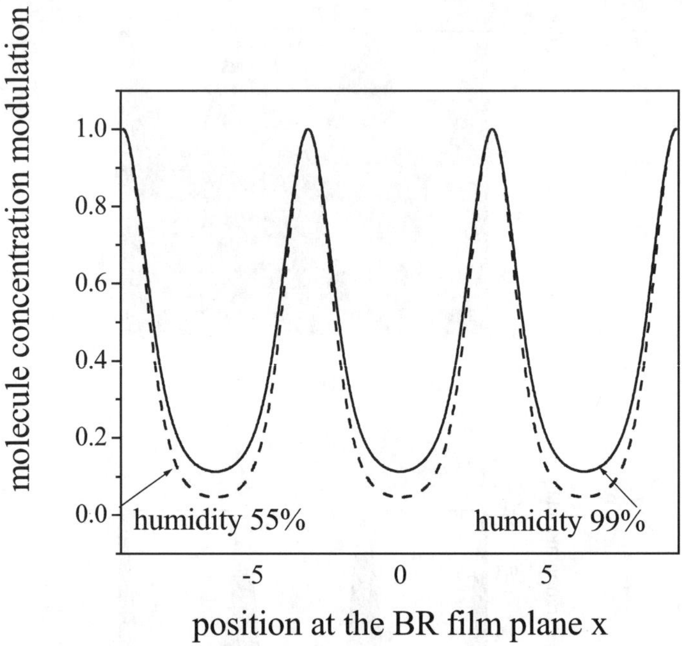

Figure 8. The spatial grating profiles recorded in BR film at environment humidity 55 and 99% by the same spatial-intensity modulation I(x) (Figure 4).

trans-retinal take part in the photocycle and reversible photobleaching (phototransition bR570⇔M412). In Figure 9, the spatial grating profiles for the beginning of holographic grating recording in the dark-adapted and light-adapted samples are shown.

3.3. Polarization Holographic Grating Recording

The photoinduced anisotropy takes place in BR film [2]. Likely, a majority of the biological photoreceptors possesses this property due to the photoreceptor molecule structure. An optical material with photoinduced anisotropy allows the holographic recording of both the usual intensity grating (only light intensity is modulated in space) and polarization grating (only light polarization is modulated in space) [4]. For polarization grating recording, an interference of the orthogonal-polarized recording beams I_1 and I_2 produces the modulation of polarization state in space, whereas the light intensity is not modulated (Figure 10). For polarization grating, it is also possible to vary a grating period Λ (a distance between the polarization state) from 0.2 to 100 micrometers. It is possible to study an influence of the light polarization on the molecule and molecular cluster formation.

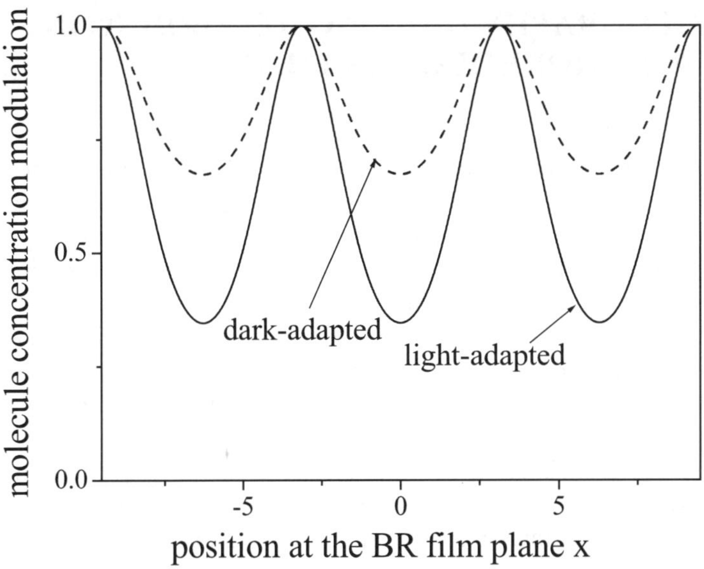

Figure 9. The spatial grating profiles for the beginning of holographic grating recording in the dark-adapted and light-adapted BR samples.

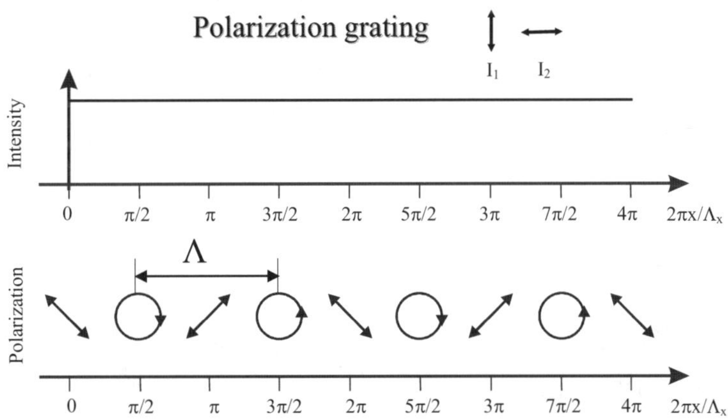

Figure 10. An illustration interpreting the polarization grating.

4. HUMIDITY EFFECT ON DYNAMIC HOLOGRAPHY GRATING RECORDING IN BACTERIORHODOPSIN FILM

It is known that biological processes depend strongly on the environment conditions. Mounting of a tiny cell containing a BR film inside the holographic set up allowed us maintaining and changing humidity in the cell for a long time (for several days) and at the same time perform holographic measurements on the film. Also a temperature changes are possible by means of tiny cell containing a BR film.

A comparison of the numerical and experimental holographic kinetics of first-order self-diffracted and diffracted probe beams in the transmission thin holographic grating (Raman-Nath) was used to study the light-induced changes in the purple membrane fragments, which occur at different humidity and are associated with proton release to the purple membrane surface [5]. An interference of the higher-order diffracted beams at the self-diffraction conditions in low-saturated BR films was taken into account. The experimental dependences of the peak and steady-state values of diffraction efficiency of the first-order self-diffracted and diffracted probe beams on humidity values are shown in Figure 11. A calculation of the diffraction

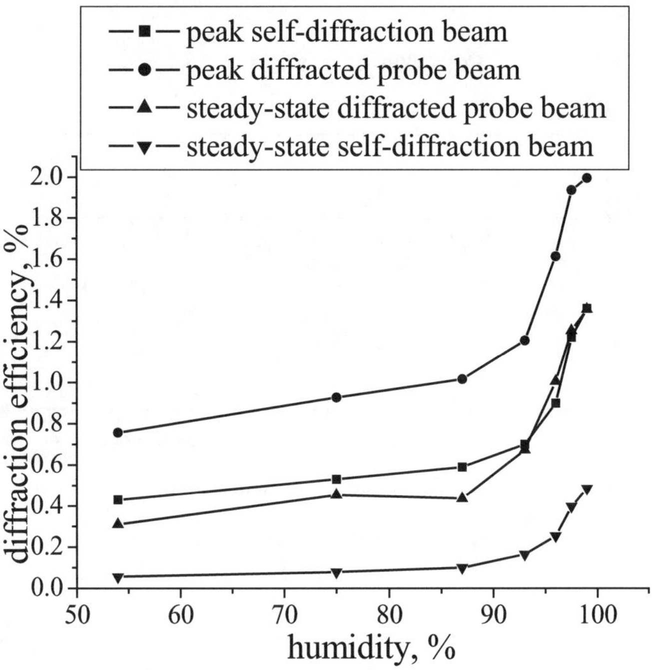

Figure 11. The experimental dependences of the peak and steady-state values of diffraction efficiency of the first-order self-diffracted and diffracted probe beams on humidity values.

efficiency of the first-order self-diffracted and diffracted probe beams at different humidity values (Figure 12) was performed on Equations (3-4) taking into account the experimental dependence of the average decay time of M412 τ in gelatin film with wild type BR on humidity and actinic light

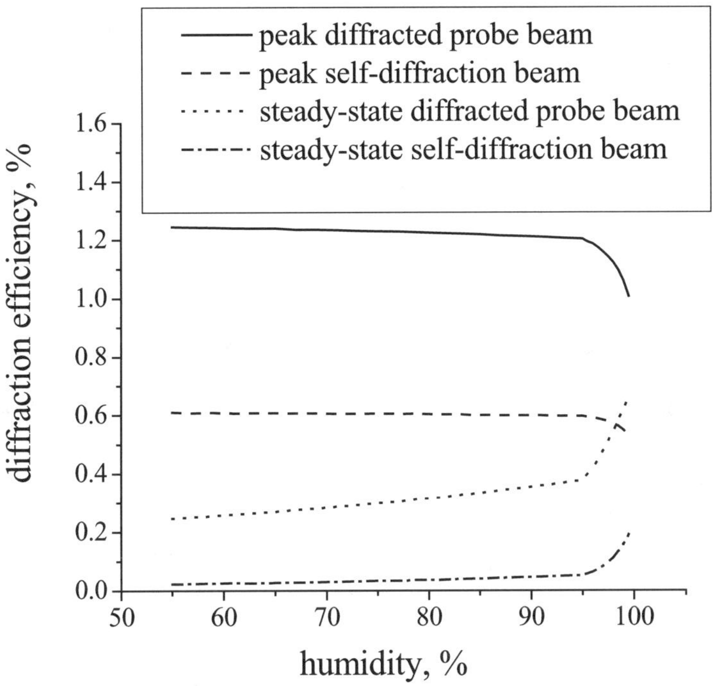

Figure 12. A calculation of the diffraction efficiency of the first-order self-diffracted and diffracted probe beams at different humidity values.

intensity [4]. As shown, the theoretical calculation of refractive index modulation due to the absorption changes (Kramers-Kroning relation) predicts a decrease in the peak value and increase in the steady-state value of diffraction efficiency at high humidity. Whereas the experimental results (Figure 11) demonstrate a drastic increase in the peak value of diffraction efficiency of the first-order self-diffracted and diffracted probe beams at high humidity. Hence large amplitude of the initial peak of diffraction efficiency at high humidity is due not only to the grating profile distortion caused by the absorption saturation (Figure 7) but also due to another processes accompanied proton release to the purple membrane surface. It thus appears that the latter process is more important in modulating the dynamic holographic recording than in the changes in pigment absorption spectrum. If it would be possible to observe the sample in microscope under the concurrent light irradiation, perhaps it would be more clearly the origin of these processes.

5. EXPLORING ANISOTROPY OF LIGHT-DARK ADAPTATION PROCESS BY DYNAMIC POLARIZARION HOLOGRAPHY

In the dark BR relaxes in the mixture with equal fractions of all-*trans*- and 13-*cis*-retinal isomers (Figure 13). Upon exposure BR becomes light-adapted and all the BR molecules convert to the initial photocycle form

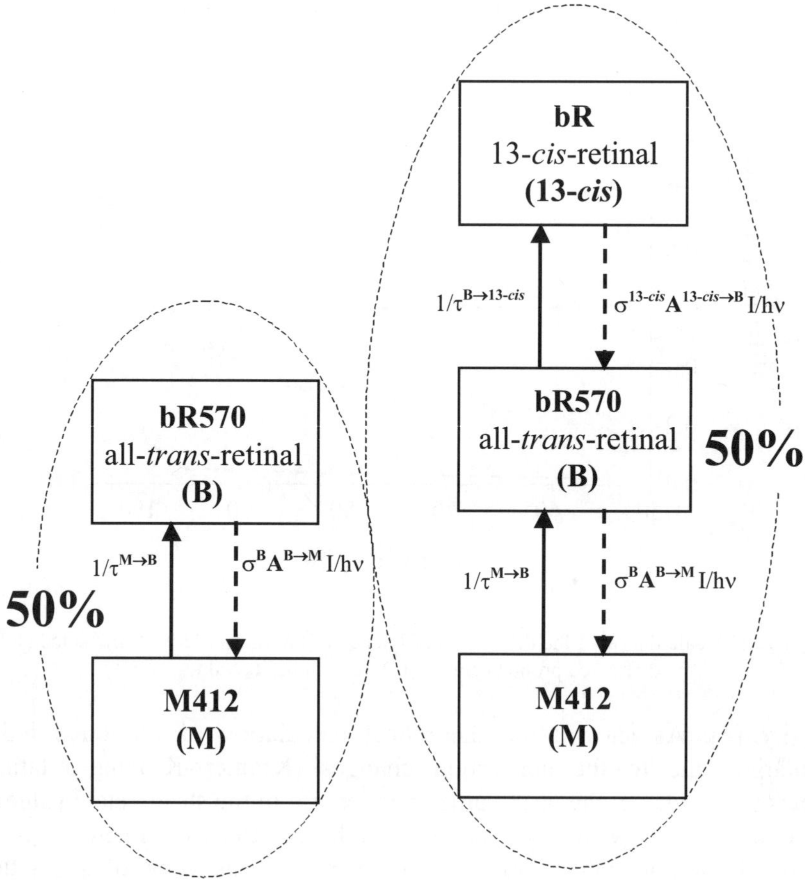

Figure 13. A simplified BR photocycle model for the light adaptation process. $\sigma^{13\text{-}cis}$ and σ^{B} are absorption cross-sections of BR molecules with 13-*cis*- and all-*trans*-retinal; $A^{13\text{-}cis\to B}$ and $A^{B\to M}$ are quantum yields of 13-*cis*→B and B→M phototransitions; $\tau^{B\to 13\text{-}cis}$ and $\tau^{M\to B}$ are average lifetimes of bR570 with all-*trans*-retinal and M412; I/hν is a photon flow density.

bR570 with all-*trans*-retinal. Only the molecules with all-*trans*-retinal take part in the photocycle and reversible photobleaching (phototransition bR570⇔M412). On the example of light-dark adaptation process, we demonstrate how dynamic holography approach might be applied for exploring multistep sequential phototransformations with different transition rates. For BR, the transition 13-*cis*→bR570 is considerably slower than transition bR570→M412.

Due to the weak divergence in spectra of dark- and light-adapted BR samples, an estimation of the molecular dichroism of the BR dark-adapted form with 13-*cis*-retinal is difficult to make from spectroscopic measurements. The anisotropy measurements on the oriented BR samples were also not able to detect the molecular dichroism changes in the plane parallel to membrane.

A molecular dichroism k : $k = \sigma_{\perp}/\sigma_{\parallel}$, where $\sigma_{\parallel}$ and $\sigma_{\perp}$ are absorption cross-sections for the light polarized parallel and perpendicular to a long absorption axis of a molecule, respectively.

The He-Ne laser, λ=633 nm, is used for the light adaptation and holography recording in BR films. Before a holography grating kinetics measurement, the sample was light adapted during one hour. The circular polarization of interfering beams was used for a holographic grating recording. For polarization grating, the two circular polarized beams rotating in the mutually opposite directions are recombined to form in space the grating of linear polarizations with no intensity modulation (Figure 14). For

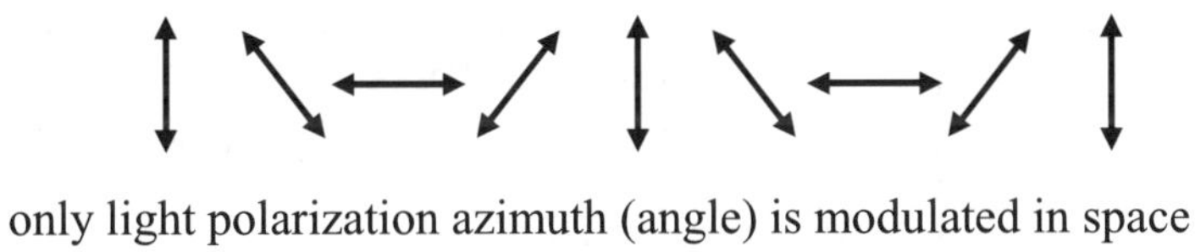

Figure 14. A space modulation patterns at circular polarization of interfering beams for polarization grating.

intensity grating, the two circular polarized beams rotating in the same directions are recombined to form the intensity modulation in space (Figure 15).

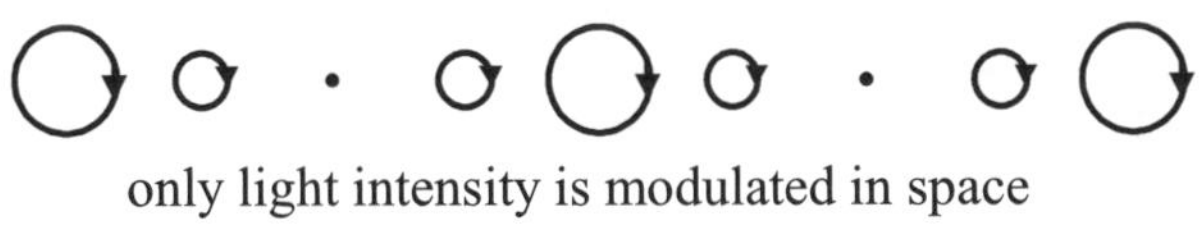

Figure 15. A space modulation patterns at circular polarization of interfering beams for intensity grating.

Two sets of experiments are performed [4]. The first set is light adaptation of the BR sample by uniform circular polarized irradiation (Figure 16). The holography recording kinetics on the dark- and light-adapted BR samples are shown in Figure 17. For this case, the holographic recording kinetics of polarization grating is coincident with one of the

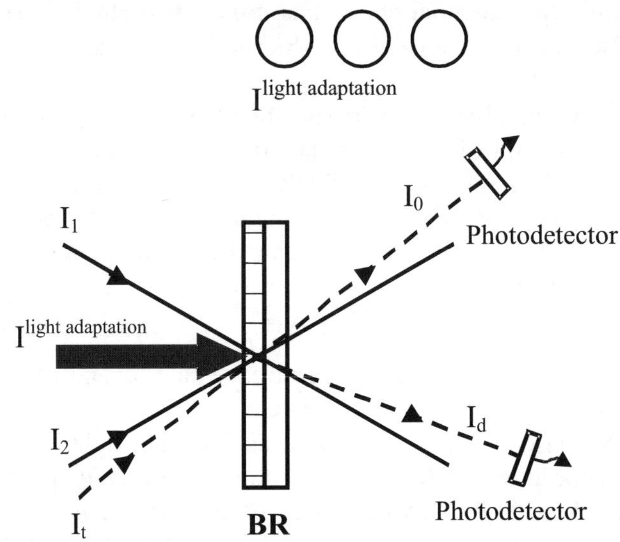

Figure 16. The scheme for light adaptation by uniform circular polarized irradiation and dynamic holography grating recording.

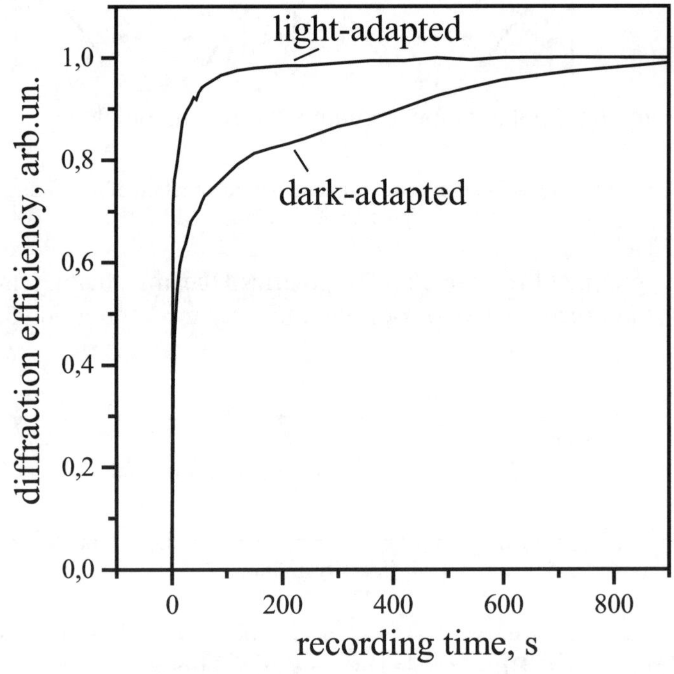

Figure 17. The diffraction efficiency recording kinetics on the dark-adapted BR film and BR film adapted by uniform circular polarized irradiation.

intensity grating. In the dark-adapted sample, initial signal is two fold smaller but due to the less depth of molecule concentration modulation (Figure 9) and slow increase takes place to a level of light-adapted sample.

The second set is the light adaptation of the BR sample by interference pattern of polarization grating (Figure 18). The holography recording

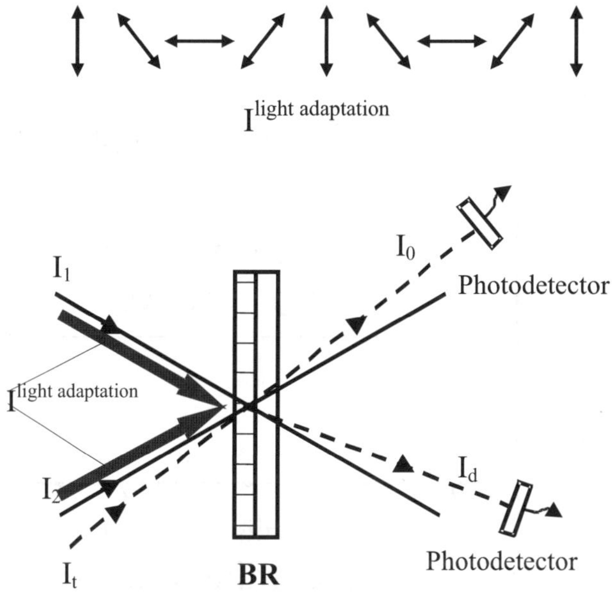

Figure 18. The scheme for light adaptation by the "fan" of linear polarizations with no intensity modulation (interference pattern of the polarization grating) and dynamic holography grating recording.

kinetics of the intensity and polarization gratings on light-adapted BR sample are shown in Figure 19. For the polarization grating (curve 1) with the same space polarization modulation as used for the light adaptation, the holographic recording kinetics is similar to the one in the sample adapted by the circular polarized irradiation. For the intensity grating (curve 2), the slow component is present corresponding to conversion of 13-cis-retinal to all-trans isomer.

The diffraction efficiency of holographic grating distinctly depends on the molecular dichroism of dark-adapted form of BR in the plane parallel to membrane. A comparison of the experimental results with calculated ones enables us to estimate a molecular dichroism of the BR dark-adapted form with 13-*cis*-15 syn retinal $k^{13\text{-}cis}$=0.14 [4], whereas a molecular dichroism of the bR570 with all-*trans*-retinal k^B=0.04 and a molecular dichroism of photocycle intermediate M412, which has 13-*cis*-15-anti retinal, *k_M*=0.11. The close molecular dichroism values *k_13-cis* and *k_M* probably indicate

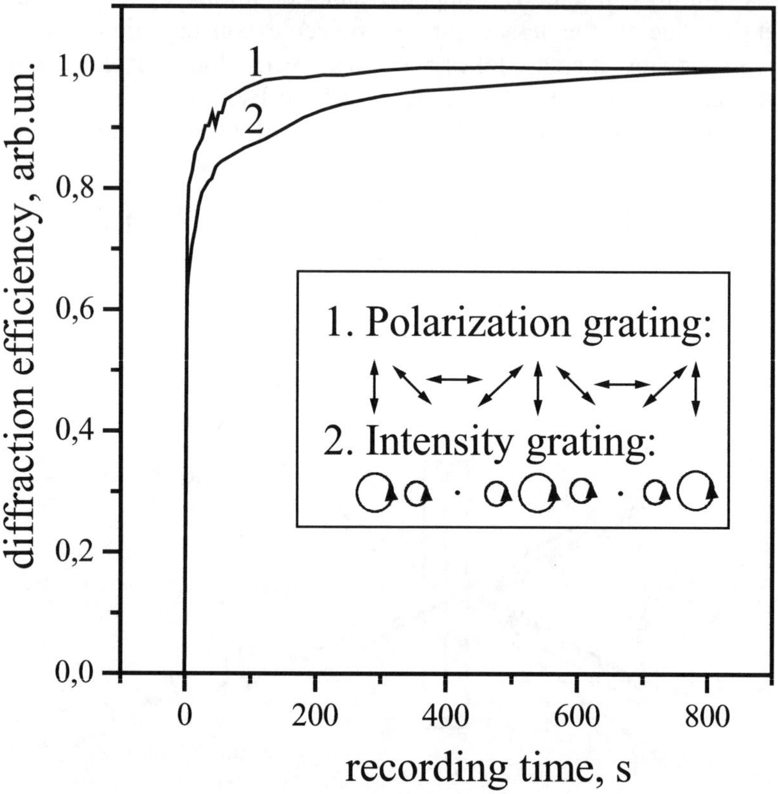

Figure 19. The diffraction efficiency recording kinetics for the intensity and polarization gratings on BR film adapted by the "fan" of linear polarizations: the curve 1 for polarization grating, curve 2 for intensity grating.

the effect of retinal chromophore isomerization on the molecular dichroism value.

6. REAL-TIME SELECTION BASED ON MOTION VELOCITY IN THIN HOLOGRAPHIC GRATING ON BR FILM

The high amplitude of the initial peak sharpness in self-diffraction holography recording kinetics in Raman-Nath thin grating is a useful feature for detection of rapid object movement [10]. Of particular interest is the situation when the movement has rapid small periodic shocks on the background of slower complex periodic vibrations (Figure 20). Among other things, closer examination of this pattern may be helpful in the context of a possibility to predict earthquakes. It is seen, that slow complex vibrations are

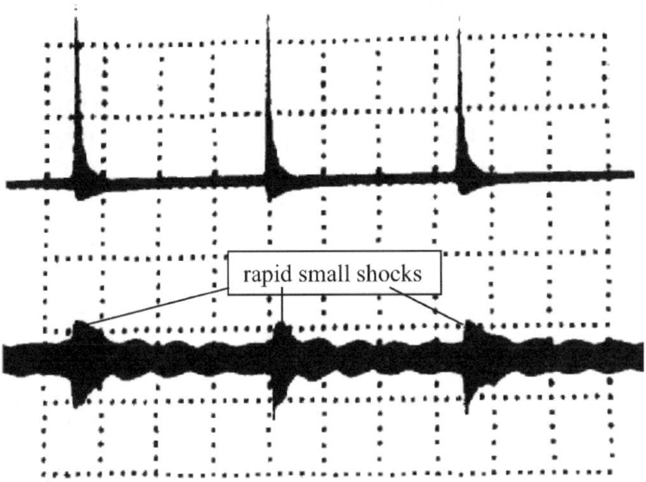

Figure 20. Real-time analysis of rapid small shocks on background of slower complex vibrations (bottom) in first-order self-diffraction beam on BR film (top).

suppressed whereas the small shocks are significantly increased in the first-order self-diffraction beam. Computational circuits were not needed with this method because it based on the natural BR ability to select the motion velocity in Raman-Nath thin grating. Utilization of the bacterium photoreceptor systems which are more evolutionary developed than Archaebacterium might considerably improve this method.

7. CONCLUSION

We demonstrate that dynamic holography is the sensitive nondestructive method for real time study of biological photoreceptor molecule. A value of the diffraction efficiency of dynamic holography grating is determined by the spatial grating profile, whereas the nonlinear optical processes are responsible for spatial grating profile. The important advantages of dynamic holography technique that it is a zero background technique, it is possible to vary a special grating period from 0.2 to 100 micrometers in real time. Polarization holographic grating recording allows one to study an influence of the light polarization on the molecule and molecular cluster formation. In combination with the light microscopy method, it opens a way to elucidate some problems in biological photoreceptor which remain unclear after a comparison of the holographic and spectroscopic results.

8. ACKNOWLEDGMENTS

We thank Mr. I. Basistiy (Institute of Physics, National Academy of Sciences, Kiev Ukraine) for a help in the result processing and Prof. S.P. Balashov (University of California, Irvine USA) for a useful discussion.

The research described in this publication was made possible in part by Award No. UB2-2427-KV-02 of the U.S. Civilian Research & Development Foundation for the Independent States of the Former Soviet Union (CRDF).

9. REFERENCES

1. Spudich J.L., Yang C.S., Jung K.H., Spudich E.N. Retinylidene proteins: structures and functions from archaea to humans. Annul Rev. Cell Dev. Biol. 2000; 16:365-392.
2. Korchemskaya E., Stepanchikov D., Druzhko A., Dyukova T. Mechanism of nonlinear photoinduced anisotropy in bacteriorhodopsin and its derivatives. Journal of Biological Physics 1999; 24:201-215.
3. Barsanti L., Passarelli V., Walne P.L., Gualtieri, P. The photoreceptor protein of *Euglena gracilis*. FEBS Letters 2000; 482:247-251.
4. Korchemskaya E., Stepanchikov D. Dynamic polarization holography for exploring photoinduced reactions in bacteriorhodopsin. Optical Memory & Neural Networks 2001; 10:137-146.
5. Korchemskaya E., Stepanchikov D., Burykin N., Basistiy I., Bugaychuk S., Ebrey T.G., Balashov S.P. Investigation of dynamic holography features accompanying phototransformation in bacteriorhodopsin. Proceeding of the International Symposium on Retinal Proteins: Experiments and Theory; 2004 September 19-21; Heidelberg, Germany.
6. Korchemskaya E., Soskin M. Polarization properties of four-wave interaction in dynamic recording material based on bacteriorhodopsin. Optical Engineering 1994; 33:3456-3460.
7. Hampp N. Bacteriorhodopsin as a photochromic retinal protein for optical memories. Chemical Reviews 2000; 100:1755-1776.
8. Dyukova T. "Properties of chemically and genetically modified bacteriorhodopsin films as optical storage media." In *Bioelectronic Applications of Photochromic Pigments*, Der A., Keszthelyi L., eds. Amsterdam - Berlin - Oxford - Tokyo - Washington, DC: IOS Press, 2001.
9. Druzhko A. "Bacteriorhodopsin and its derivatives as optical recording materials: photochromism of 4-keto bacteriorhodopsin and 14-F bacteriorhodopsin." In *Bioelectronic Applications of Photochromic Pigments*, Der A., Keszthelyi L., eds. Amsterdam - Berlin - Oxford - Tokyo - Washington, DC: IOS Press, 2001.
10. Korchemskaya E., Burykin N, Stepanchikov D. Properties of photorefractive gratings in bacteriorhodopsin films for real-time optical signal processing. OSA Trends in Optics and Photonics Series 2001; 62:370-378.

"SEEING" LIPID MEMBRANES BY SOLID-STATE NMR

Vasco Castro
Division of Physical Chemistry, Arrhenius Laboratory, Stockholm University, S-10691 Stockholm, Sweden

Abstract: Solid-state NMR has been for a long time a major tool for looking into membranes. Presently this versatile method is being used to understand other biological aspects such as the effect of anaesthetics or the mechanism of translocation of cell penetrating peptides.

Keywords: Membranes, Solid-State NMR, Lipids, Alcohols, Cell-Penetrating Peptides

1. INTRODUCTION

1.1. Membranes

Cell membranes have been subjected to very exhaustive studies in the past century, having an extremely important biological function. They serve not only as a barrier that separates the interior from the outside of the cell, but also as the structural base where transport systems and other structures crucial to the life of the cell can be found. Biological membranes are very diverse, but they all have a very thin structure (about 8 nm) constituted by lipids and proteins, containing different compositions depending on the physiological state. Membrane proteins serve as an important communication way between the exterior and the interior of the cell. However, lipids also serve as precursors for hormones or other signalling pathways like arachidonic acid derivates, and peptides are able to transfer signals over the membrane either by translocation or by binding to various receptors, which are often integral membrane proteins.

The liquid-crystalline lipid bilayer is the fundamental structure of all cell membranes, where cholesterol, proteins, ion channels and other structures may be found. In membranes, hundreds of different lipids may be found but the most common class present in membranes are the phospholipids, having a polar head group (hydrophilic part) and a hydrophobic hydrocarbon tail.

V. Evangelista et al. (eds.), From Cells to Proteins: Imaging Nature across Dimensions, 451–458.

Lipids can organize themselves in various ways, depending on their chemical structure, temperature, solvent conditions and other molecules present in the membrane. The fluid phase lipid bilayers (Lα) being the biologically most relevant has liquid crystalline properties. Lipids can also be present in the gel, cubic and the hexagonal phases, this last one being crucial for the Xanthophyll cycle enzyme activity [1].

The lipid bilayer is very elastic, having motions on timescales from femtoseconds to seconds. From rotations about chemical bonds (pico- to nanoseconds), passing by bond vibrations, torsional oscillations in the subpicosecond range, wobbling (10^{-9}s), collective undulations (10^{-6}s – 1s) and protrusion (individual movement of lipids along the bilayer normal) in the nanosecond range, the dynamics of lipid bilayers have been extensively studied by a large variety of different methods, like NMR, EPR, IR, Raman, CARS, X-Ray, DSC, neutron scattering, and on the recent years molecular dynamics simulations have become a major tool. But the motion that attracted most of the attention was the lateral diffusion in the plane of the bilayer (nanoseconds to microseconds).

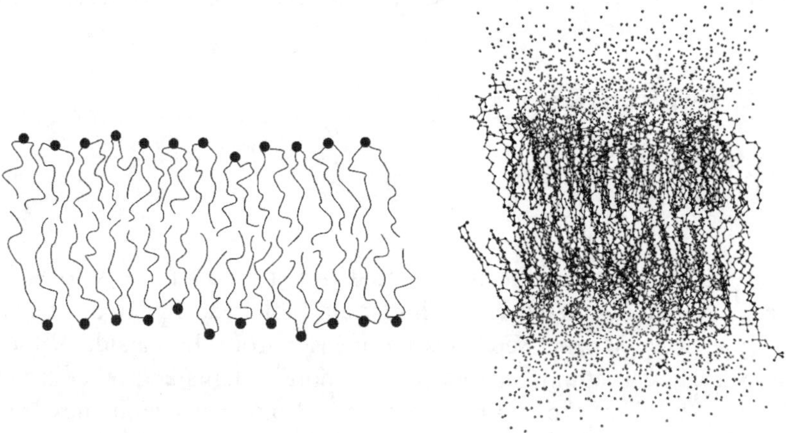

Figure 1. Schematic representation of a lipid bilayer in the Lα phase (left) and MD simulation of a lipid bilayer in the gel phase (right).

It is interesting to notice that different experimental methods (with a inherent measurement distance) result in different diffusion rates, because the lipids encounter barriers during their way. In the case of biological membranes the rate will be even lower due to the presence of regions with different molecular composition (lipid, protein, cholesterol, etc.). For the study of flip-flops (translocation of lipids, from one layer to the other, in the membrane), the difference between the model lipid bilayers and the biological one is bigger, since in the late ones there is an enzyme present called flippase that serves as a catalyzer for the flip-flop. The existence of flippases is of crucial importance, since the lipids are only synthesized in the inner leaflet of the cytoplasmic membrane and must translocate to the outer one to propagate the bilayer at a rate compatible with cell growth, which is incompatible with the flippase independent flip-flop rates ranging from milliseconds to hours.

1.2. Solid-State NMR

Nuclear magnetic resonance (NMR) is a powerful and versatile technique for the study of structure and dynamics of molecules. Although Stern and Gerlach verified that a beam of atoms under a non-uniform magnetic field undergo a deflection as they go through the field in the 1920s, it was only in the 1970s that NMR made a qualitative leap with the introduction of pulse techniques associated with mathematical analysis techniques based on Fourier transformations.

Presently, the principles of NMR techniques are based on the selective absorption of radio frequency (r.f.) waves by a sample inserted in a static magnetic field. By subjecting the sample to a magnetic field, a small majority of nuclear spins will be aligned with respect to the applied field in the lower energy state. When nuclei are irradiated with a radio frequency, energy is absorbed and nuclear spins flip from the lower energy state to the higher one (resonance). The excited nuclei will then return to the lower energy level emitting energy on the radiofrequency range. The precise determination of this energy and the time that takes for the nuclear spins to return to the starting energy level (relaxation) will give us detailed information on the molecular structure and dynamics of our object of study.

Although in the past, the field of liquid-state NMR has experienced a bigger development when compared to solid-state NMR, presently the former has seen stagnation in its rate of advancement while the later one have undergone major progress in the last decades. The reason for this delay was the broad spectra of solid-state NMR (SSNMR), but with the implementations of cross-polarization (CP), magic angle spinning (MAS), special 2D pulse sequences and enhanced probe electronics, it is now possible to obtain liquid-like spectra. But what was thought to be a barrier for the development of SSNMR, is actually a source of information about structure and dynamics of molecules. The direct observation of anisotropic interactions, as well as the non-requirement of well-formed crystals and molecular tumbling (when compared with X-ray and liquid state NMR, respectively), makes SSNMR a very versatile technique capable of looking at a variety of materials like glasses, biological molecules, zeolites, organic/inorganic complexes, semiconductors, ceramics and many more.

2. SOLID-STATE NMR OF MEMBRANES

2.1. ^{2}H NMR

Due to its small quadrupolar coupling, deuterium is the most common nucleus studied in lipids by SSNMR [2]. Quadrupolar interactions will dominate the spectra, and even dipole-dipole interactions with surrounding protons will be negligible, which is problematic when doing ^{1}H NMR experiments since this will create a featureless spectrum. Although deuterium is a very sensitive probe of its environment and reorientational motions, there are drawbacks like the need of sample labelling and the low signal to noise due to its low gyromagnetic ratio.

There are two allowed transitions for nuclei with spin = 1, which will generate a spectrum with 2 resonances separated by a quadrupolar splitting. The magnitude depends on the angle and the magnetic field. In a deuterated lipid there will be a series of doublets with an associated quadrupolar splitting that holds information about structure and dynamics of the associated CD bond.

Seelig showed that the C2 (carbon at the position number 2 in the chain) of 1,2-dipalmitoyl-*sn*-glycero-3-phosphocholine (DPPC) doesn't have the same conformation in both chains [3]. While for the C2 of the sn-1 chain only one splitting was seen, for the same carbon position on the sn-2 chain two splittings were observed. This is due to the fact the beginning of the sn-2 chain is oriented parallel to the plane of the bilayer, while the sn-1 is perpendicular.

The order parameter of CD bond, S_{CD}, can be determined from the quadrupolar splitting, where S_{CD} is the ensemble average of the function $<3 \cos^2 \theta - 1> / 2$, where θ is the angle between the CD bond and the magnetic field.

The lipids have their headgroups at the lipid-water interface, and the close packing of this region will cause constraints on the motional freedom of the individual chain molecules, causing an almost constant order parameter at the beginning of the chains. When we are heading to the end of the chains, these constraints are less pronounced, which will cause a decrease in the order parameter.

It is also interesting to notice the influence of cholesterol on the order parameters of the chains. The cholesterol "sits" in the upper/middle hydrophobic chain, as can be seen by the increase on the order parameters, but at the end of the chain it is clear that the motional restrictions are similar to the lipid alone, which indicates for an absence of the sterol at this part of the bilayer.

Although the order parameters give us an indication of the degree of freedom, it doesn't tell us much about the nature of the motions involved as a relaxation study would.

2.2. ^{31}P NMR

Since phospholipids are present in most cell membranes, phosphorus became a popular probe used for structural and dynamical information from membranes. Although ^{31}P has a 100% natural abundance and a relatively high sensitivity, the large linewidth due to chemical shift anisotropy (CSA) has limited the use of ^{31}P NMR, since the dipole-dipole interactions with the neighbouring protons are removed.

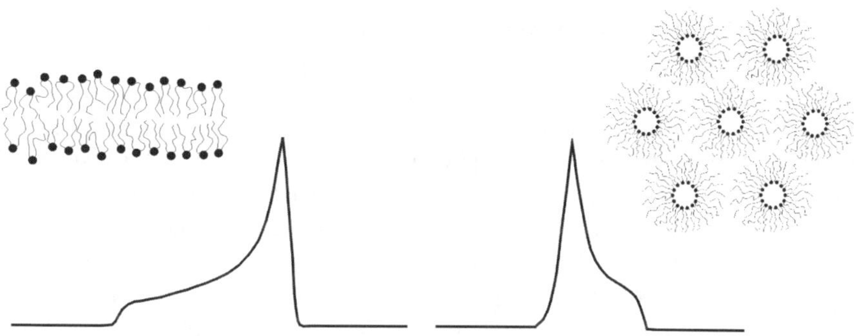

Figure 2. ^{31}P NMR spectra for the Lα (left) and hexagonal phases (right)

But CSA can help in the determination of phase transitions [3]. In the case of a phosphatidylcholine lipid powder, the linewidth will be around 130 ppm, but if we fully hydrate it (liquid crystalline phase) there will be a reducing of the broadening due to rapid axial rotation. On the hexagonal phase (H_{II}) the resulting powder pattern will have the opposite sign as well as it will be reduced by a factor of 2 due to diffusion along the cylinder structure [4,5]. The fact that the intermediates of the transition between the bilayer and the non-bilayer phase are involved in membrane fusion has raised an interest in this hexagonal and other non-bilayer phases [5-7].

The ^{31}P relaxation is governed by CSA and dipole-dipole interactions, and although it has been subject to many studies [8-10] it is still not fully understood. These measurements can however yield a good estimate of the time-averaged angle between the membrane surface and the vector connecting the phosphorus to the glycerol C3 protons [10].

2.3. Alcohols in membranes: the lipid theory of anaesthesia

Anaesthetics are extensively used in medicine, but their mechanism is not yet fully understood. Meyer and Overton reported about 100 years ago that the increase of potency of a general anaesthetic was associated with its solubility increase in olive oil. This and the fact that a different range of compounds having no apparent chemical structure similarities, as well as the observation that anaesthetic efficiency increases with the increase of alcohol chain length until 1-dodecanol, suffers a drop with larger saturated n-alkanols (which shows a very interesting relationship between the size of the anaesthetic and the lipid bilayer) made the lipids the primary candidate site of general anaesthetics. But this theory suffered two major drawbacks when it was seen that the observed differences (fluidity, thickness, area and volume of the lipid bilayer) were only observed at toxic anaesthetics concentrations and that these differences could also be observed by changing slightly the temperature. Nowadays there are 4 different candidate target sites: a) the surface and b) the interior of the membrane, c) lipids at the lipid/protein interface and d) hydrophobic pockets and amphiphilic surfaces of proteins. Thus the question is kept alive: is the primary action site the lipid bilayer or the proteins [12,13]? Contributions made by solid-state NMR trying to find a solution for this question consisted mainly by ^{2}H NMR experiments where the mode of interaction between the anaesthetics and the bilayer experiments was determined [14,15]. Alcohols showed a strong chain length dependence of its incorporation in the lipid bilayers and did not change the membrane structure dramatically [14] or change the chain ordering. Another investigation made with chloroform also could not detect any changes in the order of the lipids [15].

More insights on these interactions have been made by MD simulations of ethanol in membranes, where the alcohol was seen to translocate from one layer to the other in the membrane and to locate close to the esther oxygen in the lipid/water interface [16], as had been already suggested by investigations carried out on this system by ^{2}H and ^{31}P NMR [17].

2.4. Cell-penetrating peptides

Cell-penetrating peptides (CPPs) or in a more general terminology protein transduction domain (PTD), are a class of small peptides (less than 30 aminoacids) that enable the delivery of large biologically active molecules into most cells. Although a large amount of distinct peptides can be found (penetratin, transportan, prion protein, short sequences of arginines and lysines, etc.) they often consist of positively charged aminoacids [18]. In vivo applications of CPPs for treating diseases like ischemia, cancer as well as other diseases have been reported [19].

Although a large amount of work has been done in this area, there is a lack of consensus regarding the mechanism of penetration of CPPs [20]. An endocytosis independent translocation through the lipid bilayer was accepted some years ago, but cell fixation induced misleading artefacts. It was shown on live unfixed cells that the mechanism of cellular uptake of Tat 48-60 and $(Arg)_9$ involved an endocytic route. But a general agreement appears to reside in 3 different mechanisms depending on the peptide involved: a) endocytosis, b) physicochemical interactions between the membrane and the peptide, and c) the conjugation of the first two.

The work made in this area in solid-state NMR has mainly been the study of the interactions between the CPPs and the lipid membranes [21]. It was seen that mastoparan interacts strongly with cationic lipids and separates the membrane into a lamellar and a non-lamellar phase beyond forming pores in it induced by a peptide helix [22]. Another interesting work combined liquid-state NMR and solid-state for the determination of the structure and the positioning of bovine prion protein in bicelles. A justification for the loss of signal in the liquid state experiments was found in ^{2}H NMR where a hydrophobic mismatch in the bicelles was seen [23].

3. ACKNOWLEDGMENTS

I would like to thank Jonas Danielsson and August Andersson for their critical reading of the manuscript, as well as Martin Dahlberg for making the MD simulation picture available.

4. REFERENCES

1. Latowski D., Åkerlund H., Strzałka, K. Violaxanthin de-epoxidase, the xantophyll cycle enzyme, requires lipid inverted hexagonal structures for its activity. Biochemistry 2004; 43: 4417 – 4420
2. Seelig J. Deuterium magnetic resonance: theory and application to lipid membranes. Quart Rev Biophys 1977; 10: 353-418
3. Kleinschmidt, J.H., Tamm, L.K. Structural transitions in short-chain lipid assemblies studied by ^{31}P-NMR spectroscopy. Biophys J 2002; 83: 994-1003
4. Seelig, J., Seelig, A. Lipid conformation in model membranes and biological membranes. Quart Rev Biophys 1980; 13: 19-61
5. Cullis, P.R., De Kruijff, B. Lipid polymorphism and the functional roles of lipids in biological membranes. Biochim Biophys Acta 1979; 559: 399-420
6. Seelig, J. Phosphorus-31 nuclear magnetic resonance and the head group structure of phospholipids in membranes. Biochim Biophys Acta 1978; 515: 105-140
7. Siegel, D.P. Energetics of intermediates in membrane fusion: comparison of stalk and inverted micellar intermediate mechanisms. Biophys J 1993; 65: 2124-2140

8. Milburn, M.P., Jeffrey, K.R. Dynamics of the phosphate group in phospholipid bilayers. A ^{31}P nuclear relaxation time study. Biophys J 1987; 52: 791-799
9. Dufourc, E.J., Mayer, C., Stohrer, J., Althoff, G., Kothe, G. Dynamics of phosphate head groups in biomembranes. Comprehensive analysis using phosphorus-31 nuclear magnetic resonance lineshape and relaxation time measurements. Biophys J 1992; 61: 42-57
10. Roberts, M., Redfield A. Phospholipid bilayer surface configuration probed quantitatively by ^{31}P field-cycling NMR. Proc Natl Acad Sci USA 2004; 101: 17066-17071
11. Pope, J. M., Dubro, D.W. The interaction of n-alkanes and n-alcohols with lipid bilayer membranes: a deuterium NMR study. Biochim Biophys Acta 1986; 858: 243-253
12. Franks, N.P., Lieb, W.R. What is the molecular nature of general anesthetic target sites? Trends Pharmacol Sci 1987; 8: 169-174
13. Antkowiak, B. How do general anaesthetics work? Naturwissenschaften 2001; 88: 201-213
14. Westerman, P.W., Pope, J.M., Phonphok, N., Doane, J.W., Dubro, D.W. The interaction of n-alkanols with lipid bilayer membranes: a ^{2}H-NMR study. Biochim Biophys Acta 1988; 939: 64-78
15. Phonphok, N., Chidichimo, G., Westerman, P.W. Disposition of chloroform in phosphatidylcholine membranes: a ^{2}H- and ^{31}P-NMR study. Chem Phys Lipids 1996; 83: 25-37
16. Patra, M. Salonen, E., Terama, E., Vattulainen, I., Faller, R., Lee, B.W., Holopainen, J., Karttunen, M. Under the influence of alcohol: The effect of ethanol and methanol on lipid bilayers. (in press)
17. Barry, J.A., Gawrisch K. Direct NMR evidence of ethanol binding to the lipid-water interface of phospholipid bilayers. Biochemistry 1994; 33: 8082-8088
18. Lundberg, P., Langel, Ü. A brief introduction to cell-penetrating peptides. J Mol Recognit 2003; 16: 227-233
19. Snyder, E.L., Dowdy, S.F. Cell penetrating peptides in drug delivery. Pharmaceut Res 2004; 21: 389-393
20. Järver, P., Langel, Ü. The use of cell.penetrating peptides as a tool for gene regulation. Drug Discov Today 2004; 9: 395-402
21. Bechinger, B. The structure, dynamics and orientation of antimicrobial peptides in membranes by multidimensional solid-state NMR spectroscopy. Biochim Biophys Acta 1999; 1462: 157-183
22. Hori, Y., Demura, M., Niidome, T., Aoyagi, H., Asakura, T. Orientational behaviour of phospholipid membranes with mastoparan studied by ^{31}P solid state NMR. FEBS Lett 1999; 455: 228-232
23. Biverståhl, H., Andersson, A., Gräslund, A., Mäler, L. NMR solution structure and membrane interaction of the N-Terminal Sequence (1-30) of the bovine prion protein. Biochemistry 2004; 43: 14940-14947

VISUOMOTOR COORDINATION IN BEHAVIOUR OF PIGEONS FOLLOWING POST-HATCHING MONOCULAR EXPERIENCE: AN IMAGE ANALYSIS STUDY

Giovanni Cesaretti[1], Claudia Kusmic[2] and Daniela Musumeci[3]
[1] *Consorzio Pisa Ricerche, Piazza Ancona 7, 56127 Pisa, Italia*
[2] *Istituto di Fisiologia Clinica CNR, Area di S. Cataldo, Via G Moruzzi 1, 56126 Pisa, Italia*
[3] *Dip. di Fisiologia e Biochimica, Università di Pisa,Via S. Zeno 31, 56127 Pisa, Italia*

Abstract. We investigated the role of visual binocular inputs during the development on visuo-motor coordination in conditioned key-pecking responses to differently sized spots. Five groups of pigeons were used: monocular control group; two early groups (retinal ablated –ERA-, and eyeball enucleated –EEE) operated in two days post-hatching; two late groups (adult pigeons, retinal ablated – LRA-, and eyeball enucleated -LEE). The most substantial results occurred in early groups: the impairment of visual performance at small size spot, the loss of bill gape/spot size coordination and the permanent inaccurate reaching trajectory, particularly in the EEE group. On the contrary, *late* groups did not show any significant difference compared with the *control* group. The present data indicate that binocular interactions are crucial during the development of the visual system and, in addition, prove the differential role played by retinal and extraretinal inputs on the development of the visuomotor coordination. Once the maturation stage is accomplished (*late* groups) the eyes can work independently one from the other in conditioned key-pecking responses.

Key words: Retinal and Eye-Ball Removal, Post-Hatching Development, Visuomotor Coordination, Pigeons

1. INTRODUCTION

Some visually guided behaviours, as pigeon's pecking, require a particular precision in execution for an accurate output. The favourable outcome will rely on skill in evaluating both intrinsic (shape, dimension and colour) and extrinsic (location and distance) target features as well as on a correct motor output of both jaw and head movements. Binocular vision is generally assumed to play a significant role in assessing stimulus features rather than a monocular one.

V. Evangelista et al. (eds.), From Cells to Proteins: Imaging Nature across Dimensions, 459–469.

The pecking motor pattern consists in fast approaching movements interspersed with stationary fixation phases, in which the head is stopped with the eye-beak axis pointing the target. In this way animals evaluate all grain's features, since the last pecking phase is ballistic, performed with the eyes closed and the bill opened. Pecks to two-dimensional stimuli (i.e., keypeck responses) share the same behavioural patterns even if the target cannot be grasped, and the bill opening approximately scales stimulus dimension [1-5].

Although pigeons possess a wide panoramic monocular vision they deserve the narrow binocular window for pecking behaviours [1,5,7-9]. Even if retinal fibres are totally crossed at chiasmatic level, a few retino-recipient brain structures, involved in visual perception as well as in oculo-motor control, make binocular interactions possible [10-14]. Because pigeon's visual system is immature at hatching, retinal removal early in the life represents a useful tool to grow-up animals having only few chances of developing a binocular experience [15-16].

In a previous study, pigeons in which binocular vision was abolished in the early stages of life showed a considerable reduction of the contralateral visual field when tested in a visual perimetry experiment. However, a visual window persisted around the eye-bill axis, and it was approximately as wide as the normal binocular field [17]. These results were interpreted as a reduction of the convergence angle due to the unbalance of binocular inputs.

In the present paper, the same pigeon groups were tested for visuo-motor coordination towards targets falling in their visual windows in a key-pecking behaviour test.

2. MATERIALS AND METHODS

2.1. Animals and Surgery

The animals were treated in accordance with the guidelines to the care and use of experimental animals published by the NIH (N.80-23, revised 1978). Twenty-one pigeons were divided into five groups: monocular reversible control (control, n=5); early retinal ablated (ERA, n=4), late retinal ablated (LRA, n=4; two birds had been previously tested as control); eye-ball enucleated pigeons both early (EEE, n=6) and late (LEE, n=4) in their life.

Retina or eye-ball were gently drawn away by means of a vacuum pump under deep ethyl-ether anaesthesia [14]. The *early* groups were operated in two days post-hatching and tested one year later; the *late* groups (one year old pigeons) were tested two weeks after surgery. In operated birds we tested the blindness of the operated eye by covering the spared eye during the training. To obtain the monocular reversible condition in the *control* group, pigeons wore thin rings of velcro permanently fastened around their eyes, and a light metal cup (0.3 g) was alternatively coupled on the ring of one eye during the training.

2.2. Apparatus

Details of the conditioning box were described in a previous paper [5]. Briefly, it consisted in a Skinner box (13x50x40 cm), black inside, equipped with a VGA b/w monitor on the frontal wall and sloped 45° outward. Below the monitor there was a small container where the grain dispenser released a pellet at once. A lamp turned on for 3 seconds during pellet releasing. A TV-camera, 90° laterally placed respect to the monitor, acquired and recorded on the same frame a side view, an above view (via a 45° sloped mirror), and a 50Hz digital counter. Images were off-line analysed frame by frame at 20 ms rate by means of a digitizer card.

2.3. Procedure

Pigeons, trained according to a Go NoGo procedure, had to discriminate and peck a black spot displayed on the monitor and to refrain from pecking a white background. Spots could be displayed in one out of 3 sizes (∅ = 1.9, 3.8, and 7.6 mm) on the centre of the monitor and it lasted 600 ms on the screen. In order to get the stimulus presentation pigeons had to peck a black square (*starter*) occurring at the same co-ordinates of the spot. A single pellet was delivered after a peck to the spot (positive reward) or few seconds of off light (negative reward) followed either a peck to the white background or a missing peck to the spot. During the pre-training pigeons learned the stereotyped sequence of pecking phases, consisting in the *starter* peck, the *withdrawal* after the starter peck, the *head fixation* to discriminate the stimulus, and the *motor output* to peck the spot, or the avoiding *behaviour* for the white background, and the reset to the initial position. Pigeons performed daily sessions of 600 trials, divided into 5 blocks, in each block the same number of Go and NoGo stimuli occurred according to a pseudo-random sequence. Each session was composed of one block for every spot size and two blocks of mixed spot sizes [18]. Within each session subjects of the *control* group performed discriminations both in left and right monocular viewing conditions. When all training phases were gathered by pigeons, at least three successive sessions were planned and video-recorded to collect useful results. A personal computer drove the entire set-up.

2.4. Measurements

The parameters *fixation distance* to target, *fixation time* to target and *bill gape* were measured by selecting at least 60 video-recorded frames for each stimulus and every subject. The *fixation distance* was the length between the centre of the eye and the centre of the spot along the eye-bill line, corrected for posterior nodal point [4,19]. The *fixation time* was derived by counting frames in which the head was stopped above the stimulus. The *bill gape* was the inter tips distance measured immediately prior or at the first contact with the monitor. The percentage of correct responses (% of Go) was collected for every spot by the computer.

2.5. Data analysis

Two-way ANOVA and Fisher *post hoc* test were used for the statistical analysis using individual data. In the control group the average of the two monocular conditions was used for each pigeon because the statistical comparison between left and right indicated only marginal differences.

3. RESULTS

3.1. Anatomy

At the end of the experiments the pigeons were sacrificed for the anatomical examination of the eyes. All surgical eye-ball removals and the retinal ablations had been successful. The diameters of the intact eye of each operated group (ERA, EEE, LRA, and LEE) as well as the retina ablated eyes of ERA and LRA groups were measured and compared with the eyes of the control group. The average eye-diameter in the control group was 15,5 ± 0,3 mm and it was not significantly different from the other groups, except for the ERA pigeons in which the retina ablated eyes measured 11,2 ± 1 mm, $p<0{,}01$ vs. control, figure 1.

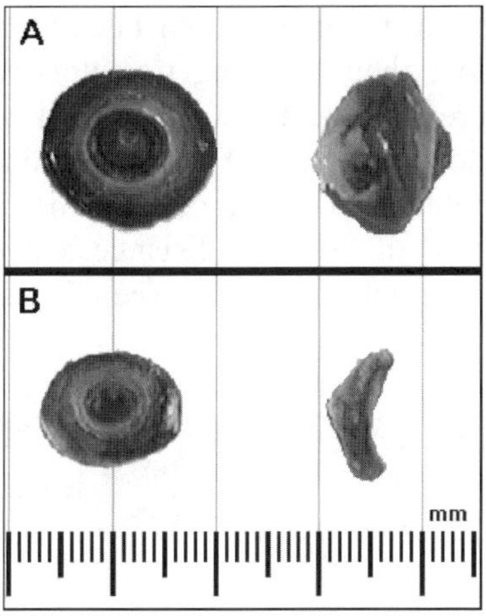

Figure 1. Frontal and side view of pigeon eye-ball. A: representative eye of an intact adult pigeon (*control group*); B: representative eye of an adult pigeon submitted to post-hatching retina removal (*ERA group*).
Minor tick scale bar: 1 mm.

3.2. Behaviour

Table 1 shows average data of each experimental group for all the parameters measured at every stimulus size.

Table 1. Mean ±S.E data of each group for all parameters at the three spot sizes

Group	*Time of fixation*			*Distance of fixation*			*% of correct GO*			*Bill gape*		
	spot size (mm)			*spot size (mm)*			*spot size (mm)*			*spot size (mm)*		
(n)	1,9	3,8	7,6	1,9	3,8	7,6	1,9	3,8	7,6	1,9	3,8	7,6
Control (5)	134±14	119±20	124±8	58±3	58±4	58±3	99±2	97±3	99±1	4,4±0,9	5±1,0	5,4±1,0
ERA (4)	128±11	104±13	99±16	62±2	64±2	61±2	65±25	93±3	91±3	7,0±1	7,4±0,9	8,0±2
LRA (4)	127±3	98±2	112±5	62±2	61±2	63±3	98±3	99±1	98±1	4,7±0,2	5,5±0,3	5,9±0,4
EEE (6)	127±24*	88±24*	98±23*	61±4	58±8	60±7	66±17	90±8	89±7	7,0±2	6,0±1	7,0±1
LEE (4)	123±14	96±13	88±12	62±4	61±4	60±4	96±4	99±2	99±1	4,5±0,4	5,1±0,7	6,7±1,3

* *indicates the mean time spent in moving during the fixation phase of EEE group*

The most important effects of the loss of binocular vision were evident in post-hatching groups, in both *ERA* and *EEE*, but with heavier abnormal behavioural outcomes in the enucleated one (*EEE*).

Fixation Time and Fixation Distance. Time and distance of fixation are reported in Table 1. As the EEE group is concerned, the measurement of the parameters requires some explanations. In fact, during the fixation phase head's sagittal plane of *EEE* pigeons was not regularly orthogonal to the monitor (figure 2) and little head's back movements (denoted by asterisks in table 1) were habitually current at the smallest spot size and occasionally present at the other spot sizes. In addition, their reaching trajectory was frequently out of the spot localisation (figure 3, white arrows) and their head was slightly tilted and/or rotated towards the seeing eye. For this reason, in this group, the time of fixation was computed as the time of head moving, and the distance of fixation was computed as the average distance between the starting and the ending head position, which ranged between 3 and 5 mm, with the tip of the bill as reference.

CONTROL LEE EEE LRA ERA

Figure 2. Fixation to the spot of a representative pigeon for each experimental group. *Control*, *LRA*, *ERA*, *LEE* animals are seeing with the left eye, whereas *EEE* pigeon is using the right eye.

The ANOVA (group vs. spot size) carried out on the *fixation time* indicated a significant effect of spot size ($F=10,7$ df $(2,46)$, $p<0,001$), it was due to the longest time spent on the small spot ($p<0,005$) by all groups. The ANOVA (group vs. spot) carried out on *fixation distance* did not evidence any difference. It is interesting to note that the average value of the *fixation time* and the *fixation distance* of even the *EEE* pigeons did not differ from the other groups.

Percentage of correct GO responses. As the percentage of correct Go responses is concerned, it was evident a difference among groups in visual discrimination performance. The ANOVA (group vs. spot size) revealed a significant effect of both factors: group ($F= 17,7$; df $(4,46)$; $p<0,001$) and spot size ($F=14,7$; df $(2,46)$, $p<0,001$). The *post hoc* test ($p<0,001$) indicated lower scores of performance of both *early* groups than *control* and *late* groups, especially on the smallest size spot. It is interesting to note that the *late* groups performed the task as the *control* group for all spot sizes.

Bill Gape opening. Figure 3 shows keypecks of representative pigeons of all experimental groups (rows) at each spot size (columns). The scaling of the beak opening amplitude to target size was evident both in *control* and *late* groups. On the contrary, *early* groups showed a loss of the gape's control and an overshooting of beak opening, that resulted about as large as the largest spot size (see Table 1). Moreover, as mentioned above, *EEE* group showed a failure in peck localisation and the head slightly tilted and rotated (arrow in figure 3). The ANOVA (group vs. spot size) carried out on bill opening values pointed out a significant effect of group ($F=9.1$, df $(4,46)$, $p<0,001$) and a marginal spot size effect ($F=3.4$, df $(2,46)$, $p<0,04$). The *post hoc* comparison confirmed the larger bill opening ($p<0,004$) of both *early* groups than in the other groups. The spot size effect was due to the ability in scaling the bill amplitude to spots of different sizes only for *control* and *late* groups. This result was supported by the significant correlation between the bill gape and spot size for *control*, *LRA* and *LEE* groups ($r=0,82$, $p<0,001$; $r=0,73$, $p<0,001$; $r=0,85$, $p<0,001$, respectively), whereas no correlation was found for *ERA* and *EEE* groups ($r=0,53$, $p=0,11$; $r=0,07$, $p=0,81$, respectively).

4. DISCUSSION

The most relevant results of the present study involve the post hatching operated birds and could be summarised in few points: the impairment in visual performance at small spot; the loss of the bill gape versus spot size coordination and, finally, the permanent impairment in the reaching trajectory in the *EEE* group.

The lack of binocular vision by reversible monocular occlusion causes only little behavioural deficits in adult pigeons. Impairments have been reported only in depth judgment capacity and in grasping efficiency [20-21].

Moreover, a little inaccuracy of target localization [22] and the slow down of reaction times and fixation times were observed in a key-pecking task, suggesting independence between the eyes even for those behaviours accomplished within the binocular visual field [23-24]. The satisfactory quality of monocular performance raises the question about the authentic benefits of binocular vision. An answer should be provided by operated monocular animals in which great anatomical and functional modifications have been induced.

However, the present results of the *late* groups do not add any supplementary knowledge, since they do not differ from *control* group. To support this claim, also the two birds enrolled in both *control* and *late* groups did not reveal any difference for all parameters considered.

Very different is the case of the *early* groups in which the compelled mono-ocular development determines permanent impairments of some visually guided motor outputs. The most evident of which is the lack of bill opening calibration with respect to the spot size in both *early* groups.

The loss of control of the bill scaling could be unpredicted because all the other parameters are similar to those of *control* group, but for visual discrimination performance at the smallest spot that, anyway, was maintained at a good quality level (Table 1). Generally, a free moving animal puts itself at the optimal viewing distance of head stop and spends its time of fixation in codifying the stimulus features and in programming the motor output. In this case, a perceptive deficit should result in a delayed time of fixation and, probably, in a different distance of fixation too, but this is not the condition of *ERA* pigeons.

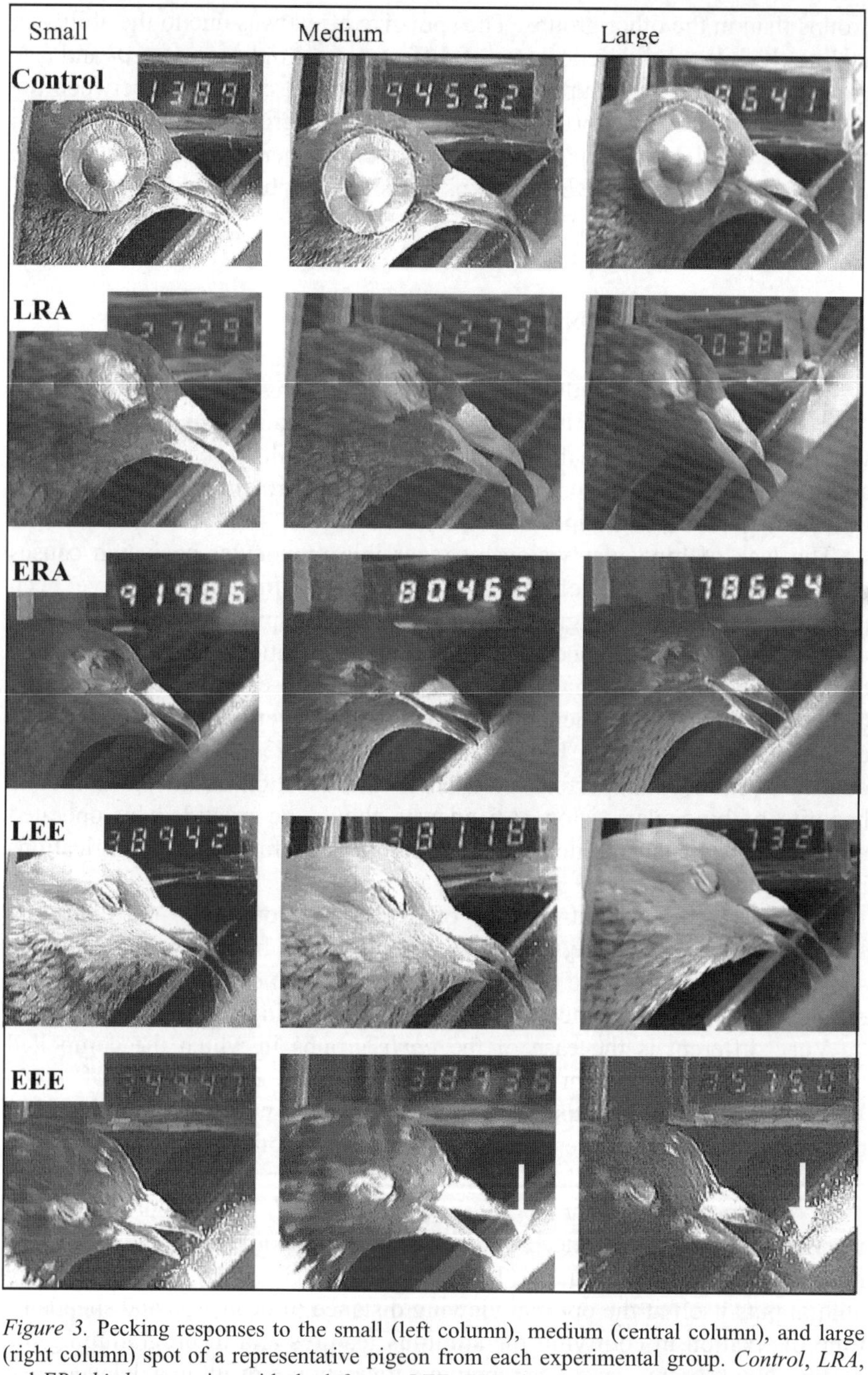

Figure 3. Pecking responses to the small (left column), medium (central column), and large (right column) spot of a representative pigeon from each experimental group. *Control*, *LRA*, and *ERA birds* are seeing with the left eye; *LEE*, *EEE* pigeons are using the right eye. White arrows point to the spot position.

At a first sight, the *EEE* group seems to behave differently, since it showed instability during the fixation phase and a misdirected peck to the target together with the loss of bill scaling. However, one could interpret the moving fixation phase as a compensation strategy to keep the image on the retina. In this case, the mean values of the time of fixation should be significantly delayed with respect to the *ERA* group and, probably, the distance of fixation should be different too, especially on the small stimulus which is the less compensated. But this is not true.

Pigeons are very sensitive to small differences in the size of projected visual stimuli [25] and the precision with which the bill opening is scaled to the pellet size during grasping is impressive (i.e., a small shift in stimulus size elicits a significant shift in opening amplitude) [6]. Thus, it is possible that the little impairment in visual perception of *early* groups, as inferred by the minor reduction in the visual discrimination performance, could be anyhow meaningful in affecting the control of bill scaling accuracy. Moreover, the large bill opening could represent a functional utility for *early* groups in increasing the chances to catch the rewarded target, which is detected but not sharply perceived [6].

As misdirected pecks to target in *EEE* group are concerned, some considerations are needed. In a key-pecking paradigm, during the fixation phase, pigeons work to the closest point to the screen with the eyelids still fully opened. In this position the eyes are working at their near point of accommodation, and they are in a high grade of convergence that increases further in the final ballistic phase, when the eyelids are closed [3-4,22]. Visual cues are critical in detecting the target presence and its extrinsic features, such as distance and location, but they loose importance in the control of the output systems, mainly neck and body musculature, that are involved in the reaching trajectory, with the eyes closed. The kernel of the matter making the difference between the two *early* groups is the role of extraretinal and/or extraocular cues involved in the moment-to-moment control of eye-ball movements. The presence of the mere ocular globe, even if reduced in size, is adequate to maintain the symmetry provided by the two eyes in some of the extraretinal reflexes, namely vestibulo-ocular and vestibulo-collic reflexes [26-27].

Thus, in the case that the extraretinal inputs along with the retinal inputs are unbalanced, EEE animals are able to develop compensating oculo-motor mechanisms and/or head postural adjustments to keep their seeing eye projecting in the frontal field [26], but unable to guide the head towards the target with the eyes closed.

5. CONCLUSIONS

The present data indicate that interactions between the two eyes are crucial during the development of the visual system and, in addition, they point out the differential role of retinal and extraretinal binocular inputs during the early life stages for the suitable setting up of visuomotor coordination. Once the developing stage is accomplished, the loss of either retinal or extraretinal information from one eye does not affect the visual and the visuomotor coordination, as shown by the *late* groups, at least when a visual conditioned key-pecking response is required.

6. ACKNOWLEDGEMENTS

The Authors wish to thank Dr. A. Gazzano for surgical assistance and Mrs. G. Bresciani for her technical assistance in animal housing. This study has been supported by MURST grant.

7. REFERENCES

1. Goodale M.A. Visually guided pecking in the pigeon (*Columba livia*). Brain Behav Ev 1983; 22: 22-41
2. La Mon B., Zeigler H.P. Grasping in the pigeon: stimulus control during conditioned and consumatory responses. Animal Learn Behav 1984; 16: 256-267
3. Martinoya C., Le Houezec J., Bloch S. Pigeon's eyes converge during feeding: Evidence for frontal binocular fixation in a lateral-eyed bird. Neurosci Letters 1984; 45: 335-339
4. Macko K.A., Hodos W. Near point of accommodation in pigeons. Vision Res 1985; 25: 1529-1530
5. Cesaretti G., Kusmic C., Musumeci D. Binocular field in pigeons: behavioral measures of stimulus detection and coding. Arch Ital Biol 1997; 135: 131-145
6. Bermejo R., Allan R. W., Deich J., Houben D., Zeigler, H. P. Prehension in the pigeon I: Descriptive analysis. Exp Brain Res 1989; 75: 569-576
7. Walls GL. *The Vertebrate Eye and its Adaptive Radiation.* New York: Hafner, 1963. : 247-367
8. Martinoya C., Rey J., Bloch S. Limits of the pigeon's binocular field and direction for best binocular viewing. Vision Res 1981; 23: 911-915
9. Nalbach H.O., Oberhollenzer F.W., Kirschfeld K. The pigeon's eye viewed through an ophthalmoscopic microscope: orientation of retinal landmarks and significance of eye movements. Vision Res 1990; 30: 529-540
10. Karten H.J. "Visual lemniscal pathways in birds." In *Neural mechanisms of behavior in the pigeon*, Granda A.M and Maxwel J.H, eds. New York: Plenum Press, 1979.
11. Gamlin P.D.R., Cohen D.H. Projections of the retinorecipient pretectal nuclei in the pigeon (Columba livia). Comp Neurol 1988; 269: 1-17
12. Bulkharter A., Cuénod M. Changes in pattern discrimination learning induced by visual deprivation in normal and commisurotomized pigeons. Exp Brain Res 1978; 31: 369-385
13. Wylie D.R., Frost B.J. Binocular neurons in the nucleus of the basal optic root (nBOR) of the pigeon are selective for either translational or rotational visual flow. Visual Neurosc 1990; 5: 489-495

14. Bagnoli P., Porciatti V., Fontanesi G., Sebastiani L. Morphological and functional changes in the retinotectal system of the pigeon during the early posthatching period. J Comp Neurol 1987; 256: 400-411
15. Wylie D.R. Binocular neurons in the nucleus lentiformis mesencephali in pigeons responses to translational and rotational optic flowfields. Neurosci Lett 2000; 291: 9-12
16. Manns M., Güntürkün O. Development of the retinotectal system in the pigeon: a cytoarchitectonic and choleratoxin study. Anat Embryol 1997; 195: 539-555
17. Musumeci D., Cesaretti G., Kusmic C. "Pigeon's visual field when binocularity is kept out at different life stages." In *Vision: The approach of biophysics and neurosciences*, Musio, ed. Singapore: Word Scientific 2001
18. Bandoni G, Cesaretti G., Kusmic C., Musumeci D. "An algorithm generating long sequences of stimuli in behavioral science: a suitable test for biosensors." In *Molecular Electronics: Bio-sensors and Bio-computers*, Barsanti L. et al., eds. Dordrecht: Kluver Academic Publishers 2003
19. Erichsen J.T., Hodos W., Evinger C., Bessette B. B., Phillips S. J. Head orientation in pigeons: postural, locomotor and visual determinants. Brain Behav Evol 1989; 33: 268-278
20. Martinoya C., Le Houezec J., Bloch S. Depth resolution in the pigeon. J Comp Physiol *A* 1988; 163: 33-42
21. Mc Fadden S.A. Eye design for depth and distance perception in the pigeon: an observer orientated perspective. Int J Comp Psychol 1989; 3: 101-130
22. Jäger R., Zeigler H.P. Visual fields and peck localization in the pigeon (*Columba livia*). Behav Brain Res 1991; 45: 65-69
23. Di Stefano M., Kusmic C., Musumeci D. Binocular interactions measured by choice reaction times in pigeons. Behav Brain Res 1987; 25: 161-165
24. Kusmic C., Musumeci D., Spinelli P. Binocular probability summation in a choice reaction time taskin pigeons. NeuroRep 1991; 2: 615-618
25. Hodos W., Weiss S.R.B., Bessette B.B. Size threshold changes after lesions of the visual telencephalon in pigeons. Behav Brain Res 1986; 21: 203-214
26. Hayman M.R., Donaldson J.P., Donaldson I.M.L The primary afferent pathway of extraocular muscle proprioception in the pigeon. Neuroscience 1995; 69: 671-683
27. Goode C. T., Maney D. L., Rubel E. W., Fuchs A. F. Visual influences on the development and recovery of the vestibuloocular reflex in the chicken. J. Neurophysiol 2001; 85: 1119-1128

INDEX

List of Contributors

Aebi Ueli
M.E. Müller Institute
BiozentrumUniversity of Basel
Klingelbergstr. 70
CH-4056 Base
Switzerland

Albrechtová Jana
Dept. of Plant Physiology
Faculty of Sciences
Charles University in Prague
Czech Republic

Apostol Simona Larisa
Valahia University
Physics Dept. Faculty of Science
24, Bd. Unirii
0200 Targoviste
Romania

Arndt-Jovin Donna
Dept. of Molecular Biology
Max-Planck-Inst. Biophysical
Chemistry
Am Fassberg 11
D- 37070 Goettingen
Germany

Aumayr Friedrich
Institut für Allgemeine Physik,
Technische Universität Wien,
Wiedner Hauptstraße 8-10/E134,
1040 Vienna
Austria

Babykin Michael M.
International Biotechnological Center
Moscow State University
Moscow 119899
Russia

Barsanti Laura
Istituto di Biofisica C.N.R.
Area della Ricerca di Pisa
Via G. Moruzzi,1
I-56124 Pisa
Italy

Bertero Mario
Dip.to di Informatica e Scienze
dell'Informazione
Università di Genova
Via Dodecaneso, 35
I-16146 Genova
Italy

Bianchini Paolo
Dip.to di Fisica
Università di Genova
Via Dodecaneso, 33
I-16146 Genova
Italy

Boccacci Patrizia
Dip.to di Informatica e Scienze
dell'Informazione
Università di Genova
Via Dodecaneso, 35
I-16146 Genova
Italy

Bugaychuk S.
Institute of Physics
National Academy of Sciences
46 Prospect Nauki
Kiev 03028
Ukraine

Burykin N.
Institute of Applied Optics
National Academy of Sciences
Kiev 01054
Ukraine

Cannone Fabio
Dip.to di Fisica
Università di Genova
Via Dodecaneso, 33
I-16146 Genova
Italy

Caorsi Valentina
Dip.to di Fisica
Università di Genova
Via Dodecaneso, 33
I-16146 Genova
Italy

Castro Vasco
Dept. Physical Chemistry
University of Stockolm
Docentbackn, 13
10405 Stockholm
Sweden

Cesaretti Giovanni
Consorzio Pisa Ricerche
Piazza Ancona 7
56127 Pisa,
Italy

Chirico Giuseppe
Dip.to di Fisica
Università di Genova
Via Dodecaneso, 33
I-16146 Genova
Italy

Coltelli Primo
Istituto Scienza e Tecnologie
Informatiche C.N.R.
Area della Ricerca di Pisa,
Via G. Moruzzi,1
I-56124 Pisa
Italy

Dallai Romano
Dip.to Biologia evolutiva
Università di Siena
Via A. Moro, 2
I-53100 Siena
Italy

Diaspro Alberto
Dip.to di Fisica
Università di Genova
Via Dodecaneso, 33
I-16146 Genova
Italy

Dubrovin Evgeniy
Physics Department
Moscow State University
Leninskie Gory
119992 Moscow
Russia

Evangelista Valtere
Istituto di Biofisica C.N.R.
Area della Ricerca di Pisa,
Via G. Moruzzi,1
I-56124 Pisa,
Italy

Ford Tom
School of Biological Sciences
Royal Holloway
University of London
Egham, Surrey,
TW20 0EX
United Kingdom

Frassanito Anna Maria
Istituto di Biofisica C.N.R.
Area della Ricerca di Pisa,
Via G. Moruzzi,1
I-56124 Pisa,
Italy

Gabrielli Simone
Dip.to Etologia Ecologia Evoluzione
Università di Pisa
Via A.Volta, 4
I-56100 Pisa
Italy

Girkin John
Institute of Photonics
University of Strathclyde
16 Richmond Street
Glasgow, G1 1XQ,
Scotland (UK)

Gebeshuber Ille
Institut für Allgemeine Physik
Vienna University of Technology
Wiedner Hauptstrasse 8-10/134
A-1040 Wien
Austria

Gualtieri Paolo
Istituto di Biofisica C.N.R.
Area della Ricerca di Pisa,
Via G. Moruzzi,1
I-56124 Pisa
Italy

Janáček Jiří
Dept. of Biomathematics
Institute of Physiology
Academy of Sciences of the Czech Republic
Videnska 1083
14220 Prague
Czech Republic

Jares-Erijman Elizabeth A
Departamento de Química Orgánica, Facultad de Ciencias Exactas y Naturales
Universidad de Buenos Aires
C1428EHA Buenos Aires
Argentina

Jovin Thomas
Dept. of Molecular Biology
Max-Planck Inst. Biophysical Chemistry
Am Fassberg 11
D- 37070 Goettingen
Germany

Karen Petr
Dept. of Biomathematics
Institute of Physiology
Academy of Sciences of the Czech Republic
Videnska 1083
14220 Prague
Czech Republic

Kirik Inessa A.
Dept. of Genetics
Moscow State University
Moscow 119899
Russia

Korchemskaya E.
Institute of Physics
National Academy of Sciences
46 Prospect Nauki
Kiev 03028
Ukraine

Kozubek Michael
Laboratory of High Resolution Cytometry, Masaryk University of Brno
Botanicka 68a
CZ-60200 Brno
Czech Republic

Kubinova Lucie
Dept. of Biomathematics
Institute of Physiology
Academy of Sciences of the Czech Republic
Videnska 1083
14220 Prague
Czech Republic

Kusmic Claudia
Istituto di Fisiologia Clinica C.N.R.
Area della Ricerca di Pisa,
Via G. Moruzzi,1
I-56124 Pisa
Italy

Lidke Diane S.
Dept of Molecular Biology
Max-Planck Inst. Biophysical Chemistry
Am Fassberg 11
D- 37070 Goettingen
Germany

Lupetti Pietro
Dip.to Biologia evolutiva
Università di Siena
Via A.Moro, 2
I-53100 Siena
Italy

Mazza Davide
Dip.to di Fisica
Università di Genova
Via Dodecaneso, 33
I-16146 Genova
Italy

Musumeci Daniela
Dip.to di Fisiologia e Biochimica
Università di Pisa
Via S. Zeno 31,
56127 Pisa
Italy

Parshykova Tatyana
Dept. Of Plant Physiology and Ecology
Kiev National University
Jaraslov Val, 11
01034 Kiev
Ukraine

Passarelli Vincenzo
Istituto di Biofisica C.N.R.
Area della Ricerca di Pisa,
Via G. Moruzzi,1
I-56124 Pisa
Italy

Pesce Mattia
Dip.to di Fisica
Università di Genova
Via Dodecaneso, 33
I-16146 Genova
Italy

Rosati Giovanna
Dip.to Etologia Ecologia Evoluzione
Università di Pisa
Via A.Volta, 4
I-56100 Pisa
Italy

Samorì Bruno
Dip.to Biochimica
Università di Bologna
Via Irnerio, 48
I-40126 Bologna
Italy

Sandal Massimo
Dip.to Biochimica
Università di Bologna
Via Irnerio, 48
I-40126 Bologna
Italy

Sbrana Francesca
Dip.to di Fisica
Università di Trento
Via Sommarive, 14
I-38050 Povo (TN)
Italy

Smith Richard A. P.
Institut für Allgemeine Physik,
Technische Universität Wien,
Wiedner Hauptstraße 8-10/E134,
1040 Vienna
Austria

Stepanchikov D.
Zhytomir State University
Zhytomir 10008
Ukraine

Stoffler Daniel
M.E. Müller Institute
BiozentrumUniversity of Basel
Klingelbergstr. 70
CH-4056 Base
Switzerland

Testa Ilaria
Dip.to di Fisica
Università di Genova
Via Dodecaneso, 33
I-16146 Genova
Italy

Usai Cesare
Dip.to di Fisica
Università di Genova
Via Dodecaneso, 33
I-16146 Genova
Italy

Verni Franco
Dip.to Etologia Ecologia Evoluzione
Università di Pisa
Via A.Volta, 4
I-56100 Pisa
Italy

Vicidomini Giuseppe
Dip.to di Fisica
Università di Genova
Via Dodecaneso, 33
I-16146 Genova
Italy

Winter Hannspeter
Institut für Allgemeine Physik,
Technische Universität Wien,
Wiedner Hauptstraße 8-10/E134,
1040 Vienna
Austria

Yaminsky Igor V.
Physics Department
Moscow State University
Leninskie Gory
119992 Moscow
Russia

Zuccheri Giampaolo
Dip.to Biochimica
Università di Bologna
Via Irnerio, 48
I-40126 Bologna
Italy